S0-BPI-529

A Checklist of the

# *Newberry Library's*

Printed Books
in Science, Medicine, Technology,
and the Pseudosciences,
ca. 1460–1750

GARLAND REFERENCE LIBRARY
OF THE HUMANITIES
(Vol. 1195)

# A Checklist of the *Newberry Library's* Printed Books in Science, Medicine, Technology, and the Pseudosciences ca. 1460–1750

*Jean S. Gottlieb*



GARLAND PUBLISHING, INC. • NEW YORK AND LONDON
1992

**Library of Congress Cataloging-in-Publication Data**

Gottlieb, Jean S.
A checklist of the Newberry Library's printed books in science, medicine, technology, and the pseudosciences, ca. 1460–1750 / Jean S. Gottlieb.
p. cm. — (Garland reference library of the humanities ; vol. 1195)
Includes bibliographical references.
ISBN 0-8240-5171-8
1. Science—History—Sources—Bibliography—Catalogs.
2. Technology—History—Sources—Bibliography—Catalogs.
3. Medicine—History—Sources—Bibliography—Catalogs.
4. Occultism—History—Sources—Bibliography—Catalogs. 5. Newberry Library—Catalogs. I. Newberry Library. II. Title. III. Series.
Z7405.H6G67 1992
[Q125]
001.3s—dc20
[016.5] 92-17941
CIP

Printed on acid-free, 250-year-life paper
Manufactured in the United States of America

To C. K. C.

## Contents

## Acknowledgments

It seems as though almost as many people deserve thanks for their help and support in the preparation of this list as the list has titles. In twelve years a sizable parade could be marshaled—of Newberry Library staff members and fellows, academic pen pals, sympathetic employers, long-suffering family members and friends, and mentors from the Newberry, the University of Wisconsin, and the University of Chicago.

John Tedeschi and Allen Debus first suggested to me that there might be quite a few interesting scientific texts in the library. They read much of the manuscript and continue to offer encouragement. My thanks to them go beyond acknowledgment of their part in this work and extend into the realm of abiding friendship. Paul Gehl has also been an unflagging supporter; he has offered many useful insights as well as translations of Latin passages. His has been a reassuring presence. To Richard Ferguson, whose patience, imagination, and vision have contributed more than the words "technical support" can convey, my fervent admiration and gratitude. To Mary Nell Hoover, who edited and typed the first list ten years ago, my heartfelt thanks. The late Richard Meredith offered early word processing help that was invaluable. The late Jon Perry's contribution is noted in the checklist (see Decembrio). Dick Brown, Margaret Brenneman, Lucille Wehner, Helga Miz, John Aubrey, Paul Saenger, Bob Karrow, Bernie Wilson, and the late Michael Kaplan represent all the Newberry administration and staff who have become my friends and valued associates over the years. It would have been impossible to compile this list without their help.

I am profoundly grateful to those who read sections of this list: William Ashworth, Jr., Kevin Berland, Bruce Bradley, Nicholas Clulee, Betty Jo Teeter Dobbs, Edward Gosselin, John Neu, John Parascandola, and Alice Stroup. Any errors that remain are mine alone.

To my dear family, all twenty-one of them, who have put up with this preoccupation, nay, obsession, all these years, *vale la pena* and hallelujah!

# Introduction

This checklist of the Newberry Library's printed books in science, medicine, technology, and pseudoscience from the Renaissance to the Enlightenment testifies to the breadth of the library's holdings. The titles, now that they are brought together, constitute a conceptual collection that reflects the library's strengths in the aspect of Renaissance studies that relates to history of printing, witchcraft and magic, exploration and discovery in the New World, geography, medicine, and other science-related fields.

The books that are listed here are dispersed throughout the library's general and special collections, and they have generally not been seen as constituting a history of science collection. Indeed, since the Newberry Library is a renowned history and humanities research institution, its emphasis on general intellectual and cultural development has tended to place "science" outside its purview. Earlier in its history, it divested itself of a considerable portion (though how far from all can be seen in the following pages) of the science holdings it had accumulated. With the publication of this list, however, science at the Newberry will no longer languish, because the books listed here will become more visible and thus more available to the community of scholars.

*Scope*

The first task in the preparation of the checklist was to decide on a cutoff date. (The earliest imprint I found became the opening date.) The year 1750 is arbitrary in the sense that it does not represent the date of a specific work, nor is it the one "watershed year" after which humanistic science ceased to exist. It does, however, mark the chronological end point of most of the library's choicest science holdings, which are concentrated in the sixteenth through the mid-eighteenth centuries, the period when the sciences and technology were still relatively undifferentiated from humanistic and philosophic inquiry. Though there are a number of interesting post-1750 works in the collections, this list is intended to document the character of science as it is represented at the Newberry Library by early imprints up to and just following the scientific revolution.

A more difficult task involved determining what subjects would, for the purposes of this checklist, qualify as science. The list reflects the bias that it is preferable to "err on the side of inclusion";[1] as a result, it doubtless contains entries of questionable relevance. "On the other hand, calling attention to the obscure, distant, offbeat or unsuspected is exactly what makes many a bibliography invaluable."[2] Thus I have included such subjects (often considered marginal) as agriculture, diet, gardening, and gymnastics when the books on these subjects inform or instruct. The engineering, chemistry, or technology that is often part of military art and science secures a place in the list for military tracts that deal with these subjects, but those discussing such topics as battle strategy or the placement of fortifications are not admitted.

Although some dramatic, satirical, or imaginative literature might deserve a place in the list and the distinction made between the didactic and the "artistic" may seem arbitrary or capricious, I have excluded literary genres (e.g., poetry, drama, satire, and fiction) except when those works use scientific subjects to observe, instruct, or inform (see the list of subject descriptors on page xv). Thus Guillaume Du Bartas's *Sepmaine* (a poem on the creation of the earth) and *The Purple Island, or the Isle of Man* (an allegorical work on human anatomy by Phineas Fletcher) are listed, but predominantly theological tracts on the Creation and descriptions of imaginary voyages are not.[3] Atlases consisting primarily of maps are also excluded because they have already been the subject of considerable published bibliographic work; further, the Newberry Library's collection is extensive enough to merit its own checklist.[4]

The library's collection of almanacs has been less well documented, and I have included many—but probably not all—of the considerable holdings in this field. A future bibliographer could perform a real service by preparing a separate list of these interesting and important ephemera. In a similar vein, popular periodicals (e.g., *The British Apollo*, the *Bureau d'Adresse*), which printed questions and observations on miscellaneous scientific topics, deserve to be included in the checklist. However, the prospect of sifting through each issue of all the serials seemed too daunting a task. Therefore, both the journals and the almanacs that appear in the checklist are a sampling of these items rather than a complete inventory of holdings.

I have included many emblem books despite their sometimes uncertain relevance because they often employ alchemical or other kinds of mystical symbols, the illustrations frequently depict science-related activities, and they can reveal patterns of transmission of images and ideas. Tracts on angling, weights and measures, and cookery that relate to natural history, physics, or other scientific subjects are also admitted to the checklist. The list of subject descriptors suggests how broad the Newberry's science collection is as a whole; a look at the entries themselves reveals the literary, stylistic, artistic, and scientific variety of the individual items.

## *Arrangement of Entries*

The checklist is arranged alphabetically by author; anonymous works are listed by title. Each entry consists of the author's name (and dates) where known, title of the item, publication information, brief subject characterization, and, in some instances, notes describing the work. The library call number follows each entry. I have standardized authors' names and have kept the original spelling but not the capitalization in titles. Names of cities in the imprint are anglicized, as are such terms as "compagni," "veuve," "fratelli." Wherever I have been able to supply them, I have included the bibliographic reference numbers for both incunables (*Gesamtkatalog*, Goff, *British Museum Catalogue*, and others) and later English imprints (Pollard and Redgrave's *Short-Title Catalogue* to 1640 and Wing's to 1700). I have not, however, supplied this bibliographic information for non-English works; see the list of descriptors on page xv for full citations to reference materials consulted and for a list of abbreviations.

When two or more separately published entities are bound together, each has its own entry in the checklist. The notes with each entry supply a cross-reference giving author and short title for the other work(s) bound into the volume. Continuously paginated collections are treated as one item, and the contents, except in a few cases (e.g., Ashmole's *Theatrum Chemicum Britannicum*), are not given in full in the notes. I base exceptions on the accessibility of the tracts in question, that is, on whether they went through many editions and whether they were widely disseminated, either as single publications or as parts of larger entities. Such items as compendia of voyages and discoveries (e.g., de Bry's *America, India Orientalis*) are treated as multivolume works, and I do not itemize the individual tracts, even though many have separate title pages, in order to limit the size and complexity of the checklist.

## *Methodology*

Once I had determined the chronological scope and the range of subjects for the checklist, I began looking for science titles, first, by subject in the library's card catalogue (guided by the subject categories devised by other bibliographers and historians of science).[5] Next, I combed the shelf lists, which involved developing some familiarity with the cataloguing systems previously employed by the library. (These included William F. Poole's adaptation, the Cutter system, and the fixed location system. Since, under the fixed location system, volumes are shelved by size rather than subject, I consulted library staff and accession lists of specific collections—such as the Silver Collection—to identify relevant materials.) I also explored the library's early accession books, large manuscript ledgers dating from the early 1890s. When the library's online catalogue was activated, a final check was made to be sure that accessions after 1980 were incorporated. I read broadly in the history of science and kept a list of names of early scientific authors that could then be searched in the card catalogue. This nearly doubled the number of titles in the checklist. Both staff and fellows at the Newberry Library offered additional names or titles of works for me to search, and I developed an active correspondence with scholars at other institutions who came up with yet more names and titles. Additional names were gleaned from the library's extensive biographical as well as bibliographical reference collection.

Striving for consistency, accuracy, and completeness in a list of works as varied in subject, language, and approach as these 2,700 titles is both humbling and challenging. Assigning subject descriptors unavoidably limits the natural breadth and subtlety of many of these books. Omissions and worse are sure to be revealed. Shortcomings notwithstanding, however, the goal of this checklist, to introduce the Newberry Library's holdings in early science, is realized because the list is now available: the books can finally speak for themselves.

## *Notes*

1. D. W. Krummel, *Bibliographies: Their Aims and Methods* (London, 1984), p. 31. For full titles of works listed in these notes, consult the selected bibliography on page xix.

2. Ibid., p. 32.

3. See Robert M. Schuler, *English Magical and Scientific Poems to 1700* (New York, 1979), pp. xii–xv, for the subjects he admits and excludes and the argument he makes in support of his selection.

4. See *Atlases in Libraries of Chicago: A Bibliography and Union Checklist* (Chicago, 1936); Philip Lee Phillips, comp., *A List of Geographical Atlases in the Library of Congress*, 4 vols. in 2 (Washington, D.C., 1909–14; reprint, Amsterdam, 1971). An up-to-date catalogue of the atlases in the Newberry Library is in preparation.

5. John Neu, ed., *Isis Cumulative Bibliography, 1966–1975* (London, 1980); Margaret Stillwell, *The Awakening Interest in Science* (New York, 1970); A. J. Walford, *Guide to Reference Material* (London, 1966); W. P. D. Wightman, *Science and the Renaissance*, vol. 2 (Aberdeen, 1962).

# List of Descriptors and Concordance with Library of Congress Subject Headings

ACADEMIES AND LEARNED SOCIETIES = learned institutions and societies
ACOUSTICS*
AGRICULTURE
ALCHEMY
ALMANACS
ANATOMY
ANCIENTS VS. MODERNS = ancients and moderns, quarrel of
ANIMAL HUSBANDRY = animal culture
ANTHROPOLOGY
APICULTURE = bee culture
ARCHAEOLOGY
ARCHITECTURE
ARISTOTLE *
ARITHMETIC
ASTROLABE = astrolabes
ASTROLOGICAL MEDICINE*
ASTROLOGY
ASTRONOMY
ATOMISM
BALNEOLOGY
BIOLOGY
BODY AND SOUL*
BOTANY
CABALA
CALENDAR
CHEMISTRY
CHIROMANCY = palmistry
CHOROGRAPHY*
CHRONOLOGY
CIPHERS
CLIMATE
COMETS
CONSILIA* (used for law, not medicine)
COSMETICS
COSMOGONY
COSMOGRAPHY
COSMOLOGY
CREATION
DEAFNESS
DEMONOLOGY
DESCRIPTIVE GEOGRAPHY*
DIAGNOSIS
DICTIONARIES = encyclopedias and dictionaries
DICTIONARY ENCYCLOPEDIAS = encyclopedias and dictionaries
DIET
DISCOVERY = discovery and exploration
DISEASE = diseases
DIVINATION
DOCTRINE OF SIGNATURES*
DREAMS
EARTH SCIENCES
EARTHQUAKES
EDUCATION
EDUCATIONAL REFORM*
ELEMENTS = chemical elements
EMBLEMS
ENCYCLOPEDIAS = encyclopedias and dictionaries
ENGINEERING
ENTOMOLOGY
EPISTEMOLOGY = knowledge, theory of
ETHNOGRAPHY = ethnology
EXORCISM
EXPERIMENTAL PHILOSOPHY*
EXPERIMENTAL SCIENCE*
EXPLORATION = discovery and exploration
FALCONRY
FOSSILS
GARDENING
GENERATION AND CORRUPTION*
GEOGRAPHY
GEOLOGY
GEOMETRY
GRAVITY
GYMNASTICS
HARMONICS = harmonics (music)
HEALTH
HERBAL MEDICINE = herbs, therapeutic use
HERBALS

HERMETICISM = Hermetism
HIEROGLYPHICS
HIEROGRAPHY*
HOROLOGY
HUMORS*
HYDROGRAPHY
HYGIENE
IATROCHEMISTRY*
INK MAKING*
INSTRUMENTATION*
INSTRUMENTS
INVENTIONS
LATITUDE AND LONGITUDE = latitude
LOGIC
MACROCOSM-MICROCOSM = microcosm and macrocosm
MAGIC
MAGNETISM
MATERIA MEDICA
MATHEMATICS
MATHEMATICS AND MUSIC*
MECHANICS
MEDICINE
MELANCHOLY
MENSURATION
METALLURGY
METAPHYSICS
METEOROLOGY
METEORS
MILITARY SCIENCE = military art and science
MILITARY TECHNOLOGY*
MINING = mineral industries
MNEMONICS
MONSTERS
MOTION
MUSIC
MUSIC AND MEDICINE = music therapy
MYSTICISM
NATURAL HISTORY
NATURAL MAGIC*
NATURAL PHILOSOPHY = physics
NATURAL RELIGION = natural theology
NATURAL SCIENCE = natural history, science
NATURAL THEOLOGY
NAVAL ART AND SCIENCE
NAVIGATION
NEO-PLATONISM = neoplatonism
NEW WORLD*
NUMEROLOGY
OCCULT = occultism
OPTICS
ORACLES
ORNITHOLOGY
PARACELSIANISM*
PASSIONS = emotions
PASSIONS OF THE MIND = emotions
PATHOLOGY
PEDAGOGY = education
PERSPECTIVE
PEST TRACTS*
PHILOLOGY
PHILOSOPHERS STONE = alchemy
PHILOSOPHY
PHILOSOPHY OF SCIENCE*
PHYSICS
PHYSIOGNOMY
PHYSIOLOGY
PLATO = Platonism (or ancient philosophy)
PLURALITY OF WORLDS
POPULAR SCIENCE*
POSSESSION = demoniac possession
PRODIGIES = omens
PROGNOSTICATION = forecasting
PROGNOSTICATIONS = forecasting
PROPORTION
PSYCHOLOGY
PYROTECHNICS = fireworks
QUADRIVIUM*
RATIONAL SKEPTICISM*
RELIGION AND SCIENCE
ROSICRUCIANS
SCIENCE
SCIENCES OF MAN*
SECRET WRITING = cryptography
SIGN LANGUAGE
SILK
SORCERY = magic, witchcraft
SOUL
SPAGYRIC MEDICINE = medicine, magic, mystic, and spagiric
SPHERE
SUPERNATURAL

SUPERSTITION
SURGERY
SURVEYING
SYMPATHETIC MEDICINE*
SYMPATHY AND ANTIPATHY*
TAXONOMY = biology, classification
TECHNOLOGY
TOPOGRAPHY*
TRAVEL
UNIVERSAL CHRONOLOGY*
VETERINARY MEDICINE
VOYAGES = voyages and travels
WATER POWER = water-power
WEATHER
WEIGHTS AND MEASURES
WIND = winds
WITCHCRAFT
ZODIAC
ZOOLOGY

Headings marked with an asterisk (*) do not appear in the Library of Congress list of subject headings.

## Selected Bibliography

### *I. Reference Bibliographies*


Allut, Paul. *Etude biographique et bibliographique sur Symphorien Champier*. Lyon, 1859.

*Atlases in the libraries of Chicago: A bibliography and union checklist*. Chicago, 1936.

Baudrier, Henri Louis. *Bibliographie Lyonnaise: Recherches sur les imprimeurs, libraires, relieurs et fondeurs de lettres de Lyon au XVIe siècle*. 12 vols. Paris, 1895–1921.

British Museum, Department of Printed Books. *Catalogue of books printed in the XVth century now in the British Museum*. 10 pts. Reprint. London, 1908–1963.

Brown University. John Carter Brown Library. *Bibliotheca Americana. Catalogue of the John Carter Brown library in Brown University*. 4 vols. in 6. Providence, R.I., 1919–73.

Caillet, Albert Louis. *Manuel bibliographique des sciences psychiques ou occultes*. Paris, 1912.

Choulant, Johann Ludwig. *History and bibliography of anatomic illustration in its relation to anatomic science*. Translated and annotated by Mortimer Frank. New York, 1945.

Choulant, Johann Ludwig. *Bibliotheca medico-historica sive catalogus librorum historicorum de re medica et scientia naturali systematicus*. Hildesheim, 1960.

Church, Elihu Dwight. *A catalogue of books relating to the discovery and early history of North and South America*. New York, 1907.

Collins, Victor. *Attempt at a catalogue of the library of the late Prince Louis-Lucien Bonaparte*. London, 1894.

Copinger, Walter A. *Supplement to Hain's Repertorium*. 2 vols in 3. London, 1895–1902.

Durling, Richard J. *A catalogue of sixteenth-century printed books in the National Library of Medicine*. Bethesda, Md., 1967. 1st suppl. compiled by Peter Krivatsy, 1971.

Ebert, Friedrich Adolf. *Allgemeines bibliographisches lexicon*. 2 vols. in 1. Leipzig, 1821–30.

*Emblemata. Handbuch zur sinnbildkunst des XVI. und XVII. jahrhunderts*. Stuttgart, 1967.

Garrison, Fielding H. *A medical bibliography*. 3d ed. London, 1970. 2d impression, 1976.

*Gesamtkatalog der wiegendrucke*. 8 vols. Leipzig, 1925–40.

Gibson, Reginald Walter. *Francis Bacon: A bibliography of his works and Baconiana to the year 1750*. Oxford, 1950. Suppl., 1959.

Goff, Frederick R., comp. and ed. *Incunabula in American libraries: A third census of fifteenth-century books recorded in North American collections*. Milwood, N. Y. 1973.

Grässe, J. G. T. *Bibliotheca magica et pneumatica*. Leipzig, 1843. Facsimile. Hildesheim, 1960.

Gray, George John. *A bibliography of the works of Sir Isaac Newton*. 2d ed. rev. and enl. Cambridge, 1907.

Green, Henry. *Andrea Alciati and his books of emblems: A biographical and bibliographical study*. London, 1872.

Hain, Ludwig Friedrich Theodor. *Repertorium bibliographicum*. 2 vols. in 4. Stuttgart, 1826–38.

Harrisse, Henry. *Bibliotheca Americana*. New York, 1866.

Houzeau, J.-C., and A. Lancaster. *Bibliographie générale de l'astronomie*. 2 vols in 3. Brussels, 1887.

Howes, Wright. *USiana*. Chicago, 1962.

La Vallière, Louis César de La Baume Le Blanc, duc de. *Catalogue des livres de la bibliothèque de feu M. le duc de La Vallière*. 3 vols. Paris, 1783.

Le Petit, Jules. *Bibliographie des principales éditions originales d'écrivains français du XVe au XVIIIe siècle*. Paris, 1888.

Mieli, Aldo. *Gli scienziati Italiani*. Vol. 1 [no more published]. Rome, 1923.

Neu, John, ed., Samuel Ives, Reese Jenkins, and John Neu, comps. *Chemical, medical, and pharmaceutical books printed before 1800 in the collections of the University of Wisconsin Libraries*. Madison, 1965.

Nissen, Claus. *Herbals of five centuries*. Zurich: L'Art Ancien, 1958.

Pellechet, Marie L. C. *Catalogue général des incunables des bibliothèques publiques de France*. Paris, 1897–1909.

Pforzheimer, Carl H. *The Carl H. Pforzheimer library, English literature, 1475–1700*. 3 vols. New York, 1940.

Phillips, Philip Lee, comp. *A list of geographical atlases in the Library of Congress*. 4 vols in 2. Washington, D.C., 1909–14. Reprint. Amsterdam, 1971.

Polain, Louis. *Catalogue des livres imprimés au 15e siècle des bibliothèques de Belgique*. 4 vols. Brussels, 1932.

Pollard, A. W., and G. R. Redgrave, comps. *A Short-title catalogue of books printed in England, Scotland, and Ireland, and of books printed abroad, 1475–1640*. 2 vols. 2d ed. revised and enlarged by W. A. Jackson and F. S. Ferguson; completed by Katharine Pantzer. London, 1986.

Pritzel, Georg. *Thesaurus literaturae botanicae*. 7 pts. in 5. Leipzig, 1872–77.

Proctor, Robert G. C. *Handlists of books printed by London printers, 1501–1556*. 4 pts. in 2 vols. London, 1913.

Proctor, Robert G. C. *An index to early printed books in the British Museum*. 4 vols. London, 1898–1903. (Reprint, 1938.)

Reichling, Dietrich. *Appendices ad Hainii-Copingeri repertorium bibliographicum*. 8 vols in 2. Milan, 1953.

Sabin, Joseph. *Bibliotheca Americana. Begun by Joseph Sabin and continued by Wilberforce Eames and completed by R. W. G. Vail*. 29 vols. New York, 1868–1936.

Sanz, Carlos. *Bibliotheca Americana vetustissima*. 7 vols. Madrid, 1960.

Sayce, R. A., and D. Maskell. *A descriptive bibliography of Montaigne's 'Essais.'* London, 1983.

Schramm, Albert. *Der bilderschmuck der frühdrucke*. 23 vols. Leipzig, 1920–43. (1924–43.)

*Short-title catalogue of books printed in Italy and of books in Italian printed abroad, 1501–1600, held in selected North American libraries*. 3 vols. Boston, 1970.

Smith, David Eugene. *Rara arithmetica: A catalogvs of the arithmetics written before the year MDCI*. Boston, 1908.

Stevens, Henry Newton. *Ptolemy's geography*. London, 1908.

Stillwell, Margaret B. *Incunabula in American libraries: A second census*. New York, 1940.

Tafuri, Michele. *Catalogo delle edizioni, e traduzioni messe a stampa delle opere di Gio. Gioviano Pontano*. Naples, 1827.

Tchemerzine, Avenir. *Bibliographie d'éditions originales et rares d'auteurs français des XVe, XVIe, XVIIe, et XVIIIe siècles*. 5 vols. Paris, 1977. Originally in 10 vols., 1927–33.

Thorndike, Lynn, and Pearl Kibré. *A catalogue of incipits of mediaeval scientific writings in Latin*. Cambridge, Mass., 1937.

Van Egmond, Warren. *Practical mathematics in the Italian Renaissance: A catalogue of Italian abacus manuscripts and printed books to 1600*. Florence, 1980.

Voulliéme, Ernst H. *Der buchdruck Kölns bis zum ende des fünfzehnten jahrhunderts*. Bonn, 1903.

Wing, Donald, comp. *Short-title catalogue of books printed in England, Scotland, Ireland, Wales, and British America, and of English books printed in other countries, 1641–1700*. 3 vols. New York, 1972.

## *II. Additional References*

*Alchemy and the occult: A catalogue of books and manuscripts from the collection of Paul and Mary Mellon*. 4 vols. New Haven, Conn., 1968–77.

Arber, Agnes. *Herbals, their origin and evolution*. Cambridge, 1912.

L'Art ancien, S.A. *Early books on medicine, natural sciences, and alchemy*. [Lugano, 1926–28?]

Ash, Lee, comp. *Subject collections*. 5th ed. New York, 1978.

[Baillet, Adrien] *Auteurs deguisez sous des noms etrangers; empruntez, supposez, feints à plaisir, chiffrez, renversez, retournez, ou changez d'une langue en une autre.* Paris, 1690. "Liste d'auteurs déguisez," pp. [523]-615.

Benzing, Joseph. *Die büchdrucker des 16. und 17. jahrhunderts in deutschen sprachgebeit.* Band 12 of *Beiträge Buch- und bibliothekwesen.* Wiesbaden, 1963.

*Bibliotheca esoterica.* Paris [1940].

Bird, D. T., comp. *A catalogue of sixteenth-century medical books in Edinburgh libraries.* Edinburgh, 1982.

Butler, Pierce. *A check list of fifteenth-century books in the Newberry Library and in other libraries in Chicago.* Chicago, 1933.

*A catalogue of printed books in the Wellcome Historical Medical Library.* Vol. 1, *Books printed before 1641.* London, 1962.

Cornell University Library. *Catalogue of the Witchcraft Collection in the Cornell University Library.* Introduction by Rossell Hope-Robbins. Milwood, N.Y., 1977.

Cosenza, Mario Emilio, comp. *Biographical and bibliographical dictionary of the Italian humanists and of the world of classical scholarship in Italy, 1300–1800.* 5 vols. and suppl. Boston, 1962.

Cosenza, Mario Emilio, comp. *Checklist of non-Italian humanists, 1300–1800.* Boston, 1966.

Debus, Allen G. *The English Paracelsians.* New York, 1966.

Debus, Allen G. *Man and nature in the Renaissance.* New York, 1978.

Díaz y Díaz, Manuel C. *Los capítulos sobre los metales de las etimologías de Isidoro de Sevilla: Ensayo de edición critica, con traducción y notas.* León, 1970.

Fierz-David, Linda. *The dream of Poliphilo.* New York, 1950.

Firpo, Luigi. *Gli scritti giovanili di Giovanni Botero, bibliografia ragionata.* Florence, 1960.

Heninger, S. K. *The cosmographical glass.* San Marino, Calif., 1977.

Henrey, Blanche. *British botanical and horticultural literature before 1800.* 3 vols. London, 1975.

Hocker, Sally Haines. *Herbals and closely related medico-botanical works, 1472–1753, in the Department of Special Collections, Kenneth Spencer Research Library, and the History of Medicine Collection, Clendening Medical Library.* Lawrence, Kans., 1985.

Hvolbeck, Russell. *Seventeenth-century dialogues: Jacob Boehme and the new science.* Ph.D. diss. University of Chicago, 1984.

Jayawardene, S. A., comp. *Reference books for the historian of science.* London, 1982.

Krummel, D. W. *Bibliographies: Their aims and methods.* London, 1984.

Lipen, Martin. *Bibliotheca realis medica . . . ordine alphabetico sic disposita, vt primo statim intvitv tituli, et svb titvlis autores medici.* Frankfurt, 1679.

[Melzi, Gaetano, conte] *Dizionario di opere anonime e pseudonime di scrittori Italiani o come che sia a venti relazione all' Italia*. 3 vols. in 2. Milan, 1848–59.

Montgomery, John Warwick. *Cross and crucible: Johann Valentin Andreae, 1586–1654, phoenix of the theologians*. The Hague, 1973.

*The natural sciences and the arts: Aspects of interaction from the Renaissance to the 20th century, an international symposium*. Acta Universitatis Upsaliensis. Series 22. Stockholm, 1985.

Neu, John, ed. *Isis cumulative bibliography, 1966–75: A bibliography from Isis critical bibliographies 91–100, indexing literature published from 1965 through 1974*. London, 1980.

Nichilo, Mauro de. *I poemi astrologici di Giovanni Pontano*. Bari, 1975.

Ortroy, Fernand Gratien van. "Bibliographie de l'oeuvre de Pierre Apian." In *Bibliographie moderne* (March-October 1901). Paris, 1901.

Picatoste y Rodriguez, Felipe. *Apuntes para una biblioteca cientifica española del siglo XVI*. Madrid, 1891.

Rohde, Eleanour S. *The old English herbals*. London, 1922.

Salvestrini, Virgilio. *Bibliografia di Giordano Bruno*. 2d [posthumous] ed., ed. Luigi Firpo. Florence, 1958.

Sarton, George. *Introduction to the history of science*. 3 vols. in 5. Baltimore, 1927–48.

Schuler, Robert M. *English magical and scientific poems to 1700: An annotated bibliography*. New York, 1979.

Stillwell, Margaret. *The awakening interest in science during the first century of printing: An annotated checklist*. New York, 1970.

Thorndike, Lynn. *A history of magic and experimental science*. 8 vols. New York, 1934–41.

Walcknaer, Charles Athanase, baron. *Catalogue des livres et cartes géographiques de la bibliothèque de feu M. le Baron Walcknaer*. Paris, 1853.

Walford, A. J. *Guide to reference material*. 2d ed. London, 1966.

Whitrow, Magda, ed. *Isis cumulative bibliography formed from bibliographies 1–90, 1913–1965*. 6 vols. London, 1971–84.

Wightman, W. P. D. *Science and the Renaissance*. Vol. 1, *The emergence of the sciences in the sixteenth century*. Edinburgh, 1962. Vol. 2, *An annotated bibliography of the sixteenth-century books relating to the sciences in the library of the University of Aberdeen*. Aberdeen, 1962.

Woodward, Gertrude L. *English books and books printed in England before 1641 in the Newberry Library*. Chicago, 1939.

Zinner, Ernst. *Geschichte und bibliographie der astronomischen literatur in Deutschland zur zeit der Renaissance*. Stuttgart: Anton Hiersmann, 1964.

# A Checklist of the *Newberry Library's* Printed Books in Science, Medicine, Technology, and the Pseudosciences, ca. 1460–1750

## List of Abbreviations

| | |
|---|---|
| BMC | British Museum Catalogue |
| C | Copinger |
| GW | Gesamtkatalog der Wiegendrucke |
| H | Hain |
| H-C | Hain-Copinger |
| IM | Incunabula Medica |
| n.a. | not applicable |
| n.d. | no date |
| n.p. | no place |
| Pell. | Pellechet |
| Pr. | Proctor |
| Still. | Stillwell |
| STC | Short-title catalogue |

## Abbot, George, abp. of Canterbury, 1562–1633.

A briefe description of the whole worlde. Wherein are particularly described al the monarchies, empires, and kingdomes of the same: with their seuerall titles and situations therevnto adioyning.
London: printed by R. B. for John Browne, 1600.
STC 25
Subject: Cosmography.
Ayer/*7/A15/1600

Running heads in Latin, text in English. History and topography of each country.

## Abbot, George, abp. of Canterbury.

*A briefe description of the whole world. Wherein is paritcularly described all the monarchies, empires, and kingdoms of the same, with their academies.*
London: Printed for W. Sheares, 1656.
STC II A61
Subject: Cosmography
Ayer/*7/A15/1656

Last leaf (pp.329–30): list of universities, with latitude and longitude of some.

## ʾAbd Al-ʾAzīz ibn ʾUthmān, al Ḳabīsī, 12th cent.

*Alchabitius cum commento.*
[Venice: Melchior Sessa, 1512.]
Subject: Astrology.
Case/B/8635/.01

Colophon: . . . ordinatum per Joannem de Saxonia in villa Parisiensi anno 1331. Correctum per artium & medicine doctorem magistrum Bartholomeum de Alte & Nusia. Translated from the Arabic by Joannes Hispalensis.

## ʾAbd Al-ʾAzīz ibn ʾUthmān, al Ḳabīsī.

*Libellvs ysagogicvs Abdil Azi. Id est servi gloriosi dei: qvi dicitvr Alchabitivs ad magisterivm ivditiorvm astrorvm: interpretatvs a Ioanne Hispalensi. Scriptvmqve in evndem a Iohanne Saxonie editvm vtili serie connexvm incipivnt.*
Venice: Erhard Ratdoldt, 1485.
H *617^ BMC (XV)V:290–91^ GW I, 844^ Pr. 4400^ Still. A330^ Pell. 418^ Houzeau 3847^ Goff A-363
Subject: Astrology.
Inc./4400

MS notes in several early hands. 5th ed. (according to Houzeau); 2nd. ed. by Ratdolt. With commentary of Johannes of Saxonia on text of Alchabitius.

## ʾAbd Al-ʾAzīz ibn ʾUthmān, al Ḳabīsī.

*Libellus ysagogicus Abdil Azi. Id est servi gloriosi dei: qvi dicitvr Alchabitivs ad magistervm ivdiciorvm astrorvm interpretatvs a Ioanne Hispalensi scriptvm que in evndem a Iohanne Saxonie editvm vtili serie connexvm incipivnt.*
Venice: Joannes and Gregorio de Gregorii, de Forlivio, brothers, 1491.
H *618^ BMC (XV)V:342^ GW I, 845^ Pr. 4519^ Still. A331^ Goff A-364
Subject: Astrology.
Inc./4519

6th ed., Houzeau 3847. Edited by Bartholomaeus de Alten. MS notes passim in several early hands. With commentary of Johannes of Saxonia on text of Alchabitius.

## Abraham Ben Meir Aben Ezra, 1089[–92]–1167.

*Liber Abraham Iudei de natiuitatibus. Henrici Bate Magistralis compositio astrolabii.*
Venice: Erhard Ratdolt, 1485.
H-C *21^ BMC (XV)V:291^ Redgrave 46
Subject: Astrolabe, astrology.
Inc./4407

## Abraham Ben Meir Aben Ezra.

*Abraam ivdaei de nativitatibvs, hoc est, de duodecim domiciliorum caeli figurarum significatione, ad iudiciariam astrologiam, non solum utilis sed & necessarius plane liber, pristino suo nitori restitutus, per Ioan. Dryandrvm medicum & mathematicum.*
Cologne: Eucharius Ceruicornus, 1537.
Subject: Astrological instruments, constellations, judicial astrology, zodiac.
Case/B/8635/.013

Bookplate mounted on back flyleaf: (Hebrew words) Sicut lilium inter spinas, G. V. L. S. I. D.

## Abraham ben Mordecai Farissol, 1451–1528.

*Itinera mundi, sic dicta nempe cosmographia, autore Abrahamo Peritsol. Latina versione donavit & notas passim adjecit Thomas Hyde.*
Oxford: Sheldonian Theatre, printed for Henry Bonwick, 1691.
Subject: Cosmography, descriptive geography, travel.
Ayer/7/A2/1691

BOUND WITH Bobovius, Albertus, *Tractatus Alberti Bobovii*, q.v.In *Itinera mundi* (p. 178, ch. 29): "De inventione mundi novi magni & terribilis & de dispositione ejus, qui repertus est ultra aequatorem diei: & an possibilis sit res ut detur habitatio in meridie ultra istum aequatorum, ubi vocatur Torrida Zona lingua eorum Christiani & Graeci." Author's preface: mentions climate, habitation, desert sea, river, mountain, woods, aromatics and pepper (herbs) among subjects dealt with. Farissol son of Mordecai of Avignon.

* * *

*Abriss der blü und frucht die in den nidergängigen Indien wächst und newlichen uber sandt worden ist ihrer bäbst: heyl: Paulo dem fünfften zu praesentieren.*
Augsburg: Christoff Mang [1610?].
1 leaf.
Subject: Descriptive botany.
Ayer/109.9/B6/A16/1610
Mentions Cuzco and Peru. Description of native plant, similar to passion flower, "the flower of the Granadilles."

* * *

*Academia Veneta. Svmma librorvm, qvos in omnibvs scientiis, ac nobilioribvs artibvs, variis lingvis conscriptos, vel antea nvnqvam divvlgatos, vel vtilissimis, et pvlcherrimis scholiis, correctionibvs'qve illvstratos, in lvcem emittet Academia Veneta.*
[Venice] In Academia Veneta, 1559.
Subject: Bibliography, metaphysics, science.
Case/ā/9/.014

## Accademia Fiorentina.

*Novae Academiae florentinae opvscvla. Aduersus Auicennam, & medicos neotericos, qui Galeni disciplina neglecta, barbaros colunt.*
Lyon: Seb. Gryphius, 1534.
Subject: Bloodletting, medicine, pharmacology.
Case/Q/.013

Contents: Barbaromastix, seu medicus dicitur Petri Francisci Pauli medici Galenici aduersus Auicennam de venae sectione tractatus. Leonardi Giachini aduersus Mesuem & uulgares medicos omnes tractatus.

* * *

*Accademici occulti, Brescia carmina Acad. Occvltorvm Io. Francisco Commendono Card. ampliss.*
Brescia [Vincentivs Sabiensis] 1570.
Subject: Academies and learned societies, dedicatory poetry, occult.
Case/Y/6809/.014

Poems by [?] Thomae Porcacchi.

## Achillini, Alessandro, 1463–1512.

*Alexandri Achillini Bononiensis philosophi celeberrimi opera omnia in vnvm collecta. De intelligentijs. De orbibus. De vniuersalibus. De physico auditu. De elementis. De subiecto physionomiae & chiromantiae. De subiecto medicinae. De prima potestate syllogismi. De distinctionibus. De proportione motuum. Cvm annotationibvs excellentissimi doctoris Pamphili Montij Bononiensis scholae Patauinae publici professoris.*
Venice: Hieronymvs Scotvs, 1568.
Subject: Medicine, natural philosophy, natural science.
Case/fB/235/.014

MS inscription on verso of back flyleaf (in Italian?) signed A. B. C.

## Achillini, Alessandro.

*Alexander Achillinus Bononiensis de elementis tripartitum opus. In primo libro de intrinsecis elementorum principiis. In duo de elementis. In tertio de ipsorum accidentis differendum.*
Bologna: J. A. de Benedictis, 1505.
Subject: Natural history, natural philosophy, matter, soul, motion
Case/fB/235/.016

MS marginalia in early hands. Section on heaven-earth Averroes, Avicenna, leaf 44. Substances, qualities, accidents. Anatomical research. Demonstrated errors of Galen. Not much influence on pre-Vesalian anatomists because not illustrated, used obscure medieval terminology, sprinkled with Arabic words. Language derived from Mondino.

## Achillini, Alessandro.

*Secreta secretorvm Aristotelis. Philosophorum maximi Aristotelis secretum secretorum: alio nomine liber moralium de regimine principum ad Alexandrum.*
Lyon: Antonii Blanchard, 1528.
Subject: Astrological medicine, hygiene.
Case/Y/642/.A8998

Contents: *Secretum secretorum.* Aristotle, *De signis aquarum ventorum & tempestatum.* Aristotle, *De mineralibus.* A. Aphrodisei, *De intellectu.*Averrois, *De beatudine anime.* A. Achillini, *De universalibus.* A. Macedonis, *Ad Aristotelem de mirabilibus Indie.*

## Acosta, Cristoval, 1515–1580.

*Tractado de las drogas, y medicinas de las Indias orientales, con sus plantas debuxadas al biuo por Christoual Acosta medico y cirujano que las vio ocularmente. En el qual se verifica mucho de lo que escriuio el Doctor Garcia de Orta.*
Bvrgos: Martin de Victoria, 1578.
Subject: Herbal medicine, materia medica.
Ayer/*109.9/B6/A2/1578

Drugs mentioned include: cinnamon, camphor, cardamon, tamarind, pepper, assafoetida, palm and its fruit. List of authors mentioned: L. Fuchs, Amatus Lusitanus, Aristotle, Avicenna, Averroës, Gerardus Cremonensis, Ptolemy, Dioscorides. P. 417: *Tractado del elephante y sus calidades.* All copiously illustrated with woodcuts.

## [Acosta, José de], ca. 1539–1600.

*Geographische vnd historische beschreibung der oberauss grosser landschafft America: welche auch West India, und ihrer grosse halben die New Welt genennet wirt.*
Cologne: Johann Christoffel, 1598.
Subject: Atlas, cosmography.
Case/5A/547/no./5

Vellum cover stamped with coat of arms, "Anthoni Fvgger 1586." A combination of Acosta's *De natura novi orbis* and Wytfliet's *DescriptionisPtolemaicae augmentum.* First German translations of these 2 works. BOUND WITH J. M. Metellus, *Germaniae superior*, and G. Botero, *Theatrvm principum*, qq.v.

## Acosta, José de.

*De natura novi orbis libri dvo et promvlgatione evangelii apvd barbaros, siue, de procvranda indorum salute, libri sex.*
Cologne: printed for Arnoldi Mylij, in officina Birckmannica, 1596.
Subject: Climate, cosmography, natural history, New World.
Ayer/*108/A2/1596d

Chap. XVI, p. 39: Quomodo primi homines ad Indos venire potuerint & quod venerint certa nauigatione. Chap. XVII, p. 43: De magnetis admirabili efficacia & vsv ad nauigandum veteribus ignoto. Bk. 2: De aequinoctialis natura differendum esse. P. 319: De metallorum operatione.

## Acosta, José de

*The natvrall and morall historie of the east and west Indies. Intreating of the remarkeable things of heaven, of the elements, mettalls, plants and beasts which are proper to that country: . . . Written in Spanish by the R. F. Ioseph Acosta, and translated into English by E. G.* [Edw. Grimston?]
London: printed by Val. Sims for Edward Blount and William Aspley, 1604.
STC 94
Subject: Natural history of East and West Indies
Case/F/96/.01

Elements [earth, air, fire, water] wind, earthquakes, volcanoes. Mixtures, compounds [metals, plants, beasts].

## Acuña, Cristóbal, 1597–1680.

*Nvevo descvbrimiento del gran rio de las Amazonas. Por el padre Christoval de Acuña.*
Madrid: Impr. del reyno, 1641.
Subject: Amazon River, discovery, medicinal plants.
Ayer/*1300.5/B8/A18/1641

1st ed. Spanish government ordered it suppressed to prevent Portuguese explorers from benefiting from information about Amazon River. Brunet I, col. 45. Graesse I, p. 17.

## Ady, Thomas.

*A candle in the dark: or, A treatise concerning the nature of witches & witchcraft: being advice to judges, sheriffes, justices of the peace, and grand-jury-men, what to do, before they passe sentence on such as are arraigned for their lives, as witches.*
London: printed for R. I., 1656.
STC II A674
Subject: Witches, witchcraft.
Case/B/88/.016

Definition of witch and witchcraft. Use of divination qualifies person as a witch. Prophesying, through dreams, or as a divine forewarning. Witches related to "planetarians," i.e., astrologers. Conjecturing—use of animals for predicting. Magic and witches ("juglers"). There *are* legitimate astrologers. Witches abuse astrology. Albertus Magnus tells of wonderful things that can be done through knowledge of natural causes and secrets of nature. Use of charms and incantations. Oracles (associated with word "python"). Soothsayers. Necromancers (seek counsel of dead). Witches can infect air and cause diseases. Bk. 3 refutation of King James, *Demonology*.

## Aegidius (Columna) Romanus [Colonna, Egidio, abp.], 1247–1316.

*In Aristotelis analytica posteriora commentum. Excellentissimi artium & sacre theologie doctoris domini Egidu Romani . . . ordinis heremitarum sancti Augustini libros posteriorum Arist. expositio.*
Venice: Bonetus Locatellus for Octavianus Scotus, 1488.
GW VI, 7192^ Goff A-65^
Subject: Mathematics, posterior analytics.
Inc./f 5017/.4

Bound in vellum MS with MS marginalia passim in early hands. Passions, eclipses, liquids, solutions, complexions, temperament.

## Aelianus, Claudius.

*Aeliani de historia animalivm libri xvii. Quos ex integro ac veteri exemplari Graeco, Petrus Gillius vertit. Vna cum noua elephantorum descriptione. Item Demetrii de cvra accipitrum, & de cura & medicina canum, eodem Petro Gillio interprete.*
Lyon: Gvliel. Rovillivm, 1565.
Subject: Descriptive zoology, elephants, veterinary medicine (esp. hawks and dogs).
Case/Y/642/.A1605

## Aeneas Sylvius Piccolomini, Pius II, pope, 1405–1464.

*La discrittione de l'Asia et Evropa di Papa Pio II e l'historia de le cose memorabili fatte in quelle, con l'aggionta de l'Africa, secondo diuersi scrittori, con incredibile breuità e diligenza.*
Vinegia: Vincenzo Vaugris, 1544.
Subject: Cosmography, Geography, Asia and Europe.
Greenlee/5100/P69/1544

Asia, Asia Minor, Discrittione de l'Asia, Evropa, et Africa Il sito, forma, distanza confini, prouincie, cittado, castella, villagi, mari, fonti, fiumi, stagni, laghi, paludi, monti, selue, boschi, pascoli, solitudini.

## Aevoli, Caesare.

*Caesaris Aevoli, Neapolitani de cavsis antipathiae & sympathie rerum naturalium.*
Venice: Francisco Ziletti, 1580.
Subject: Sympathy and antipathy in elements, compositions, planets.
Case/L/O/.018

Intelligence, soul, matter, form, composition, qualities. Universality of system of sympathy and antipathy. See Thorndike VI:414.

## Agricola, Franciscus, d. 1621

*Gründtlicher bericht, ob Zauber und hexerey die ärgste und grewlichste sünd auff erden sey. Zum andern ob die zauberer noch büsz thun vnd selig werden mögen.*
Ingolstadt: Gregorius Hänlin, 1618.
Subject: Magic, witchcraft.
B/88/.018

3d printing.

## Agricola, Georg, 1494–1555.

*De re metallica libri XII quibus officia, instrumenta, machinae, ac omnia denique ad metallicam spectantia, non modo luculentissime describuntur, sed & per effigies, suis locis insertas, adiunctus Latinis, Germaniscisque appellationibus ita ob oculos ponuntur, ut clarius tradi non possint. Eivsdem de animantibvs svbterraneis liber ab autore recognitus: cum indicibus diuersis, quicquid in opere tractatum est, pulchrè demonstrantibus.*
Basel: Hieronymus Frobenius & Nicolaus Episcopius, 1556.
Subject: Chemistry, fossils, metallurgy, mining, occult, smelting, technology.
Wing/fZP/538/.F9283

Metals related to philosophy, subterranean and natural causes, medicine for curatives, astronomy to comprehend the heavenly bodies (?), mensuration. Contains index translating technical words from Latin and Greek into German. 292 woodcuts executed by Blasius Weffring. Dedicated to Augustus I of Saxony. *De animantibus* usually published with *De re metallica*. Deals with subterranean-dwelling animals. In *De re metallica* Agricola brings together an accumulation of information on methods and processes of mining and metallurgy from previous generations. One of first to use observation and research as foundation for a natural science. Deduction from observed phenomena as in the case of attributing mountain sculpture to erosion from wind and water.

## Agricola, Geo. Andreas, 1672–1738.

*A philosophical treatise of husbandry & gardening: being a new method of cultivating and increasing all sorts of trees, shrubs, and flowers. A very curious work: containing many useful secrets in nature, for helping the vegetation of trees and plants, and for fertilizing the most stubborn soils. by G. A. Agricola, M. D. & doctor in philosophy at Ratisbonne . . . The whole revised and compared with the original, together with a preface, confirming the new method, by Richard Bradley, Fellow of the Royal Society.*
London: printed for P. Vaillant, W. Mears, & F. Clay, 1721.
Subject: Botany, diseases of plants.
Case/R/50/.02

## Agrippa von Nettesheim, Heinrich Cornelius, 1486?–1535.

*Opera.*
Lyon: Beringos brothers, n.d.
Subject: Alchemy, chemistry, occult.
Case/B/247/.014

2 vols.

## Agrippa von Nettesheim, Heinrich Cornelius.

*Henrici Cornelii Agrippae ab Nettesheym, De incertitudine & vanitate scientiarum declamatio inuectiua, qua vniuersa illa sophorum gigantomachia plus quam Herculea impugnatur audacia: doceturque nusquam certi quicquam, perpetui, & diuini nisi in solidis eloquens atque eminentia verbi dei latere.*
[Cologne: Melchior Noresianus] 1531.
Subject: Alchemy, arithmetic, divination, geology, geometry, magic, medicine, witchcraft.
Case/B/247/.0186

Includes Art of Raymond Lull; mathematics; chiromancy; geomancy; "De sorte Pythagorica"; optics; "De metallaria"; judicial astrology; metoposcopia [art of discovering character from markings of forehead]; magic; cabala; medicine; surgery; anatomy; veterinary medicine; dietetics. 1st published 1530. BOUND WITH Pirckheimer, Bilibald. *Germaniae ex variis scriptoribus per breuis explicatio.* Nuremberg: Io: Petreius, 1532.

## Agrippa von Nettesheim, Heinrich Cornelius.

*Henrici Cornelii Agrippae ab Nettesheym, de incertitvdine et vanitate scientiarum declamatio inuectiua, ex postrema authoris recognitione.*
Cologne: Theodorus Baumius, 1583.
Subject: Alchemy, arithmetic, divination, geology, geometry, magic, medicine, witchcraft.
Case/B/247/.0188

102 Chaps.

## Agrippa von Nettesheim, Heinrich Cornelius.

*De occvlta philosophia, libri III. Quibus accesservnt, spurius Agrippae liber de ceremonijs magicis. Heptameron Petri de Albano. Ratio compendiaria magiae naturalis, ex Plinio desumpta. Disputatio de fascinationibus. Epistola de incantatione & adiuratione, colllique suspensione. Iohannis Tritemij opuscula quaedam huius argumenti metoposcopia, physiognomia, chiromatitia.*
Paris: Jacob Dupuys, 1567.
Subject: Astrology, cryptography, elements, macrocosm–microcosm, medicine, natural magic, occult, optics.
Case/B/247/.016

Bk. 3: Religion and magic/occult, witches, supernatural, demons. Pp. 505 ff.: "Capita censvrae sive retractionis de magia cum reliquis." Divination by means of augury, by earth, air, fire, water (called geomantia, aeromantia, pyromantia, hydromantia); divination by dreams; passions of the soul; incantations; mathematics, numerology, music, harmony, magic of sun, moon, stars, planets, chance; steganographia. Vol. I published 1531; complete ed. (3v.) 1533. Agrippa follows speculative tradition of natural philosophy.

**Agrippa von Nettesheim, Heinrich Cornelius.**

*Henrici Cornelii armatae militiae eqvitis avrati, & utriusque; iuris doctoris in artem breuem Raymundi Lullij commentaria.*
Cologne: Ioannes Soter, 1533.
Subject: Incantation, magic, mysticism, occult symbolism.
Case/B/240/.52

**Agrippa von Nettesheim, Heinrich Cornelius.**

*Of the vanitie and uncertaintie of artes and sciences: Englished by Ia. San. Gent.*
London: Henry Bynneman, 1575.
STC 205
Subject: Alchemy, arithmetic, divination, geology, geometry, magic, medicine, witchcraft.
Case/B/247/.019

"To the reader: [author's] intent is not to deface the vvoorthiness of arts and sciences but to reprove and detect their euill uses. . . . " Chapters include: Arithmetic, mathematics, Raymond Lull, perspective, [eye]glasses, Pythagoras, geometry, cosmography, finding metals,astronomy, judicial astrology, divination, metoposcopy, physiognomy, palmistry, interpretation of dreams ["onocritica"]. He calls practitioners of dream interpretation "conjectors" and quotes Euripides, "He that does not conject[ure] amiss, a perfect prophet compted is." Natural magic, witchcraft, alchemy, Cabala, theurgy, aurispicy [a kind of soothsaying] or augury (denigrated by Agrippa as being believed in in *his* time by the "superstitious and common sort of men . . . as wordes procaeding out of Gods owne mouth"). Chaps. 82–88 "Of physicke," including practice of medicine, apothecaries, surgery, anatomy, veterinary medicine, dietetics. Agrippa takes negative view of poetry (akin to secret writing) as needful to bawdry. Painting can also lead people astray via "fascination."

**Agrippa von Nettesheim, Heinrich Cornelius.**

*The vanity of arts and sciences. by Henry Cornelius Agrippa, knight, doctor of both laws, judge of the prerogative-court, and counsellour to Charles the fifth, Emperour of Germany.*
London: printed by J. C. for Samuel Speed, 1676.
STC II A790
Subject: Alchemy, arithmetic, divination, geology, geometry, magic, medicine, witchcraft.
Case/B/247/.02

Pp. 113–126 on witchcraft. Biographical preface (sig. A4v) describes book as "his satyrical invective or cynical declamation against the vanity of arts and sciences." Has alphabetical table of contents following dedicatory poem to author.

**Aiguillon, François d', 1566–1617.**

*Opticorum libri sex philosophis iuxtà ac mathematicis vtiles.*
Antwerp: Plantiniana, widow & sons of I. Moreti, 1613.
Subject: Anatomy of the eye, optics, rainbows, reflection–refraction, vision.
Wing/fZP/6465/.P694

Contents: The organ of sight. Object of sight, light, color. Nature of sight. Projections: sphere stereographics, scenographics. Mathematical projection of optics. Organized as propositions, theorems, problems.

**Ailly, Pierre d', 1350–1420.**

*Imago mundi.*
[Incipit] "Imago mundi seu eius ymaginaria descriptum ipsum velut in materiali quodam speculo representans non parum utilis esse videtur ad divinarum elucidationem scripturarum. Cum in eis de partibus ipsius & maxime de locis terre habitabilis mentio sepius habeatur."
[Louvain: Johann von Paderborn, ca. 1483.]
H-C 836, 837^ Pr. 9258^ Pell. 548
Subject: Astrology, astronomy, geography.
Inc./f9258

16 Tracts on Geography, astronomy, calendar, cosmography by Ailly, on concordance of astrology, astronomy, and theology with historical events. Also on astrology, superstition, by Joannes Gerson. Some MS marginalia. Not in Goff. Titles of some of the tracts: *Tractatus de ymagine mundi. De correctione kalendarii. De vero ciclo lunari. Vigintiloquum de concordantia astronomice veritatis cum theologia. Tractatus elucidarius astronomice concordie cum theologia & cum hystorie narratione.*

**Alanus de Insulis [Alain de Lille], 1114?–?1203.**

*Prophetia anglicana Merlini Ambrosii Britanni ex ncvbo [sic] olim (vt hominvm ama est) ante annos mille ducentos circiter in Anglia nati, vaticinia & praedictiones: à Galfredo Monumetensi Latinè conversae: una cum septem libris explanationvm in eandem prophetiam, excellentissimi sui temporis oratoris, polyhistoris & theologi, Alanus de Insvlis, Germani doctoris.*
Frankfurt: Joachim Bratheringius, 1603.
Subject: Astronomy, natural history.
Case/Y/A227/.56

Bk. VI: natural history. Bk. VII: signs of zodiac, astronomy. Typed note on back flyleaf: "This is an extract from Geoffrey of Monmouth's *Historia regum Britanniae*, bk. VII, pts. 3–4." 7 chaps. of explanatory material added by A. de Insulis.

**[Albergati, Vianesio]**

*La pazzia.*
[Venice?] 1546.
Subject: Alchemy (?) [quintessence, sig. iii c], disease, insanity.
Case/B/569/.02

Erroneously attributed to Ortensio Landi. Cf. Melzi.

**Alberti, Leone Battista, 1404–1472.**

*De re aedificatoria.*

Florence: Nicolaus Laurentius, 1485.
H-C 419*^ BMC (XV)VI:630^ GW I, 579^ Pr. 6131^ Pell. 266^ Goff A-215
Subject: Architecture.
Inc./f6131

Edited by Angelo Poliziano. Klebs 32.1.

## Alberti, Leone Battista

*Libri de re aedificatoria decem. Opus integrum et absolutum: diligenterque recognitum.*
Edited by Geoffroy Tory.
Paris: B. Rembolt & Lodovicus Hornken, 1512.
Subject: Architecture, building materials.
Case/*W/2/.024

Discusses rarity of earthquakes, floods, pests (snakes, scorpions) and how to eradicate them (bk. X, Fol. CLXXI).

## Alberti, Leone Battista.

*Opvscoli morali di Leon Batista Alberti gentil'hvomo Firentino* [*sic*]; *nel quali si contengono molti ammaestramenti, necessarij al viuer de l'huomo, cosi posto in dignità, come priuato. Tradotti, & parte corretti da M. Cosimo Bartoli.*
Venice: Francesco Franceschi, Senese, 1568.
Subject: Mathematics, natural history.
Case/Y/712/.A334

Included in contents: "La cifera"; "Piaceuolezze mathematiche"; "Della mosca"; "Del cane." "Piaceuolezze matematiche" [spelling varies thus in text, p. 225] is illustrated. Measurement of heights and distances by triangulation.

## Albertus Magnus, ca. 1200–1280.

*Aureus liber metaphysice diui Alberti Magni epi. Ratisponensis divisus in libros xiii.*
Venice: Johannes & Gregorius de Gregoriis, 1494.
H–C 501*^ BMC (XV)V:345–46^ GW I, 683^ Pr. 4540^ Pell. 322^ Goff A-276
Subject: Metaphysics, mathematics, Pythagoras, esp. bk. XII
Inc./f4540

Some MS marginalia.

## Albertus Magnus.

*Diui Alberti Magni de anima libri tres. De intellectu et intelligibili libri duo.*
Venice: Joannes & Gregorius de Forlivio, 1494.
H-C add. 494^ Pell. 320^ Pr. 4539^ BMC (XV)V:345^ GW 586^ Goff A-222
Subject: Epistemology, natural philosophy, psychology, sense perception.
Inc./f4539

*De anima* (pt. 1) in 3 bks., deals with the senses and perceptions of color, odor, vision, hearing, respiration, taste, touch. Mentions Avicenna and Averroës. Is flesh an organ of touch? "Cap. 4. In quo probatur non esse nisi quinque sensus per naturam organorum." " Tractatus primus est de modo quo cognoscenda est anima. Tractatus secundus secundi libri in quo agit de potentijs anime vegetabilis sigillatim & primo de ordine precedendi & quot sint & que potentiae anime vegetabilis. Liber tertius de anima qui est de apprehensiuis viribus de intus cuius tractatus primus est de viribus anime sensibilis continet infra ascripta capitula." 2d part compares "intellectus" and "intelligibile."

## Albertus Magnus.

*De animalibus.*
Mantua: Paul Johannes von Butzbach, 1479.
H *546^ BMC (XV)VII:931^ GW I,588^ Pr. 6895^ Pell. 340^ Goff A-224^
Subject: Comparative anatomy, natural history, zoology.
Inc./f6895

Also Klebs 14.2^ Osler (IM) 182. Incipit liber Alberti Magni animalium primus qui est de communi diuersitate animalium. Physical descriptions of animals, including man. In 26 Bks. Bk. I: various animals (diversity); II: comparison of humans & other animals; III: similarity of. of, e.g., venous systems of man and other animals (Galen, Aristotle, Avicenna, Averroës); IV: marine animals; V: generative functions; VI: vivparous animals; VII: life and death of animals, illnesses; X: causes of sterility.

## Albertus Magnus.

*De duabus sapientiis et de recapitulatione omnium librorum astronomie.*
[Nuremberg: Kaspar Hochfeder, ca. 1493–96]
H-C *485^ BMC (XV)II:474^ GW I, 718^ Pr. 2300^ Goff A-243
Subject: Astronomy, judicial astrology, necromancy.
Inc./2300

"Scientia iudiciorum astrorum," necromancy, chiromancy. According to *BMC*, watermarks connect this work with *De natura et immortalitate animae cum commento*, Nuremberg, 1493, whose colophon contains Hochfeder's name.

## Albertus Magnus.

*De natura locorum.*
[Vienna: H. Vietor & I. Singrenius for Leonhard and Lucas Alantse] 1514.
Subject: Effect of climate on temperament, geography, natural science.
Ayer/*6/A33/1514

Edited by G. Tanstetter."Habes in hac pagina amice lector Alberti Magni Germani principis philosophi. De natura locorum. Librum mira eruditione & singulari fruge refertum & iam primum, summa diligentia reuisum, in lucem aeditum, quem leges diligentius si uel cosmographia uel phisica profecise te uolueris." Bk. I, de longitudine & latitudine loci per distantias ab orbe. Bk. II, de natura generatorum causata a locis in quibus generantur. Bk. III, distinctio tertia libri de natura locorum habitabilium in quae est cosmographia.

## Albertus Magnus.

*De sex principiis Gilberti Porretani Logica Liber III.*
[Pavia: Christophorus de Canibus, ca. 1490.]
HR 489^ BMC (XV)VII:1010^ GW I,676^ Goff A-269
Subject: Epistemology, physical component of emotions.
Inc./f7093.5

"Incipit commentum summi philosophi Alberti Magni super sex principiis Gilberti Porretani." Leaf b ii verso: on the passions and heat and cold and action of heart: physical component of emotions. Leaf b vi: complexion, heat, cold, "seasons" of the body, melancholy, motions of the spirit, generation and corruption.

## Albertus Parvus Lucius [pseud.]

*Secrets merveilleux de la Magie Naturelle & cabalistique du Petit Albert, traduits exactement sur l'original Latin, intitulé Alberti Parvi Lucii libellus de mirabilibus Naturae Arcanis. Enrichi de Figures mistérieuses, & de la manière de les faire.*
Lyon: heirs of Beringos brothers, 1743.
Subject: Alchemy, astrological medicine, cabala, iatrochemistry, occult, Paracelsianism.
B/85/.027

"Nouvelle edition corigée & augmentée." Among subjects dealt with: love potions, chastity in women, chance, prediction, removal of wrinkles [cosmetics?], improve quality of farm produce, charms to prevent "accidens nuisibles à l'homme," cabalistic method of fixing mercury, Paracelsus, perfumes of the seven planets, mandragora, magic candle, talisman of Mercury, how to make Holy Water, balm made from Holy Water a protection against plague, sympathetic powder, how to make gold artificially, alchemical processes, secret for keeping health, how to prevent gout, smallpox, stones (de la vessie), colic, urinary blockage, hydropsy, stomach problems, talismans of Paracelsus, Cabalistic talismans, astronomy (table of hours of sunrise), pharmacological recipes, simultaneous communication (how to receive instantaneous reply to a communication): "le cadran ou boussole sympatique," alchemical medicine.

## Albin, Eleazar, fl. 1713–1759.

*Insectorum angliae naturalis historia: illustrata iconibus in centum tabulis Aeneis eleganter ad vivum expressis, et istis, qui id poscunt, accuratè etiam coloratis ab authore, Eleazare Albin, pictore. His accedunt annotationes amplae, & observationes plurimae insignes, a Guil. Derham, R. S.*
London: Guilielmi Innys, 1731.
Subject: Entomology.
Case/*W/764/.02

100 colored plates, each dedicated to an individual. Pl. 74, only one not signed, dedicated to an Italian. Descriptions include developmental stages of insect, anatomy, and color, but no taxonomic material.

## Albizzini, Bartolommeo.

*Trattato astrologico di quanto influiscono le stelle dal cielo a prò, e danno delle cose inferiori per tutto l'anno bis 1704. Calcolato alla longit. e latit. della città di Firenze.*
Florence, Anton Maria Albizzini [n.d.]
Subject: Astrology.
UNCATALOGUED

## Albrecht, Johann Wilhelm, 1703–1736.

*Tractatus physicus de effectibus musices in corpus animatum.*
Leipzig: Joann. Christian Martini, 1734.
Subject: Influence of music on body and spirit of man and other animals.
Case/3A/757

## Alciati, Andrea, 1492–1550.

[*Emblemata*]
*Andreae Alciati emblematvm libellvs.*
Paris: C. Wechel, 1536.
Subject: Astrology, emblems.
Case/W/1025/.0165

P. 25: Anchor and dolphin emblem; p. 30: "In parasitos" (lobsters? crawfish?); p. 57: "In astrologos"; p. 109: Map of Tuscany?

## Alciati, Andrea.

[*Emblemata*]
*Les emblemes de Maistre Andre Alciat, mis en rime francoyse* [par Jean Lefevre], *et pvis nagueres reimprime auec curieuse correction.*
Paris: Chrestien Wechel, 1540.
Subject: Emblems.
Case/*W/1025/.019

Text in Latin and French. Emblems not numbered. Colophon lacking. "Contre astrologues" pp. 116 (Latin), 117 (French). Green no. 17.

## Alciati, Andrea.

*Les emblemes de Maistre Andre Alciat, puis nagueres augmentez par ledict Alciat & mis en rime francoise.*
Paris: Chrestien Wechel, 1542.
Subject: Astrology, emblems.
Case/W/1025/.02

Emblems numbered to 115. Penciled note on front flyleaf: 1st ed. with the Latin and French text . . . Over 50 eds. throughout this period, certifying to the great popularity of the work. This is the 2nd ed. with the additional plates, the 1st being published in 1540. See [illegible] *Gravure sur bois*, p. 153. Contre astrologues, no. 53.

## Alciati, Andrea.

[*Emblemata*]
*Andreae Alciati emblematvm libellvs, nvper in lvcem editvs.*
Venice: Aldus sons, 1546.
Subject: Emblems.
Case/*W/1025/.0168

Emblems unnumbered. P. 28 verso is hand with eye in palm. 47 leaves. No "In astrologos."

**Alciati, Andrea.**

*Los emblemas de Alciato; traducidos en rhimas españolas. Añadidos de figuras y de nueuos emblemas en la tercera parte de la obra.* [por Bernardino Daza].
Lyon: Gvulielmo Rovillio, 1549.
Subject: Emblems.
Case/W/1025/.0205

Emblems not numbered. Bk. II begins p. 151, sig. K4: "Contra los que enrriqueçen con el mal commun." Occasional marginal italic glosses e.g., "ottaua rhima." Dialogues, type of rhyme scheme (tercetos, soneto, media rhima) printed beneath title of emblem. P. 167: El vaso de Nestor. P. 186 begins section on trees. P. 210: Los colores has marginal gloss: "A los astrologos."

**Alciati, Andrea.**

[*Emblemata*]
*Clarissimi viri D. Andreae Alciati emblematvm libri dvo.*
Lyon: Ioan. Tornaesius & Gulielmus Gazius, 1554.
Subject: Emblems
Case/*W/1025/.0169

Bk. II, no woodcuts. Emblems numbered to 103.

**Alciati, Andrea.**

[*Emblemata*]
*Diverse imprese accommodate a diverse moralità, con versi che i loro significati dichiarano insieme con molte altre nella lingua italiana non piu tradotte. Tratte da gli emblemi dell' Alciato.*
Lyon: Gvlielmo Rovillio, 1564.
Subject: Emblems.
Case/W/1025/.01692

Reprint of 1551 ed.

**Alciati, Andrea**

[*Emblemata*]
*D. And. Alciati Emblemata denvo ab ipso autore recognita, ac, quae desiderabantur, imaginibus locupletata.*
Lyon: Gvlielmvm Rovill., 1564.
Subject: Emblems.
Case/W/1025/.01693

Pages numbered; emblems numbered in MS. Some emblems have subject headings. P. 104, "Doctorvm," no. 1054, mentions Pico. "Natvra": pp. 106–9 [seasons]. "Astrologia": pp. 110–14. "Arbores": pp. 213–26, gives name of tree, description, symbolism or use [as cypress, which is funerary], uses [if it has edible fruits, or if its wood is useful]. Has extensive commentaries and explanation of emblems.

**Alciati, Andrea.**

[*Emblemata*]
*Omnia Andreae Alciati v.c. emblemata, cvm commentariis qvibvs emblematum omnium aperta origine per Clavdivm Minoem.*
Antwerp: C. Plantin, 1577.
Subject: Emblems.
Case/W/1025/.01694

Commentary of Claude Mingault. Bookplate Theo. L. De Vinne. Has some MS notes in early hand in preliminaries and on back flyleaf. Emblems numbered to 213. "In astrologos," no. 103. Includes section on trees. Has "Syntagma de symbolis." Woodcuts similar to those in early eds. but show a somewhat more refined technique.

**Alciati, Andrea.**

[*Emblemata*]
*Emblemata V. P. Andreae Alciati Mediolanensis ivrisconsvlti; cum facili & compendio explicatione, qua obscura illustrantur, dubiàque omnia soluntur. Per Clavdivm Minoem Diuionensem.*
Antwerp: Christophorus Plantin, 1584.
Subject: Emblems.
Case/*W/1025/.01695

**Alciati, Andrea.**

[*Emblemata*]
*Andreae Alciati emblemata elucidata doctissimus Claudij Minois commentarijs: quibus additae sunt eiusdem auctoris notae posteriores.*
Lyon: heirs of Gulielmi Rouillius, 1614.
Subject: Emblems.
Case/W/1025/.01697

"In astrologos" emblem 103, p. 361.

**Alciati, Andrea.**

[*Emblemata*]
*Andreae Alciati Emblemata, cvm commentariis Clavdii Minois i. c. Francisci Sanctii Brocensis & notis Lavrentii Pignorii Patavini . . . opus copiosa sententiarum, apophthegmatum, adagiorum, fabularum, mythologiarum, hieroglyphicorum, nummorum, picturarum & linguarum varietate instructum & exornatum: . . . cum indice triplici.*
Padua: Petrus Paulus Tozzius, 1621.
Subject: Astrology, emblems, hieroglyphics, materia medica, natural history.
Case/W/1025/.0206

Pp. xiii–xvi, preliminaries: emblem, "Nullus indiga virtus." Ioannis Thuilius to Rev. Father D. Nicolaus de Oddis, abbot. The physician prescribes pills to purge the viscera (with more about pharmacology in commentary). Is medicine a metaphor for emblems and their application? 3 kinds of symbols: historical, physical, ethical. Subject index includes under Astrologia, p. 102, Scyphus Nestoris; p. 103, quae supra nos nihil ad nos; p.104, in astrologos; p. 105, qui alta contemplantur, cadere. Includes index rerum et verborum. Woodcuts larger than (and different from) those in Spanish (1615), French (1540, 1542), Italian (1564), and Latin (1554, 1564). "Corollaria et monita," Frederick Morelli.

**Alciati, Andrea.**

[*Emblemata*]
*V. C. Andreae Alciati Mediolanensis jvrisconsvlti, Emblemata, cum facili, & compendiosa explicatione, qua obscura*

***illustrantur, dubia que omnia solvntur,*** *Clavdivm Minoem Divionensem.*
Antwerp: Henricvs & Cornelivs Verdvssen, 1715.
Subject: Astrology, emblems.
Case/*W/1025/.017

Editio novissima.

## Alciati, Andrea.

*And. Alciati libellvs, De ponderibus et mensuris. Item. Philippi Melanchthonis, de ijsdem, ad Germanorum usum, sententia. Alciati quoque, & Philippi Melanchthonis, in laudem iuris ciuilis, orationes duae elegantissime.*
Hagenau: Iohan Sec, 1530.
Subject: Weights and measures.
Case/R/22/.024

Coloph.: Iohan. Secerii 1531.

## Alciati, Andrea.

*Declaracion magistral sobre las emblemas de Andres Alciato con todas las historias, antiguedades, moralidad, y doctrina tocante a las buenas costumbres, por Diego Lopez.*
Najera: printed by J. de Mongaston at author's expense, 1615.
Green no. 142.
Subject: Emblems, magic, natural history.
Wing/ZP/640/.M743

"Signed by the author" [Lopez] (there is an autograph in an early hand at end of dedication.) Emblems numbered to 210, with explanatory material following each one. "In astrologos" no. 102. "Inviolabiles telo cupidinis," no. 78, a bird in a sort of gyroscope, has commentary, a remedy against "hechizos," bewitchment, and magic arts, according to Paulo Geometria Florentino, whose source was Albertus Magnus.

## Alciati, Andrea.

*Francisci Sanctii Brocensis in inclyta Salmaticensi Academia rhetoricae, Graecaeque linguae professoris, comment. in And. Alciati emblemata.*
Lyon: Gvliel. Rovillivm, 1573.
Subject: Astrology, emblems, natural history.
Case/W/1025/.021

Emblems numbered to 211. In astrologos, no. 103.

## Alexander, of Aphrodisias, fl. 193–217.

*In Aristotilis* [*sic*] *institutione de anima: Hieronymo Donato interprete.*
Brescia: Bernardinus de Misinta, 1495.
H*656^ BMC (XV) VII:989^ GW I:859^ Pr. 7030^ Pell. 441^ Goff A-386.
Subject: Soul, genesis and parts of, senses, body and soul, order of animals according to their powers.
Inc./7030

Leaf c: " ut simul cum corpore permaneat anima." [the soul endures at the same time with the body.] c ii verso, chapter titled "quod inseparabilis est anima a corpore cuius en anima."

## Alexander, of Aphrodisias.

*Alexandri Aphrodisiensis maximi peripatetici, in quatuor libros meteorologicorvm Aristotelis, commentatio lucidissima, Alexandro Piccolomineo interprete. Huc insuper accessit de iride breuis tractatus, eodem Alexandro Piccolomineo authore.*
Venice: Hieronymus Scotus, 1548.
Subject: Meteorology, motion, rainbows.
Case/*Y/642/.025/v.2

In 2 vols. Vol. 1 is in Greek and consists of Olympiodorus's commentary on Aristotle's *Problemata* and Ioannis Grammaticus on book 1 of Aristotle, *Meteorum.* BOUND WITH Alexander of Aphrodisias in vol. 2: *Olympiodori Philosophi Alexandrini in meteora Aristotelis commentarii. Ioannis Grammatici Philoponi Scholia in I. Meteorvm Aristotelis. Ioanne Baptista Camotio philosopho interprete.* Venice: Aldus, 1551.

## Alexander, of Aphrodisias.

*Problemata. In hoc volumine continentur Alexandri Aphrodisei problemata per Georgium Ualla in Latinum conversa. Aristotelis problemata per Theodorum Gazam. Plutarchi problemata per Iohannem Petrum Incensem impressa per Antonium de Strata Cremonensem.*
Venice: Antonius de Strata, 1488/89.
H-C (Add.) 658*^ BMC (XV) V:295^ GW 860^ Goff A-387^ Pell. 439^ Pr. 4594^
Subject: Anatomy, astrology, biology, medicine, natural history.
Inc./f4594

BMC: edited by Joannes Calphurnius. Section on Plutarch: historical and philosophical problems. Effect on man of sounds, odors, other sensory stimuli, geographic locations, study of literature,, mathematics, plants, fruits, saltiness of sea, climate, seasons.

## Alexander, of Aphrodisias.

*Quaestiones Alexandri Aphrodisei naturales, de anima, morales: siue difficilium dubitationum & solutionum libri IIII. . . . Gentiano Herueto Aureliano interprete.*
Basel: Johannes Oporinus, 1548.
Subject: Astronomy, corruptibility of matter, magnetism, motion, optics, soul.
Case/B/181/.A64

Bk. II, ch. 13, opposes Platonic view that all corporeal things ultimately relate or are reducible to the triangle. Corruptibility of matter, p. 25.

## Alexander, of Aphrodisias.

*Qvaestiones natvrales et morales et de fato, Hieronymo Bagolino Veronensi patre, et Ioanne Baptista filio interpretibus. De anima liber primus, Hieronymo Donato patritio Veneto interprete.*
Venice: Hieronymus Scotus, 1549.
Subject: Corruptibility of matter, dreams, epistemology, magnetism, motion, physics, soul.
Case/fB/181/.A642

Also contains *De anima, liber secundus,* on fate, fortune—and Alexander Aphrodisias, *De mistione,* on fate, fortune, 4 elements —both trans. by Angelo Caninio Anglariensi.

## Alfonso X, el Sabio, King of Castile and Leon, 1226?–1284.

*Tabule tabularum celestium motuum diui Alfonsi regis Romanorum et castelle ilustrissimi. Necnon stellarum fixarum longitudines ac latitudines ipsius tempore ad motus veritatem mira diligentia reducte: ac in ipsas primo tabulas Alfonsi canones hue propositiones ordinatissime incipiunt felici sidere.*
Venice: Johannes Hamman, 1492.
H *869^ BMC (XV)V:424^ GW II, 1258^ Pr. 5188^ Still. A474^ Pell. 558^Goff A-535
Subject: Astrology, astronomy.
Inc./5188

Colophon: Exp[plici]unt tabulae tabularum astronomice diui Alfonsi Romanorum & caste[lle] regi ilustrissimi: opera & arte mirifica viri solertis Johannis Hamman de Landoia dictus Hertzog. Curaque sua non mediocri: impressione complete existunt felicibus astris. Anno a prima reorum etherearum circuitione. 8476. Sole in parte. 18. gradiente Scorpij sub celo Ueneto. Anno salutis. 1492. Ed. by Johannes Lucillus Santritter with additions by Augustinus Moravus.

* * *

*Almanach auf das jahr 1481.*
Augsburg: Johann Blaubirer, 1480.
GW Ergänzungen und verbesserungen, 1348/20.
Subject: Almanacs, bloodletting.
Inc./+1580/.8

Folio broadside. Possibly an advertisement for an almanac. Best days for bloodletting; what will occur on day following bloodletting. Cost of almanac, 19 gulden. Joseph Blaubirer, printer.

## Almanacs.

*English almanacs for the year 1674.*
Various authors. A group of 11 almanacs, bound together.
Case/3A/1338/nos. 1–11

## Almanacs.

*English almanacs, 1687.* (Binders title)
A group of 13 almanacs, by various authors, for the year 1687. From the library of James II, king of England, with his monogram gold stamped on front and back covers.
Case/A/1/.256

## Almanacs.

*Almanacs 1691–1700.* (Binders title)
A group of 13 almanacs, by various authors, bound together.
Case/A/1/.0175

## Almanacs.

*Almanacs for the year 1694.*
A group of 12 almanacs, by various authors, bound together.
Case/A/1/.258

## Almanacs.

*English almanacs for the year 1696.*
A group of 13 almanacs, by various authors, bound together.
Case/3A/1342/nos. 1–13

## Almanacs.

*English almanacs for the year 1699.*
A group of 15 almanacs, by various authors, bound together in a vol. with possible ownership mark of King William III on cover and spine.
Case/3A/1343/nos. 1–15

## Almanacs.

*English almanacs, 1717.*
A boxed collection of 17 unbound almanacs by various authors.
Case/A/1/.26

## Almeloveen, Theodor Jansson van, 1657–1712.

*Inventa nov-antiqua. Id est brevis enarratio ortus & progressus artis medicae.*
Amsterdam: Janssonio-Waesbergios, 1684.
Subject: Medicine and science, discoveries and inventions.
R/2/.029

2 vols. in 1. BOUND WITH *Rerum inventarum onomasticon.* Lists inventions alphabetically by subject, e.g., astrologia, astronomia. Names inventor, gives date of invention, offers reference sources. Inventors of medicine include Apis Aegyptius, Aesculapius, Arabs, Babylonians, Chaldeans, Apollo, Simonides first to demonstrate art of memory.

## Alpini, Prosper, 1553–1617.

*De plantis Aegypti liber. In qvo non pauci, qvi circa herbarum materiam irrepserunt, errores, deprehenduntur, quorum causa hactenus multa medicamenta ad vsum medicine admodum expetenda, plerisque medicorum, non sine artis iactura, occulta, atque obsoleta iacerunt.*
Venice: Franciscus de Franciscis, Senese, 1592.
Pritzel 111^ Mieli 88^ Arber 88–90
Subject: Botany, herbal medicine.
Wing/ZP/535/.F85

1st ed. Woodcuts of most plants, with description, habitat, practical uses, especially medicinal. With [not illustrated] his *De balsamo.dialogvs. In qvo uerissima balsami plantae, opobalsami, carpobalsami, & xilobalsami cognitio, plerisque antiquorum atque iuniorum medicorum occulta, nunc elucesit.* Dedication: "Ad illvstrissimos et sapientissimos Patauinae academiae curatores."

## Alpini, Prosper.

*Historiae Aegypti naturalis. Pars prima qua continentur rerum Aegyptiarum libri quatuor. Opus postumum.* [second part]

*Historia naturalis Aegypti, sive, de plantis Aegypti . . . cum observationibus & notis Joannis Veslingii.*
[Leyden]: Gerardus Potvliet, 1735.
Subject: Anthropology, descriptive geography, medicine, natural history, public health, topography of Egypt.
Case/M/071/.03

2 vols. in 1. Bk. 1: the Nile; bk. 2: Knowledge, medicine (chap. ix), public health, sanitation; bk. 3: Geology, stones, minerals, shells, metals, including those used for medicinal purposes, plants, edible and nonedible, fruits, trees; bk. 4: animals, fish, insects, serpents, domesticated and wild quadrupeds, elephants, monkeys, hippos. Books 1 and 2 copiously illustrated.

## Alsario Dalla Croce, Vincenzo, b. 1576.

*Vincentii Alsarii . . . De inuidia, et fascino veterum libellus.*
Lucca: V. Busdrachius, 1595.
Subject: Sympathy and antipathy, witchcraft and occult.
Case/B/8812/.032

End of leaf. 27: "Index rerum magis insignivm, quae in hoc paruo volumine continentur."

## Alsted, Johann Heinrich, 1588–1638.

*Thesaurus chronologiae in quo universae temporum & historiarum series omni vitae genere ita ponitur ob oculos, ut fundamenta chronologiae ex s. literis & calculo astronomico eruantur, & deinceps tituli homogenei in certas classes memoriae causâ digerantur.*
Herborn, 1650.
Subject: Alchemy, astronomy, universal chronology, comets, eclipses, medicine, witchcraft.
F/017/.03

4th ed. Includes "chronologia praecipuarum eclipsum, & observationum astronomicarum" (p. 53); "chronologia medicorum" (p. 460); "chronologia sibyllarum"; "chronologia alchymiae" (Hermes Trismegistus, Moses, expedition of Argonauts, Cumaean sibyll, Diocletian, English Merlin, philosophers stone, King Arthur); chronology of comets, of planetary conjunctions, and synchronic chronology, beginning with year 1 when God created the world. The year corresponding to 1603 A.D. is 5592 by Alsted's reckoning.

## Altomarus, Donatus Antonius.

*Nonnulla opuscula nunc primum in unum collecta, & recognita, cum locis omnibus in margine additis. Quibus vltimo accedit de sanitatis latitudine tractatus, vna cum eiusdem latitudinis tabula denuo in lucem aeditus.*
Venice: Marci de Maria, Salernitani, 1561.
Subject: Galen, gynecology, medicine.
Case/4A/964

" Quod utero pro praeseruatione aborsus venae sectio non competat. Qod ars medica & medici sunt necessarij. De alteratione, de concoctione, de digestione, de purgatione, de sedimento in vrinis." With MS marginalia.

## Altoni, Giovanni, fl. 1604.

*Il soldato di Giov. Altoni Fiorentino della scienza, et arte della gverra.*
Florence: printed by Volcmar Timan German, 1604.
Subject: Arithmetic, geometry, military science.
Case/fU/O/.032

P. 3: "Hauendo queste in pratica, e facile trouare tutto quello, che apporta alla suddetta scienza, dalla quale si viene in cognizione della geometria, e misure matematiche, dell' altezze, e distanze della cosmografia, e geografia, e cose sferiche appartenenti alle regioni celesti, e misure terrestri." BOUND WITH Capobianco, Alessandro. *Corona e palma militare*, q.v.

## Amboise, François d', ca. 1550–1620.

*Discovrs ov traicté des devises. ov est mise la raison et difference des emblemes, enigmes sentences & autres. Pris & compilé des cahiers de feu Messire François d'Amboise . . . par Adrian d'Amboise, son fils.*
Paris: Rolet Bovtonne, 1620.
Subject: Emblems, rules for making, meaning of.
Case/F/0711/.03

P. 66 emblem of a Dr. Antoine Valetus who broke a rule by using human figure.

## Amboise, Jacques Marius d', 1538–1611.

*De rebus creatis & earum creatore liber tripertitus.*
Paris: Federicus Morell, 1586.
Subject: Astrology, astronomy, cosmography, created things, their creator, geometry, seasons, zodiac.
Case/B/72/.032

Chap. 1: "Iacobi Marii Ambosii philosophiae professoris regii disputatio, quam de rebus creatis & earum creatore instituimus, est omnis diuisa in partes tres: quarum vna est de sphaera mundi, altera de aeterno eius parente effectoréque Deo, tertia de idea quam mundam efficere moliens Deus sibi proposuit exemplar."

## Ambrogio, Teseo, 1469–1540.

*Introdvctio in Chaldaicam linguam, Syriacam, atque Armenicam, & decem alias linguas. Charactervm differentium alphabeta, circiter quadraginta, & eorundem inuicem conformatio. Mystica et cabalistica quamplurima scitu digna. Et descriptio ac simulachrum Phagoti Afranij.*
[Pavia: Gio. Maria Simonetta, 1539]
Subject: Cabala, secret and mystical writing.
Wing/ZP/535/.S611

From bookseller's catalogue copy pasted on front flyleaf: "Forty exotic alphabets, including runic alphabet [fol. 206], account of the first printing of the Koran in Arabic" (fol. 200v).

## Amerbach, Vitus

*Poemata Pythagorae et Phocylidis: cum dvplici interpretatione Viti Amerbachij.*
Strassburg: Cratonem Mylius, 1545.

Subject: Philosophy, Pythagoras, Pythagoreans, sorcery.
Case/Y/642/.P997

## Amerbach, Vitus.

*Poemata Pythagorae et Phocylidis. Cum duplici interpretatione Viti Amerbachii.*
Strassburg: heirs of Christian Mylius, 1570.
Subject: Philosophy, Pythagoras, Pythagoreans, sorcery.
Case/Y/642/.P998

To the reader: "Quod autem ita huic enarrationi aliquando inservi theologica, factum est hoc deteriore eventu . . . Quis enim iurisconsultus, quis medicus." P. 112: "Et quod Galenus de sectis medicorum dixit, huc torqueri commodissime potest." P. 114: "A pharmacia & magia esse abstinendum hic versiculus monet."

* * *

*The American Magazine & Historical Chronicle.*
Boston: printed by Rogers & Fowle, 1743–46.
Subject: Medicine, popular science, tar-water.
Case/A/5/.0385

Vols. 1–3 (Sept. 1743–Dec. 1746). Nov. 1746, pp. 489–90 (from a set of boxed pamphlets): "The surprizing effects of tar water in the cure of the small pox: extracted from Mrs. Prior's authentick narrative of the success of tar-water, in above 300 cases, in the asthma, fevers, gout, scurvy, small-pox, and a great many other diseases." Also, in a bound volume, articles on the longitude (p. 263), electricity (pp. 530-37), the spleen (p. 302) in other numbers in this collection (see also pp. 498–503;145–152).

## Amman, Jost, 1539–1591.

*Gynaeceum, siue theatrvm mvliervm, in qvo praecipvarvm omnivm per Evropam in primis, nationvm, gentivm, popvlorvmqve, cvivscvnqve dignitatis, ordinis, status, conditionis, professionis, aetatis, foemineos, habitus videre est, artificiosissimis nvnc primvm figuris, neque usquam ante hac pari elegantia editus, expressos à Iodoco Amano.*
Frankfurt: Sigismund Feyrabend, 1586.
Subject: Women, occupations and dress.
Wing/ZP/547/.F43

"Additis ad singvlas figvras singvlis octostichis Francisci Modii Brvg. Opus cvm ad foeminei sexvs commendationem, tum illorum maximè gratiam adornatum: qui à longinquis peregrenationibus institutae vitae ratione, aut certis alijs de causis exclusi, domi interim variorum populorum habitu, qui est morum indicium tacitum, delectantur.

## Anania, Giovanni Lorenzo d'.

*De natura daemonvm Io. Lavrenti Ananiae, Tabernatus theologi libri quatuor.*
Venice: Aldus, 1589.
Subject: Witchcraft, birth of demons from women, giants, kinds of demons.
Case/B/88/.035

I. Agit de origine & differentia daemonum. II. De eorundem in homines potestate. III. De his, quae daemones per se operantur in nobis. IV. De his, quae hominum auxilio peragunt.

## Anania, Giovanni Lorenzo d'.

*L'vniversale fabrica del mondo, overo cosmografia di M. Gio. Lorenzo d'Anania, diuisa in quattro trattati, ne i quali distintamente si misura il cielo, e la terra, & si discriuono particolarmente le prouincie, città, castella, monti, mari, laghi, fiumi, & fonti, et si tratta delle leggi, & costumi di molti popoli: de gli alberi, & dell' herbe, e d'altre cose pretiose, & medicinali, & de gli inuentori di tutte le cose.*
Venice: Iacomo Vidali for Aniello San Vito di Napoli, 1576.
Subject: Cosmography, encyclopedia.
Ayer/7/A65/1576

2d ed. Following dedication: "Nomi de libri di che s'ha servito in questa fabrica l'auttore." Lists authors whose works were source materials: Vespucci, B. della Casa, Mercator, Saxo Grammaticus, Boethius, Hagi Memet of Persia, and anonymous works: Croniche d'Aragon, Comentarii della Cina, and of East Indies.

## Andreae, Antonius, d. ca 1320.

*Quaestiones Famosissimi doctoris Antonii Andree de tribus principiis rerum naturalium.*
Padua: Laurentius Canozius, 1475.
H-C 990^ BMC (XV)VII:908^ GW II, 1667^ Pr. 6769^ Pell. 634^ Goff A-588
Subject: Form and matter, generation and corruption, natural philosophy.
Inc./*f6769

1st ed. Edited by Thomas Penketh. Leaves 53–60: Andreae, *Formalitates secundum viam.* Leaves 61–66: Aquinas, *Tractatus de ente et essentia.*

## Andreae, Antonius.

*Quaestiones super duodecim libros methaphysice Aristotelis.*
Venice: Bonetus Locatellus for Octavianus Scotus, 1491.
H *979^ BMC (XV)V:439^ GW II, 1662^ Pr. 5026^ Goff A-584
Subject: Form and matter, generation and corruption, metaphysics, motion, physics.
Inc./f5026

Edited by Lucas de Subereto. A few MS marginalia. Incipit: Altissimi doctoris Antonij Andree seraphici ordinis minorum questiones subtilissime super duodecim libros methaphisice Aristotelis feliciter incipiunt. Leaf g (verso): "Item animalis habentis pedes differentiarum scire oportet in quantum habens pedes: quare non en dicendum habentis pedes aliud alatum: aliud non alatum: si quid en bene dicit: sed propter non posse facit hoc."

## Andreae, Antonius.

*Tria principia naturalia.*
Ferrara: Lorenzo Rossi, [Laurentius Rubeus, de Valentia] 1490.
H-C-R 989^ BMC (XV)VI:612^ GW II, 1668^ Pr. 5759^ Goff A-589

Subject: Mathematics, motion, natural philosophy.
Inc./5759

2d ed. Edited by Petrus Malfeta. [Sigs. n-r wanting.] Incipit: Tria principia clarissimi doctoris Antonii Andree secundum doctrinam doctoris subtilis Scoti. Nec non & expositio Francisci Mayronis doctoris illuminati super octo libros physicorum valde vtil & breuis iuxta Ari[stotelis] propositiones & demonstrationes & formalitates eiusdem. 2d tract begins on leaf lij. Has some MS marginalia. Subject of 2d pt.: motion, place, vacuum, body, time, transmutation, change.

## Andreae, Johann Valentin.

*Sereniss. domus Augustae Selenianae princip. juventutis utriusque sexus pietatis, eruditionis, comitatisque exemplum sine pari in perfectae educationis & institutionis normam expositum a Johanne Valentino Andreae.*
Ulm: Balthasar Kühn, 1654.
Subject: Science, education, Rosicrucians.
Case/-Y/682/.A557

From catalogue card: "Correspondence with members of the house of Brunswick and Luneburg between 1649 & 1654, dealing with theology, literature, & science." Letter no. 229 (p. 273), 1653, on medicine or hygiene. P. 331: Zodiacus Augustalis. P. 350: "Joh. Valentini Andreae scripta," a bibliography of his works arranged by subject, with place and year of publication, short title, and format (e.g., 8 [octavo?]). Subjects: theologia, memorialia, funebria, philologia, germania, in lucem promota, propago [genealogy].

## Andrelinus, Publius Faustus, 1450–1518.

*De influentia siderum et querela.*
[Paris] Felix Baligault [for Denise Roce, ca. 1497].
C 467^ BMC (XV)VIII:178^ GW II, 1872^ Pr. 8278^ Still. A612^ Pell. 739^ Goff A-698
Subject: Astral influence, astrology.
Inc./8278

Carmen [a poem or song], a complaint on [or to?] the star struck.

## Angelis, Alexander de, d. 1620.

*In astrologos coniectores libri qvinqve.*
Lyon: Horatius Cardon, 1615.
Subject: Astral influence on conception, fetus, judicial astrology, birth, lower animals.
Case/B/8635/.052

## Anghiera, Pietro Martire d', 1455–1526.

*Opera.*
Seville: Jacobus Corumberger Alemanus, 1511.
Harrisse, Additions no. 41
Subject: Discovery, geography, medicine, natural history, travel.
Ayer/*111/A5/1511

2d issue. *Legato babilonica*: embassy to Babylon, describes pyramids, crocodiles, chirography [hand signals? code?]. *Oceana decas*: natural history, medicine, pharmocopoeia. *Poemata*, leaf gii, ff. mentions magnet.

## Anghiera, Pietro Maritre d'.

[*De insulis*]
*De nvper svb D. Carolo repertis insulis, simulque incolarum moribus, R. Petri Martyris, enchiridion, Dominae Margaritae, Diui Max. Caes filiae dicatum.*
Basel, 1521.
Harrisse 110
Subject: Discovery, natural history, New World, travel.
Ayer/*111/A5/1521

## Anghiera, Pietro Martire d'.

[*De insulis*]
*Petrus Martyr de insvlis nuper repertis. & de moribus incolarum earundum. Descriptio terrae sanctae exactissima, autore Brocardo Monacho, libellus diuinarum scripturarum studiosis, multò vtilissimus. De nouis insulis nuper repertis, & de moribus incolarum earundem, per Petrum Martyrem, res lectu digna.*
Antwerp: Ioannes Steelsius, 1536.
Subject: Astrology, discovery, natural history, New World, travel.
Case/G/61007/.13

BOUND WITH Palaephati, *De non credendis historijs*; Phornuti, *De natura deorum*; *Epitaphivm Isabelle*; Luciani, *De astrologia oratio*. Antwerp: Gregorio Bontio, 1528.

## Anghiera, Pietro Martire d'.

[*De insulis*]
*De nouis insulis nuper repertis & de moribus incolarum earundem, per Petrum Martyrem.*
Antwerp: I. Steelsius, 1536.
Harrisse 218
Subject: Discovery, natural history, New World, travel.
Ayer/*109/B8/1536

In Burchardus de Monte Sion, 13th cent., *Descriptio terrae sanctae exactissima autore Brocardo monacho.*

## Anghiera, Pietro Martire d'.

*Opvs epistolarum Petri Martyris Anglerij Mediolanensis protonotarij apostolici atque a consiliis rerum indicarum nunc primum et natum & mediocri cura excusum quod quidem presenter stili venustatem nostrorum quaeque temporum historie loco esse poterit.*
Madrid: Michael de Eguia, 1530.
Subject: Discovery, exploration, letters of Columbus, navigation, New World.
Ayer/*111/A5/1530a

1st ed. Includes "Index historialium" and Index moralium."

## Anghiera, Pietro Martire d'.

*Opus epistolarum Petri Martyris. . . .cui accesserunt epistolae Ferdinandi de Pulgar.*

Amsterdam: printed by Elzeviriana; Paris: Frederic Leonard, 1670.
Subject: Discovery, natural history, New World.
Ayer/111/A5/1670

Selected letters on scientific subjects: 232 (storm), 237 (cosmographica), 545 (Haeretici Boemi), 610 (de monstris), 619 (de dicessu infantis), 650 (de orbe novo), 656 (de recessu maris), 660 (de insania Valentinus), 675 (de prodigiosis cometis), 682 (de gelu in Augusto), 747 (de terra munda), 769 (de terra motu), 774 (Romana pestis), 783 (de monstris). Bk. XV has some letters on Egypt. Reprint of 1st ed. (1530).

### Anghiera, Pietro Martire d'.

*De orbe nouo Petri Martyris . . . protonotarii Cesaris senatoris decades.*
Madrid: Michael d'Eguia, 1530.
Harrisse 154^ Church 61b^ John Carter Brown: 101
Subject: Discovery, natural history, navigation, New World.
Ayer/*111/A5/1530

### [Anghiera, Pietro Martire d']

[*De orbe novo*]
*Libro primo della historia de l'Indie Occidentali. Svmmario de la historia de l'Indie occidentali cavato da libri scritti dal signor Don Pietro Martyre.*
[Venice, 1534]
Subject: Cannibals, discovery, natural history, navigation, New World.
Ayer/*111/A5/1534

Libro secondo delle Indie occidentali, 1534. Svmmario de la natvrale et general historia de l'Indie occidentali, composta di Gonzalo Ferdinando del Ouiedo. Libro vltimo del svmmario delle Indie occidentali, 1534. De cose de le Indie occidentali, doue si narra di tutto quello ch'è stato fatto nel trouar la prouincia de Peru, ouer de Cusco.

### Anghiera, Pietro Maritre d'.

[*De orbe novo*]
*The decades of the newe worlde or west India. Conteyning the nauigations and conquestes of the Spanyardes, with the particular description of the moste ryche and large landes and ilandes lately founde in the west ocean perteyning to the inheritaunce of the kinges of Spayne.In which the diligent reader may not only consyder what commoditie may hereby chaunce to the hole christian world in time to come, but also learne many secreates touchynge the lande, the sea, and the starres, very necessarie to be knowen to al such as shal attempte any nauigations, or otherwise haue delite to beholde the strange and woonderfull woorkes of God and nature. Wrytten in the Latine tounge by Peter Martyr of Angleria, and translated into Englysshe by Rycharde Eden.*
London: Guilhelmi Powell, 1555.
STC 645
Subject: Discovery, natural history, navigation, New World.
Ayer/*110/E2/1555

Contents: "3 Decades of Peter Martyr. Gonzalus Ferdinandus Ouiedus, The history of the west Indies. The vyage round about the vvorlde by Ferdinand Magalianes" (Magellan). Prices of gems and spices, and where they can be found. Contention for the Indies between Spain and Portugal. The antarctic. "Of Moscouie and Cathay. Letters, etc. Of the generation of metalles and their mynes with the manner of fyndinge the same: written in the Italian tounge by Vannuccius Biringuczius. The discription of the two viages made owt of England into Guinea in Affrike at the charges of certayne marchauntes aduenturers of the citie of London, 1553. The longitudes of regions." 4th decade, under heading, "Of the landes and ilandes lately founde," is included.

### [Anghiera], Pietro Martire d', 1455–1526.

[*De orbe novo*]
*Ander theil, der newen welt vnd indianischen nidergängischen königreichs darinn nicht allein alle warrhaffte thaten und geschichten so sich von der ersten erfindung an ordenlich haben zugetragen verzeichnet erstlichdurch Petrum Martyrem.*
[Basel: Sebastian Henricpetri, 1582.]
Subject: Discovery, natural history, navigation, New World.
Ayer/108/B4/1582

Erste theil, der newen welt vnd indianischen nidergängischen königreichs newe und wahrhaffte history von allen geschichten handlungen thaten, strengem und sträfflichem regiment der spanier gegen den indianern ungläublichem grossem gut von goldt, sylber, edelgestein . . . durch Hieronymum Benzon.
1st 3 decades of Anghiera's *De orbe novo* and his *De nuper repertis insulis.*

### Anghiera, Pietro Martire d'.

*De orbe novo Petri Martyris Anglerii Mediolanensis, protonotarij, & Caroli quinti senatoris decades octo, diligenti temporum obseruatione & vtilissimis annotationibus illustrate, suöoque nitori restitutae, labore & industria Richardi Haklvyti Oxoniensis Angli.*
Paris: Gvillelmvs Avvray, via D. Ioannis Bellouacensis, 1587.
Subject: Discovery, natural history, navigation, New World.
Ayer/*111/A5/1587

Dedication letter by Hakluyt to Walter Ralegh. World map in facsimile.

### Anghiera, Pietro Martire d'.

[*De orbe novo*]
*De nouo orbe, or the historie of the West Indies. Contayning the actes and aduentures of the Spanyardes, which haue conquered and peopled those countries, inriched with varietie of pleasant relation of the manners, ceremonies, lawes, gouernments, and warres of the Indians. Comprised in eight decades. Written by Peter Martyr a Millanoise of Angleria whereof three have been formerly translated into English by R. Eden, whereunto the other fiue, are newly added by the industrie, and painefull trauaile of M. Lok Gent.*
London: printed for Thomas Adams, 1612.
STC 650

Subject: Discovery, natural history, navigation, New World.
Ayer/*111/A5/1612

1st complete English translation of the 8 decades.

## [Anghiera, Pietro Martire d']

[De orbe novo]
*The historie of the VVest-Indies, containing the actes and adventures of the Spaniards, which haue conquered and peopled those countries, inriched with varietie of pleasant relation of the manners, ceremonies lawes, governments, and warres of the Indians. Published in Latin by Mr. Hakluyt, and translated into English by M. Lok.*
London: printed for Andrew Hebb [1625?].
STC 651^ John Carter Brown II, 1:194
Subject: Discovery, natural history, navigation, New World.
Ayer/*111/A5/1625

8 decades. "The fyrst booke of the decades of the ocean, written by Peter Martyr, of Angleria Milenoes."

## Anghiera, Pietro Martire, d'.

[*De orbe novo*]
*The famovs historie of the Indies: declaring the aduentures of the Spaniards, which haue conquered these countries, . . . Comprised into sundry decads. Set forth first by Mr. Hackluyt, and now published by L.M. Gent.*
London: printed for Michael Sparke, 1628.
STC 652
Subject: Discovery, natural history, navigation, New World.
Ayer/*111/A5/1628

2d ed. Peter Martyr's 8 decades.

## Anghiera, Pietro Martire d'.

*De rebus, et insulis nouiter repertis . . . et variis earum gentium moribus.*
Nuremberg: Frideric Peypus, 1524.
Subject: Discovery, natural history, navigation, New World.
Ayer/*655.51/C8/1524b

## Anghiera, Pietro Martire d'.

[*De rebus oceanicis*]
*Petri Martyris ab Angleria, Mediolanen. oratoris clarissimi. Fernandi & Helisabeth Hispaniarum quondam regum consilijs, de rebus oceanicis & orbe nouo decades tres.*
Basel: Ioannes Bebelius, 1533.
Subject: Discovery, natural history, navigation, New World.
Ayer/*111/A5/1516

With this is bound 1) *De orbe nouo decades.* Alcalà, 1516. Contains 3 decades, and preface by Antonio Nebrissensis. 2) *Petrvs Martyr de insvlis nvper inventis, et de moribus incolarum earundem.* Ayer copy does not contain *Legationis babylonicae libri tres.*

## Anghiera, Pietro Martire d'.

[*De rebus oceanicis*]
*Extraict ov recveil des isles nouuellement trouuees en la grand mer oceane ou temps du roy Despaigne Fernand & Elizabeth sa femme, faict premierement en latin par Pierre Martyr de Millan, & depuis translate en languaige francoys. Item trois narrations: dont la premiere est de Cuba . . . La seconde . . . est de la mer oceane . . . La tierce . . . est de la prinse de Tenustitan.*
Paris: Simon de Colines, 1532.
Subject: Discovery, natural history, navigation, New World.
Ayer/*111/A5/1532

Order of first and second narrations is transposed. Abridgement of 1st 3 decades, and also decade 4, plus translations of Cortes material. MS notes on back flyleaf.

## Anghiera, Pietro Martire d'.

*De rebus oceanicis et novo orbe, decades tres, Petri Martyris ab Angleria, Mediolanensis. Item eivsdem, de babylonica legatione, libri III. Et item de rebvs aethiopicis, indicis, lusitanicis & hispanicis opuscula quaedam historica doctissima, quae hodiè non facilè alibì reperiuntur, Damiani a Goes equitis lusitani.*
Cologne: Gervinus Calenius & heirs of Quentel, 1574.
Subject: Discovery, natural history, navigation, New World.
Case/G/131/.03

Marine monsters, p. 114. Cannibals. *De insvlis*, 4th decade, pp.329–66.

## Anghiera, Pierto Martire d'.

*De rebvs oceanicis.* [Another copy]
Ayer/*111/A5/1574

## Anghiera, Pietro Martire d'.

*De rebvs oceanicis.* [Another copy]
Greenlee/A75/A55/1574

## Anianus, fl. 1300?

*Compotus cum commento.*
Rouen: Pierre Regnault, Pierre Violette, and Noel de Harsey, 1500.
GW II, 1979^ Goff A-740^ Still. A652
Subject: Astrology, astronomy, calendar, medicine, palmistry.
Inc./8784.5

Incipit: "Liber qui compotus inscribitur una cum figuris et manibus necessariis tam in suis locis que in fine libri positis." In 4 parts: 1) solar cycle, 2) lunar cycle, 3) de festis mobilibus, 4) four seasons.

## Annius, Joannes, ca. 1432–1502.

*Antiquitatum variarum volumina XVII.*
Joannes Paruo [Jehan Petit] & Jodoco Badio [1515]
Subject: Chronology.
Case/F/009/.7

Bk. 13: Chronographia Etrusca; bk. 15, pt. 3: "Que docuit astronomiam: ob quam illi dicauerunt vrbes; & cognominauerunt coelum & solem." BOUND WITH Polybius, *De primo bello Punico* and with Plutarch, *Parallelia Guarino Veronensi paraphraste opusculum aureum.*

### [Anthonisz, Cornelis], b. ca. 1499.

*The safeguard of sailers, or great rutter. Containing the courses, distances, soundings, floods, and ebbs; with the marks for the entering of sundry harbours of England, Scotland, France, Spain, Ireland, Flanders, Holland, and the sounds of Denmark; also the coast of Jutland and Norway; with other necessary rules of common navigation.*
London: printed by W. G. for Wil. Fisher, 1671.
STC II A3476A
Subject: Navigation.
Case/L/995/.772

### Antonio, Milanese.

*Specchio Vniversale, dell' eccellente Signor Antonio Milanese. Opera à chi brama la sanità vtilissima, & necessaria à i corpi humani Doue si tratta della generatione dell' huomo, e proportione sua, con il modo di conseruarsi in sanità, e i rimedij di tutte l'infirmità dal capo sino à piedi con il pronostico, che comincia l'anno del 1590 & dura in perpetuo.*
[printed in Verona & Brescia, and reprinted in] Genoa, 1590.
Subject: Hygiene, materia medica.
Wing/ZP/535/.M463

"Secreto per la renella & dolor di fianco. Secreto à far moltiplicar il latte alle donne. Secreto à chi sputa il sangue per vena rotta nel petto. Pronostico calcvlato l'anno 1590." BOUND WITH Verini, Giovanni Battista, q.v.

### [Apianus, Petrus] 1495–1552.

[*Cosmographiae introductio*]
*Cosmographiae introdvctio: cvm quibusdam gaeometriae ac astronomiae principijs ad eam rem necessarijs.*
Ingolstadt, 1529.
Subject: Astronomy, cosmography, geometry, hydrography, latitude and longitude, meteorology.
Ayer/*7/A7/1529/[1532]

Title page: Excvsvm Ingolstadii 1529; colophon: Ingolstadij Anno M.D.XXXII.

### [Apianus, Petrus]

[*Cosmographiae introductio*]
*Cosmographiae introdvctio: cvm quibusdam gaeometriae ac astronomiae principijs ad eam rem necessarijs.*
Ingolstadt, 1529.
Subject: Astronomy, cosmography, geometry, hydrography, latitude and longitude, meteorology.
Ayer/*7/A7/1529/[1533]

Title page: Excvsvm Ingolstadii 1529; colophon: Excusum Iogolstadij [*sic*] 1533.

### [Apianus, Petrus]

[*Cosmographiae introductio*]
*Cosmographiae introdvctio cum quibusdam geometriae ac astronomiae principijs ad eam rem necessarijs.*
[Venice], 1533.
Subject: Astronomy, cosmography, geometry, hydrography, latitude and longitude, meteorology.
Ayer/*7/A7/1533b

In quo differat geographia à chorographia. Zodiac, meridians, armillary sphere, planets.

### [Apianus, Petrus]

[*Cosmographiae introductio*]
*Cosmographiae introdvctio cum quibusdam geometriae ac astronomiae principijs ad eam rem necessarijs.*
[Venice: Ioannes de Nicolinis de Sabio for D. Melchior Sessa, 1535]
Subject: Astronomy, cosmography, geometry, hydrography, latitude and longitude, meteorology.
Ayer/*7/A7/1535

### [Apianus, Petrus]

[*Cosmographiae introductio*]
*Cosmographiae introdvctio cum quibusdam geometriae ac astronomiae principijs ad eam rem necessarijs.*
[Venice: Franciscus Bindonis & Mapheus Pasini], 1537.
Subject: Astronomy, cosmography, geometry, hydrography, latitude and longitude, meteorology.
Ayer/*7/A7/1537

Also deals with instruments.

### [Apianus, Petrus]

[*Cosmographiae introductio*]
*Cosmographiae introdvctio cvm qvibvsdam geometriae ac astronomiae principiis ad eam rem necessariis.*
Venice [Io. Antonius de Nicolinis de Sabio for D. Melchior Sessa, 1541].
Subject: Astronomy, cosmography, geometry, hydrography, latitude and longitude, meteorology.
Ayer/*7/A7/1541

Also planets, climate.

### [Apianus, Petrus]

[*Cosmographiae introductio*]
*Cosmographiae introdvctio: cum quibusdam geometriae ac astronomiae principiis ad eam rem necessariis.*
Paris: Gulielmus Cauellat, 1550.
Subject: Astronomy, cosmography, geometry, hydrography, latitude and longitude, meteorology.
Ayer/*7/A7/1550a

No colophon. Running head: Rvdimenta cosmographiae.

### [Apianus, Petrus]

[*Cosmographiae introductio*]
*Cosmographiae introdvctio cvm qvibvsdam geometriae ac astronomiae principiis ad eam rem necessariis.*
[Venice: Petrus de Nicolinis Sabiensis for Melchior Sessa, 1551.]
Subject: Astronomy, cosmography, geometry, hydrography, latitude and longitude, meteorology.
Greenlee/4890/A64/1551

Bound in vellum MS.

**Apianus, Petrus.**

[*Cosmographiae introductio*]

*Cosmographiae introdvctio cum quibusdam geometriae ac astronomiae principiis ad eam rem necessariis.*

[Venice: Franciscus Bindonis, 1554]

Subject: Astronomy, cosmography, geometry, hydrography, latitude and longitude, meteorology.

Ayer/*7/A7/1554

**Apianus, Petrus.**

[*Cosmographicus liber*]

*Cosmographicus liber Petri Apiani mathematici studiose collectus.*

[Landsshut: D. Joannis Weyssenburgers for P. Apian, 1524.]

Subject: Astronomy, cosmography, chorography, geometry, zodiac.

Ayer/*7/A7/1524

1st ed. MS marginalia. Contains volvelles. 2nd part: "De abaco: hoc est partili seu radicali orbis descriptione." (A sort of geographical index?) Appendix: "Horam usualem noctu ex radiis lunaribus mediante compasso prope verum cognoscere."

**Apianus, Petrus.**

[*Cosmographicus liber*]

*Cosmographicvs liber Petri Apiani mathematici, studiose correctus, ac erroribus vindicatus per Gemmam Phrysium.*

Antwerp: Ioannis Grapheus for Roland Bollaert [1529].

Subject: Astronomy, cosmography, chorography, geometry, zodiac.

Ayer/*7/A7/1529

With volvelles.

**Apianus, Petrus.**

[*Cosmographicus liber*]

*Cosmographicvs liber Petri Apiani mathematici, iam denuo integritati restitutus per Gemmam Phrysium.*

[Antwerp: Ioan. Grapheus, 1533]

Subject: Astronomy, cosmography, chorography, geometry, zodiac.

Ayer/*7/A7/1533a

With volvelles.

**Apianus, Petrus.**

[*Cosmographicus liber*]

*Cosmographicvs liber Petri Apiani mathematici, iam denuo integritati restitutus per Gemmam Phrysium. Item eiusdem Gemmae Phrysij libellus de locorum describendorum ratione, & de eorum distantijs inueniendis, nunque ante hoc visus.*

[Antwerp: printed by Ioan. Grapheus, to be sold by Arnold Birckmann, 1534.]

Subject: Astronomy, cosmography, chorography, geometry, zodiac.

Ayer/*7/A7/1534

With volvelles. Includes definition of cosmography, and how it differs from geography and chorography.

**Apianus, Petrus.**

[*Cosmographicus liber*]

*Petri Apiani Cosmographia, per Gemmam Phrysium, apud Louanienses medicum ac mathematicum insignem, denuo restituta. Additis de eadem re ipsius Gemmae Phry. libellis, quos sequens pagina docet.*

Antwerp: Arnold Berckmann [sic], 1540.

Subject: Astronomy, cosmography, chorography, geometry, zodiac.

Ayer/*7/A7/1540

With volvelles. Colophon: "Excusum Antuerpiae opera Aegidij Copenij, 1540. Contains "De tabularum Ptol. et qualiter vnicuiusque regionis, loçiaut oppidi situs, in illis sit inueniendus. Difference between island, peninsula, isthmus, and continent. Constellations. MS notes, fols. 19v, 25.

**Apianus, Petrus.**

[*Cosmographicus liber*]

*Cosmographia Petri Apiani, per Gemmam Frisivm apud Louanienses medicum & mathematicum insignem, iam demum ab omnibus vindicata mendis, ac nonnullis quoque locis aucta. Additis eiusdem argumenti libellis ipsius Gemmae Frisii.*

Antwerp: Gregorio Bontio, 1545.

Subject: Astronomy, cosmography, chorography, geometry, zodiac.

Ayer/*7/A7/1545

**Apianus, Petrus.**

[*Cosmographicus liber*]

*Libro dela cosmographia de Pedro Apiano, el qual trata la descripcion del mundo, y sus partes por muy claro y lindo artificio, augmentado por el doctissimo varon Gemma Frisio . . . con otros dos libros del dicho Gemma, de la materia mesma. Agora nueuamente traduzidos en romance castellano.*

Antwerp: Gregorio Bontio, 1548.

Subject: Astronomy, cosmography, chorography, geometry, zodiac.

Ayer/*7/A7/1548

With volvelles. Works by Gemma: 1) "Libritto dela manera de descriuir o situar los lugares, y de hallar las distancias de aquellos nunca vista hasta agora por Gemma Frison." 2) "Vso del anillo astronomico compuesto por Gemma Frisio." Subjects include mensuration, instrumentation, navigation, astrology; principles of cosmography and geography.

**Apianus, Petrus.**

[*Cosmographicus liber*] [Another copy]

Case/G/117/.04

**Apianus, Petrus.**

[*Cosmographicus liber*]

*Cosmographia Petri Apiani, per Gemmam Frisivm apud Louanienses medicum & mathematicum insignem, iam demum ab omnibus vindicata mendis, ac nonnullis quoque*

*locis aucta. Additis eiusdem argumenti libellis ipsius Gemmae Frisij.*
Antwerp: Gregorio Bontio, 1550.
Subject: Astronomy, cosmography, chorography, geometry, zodiac.
Ayer/*7/A7/1550b

With volvelles.

## Apianus, Petrus.

[Cosmographicus liber]
*Cosmographia Petri Apiani, per Gemmam Frisium apud Louanienses medicum & mathematicum insignem iam demum ac omnibus vindicata mendis, ac nonnullis quoque locis aucta, figurisque nouis illustrata: additis eiusdem argumenti libellis ipsius Gemmae Frisii.*
Paris: Vivantius Gaultherot for D. Martin, 1551.
Subject: Astronomy, cosmography, chorography, geometry, zodiac.
Ayer/*7/A7/1551

With volvelles.

## Apianus, Petrus.

[Cosmographicus liber]
*Cosmographia Petri Apiani, per Gemmam Frisium apud Louaneinses medicum & mathematicum insigne, iam demum ab omnibus vindicata mendis, ac nonnullis quoque locis aucta figurisque nouis illustrata: additis eiusdem argumenti libellis ipsius Gemmae Frisii.*
Paris: Vivantius Gaultherot for D. Martin, 1553.
Subject: Astronomy, cosmography, chorography, geometry, zodiac.
Ayer/*7/A7/1553/[1551]

With volvelles. " Constellations.

## Apianus, Petrus.

[Cosmographicus liber]
*Cosmographia Petri Apiani, per Gemmam Frisivm apud Louanienses medicum & mathematicum insignem, iam demum ab omnibus vindicata mendis ac nonnullis quoque locis aucta. Additis eiusdem argumenti libellus ipsius Gemmae Frisij.*
Antwerp: heirs of Arnold Birckmann, 1564.
Subject: Astronomy, cosmography, chorography, geometry, zodiac.
Ayer/*7/A7/1564

Volvelles. Some MS marginalia.

## [Apianus, Petrus]

[Cosmographicus liber]
*Cosmographia Petri Apiani, per Gemmam Frisium apud Louanienses medicum & mathematicum insignem, iam demum ab omnibus vindicata mendis, ac nonnullis quoque locis aucta, & annotationibus marginalibus illustrata. Additis eiusdem argumenti libellis ipsius Gemmae Frisij.*
Antwerp: Christopher Plantin, 1574.
Subject: Astronomy, cosmography, chorography, geometry, zodiac.
Ayer/*7/A7/1574a

With volvelles. Spheres, planets, chorography, instruments (compass), hydrography, winds, navigation.

## Apianus, Petrus.

[Cosmographicus liber]
*Cosmographia Petri Apiani, per Gemmam Frisium . . . iam demum ab omnibus vindicata mendis, ac nonnullis quoque locis aucta, & annotationibus marginalibus illustrata. Additis eiusdem argumenti libellis ipsius Gemmae Frisij.*
Antwerp: Ioannes Bellerus, 1574.
Subject: Astronomy, cosmography, chorography, geometry, zodiac.
Ayer/*7/A7/1574b

With volvelles. Colophon: Antverpiae ex officina typograph. Ioan. VVithagij. Anno. 1574.

## Apianus, Petrus.

[Cosmographicus liber]
*Cosmographia Petri Apiani, per Gemmam Frisivm apvd Lovanienses medicvm et mathematicvm insignem, iam demvm ab omnibus vindicata mendis, ac nonnullis quoque locis aucta, & annotationibus marginalibus illustrata. Additis eiusdem argumenti libellis ipsius Gemmae Frisij.*
Cologne: heirs of Arnold Birckmann, 1574.
Subject: Astronomy, cosmography, chorography, geometry, zodiac.
Ayer/*7/A7/1574c

Verso of title page: "Contenta in hoc libello *Petri Apiani,* Liber cosmographicus de principijs astrologiae & cosmographiae. Eiusdem partilis descriptio quatuor partiam terrae, videlicet Europae, Asiae, Africae, & Americae. Cui adiecta est descriptio regionis Perv nuper inuentae. Eiusdem de horarum noctis obseruatione. *Gemmae Frisii* De locorum describendorum ratione, deque distantijs eorum inueniendis. Eiusdem Gemmae Frisij de vsv annuli astronomici, in multis locis recenter aucti. *Didaci Pyrrhi Lvsitani* carmen. *Distichon.*

## Apianus, Petrus.

[Cosmographicus liber]
*La cosmographia de Pedro Apiano, corregida y añadida por Gemma Frisio, 1575 medico y mathematico. La manera de descriuir y situar los lugares, con el vso del anillo astronomico, del mismo auctor, Gemma Frisio. El sitio y descripcion de las Indias y mundo nueuo, sacada dela historia de Francesco Lopez de Gomara, y dela cosmographia de Ieronymo Girana Tarragonez.*
Antwerp: Juan Bellero, 1575.
Subject: Astronomy, cosmography, chorography, geometry, zodiac.
Ayer/*7/A7/1575

With volvelles. Fol. 26 v.: Añadidura de Gemma Frisio en laqual cuenta los vientos, segun los marineros modernos. Mas del arte de nauegar y regir la nao por la aguja, y hallar la differencia de longitud y latitud.

## Apianus, Petrus.

[Cosmographicus liber]

*Cosmographie, ou description des quatre parties du monde contenant la situation, diuision, & estendue de chacune region & prouince d'icelles, escrite en latin par Pierre Apian. Corrigée & augmentée par Gemma Frison excellent geographe & mathematicien auec plusieurs autres traitez concernans la mesme matiere, composez par le susdit Gemma Frison, & autres autheurs, nouuellement traduits en langue françoise.*
Antwerp: Iean Bellere, 1581.
Subject: Astronomy, cosmography, chorography, geometry, zodiac.
Ayer/*7/A7/1581

With volvelles. Includes principles of astrology and cosmography; 4 parts of the world: Europe, Asia, America, Africa, with wonders and monsters found there. Apian: hours of night, how to know time at night using simple apparatus. Usage of astronomical ring, astronomical globe, astronomical ray, gnomonic tables of Peurbach, "la fabrique du baston astronomique, par le vulgaire, dit le Baston de Iacob, escrite en Latin par Iean Spang." Extract from Sebastian Münster from his book on mathematical principles dealing withe usage of "baston astronomique."

## Apianus, Petrus.

[*Cosmographicus liber*]
*Cosmographia, siue Descriptio vniuersi orbis, Petri Apiani & Gemmae Frisij, mathematicorum insignium, iam demum integritati suae restituta. Adiecti sunt alij, tum Gemmae Frisij, tum aliorum auctorum eius argumenti tractatus ac libelli varij, quorum seriem versa pagina demonstrat.*
Antwerp: Ioan. Bellerus, 1584.
Subject: Astronomy, cosmography, chorography, geometry, zodiac.
Ayer/*7/A7/1584a

With volvelles. Has added material: p. 223, "Gemmae Frisij medici ac mathematici de vsu globi ab eodem editi"; p. 249, "Coelestis globi. Compositio J. S. auctore"; p. 282, "Gemmae Frisij medici & mathematici, de radio astronomico & geometrico liber"; p. 341, "Tabula gnomonica Georgii Peurbachij"; p. 348, "Fabrica Baculi astronomica, vulgo Baculi Iacobi Ioan. Span. auctore"; [Baculus, Iacobi] "De arte mensvrandi. De mensvratione altitvdinis"; p. 350, "Baculo astronomico, ex lib. I. S. M. e principiis geometriae"; p. 354, "Gemmae Frisij medici ac mathematici astrolabo catholico liber."

## Apianus, Petrus.

[*Cosmographicus liber*]
*Cosmographia sive descriptio vniversi orbis, avctoribvs Petro Apiano, et Gemma Frisio, mathematicorvm insignivm. Cuius huic est de astrolabio catholico libellus, nunc primum à Martino Euerartho in epitomen contractus.*
Antwerp: Arnold Coninx, 1584.
Subject: Astronomy, cosmography, chorography, geometry, zodiac.
Ayer/*7/A7/1584b

With volvelles.

## Apianus, Petrus.

[*Cosmographicus liber*]
*Cosmographie. Ofte beschrijvinge der gheheelder werelt begrijpende de gelegentheyt ende bedeelinghe van elck landtschap ende contreye der selver, gheschreven in Latijn door Petrus Apianus. Ge corrigeert ende vermeerdert door M. Gamma Frisius, excellent geographijn ende mathematicijn, met sommighe andere tractaten van de selve materie, ghemaect van der voorseyden Gamma, ende hier by gevoeght, waer af d'inhout staet 'in't navolgende bladt.*
Amsterdam: Cornelis Claesz, 1609.
Subject: Astronomy, cosmography, chorography, geometry, zodiac.
Ayer/*7/A7/1609

With volvelles. Title page vignette map of world (with Jehovah above. 2 spheres, representing old and new worlds) signed Iodocus Hondius.

## Apianus, Petrus.

*Horoscopion Apiani. generale dignoscendis horis cvivscvmqve generis aptissimum, nêque id ex sole tantum interdiu, sed & noctu ex luna aliisque planetis & stellis quibusdam fixis, quo per vniuersum Rhomanum imperium atque adèo vbiuis gentium vti queas, adiuncta ratione, qua vtaris, expeditissima, nunc ab illo primum inuentum & aeditum.*
[Coloph.: Excusum Ingolstadij, 1555.]
Subject: Astronomical observation, instrumentation, mensuration, surveying.
Ayer/*7/A7/1533i

BOUND WITH Apian's *Introdvctio geographica*, q.v.

## Apianus, Petrus.

*Introductio geographica Petri Apiani in doctissimas verneri annotationes, continens plenum intellectum & iudicium omnis operationis, quae persinus & chordas in Géographia confici potest, adiuncto radio astronomico cum quadrante nouo meteoroscopii loco longe vtilissimo. Hvic accedit translatio noua primi libri Géographiae Cl. Ptolemaei, translationi adiuncta sunt argumenta & paraphrases singulorum capitum: libellus quoque de quatuor terrarum orbis in plano figurationib. Authore Vernero. Locvs etiam pulcherrimus desumptus ex fine septimi libri eiusdem Géographiae Claudii Ptolemaeii de plana terrarum orbis descriptione iam olim & à veterib. instituta Géographis, vna cum opusculo Amirucii Constantinopolitani de iis, quae géographiae debent adesse. Adivncta est & epistola Ioannis de Regiomonte ad Reuerendissimum patrem & dominium D. Bessarionem Cardinalem Nicenum atque patriarchum Constantinopolitanum, de compositione & vsu cuiusdam meteoroscopii armillaris, cui recens iam opera Petri Apiani accessit Torquetum instrumentum pulcherrimum sanè & vtilissimum.*
Ingolstadt, 1533.
Subject: Geography, geometry, instrumentation, latitude and longitude, mathematics, navigation.
Ayer/*7/A7/1533i

Introduction and appendix to book 1 of Ptolemy's *Geographia*, translated by Joannes Werner. Nuremberg,

1514. See Ortroy, no. 101. BOUND WITH Apian's *Horoscopion* and *Quadrans*, qq.v.

## Apianus, Petrus.

*Isagoge in typum cosmographicum seu mappam mundi (vt vocant) quam Apianus sub illustrissimi Saxoniae ducis auspicio praelo nuper demandari curauit.*
[Landshut: Joannes Weyssenburger, ca. 1520.]
Subject: Cosmography, maps.
Ayer/*7/A7/1520

"De diuersis vsibus huius mappae: potissimum his: qui pro limborum declaratione videntur conducere." Uses of a map: geographical chart, climate (wind direction), stars, education, land masses.

## Apianus, Petrus.

*Qvadrans Apiani astronomicvs et iam recens inventvs et nvnc primvm editvs. Huic adiuncta sunt & alia instrumenta obseruatoria perinde noua, ad commodata horis discernendis nocturnis simul & diurnis, idque ex sole, luna, stettisque tum erraticis tum fixis, ad quoque tamen cognitionem cuique citra omnem preceptoris operam facile peruenire licebit. Deinde altitudinis etiam, distantiae, profunditatisque puteorum, turriumque seu aedificiorum adnexae sunt dimensiones, & aquarum quoque ex monte vno in alium deductiones.*
[Ingolstadt], 1532.
Subject: Astronomy, constellations, instrumentation, mensuration.
Ayer/*7/A7/1533i

BOUND WITH Apian's *Introdvctio geographica*, 1533, q.v.

## Apicius, Caelius.

*De re culinaria libri X.*
Lyon: S. Gryphius, 1541.
Subject: Cookery, diet, herbs, hygiene, medicine.
Case/3A/290

Tracts contained in this vol.: Q. Sereni, *Liber de medicina*; Q. Rhemnii Fannij, *Palaemonis de ponderibus atque mensuris*; Caelii Apitii, *Svmmi advlatricis medicinae artificis, de re culinaria libri decem* (separate pagination); B. Platinae Cremonensis, *De tuenda ualetudine, natura rerum & popinae scientia libri X*; Pavli Aeginetae, *De facultatibus alimentorum tractatus, Albano Torino interprete; Appendicvla De conditvrus varijs, ex Ioanne Damasceno, Albano Torino paraphraste; Condimentvm malorvm medicorum, que citria dicuntur; De facvltatibvs alimentorvm, ex Pavlo Aegineta, Albano Torino interprete.* BOUND WITH Celsus, Aulus Cornelius, *De re medica libri octo*, 1542, q.v.

## Apicius, Caelius.

*Caelii Apitii svmmi advlatricis medicinae artificis de re cvlinaria libri X recens è tenebris eruti, & à mendis uindicati, typisque summa diligentia excusi. Praeterea, P. Platinae Cremonensis viri vnde cvnqve doctissimi, de tuenda ualetudine, natura rerum, & popinae scientia libri X, ad imitationem C. Apitii ad unguem facti. Ad haec Pavli Aeginetae de facvltatibvs alimentorvm tractatvs, Albano Torino interprete.*
Basel, 1541.
Subject: Cookery, diet, herbs, hygiene, medicine.
Case/Y/672/.A3

## Apollonius Pergaeus, ca. 262 B.C.–190 B.C.

*Locorum planorum lib. II.*
Glasgow: Rob. & And. Foulis, 1749.
Subject: Plane geometry.
Greenlee/5100/A/64/1749

## Appier Hanzelet, Jean.

*La pyrotechnie de Hanzelet Lorrain ou sont representez les plus rare & plus appreuuez secrets des machines & des feux artificiels propres pour assieger battre surprendre & deffendre toutes places.*
Pont a Mousson: I. & Gaspard Bernard, 1630.
Subject: Engineering, military medicine, military science, pyrotechnics, surveying.
Case/U/444/.042

Leverage, motion, inertia, force, trajectory, cures for burns, compounding of gunpowder, mechanics, triangulation, surveying, geometry.

## Apuleius, Lucius, Madaurensis, 2d cent.

*Opera.* Edited by Joannes Andrea.
[Rome: in the house of Peter Maximus (Conrad Sweynheym and Arnold
Pannartz), 1469.]
H-C 1314*^ BMC (XV)IV:6^ GW II, 2301^ Goff A-934 ^Pr. 3297^ Pell. 923^
Subject: Hermeticism, magic, natural philosophy.
Inc./f3297

Traces of manuscript notes. Colophon: Lucii Apuleii Platonici madaurensis philosophi metamorphoseos liber: ac nonnulla alia opuscula eiusdem: necnon epitoma Alcinoi in disciplinarum Platonis desinunt. Contents: "Apuleii madaurensis philosophi platonici metamorphoseos, siue de asino aureo [10 bks.]. Philosophi platonici floridorum [4bks.]. Apologie siue defensionis magie. De deo Socratis. De dogmate Platonis. Moralis philosophia. Cosmographia, siue de mundo. Trimegisti dialogus . . . Asclepius iste. Alcinoi disciplinarum Platonis epitoma. Episcopi Tropien ad Nicolaum Cusensem cardinalem conuersio cum te intelligam."

## Apuleius, Lucius, Madaurensis.

*Opera.*
Vicenza: Henricus de Sancto Ursio, Zenus, 1488.
H-C*1316^ BMC (XV)VII:1047^ GW II, 2302^ Goff A-935^ Pr. 7172^ Still. A834.
Subject: Hermeticism, magic, natural philosophy.
Inc./f7172

Ed. Joannes Andrea. MS marginalia passim. Contents same as Apuleius, 1469, above.

## Apuleius, Lucius, Madaurensis.

*Opera.*
Venice: Philippus Pincius, 1493.

H *1317^ BMC (XV)V:495^ GW I, 2303^ Goff A-936^ Pr. 5297^ Still. A835.
Subject: Hermeticism, magic, natural philosophy.
Inc./f5297

Ed. Joannes Andrea. Bound in German black-letter rubricated vellum MS. Contents same as Apuleius, 1469, above.

## Apuleius, Lucius, Madaurensis.

*Quae in toto opere continentvr 1) Apuleii Madaurensis, Metamorphoseon siue de asino aureo libri XI. 2) Floridorum libri III. 3) De deo Socratis. 4) Apologie sive defensionis magiae apvd Clavdivm Max. Procons. de dogmate Platonis. 5) Trismegisti dialogus. 6) De mundo siue de cosmographia.*
Florence: heirs of P. Iunta, 1522.
Subject: Hermeticism, magic, natural philosophy.
Case/Y/672/.A403

Ad Favstinvm filivm de dogmate Platonis liber tertius secundus enim desyderatur L. Apvleii Madavrensis philosophi Platonici dialogi, cui titulus Hermes Trismegistus in latinum conuersio.

## Aratus, Solensis, ca.315–ca.245 B.C.

*Phainomena accesserunt annotationes in Eratosthenem et hymnos Dionysii.*
Oxford: Theatro Sheldoniano, 1672.
Madan 2919^ STC II A3596
Subject: Astronomy, constellations.
Wing/ZP/645/.0979

1st ed. of Eratosthenes.

## Arbolaire.

*Le grant herbier en francois. Contenant les qualitez vertuz & proprietez des herbes: arbres: gommes: & semences & pierres precieuses extrait de plusieurs traictez de medicine: comme de Auicenne/ de Rasis/ de/ Constantin/ de Isaac: & Plateaire, selon le commun usage: bien correct.*
Paris: Pierre Le Caron [ca. 1498].
BMC (XV)VII:142–143^ GW II, 2313^ Goff A-945^ Still. A844^ Klebs 508
Subject: Herbals, herbal medicine.
Inc./*8142.5

No other copies in U.S. reported in Goff or *NUC*. Herbs arranged alphabetically. Remedies listed by part of body for which they are specifics. Also remedies for specific diseases, e.g., gout, jaundice, paralysis.

## Arbuthnot, John, 1667–1735.

*An examination of Dr. Woodward's account of the deluge with a comparison between Steno's philosophy and the doctor's, in the case of marine bodies dug out of the earth. by J.A.* [John Arbuthnot] *with a letter to the author concerning an abstract of Agostino Scilla's book on the same subject, printed in the* Philosophical Transactions. *By W. W. [William Wotton] F.R.S.*
London: printed for C. Bateman, 1697.
STC II A3601
Subject: Earth sciences, fossils, geology, natural history, natural theology.
Case/Y/109/.79

Binder's title: *Scarce tracts.* Letter compares Woodward (unfavorably) with Steno.

## Arbuthnot, John.

*Tables of ancient coins, weights, and measures.*
London: printed for J. Tonson, 1727.
Subject: Medicine, navigation, numismatics, weights and measures.
folio/oHC/31/.A8

"A dissertation concerning the navigation of the ancients" (p. 214); "A dissertation concerning the doses of medicines given by ancient physicians" (p.282); other references to pharmacopoeia of the ancients on pp. 293, 300, 311, 318, 322, 325.

## Arcussia, Charles d', ca. 1545–1617.

*La favconnerie de Charles d'Arcvssia de Capre, seignevr d'Esparron, de Pallieres, et dv Revest, en Prouence. Divisee en dix parties . . . auec les portraicts au naturel de tous les oyseaux.*
Rouen: F. Vavltier . . . & Iacqves Besonge . . . 1643.
Subject: Falconry, ornithology, veterinary medicine.
Case/*V/116/.04

"De leurs maladies communes & accidentales, auec les remedes. De l'anatomie d'iceux, par discours et par figure."

## Ardoino, Antonio.

*Examen apologetico de la historica narracion de los naufragios* . . . Part of vol. 1 of a 3 vol. work on discovery in New World. Contents listed 1749under Barcia [Carballido y Zuñiga], Andrés Gonzalez de.
Ayer/108/B2/1749

## Ardoino, Antonio.

*Examen apologetico de la historica narracion de los naufragios, peregrinaciones, i milagros de Alvar Nuñez Cabeza de Baca, en las tierras de la Florida, i del Nuevo Mexico.*
Madrid: J. de Zuñiga, 1736.
Subject: Discovery, natural history, navigation, New World, voyages.
Case/F/96/.073

Bound separately, this appears to be a duplicate of the above tract.

## Argoli, Andrea, 1570–1657.

*Andreae Argoli serenissimi senatvs veneti eqvitis, et in Patauino lyceo mathematicas profitentis pandosion sphericvm. In quo singula in elementaribus regionibus, atque aetherea, mathematicè pertractantur.*
Padua: Paul Frambotti, 1653.
Subject: Alchemy, astronomy, geometry, occult, trigonometry.

Case/L/9116/.044

2d ed. Preface suggests occult: Delphic oracle, Plato, *Timaeus*, astrology.

## Argoli, Andrea.

*Ptolemaevs Parvvs in genethliacis junctus Arabibus auctore Andrea Argolo D. Marci.*
Lyon: Joseph & Peter Vilort, 1652.
Subject: Astrological medicine, astronomy, psychology.
Case/B/8635/.054

## Aristotle, 384–322 B.C.

*Works*, 5 vols. [Greek]
V.1. Venice: Aldus Manutius, 1495.
H-C *1657(1)^ BMC (XV)V:553^ GW II, 2334(1)^ Goff A-959^ Pr. 5547^ Pell. 1175(1)
Subject: Categories, logic.
Inc./*f5547

## Aristotle.

*Works*, vol. 2. *Physica*. [Greek]
Venice: Aldus Manutius, 1497.
H-C *1657 (2)^ BMC (XV)V:556^ Goff A-959^ GW II, 2334^ Pr. 5555^ Pell. 1175
Subject: Medicine, natural history, natural science.
Inc./*f5555

Galen, the heavens, generation and corruption, weather,climate, Theophrastus on fire, wind, stones, unknown author on water and wind.

## Aristotle.

*Works*, vol. 3. *Peri zoon historia*. [Greek]
[With other tracts.] Venice: Aldus Manutius, 1497.
H-C *1657(3)^ BMC (XV)V:555–56^ GW 2334(3)^ Goff A-959^ Pr. 5553^ Pell. 1175(2)
Subject: Natural history.
Inc./5553

Contains 19 tracts of Aristotle. This copy has 259 of 467 leaves.

## Aristotle.

*Works*, vol. 4 *Problemata*. [Greek]
[With other tracts.] Venice: Aldus Manutius, 1497.
H-C *1657(4)^ BMC (XV)V:556–57^ GW 2334(4)^ Goff A-959^ Pr. 5556^ Pell. 1175(4)
Subject: Metaphysics, natural science.
Inc./*f5556

MS notes, esp. in parts of *Metaphysica*. Treatises of Theophrastus usually bound with this work are wanting in this copy. Vol.5 of the *Works* is out of scope for this checklist.

## Aristotle.

*Aristotelis stagiritae omnia qvae extant opera nunc primum selectis translationibus, collationisque cum graecis emendatissimis exemplaribus, margineis scholiis illustrata, & in nouum ordinem digesta: additus etiam nonnullis libris nunquam antea latinitate donatus: Averrois Cordvbensis in ea opera omnes qvi ad nos pervenere commentarii, aliisque ipsius in logica, philosophia, & medicina libri:... Io. Baptistae Bagolini Veronensis labore, ac diligentia.*
Venice: Giunta, 1550–52.
Subject: Metaphysics, natural science.
Case/fY/642/.A81/978

11 vols. Newberry has 8 vols. in 6. Vol. 1: "Averroës in librum Porphyrii." Vols. 2, 3 out of scope. Vol. 4: "De physico avditv . . . cvm Averrois Cordvbensis variis in eosdem commentariis." Vol. 5: "De coelo, de generatione et corruptione, meteorologicorvm, de plantis, cvm Averrois . . . commentariis." Vol. 6: "Ad animalium cognitionem cvm Averrois . . . commentariis" [MS notes on front flyleaf]. Vol. 7: "Extra ordinem natvralivm." Vol. 8: "Metaphysicorvm."

## Aristotle.

*Opervm Aristotelis. . . .nova editio, Graecè & Latinè . . . in quibus plurimae nunc in lucem prodeunt, ex bibliotheca Isaaci Casavboni.*
Geneva: Petrvs de la Roviere, 1605.
Subject: Aristotle, Metaphysics, natural science.
Case/fY/642/.A806

2 vols. In Greek and Latin. Subjects include physics, medicine, biology, optics.

## [Aristotle] Sunczel, Fridericus.

*Collecta et exercitata Friderici Sunczel Mosellani liberalium studiorum magistri in octo libros physicorum Arestotelis [sic] in almo studio Ingolstadensi.*
Hagenau: Heinrich Gran for Johannes Rynman, 1499.
H* 15186^ BMC (XV)III:686^ Pr. 3198^ Still. S770^ Goff S-869
Subject: Natural history.
Inc./3198

BOUND WITH *Textus veteris artis S. isagogarum Porphirii, predicamentorum. Aristotelis simulcum duobus libris perihermenias eiusdem*, ed. Johannes Parrent. Ingolstadt, 1501. Call no. Wing AP 547 .G756.

## Aristotle.

*De natura animalium.*
Venice: Bartholomaeus de Zanis de Portesio for Octavianus Scotus, Modoetiensis, 1498.
H-C *1703^ BMC (XV)V:433^ GW II, 2353^ Pr. 5341^ Pell. 1208^ Goff A-976
Subject: Natural history, zoology.
Inc./f5341

Translated by Theodoro Gaza. *Aristotelis de natura animalium: libri nouem.De partibus animalium: libri quattuor. De generatione animalium: libri quinque.*

## Aristotle.

*Habentvr in hoc volvmine haec Theodoro Gaza interprete. Aristotelis de natura animalium lib. IX; eiusdem de partibus animalium lib. IV; eiusdem de generatione animalium lib. V.*

[Venice: Aldus Manutius, the elder, and Andreas Torresanus, de Asula, 1513.]
Subject: Natural history, zoology.
Case/fY/642/.A8479

BOUND WITH Codrus and Theophrastus, qq.v. Also includes *ProblematvmAristotelis* and *Problematvm Alexandri Aphrodisei.*

## Aristotle.

*Organum* [Greek]
Basel: Isingrin [1549]
Subject: Scientific logic.
Case/3A/286

Preface by Porphyrius. Includes "Categories," "Prior Analytics," and "Posterior Analytics."

## Aristotle.

*Paruulus philosophiae naturalis.*
[Vienna: Hieronymus Vietor Philovallis] 1510.
Subject: Natural philosophy.
Case/Y/642/.A8678

Edited by Joachim Vadianus. MS notes. A compendium, the *Physica* abridged by an unknown scholar.

## Aristotle.

*Parva naturalia.*
Padua: Hieronymus de Durantis, 1493.
Subject: Colors, dreams, imagination, life and death, memory, natural history.
H-C 1719*^ BMC (XV)VII:925^ GW 2430^ Goff A-1019^ Pell. 1216
Inc./f6830.5

Edited by Onofrius de Funtania with commentary of Thomas Aquinas. Includes "De sensu & sensato, de memoria & reminiscentia, de somno & vigilia, de motibus animalium, de longitudine & breuitate vitè, de respiratione & inspiratione, de morte & vita."

## Aristotle.

[*Physica.* (Greek)] *Aristotelis de natvrali avscvltatione, seu de principiis, cum praefatione doctoris Zanchi.*
Strassburg: Vuendelinus Rihelius, 1554.
Subject: Natural science, Physics.
Case/3A/1171

Preface in Latin. Some MS marginalia. Bound with Melanchthon, *Ethicae doctrinae elementa* . . . Wittenberg, 1554.

## [Aristotle Spuria]

*Alexandri Piccolominei in mechanicas quaestiones Aristotelis. Eivsdem commentarivm de certitvdine, mathematicarvm disciplinarum: in quo, de resolutione, diffinitione, & demonstratione: necnon de materia, & fine logica facultatis, quamplura concinentur ad rem ipsam, tum mathematicam tum logicam, maximè pertinentia.*
[Rome: Bladus] 1547.
Subject: Geometry, levers, mechanics, motion.
Case/Y/642/.A8533

Part II has material on mathematics, passions (leaves 92, 93), on fulcrum, leverage (leaf 45).

## [Aristotle Spuria]

*Parafrasi di monsignor Alessandro Piccolomini . . . sopra le mecaniche d'Aristotile, tradotta da Oreste Vanocci Biringucci, gentilomo senese.*
Rome: Francesco Zanetti, 1582.
Subject: Geometry, levers, mechanics, motion.
Case/Y/642/.A853

## [Aristotle Spuria]

*Problemata Aristotelis ac philosophorum medicorumque; com plurium Marci Antonij Zimare Sancti petrinatis problemata, una cum trecentis Aristotelis & Auerrois propositionibus. Item Alexandri Aphrodisij super quaestionibus nonnullis physicis, solutionum liber, Angelo Politiano interprete.*
Venice: Andrea Muschio, 1568.
Subject: Human development, medicine, sciences of man.
Case/-Y/.A/84794

Material on monsters, spleen. P. 117: Verse in Italian on why death is so terrible.

## Aristotle.

*Secreta secretorvm Aristotelis.*
[Lyon: Antonius Blanchard, 1528]
Subject: Alchemy, astrology, meteorology, metaphysics, mineralogy.
Case/Y/642/.A8998

MS marginalia and MS notes on back flyleaf. Contents: "Philosophorum maximi Aristotelis secretum secretorum alio nomine liber moralium de regimine principum ad Alexandrum"; "De signis aquarum ventorum & tempestatum"; "De mineralibus"; "Alexander Aphrodisei de intellectu"; "Auerrois de beatitudine anime"; "Alexandri Achillini bononiensis de vniuersalibus"; "Alexandri macedonis ad Aristotelem de mirabilibus Indie."

* * *

*An arithmetical copy-book, containing the fundamental rules of practical arithmetick, in whole numbers and fractions, vulgar & decimal. Engraven upon copper plates in the hand now commonly used in business for the benefit of youth; that while they may at ye same time be acquainted with, and commit to memory, all ye most useful rules in arithmetick. Intended for the use of the free-school founded by Sr. Joseph Williamson at Rochester; where youth may be boarded and instructed in all parts of mathematical or other learning.*
[London, ca. 1710]
Subject: Tradesmen's arithmetic.
Wing/ZW/7451/.1

* * *

*Arithmetic vulgar and decimal; with the application thereof, to a variety of cases in trade and commerce.*
Boston, N.E.: printed by S. Kneeland and T. Green for T. Hancock, 1729.

Subject: Tradesmen's arithmetic.
Case/4A/460

## Armstrong, John, 1709–1779.

*The art of preserving health: a poem in four books.*
London: printed for A. Millar, 1745.
Subject: Diet, health, hygiene, psychology.
oPR/3316/.A6A8/1745

2d ed. BOUND WITH works by Akenside & Mason (n.a.).

## Arnaldus de Villanova, d. 1313?

*Arnaldi Villanovani philosophi et medici svmmi Opera Omnia. Cum Nicolai Tavrelli medici & philosophi in quosdam libros annotationibus.*
Basel: ex officina Pernea per Conradus VValdkirch, 1585.
Subject: Alchemy, astrological medicine, materia medica.
Case/fY/682/.A743

Among chap. headings in bk. I: "De simplicibvs, De phlebotomia, De dosibvs theriacalibvs, De vinis, Aquis medicinalibus, Regimine sanitatis, Regulae generalis de febribvs, Sterilitate ex parte viri qvam ex parte mvlieris, Remedia contra maleficia, Contra calcvlvm, De epilepsia"; title, bk. II: *Continens exoterica . . . omnium secretorum maximum secretum de verissima compositione naturalis philosophiae, qua omne diminutum reducetur ad soliscum, & luniscum verum,* including: "De modo generationis metallorum, argentum vivum; sulfur extraneum; capitula astrologiae."

## [Arnauld, Antoine, 1612–1694, and Nicole, Pierre]

*Logic: or the art of thinking. Containing . . . many new observations, not only of great use in forming an exactness of judgment in the speculative sciences, but also full of fine reflections, for the common service of life. In four parts.*
London: printed for William Taylor, 1723.
Subject: Geometry, scientific logic.
B/49/.05

Translated by John Ozell. Pt. I: Reflections upon the ideas, or first operation of the mind, called conception. II: Reflections men have made upon their judgment. III: Of argumentation, or reasoning. IV: Of method (p. 365). Chap. 3 (p. 385): Of the method of composition, & particularly that which is observ'd by the geometricians. Chap. 11: The method of sciences reduced to eight principal rules. P. 190: Of two sorts of propositions which are of great use in the sciences, division and definition. Pt. I chap. 9: definition of name & thing from Pascal, of the geometrical mind.

## Arrianus, Flavius, fl. 2d cent.

*Arriani ars tactica, acies contra Alanos periplvs ponti Euxini, periplvs Maris Erythraei, liber de venatione, Epicteti Enchiridion, ejusdem apopthegmata et fragmenta, qvae in Joannis Stobaei florilegio, et in Agelli noctibvs Atticis svpersvnt. Ex recensione & museo Nicolai Blancardi.*
Amsterdam: Janssonio-Waesbergios, 1683.
Subject: Chorography, natural history, navigation.
Case/Y/642/.A/906

Trans. by Joh. Guilielmi Stackius. In Greek and Latin, double columns. Epictetus and Stoics n.a.

* * *

*Ars Memorandi. Rationarum euangelistarum omnium in se euangelia prosa, uersu, imaginibusquè quem mirifice complectens.*
[Phorcae] 1510
Subject: Emblems, mnemonics, mystical symbolism as an aid to memory.
Case/W*/0143/.045

"To the reader" written by George Simler.

## Artemidorus, Daldianus, fl. 2d. cent.

[title in Greek] *Artemidori de somniorum interpretatione libri quinque. De insomniis, quod synesii cuiusdam nomine circumfertur.*
[Venice: Aldus & Andreas Socerus, 1518]
Subject: Interpretation of dreams, sciences of man.
Case/B/578/.036

Title and text in Greek.

## Artemidorus, Daldianus.

*Artemidori Daldiani philosophi excellentissimi de somniorvm interpretatione libri qvinqve. A Iano Cornario medico physico francofordiensi Latine lingua conscripti.*
Lyon: Seb. Gryphivs, 1546.
Subject: Interpretation of dreams, sciences of man.
Case/*B/578/.04

External and internal influences on dream creation (e.g., diet).

## Artemidorus, Daldianus.

*Dell' interpretatione de sogni nouamente di Greco in volgare tradotto per Pietro Lauro Modonense.*
Venice: Gabriel Iolito de Ferrarii, 1542.
Subject: Interpretation of dreams, sciences of man.
Case/B/578/.038

Subjects arranged in alphabetical order at beginning of each vol.

## Articella.

*A collection of ancient medical treatises, chiefly by Hippocrates and Galen.*
Venice: Baptista de Tortis, 1487.
H-C 1870^ BMC (XV)V:325^ Goff A-1144^ Pr. 4633^ Still. 1012
Subject: Medicine.
Inc./f4633

Ed. Franciscus Argilagnes. Contents: "Isagoge Joannitii ad tegni Galieni. Liber pulsuum Philareti. Liber urinarum Theophili. Libri aphorismi Hyppocratis: cum commentis Galieni. Libri pronosticorum Hyppocratis. Commenti Galieni morborum Hyppocratis. Epidimie Hyppocrates &

commentaria Joannis Alexandrini [sic]. Liber Hyppocratis de natura puerorum [etc.]. Commentum Hali supra tegni Galieni. Libellus de diuisione librorum Galieni. Hyppocratis de lege: qui introductoribus dicitur tractatus per Arnaldum de Villa noua de greco in latinum. Coloph.: Explicit Hyppocratis iusiurandum in cuiusdam sui libri principio iuventum: et e greco in latinum conuersum per Petrum Paulum Vergerium faciatum.

## Asclepius.

*Definitiones ad ammonem regem.*
Paris: Adr. Turnebus, 1554.
Subject: Cognitive psychology, Hermeticism, medicine, natural philosophy, occult, mysticism.
Wing/ZP/539/.T852

Greek followed by Latin. Also contains *Mercvrii Trismegisti Poemander, seu de potestate ac sapientia diuina*, q.v.

## Ashmole, Elias, ed., 1617–1692.

*Theatrum Chemicum Britannicum. Containing severall poeticall pieces of our famous English philosophers, who have written the Hermetique mysteries in their owne ancient language. Faithfully collected in one volume, with annotations thereon, by Elias Ashmole, Esq.*
STC II A3987
London: printed by J. Grismond for Nath. Brooke, 1652.
Subject: Alchemy, Hermeticism, magic, natural philosophy, occult.
Case/Y/184/.05

Manuscript notes on last page. Contents: Thomas Norton, *The ordinall of alchimy*; Sir George Ripley, *The compovnd of alchymie*; *Liber patris sapientiae* (in verse); *Hermes bird* (a poem); Geoffrey Chaucer, *The tale of the Chanon's yeoman*; John Dastin, *Dastin's dreame*; *Pearcke the black monke vpon the elixir*; *The worke of Rich[ard] Carpenter*; Abraham Andrewes, *The hunting of the greene lyon*; Thomas Charnock, *The breviary of naturall philosophy*; William Bloomefield, *Bloomefield's blossoms*; Sir Edward Kelle, *Worke*; *Kelley to G. S. gent. concerning the philosophers stone*; *Testamentum John Dee*; Thomas Robinson, *De lapide philosophorum*; *Experience and philosophy* (in verse); W. B., *The magistery*; *Anonymi or severall workes of unknowne authors*; John Gower, *Concerning the philosophers stone*; the following 5 works by Sir George Ripley: 1) *The vision of Sir George Ripley*, 2) *Verses belonging to an emblematicall scrowle supposed to be invented by Geo. Ripley*, 3) *The mistery of alchymists*, 4) *The preface prefixt to Sir Geo. Ripley's Medvlla*, 5) *A short worke*; John Lydgate, *Translation of the 2d epistle of Alexander to Aristotle out of Aristotle's Secreta secretorum*; the following 4 anonymi: 1) *1st to the 6th chapters of the worke*; 2) *The hermet's tale*; 3) *A discription of the stone*; 4) *The standing of the glasse for the tyme of the putrifaction and the congelation of the medicine*; D. D. W. Bedman [Redman], *Aenigma philosophicum*; *Fragments coppied from Thomas Charnock's owne hand writing* ("In some coppies I have fovnd these verses placed before Pearce the black monk vpon the elixir"); *Fragments* (pp. 434–36, with the initials R. B. [Robert Boyle]; *Annotations and discovrses vpon some part of the preceding worke.*

## [Ashmole, Elias, ed.]

*The way to bliss. In three books made publick, by Elias Ashmole, esq., qui est Mercuriophilus anglicus.*
London: printed by J. Grismond for N. Brook, 1658.
STC II A3988
Subject: Alchemy, iatrochemistry.
Case/B/8633/.052

"Of long life; "Of health"; "Of youth"; "Of riches [minerals, metals]"; "Of mending and bettering the state of man's body"; "That the philosopher's stone is able to turn all base metals into silver and gold"; "That the philosopher's stone will turn base metalls with as much advantage as we will." "To the reader: I lately met with a pretended copy of the following discourse . . . intended, that the world should take it for the child of one Eugenius Theodidactus, being (by re-baptization) called the Wise Man's Crown, or Rosie–Crucian Physick; . . . loth I was any longer to keep my perfect copy by my side. . . . and resolved . . . to venture it abroad. . . . As for our author, he was without doubt an Englishman. Onely this, I can modestly aver, that my copy was a transcript of the original. The work seems to be written about the beginning of the last (or end of the former) century. With marginal notes of Dr. Everard (they being added to a transcript of this work . . . ) [obtained from] a very intimate friend."

## Assin y Palacios de Ongoz, José.

*Florilegio theorico-practico, nuevo curso quimico, en que se contienen quatro reflexiones generales. La primera sobre la fisico-mecanica formacion de los principios inmediatos, ò proximos de los mixtos. Y las otras tres sobre los reynos mineral, vejetal, y animal, con muchas curiosas nuevas operaciones quimicas, y sobre cada vna su particular reflexion. Compuesto por D. Joseph Assin y Palacios de Ongoz, botocario colegial del antiguo colegio de la ciudad de Zaragoça.*
Madrid: Antonio Gonçalez de Reyes, 1712.
Subject: Alchemy, iatrochemistry, organic chemistry, materia medica.
Ayer/656.8/B8/A84/1712

## Astros, I. G. d'.

*Lov trimfe de la lengovo gascovo. Avs playdeiats de las quoüate sasous, & deous quoüate elemens, daoüant lou pastou de Loumaigno.*
Toulouse: Ian Bovdo, 1643.
Subject: Elements, seasons.
Bonaparte/Coll./No./4070

In verse. P. 57: "A Monsieur d'Astros sur ses oeuures des quatre saisons & des quatre elemens. . . . Dessous le ciel gascon, on n'a veu rien si beau/ Que tes quatre saisons rejointes de nouveau/ A tes quatre elemens en meme sympathie."

## Astruc, Johanne, 1684–1766.

*De morbis venereis libri novem in quibus disseritur tum de origine, propagatione & contagione horumce affectuum in genere: tùm de singulorum naturâ, aetiologiâ & therapeiâ,*

*cum brevi analysi & epicrisi operum plerorumque, quae de eodem argumento scripta sunt.*
Paris: Guillelmus Cavelier, 1740.
Subject: Medicine, venereal disease.
UNCATALOGUED

Editio altera. 2 vols., of which the library holds vol. 2 only.

## Atwell, George.

*The faithfull surveyor: teaching how to measure all manner of ground exactly, by the chain onely; also thereby to take distances of a mile space, and the situation of any building. Shewing likewise the making and use of a new instrument, called a pandoron; which supplies the use of the plain-table, theodelite, quadrant, quadrat, circumferentor, and any other observing instrument.*
Cambridge: printed for William Nealand, 1662.
STC II A4164
Subject: Arithmetic, instrumentation, mathematics, mensuration, plane and solid geometry, surveying.
Case/3A/2290

Atwell was "late teacher of mathematicks in Cambridge."

## Aubery, Jean.

*L'antidote d'amovr. Auec vn ample discours, contenant la nature & les causes d'iceluy, ensemble les remedes les plus singuliers pour se preseruer & guerir des passions amoureuses. Par Iean Avbery docteur en medecine.*
Paris: Clavde Chappelet, 1599.
Subject: Mental and physical symptoms of love, remedies.
Case/B/529/.054

Passions, physiological manifestations of love (heartbeat), love and numbers, natural philosophy, love philtres, eyes (hypnotism?).

* * *

*Auctoritates Aristotelis et aliorum philosophorum.*
Bologna: Ugo Rugerius, 1488.
H-C 1930*^ BMC (XV)VI:808^ Goff A-1200
Subject: Aristotle.
Inc./6567A

"Incipit propositiones vniuersalies Aristotelis."
Bookplate: "Edward Gibbon esq."

* * *

[*Auctoritates Aristotelis*]
Autoritates [*sic*] *Aristotelis olim recte philosophantum facile principis. insuper & Platonis. Boetii Senece. Apulei Aphricani. Porphirii. Auerroys Gilberti Porritani necnon quorundam aliorum nouissime castiori studio recognite et pigmentate.*
Cologne: Henricus Quentel, 1504.
Subject: Aristotle.
Case/Y/642/.A8088

## Avellar, Luiz do.

[*Nox Atticae hoc est, dialogvs de impressione methereologica, et cometa anni domini 1618*]
[Coimbra: N. Carvalho, 1619.]
Subject: Comets.
Greenlee/4504/.P855

Missing title page supplied from Barbosa Machado, Diego, *Biblioteca Lusitana.* Dialogue between Philosophus, Scholasticus, and Astrologus on comets: where, when, and why.

## Aventinus, Joannes, 1477?–1534.

*Annalivm Boiorvm libri septem.*
Ingolstadt: Alexander & Samuel VVeissenhorn, 1554.
Subject: Topography.
Case/fF/47908/.046

"Boioariae descriptio, situs, telluris, qualitas, & gentis mores.

## Aventinus, Joannes.

*Johannis Auentini des hoch gelerten weitberümbten Beyerischen geschichtschreibers chronica darinn mit allein dess gar alten hauss Beyern, keiser, könige, herzogen, fürsten, graffen, freyherrn geschlechte herkommen, stamm und geschichte, sondern auch der vralten Teutschen ursprung herkommen, sitten, gebreuch, religion, mannliche vnd treffliche thaten, so sie fast biss zu dieser zeyt allen thalben nit allein im Teutschland vnd Europa, sondern auch in Asia vnd Africa, auch vor Christi vnsers seligmachers geburt, gethan haben zuin fleissigsten beschriben vnn auss allerley chronicken handschrifften alten freyheiten, ubergaben, brieffen, salbüchern, reimen, liedern, vnd andern glaubwirdigen monumenten vnd schrifften zusammen getragen vnd in acht bücher getheilt.*
Frankfurt, 1566.
Subject: Chorography, topography.
Case/fF/47908/.048

## Aventinus, Joannes.

*Ioannes Aventini viri Cl. Annalivm Boiorvm, sive veteris Germaniae libri VII. In qvibvs non solvm Boiariae, sive Bavariae regionvm, vrbivm, flvminum, & syluarum, sed etiam Germaniae veteris descriptio chorographica populorum, religionis, legum, constitutionum & morum, vt & heroum, ducum & regum veterum & recentiorum Germaniae, bellorum & verum gestarum, migrationum & expeditionum historia aedo luculenta & fidelissima habetur, vt non tam Bavariae, quam totius Germaniae chronicon dici mereatur.*
Frankfurt: printed for King Ludwig, 1627.
Subject: Chorography, topography.
Case/fF/49708/.05

## Avenzoar, 1091?–1162.

*Rectificatio medicationis.*
Venice: Bonetus Locatellus for Octavianus Scotus, 1496.
H-C 2187^ BMC (XV)V:446^ GW III, 3104^ Goff A-1409^ Pell. 1653
Subject: Diagnosis, disease, medicine.
Inc./f5070.5

Ed. Hieronymus Surianus. Leaf 40: "Incipit antidotarum Abumeronis Auenzoar compositum." Diseases (abcesses, baldness). With Averroës *Colliget*. "Incipit liber de medicina Auerroys: que dicit colliget: & totalis liber iste."

### Averroës (Muhammad Ibn Ahmad, called Ibn, Rushd [Ibn Rosd]), 1126–1198.

*Destructiones destructionum Auerroys cum Augustini Niphi de Suessa expositione. Eiusdem Augustini questio de sensu agente.*
Venice: Bonetus Locatellus for Octavianus Scotus, 1497.
H-C 2190*^ GW 3106^ Goff A-1412
Subject: Metaphysics (?)
Inc./f 5078

### Averroës.

*Averrois Cordvbensis de svbstantia orbis absolvtvs ac pervtilis sermo.*
Venice: Giunta, 1552.
Subject: Earth sciences, heavenly bodies, prime mover, metaphysics.
Case/fY/642/.A83932

With Joannes de Janduno. *In libros Aristotelis de coelo et mundo*, q.v.

### Avicenna, 980–1037.

*Ugonis Senensis super quarta fen primi Aui. preclara expositio. Cum annotationibus Jacobi de partibus nouiter perque diligentissime correcta.*
[Venice: Bonetus Locatellus for Octavianus Scotus, 1502.]
NLM 395
Subject: Disease, hygiene, medicine.
Case/fQ/.058

Commentary by Ugo Benzi (1376–1439) ends in the middle of chap. 17. See C. P. Lockwood, *Ugo Benzi* (Chicago, 1951). "Fen quarta de diuisione modorum medicationis est locutio vniuersalis de medicatione. Trium rerum completur: 1) regimen & nutritiae; 2) alia medicinarum exhibitio; 3) operatio manualis." Evacuation, vomiting, poultices, bloodletting, painkillers.

### Aviler, Augustin Charles d', 1635–1700.

*Ausfürliche anleitung zu der gantzen civil-bau-kunst, worinnen nebst denen lebens beschreibungen, und den fünf ordnungen von J. Bar. de Vignola . . . und mit vielen anmerckungen auch dazu gehörigen rissen vermehret von Leonh. Christ. Sturm.*
Augsburg: Johann Georg Hertel, 1747.
Subject: Architecture, solid geometry.
W/2/.058

Trans. Leohnard Christoph Sturm.

# B

## Babington, John, fl. 1635.

*Pyrotechnia, or a discourse of artificiall fire-work's: in which the true grounds of that art are plainly laid downe: Together with sundry such motions, both straight and circular, performed by the helpe of fire, as are not to be found in any other discours of this kind, extant in any language. Whereunto is annexed a short treatise of geometry . . . Written by John Babington gunner, and student in mathematicks.*
London: T. Harper for Ralph Mab, 1635.
STC 1099–1100
Subject: Geometry, instrumentation, pyrotechnics.
Case/fR/83/.062

With *A short treatise of geometrie, contayning certaine definitions and problemes, for the mensuration of superficies and sollids, as also the use of the quadrat and quadrant in measuring of altitudes, latitudes and profundities with sundry mechanicall wayes for performing the same.* [separate pagination] Imprint as above.

## Bacci, Andrea, d. 1600.

*Del Tevere di M. Andrea Bacci medico et filosofo libri tre, ne' quali si tratta della natura, & bontà dell' acque, & specialmente del Teuere, & dell' acque antiche di Roma, del Nilo, del Pò, dell' Arno, & d'altri fonti, & fiumi del mondo.*
Venice, 1576.
Subject: Balneology, water power.
Case/G/3588/.062

Topics treated include volcanoes, steam, medicinal properties and values of water (drinking, bathing). Dell' vso dell' acque, & del beuere in fresco, con neui, con ghiaccio, & con salnitro.

## Bacon, Francis, 1561–1626.

*Opera Omnia, quae extant; philosophica, moralia, politica, historica.*
Frankfurt: printed by Mattheus Kempffer for Johannes Baptista Schönwetter.
Subject: Epistemology, experimental science, natural history, philosophy of science.
Case/B/245.05

## Bacon, Francis.

*Opera Omnia.*
Leipzig: [printed?] by C. Goezius, [for] J. J. Erythropili, 1694.
Subject: Epistemology, experimental science, natural history, philosophy of science.
Case/fB/245/.051

## Bacon, Francis.

*De dignitate et avgmentis scientiarvm libri IX.*
Paris: Petrus Mettayer, 1624.
Subject: Human faculties (imagination, memory, reason), magic, mathematics, mechanics, metaphysics, natural history, natural philosophy, natural theology.
Case/B/245/.045

Begins with bk. II, Natural history in 2 parts; bk. IV, chap. 2: medicine, curing of ills. Inscription indicates that this was a prize awarded to A. C. Dubois for Latin oration, 1650.

## Bacon, Francis.

*De dignitate et augmentis scientiarum Libri IX.*
Strassburg: heirs of Lazarus Zetzner, 1635.
Subject: Human faculties (imagination, memory, reason), magic, mathematics, mechanics, metaphysics, natural history, natural philosophy, natural theology.
Case/B/245/.047

Edited by William Rawley. Begins with bk. II. Bk. III, chap. 5: magic.

## Bacon, Francis.

*Histoire des vents, ov il est traitté de leur causes, & de leurs effets; composée par Messire François Bacon . . . et fidellement tradvitté par I. Bavdoin.*
Paris: Cardin Besongne, 1650.
Subject: Wind generation, qualities, wind energy.
Case/3A/474

Silver Coll. Poem on title page: "L'homme, ce maistre de la terre, / Oú, l'amuse un bien deceuant: / Qu'est il, qu'un ouurage de verre/ Puis qu'il se casse au moindre vent?" Includes materials on generation of winds, sudden winds, wind energy (windmills, sailboats), limits of winds.

## Bacon, Francis.

*Historia naturalis et experimentalis ad condendam philosophiam sive, phaenomena vniversi: quae est instaurationis magnae pars tertia.*
London: I. Haviland, 1622.
STC 1155
Subject: Natural history, wind.
Case/B/245/.0618

"Historia ventorum" is 1st part, to p. 246, followed by "Additvs ad titvlos proximos quinque menses destinatos."

## Bacon, Francis.

*Historia naturalis & experimentalis, de ventis, &c.*
Amsterdam: Officina Elzeviriana, 1662.
Subject: Natural history, wind.
Case/B/245/.048

Contains also "Aditus titulos in proximos quinque menses destinatos. Historia densi & rari, Historia gravis & levis, Historia sympathiae & antipathiae rerum aditus, Historia sulphuris mercurii, & solis, aditus, Historia vitae & mortis." These are brief (abstracts?). Also "Francisci Baconis de Verulamis Historia naturalis, et experimentalis de forma calidi," and "De motus, sive virtutis activae variis speciebus. BOUND WITH *Scripta in natvrali*, q.v.

**Bacon, Francis.**

*Historia vitae & mortis. Sive titvlvs secvndvs in historia naturali & experimentali ad condendam philosophiam: quae est instavrationis magnae pars tertia.*
London: Io. Haviland for Matthaeus Lownes, 1623.
STC 1156
Subject: Natural history.
Case/B/245/.06182

See Graham Rees in Vickers, *Occult and scientific mentalities in the Renaissance*, p. 307.

**Bacon, Francis.**

*History naturall and experimentall, of life and death. Or of the prolongation of life.*
London: printed by I. Haviland for Wiliiam Lee and Humphrey Mosley, 1638.
STC 1158
Subject: Experimental science.
Case/Y/145/.B142
Imprimatur on leaf before title page: Tho. Wykes R.P. Episc. Lond. Cap. domest. Dec. 29 1637.

**Bacon, Francis.**

*History naturall and experimental of life and death; or, of the prolongation of life.*
London: printed for William Lee, 1664.
STC II B311
Subject: Experimental science.
Case/fB/245/.049
BOUND WITH *Sylva sylvarum, New Atlantis, Novum organum*, qq.v.

**Bacon, Francis.**

*New Atlantis. A vvorke unfinished. Written by the right honourable, Francis, Lord Verulam, Viscount St. Alban.*
London: printed by J. Haviland for W. Lee, [1635].
STC 1172, pt.2
Subject: Alchemy, experimental science, natural magic, natural philosophy, science.
Case/5A/124

"Magnalia natvrae, praecipue quo ad usus humanos" (sig. g3).

**Bacon, Francis.**

*New Atlantis. A VVork unfinished.*
London: printed by S. Griffin for W. Lee [n.d.]
STC II B307
Subject: Alchemy, experimental science, natural magic, natural philososphy, science.
Case/fB/245/.049

BOUND WITH *Sylva sylvarum* (1664), and *Novum organum* (1677), qq.v.

**Bacon, Francis.**

*[Novum organum] Instauratio magna.*
London: John Bill, 1620.
STC 1162
Subject: Epistemology, experimental science, natural history.
Case/fB/49/.059

1st ed., 1st issue of large paper ed. See Pforzheimer I, Appendix 1:xix.

**Bacon, Francis.**

*Novum organum.*
London: J. Bill, 1620.
STC 1163^ Gibson 1036^ Pforzheimer I, Appendix 1:xix
Subject: Epistemology, experimental science, natural history.
Case/fB/49/.06

**Bacon, Francis.**

*The novum organvm of Sir Francis Bacon, Baron of Verulam, Viscount St. Albans. Epitomiz'd: for a clearer understanding of his natural history. Translated out of the Latin by M.D.B.D.*
London: printed for Thomas Lee, 1677.
STC II B311
Subject: Epistemology, experimental science, natural history.
Case/fB/245/.049

BOUND WITH *Historia vita et mortis, New Atlantis*, and *Sylva sylvarum*, qq.v.

**Bacon, Francis.**

*De sapientia vetervm liber.*
London: Robert Barker, 1609.
STC 1127
Subject: Experimental science, wisdom (hidden) of the ancients.
Case/3A/473

**Bacon, Francis.**

*De sapientia vetervm liber.*
London: Felix Kyngston, Jocosa Norton, Richard Whitaker, 1634.
STC 1129
Subject: Experimental science, wisdom (hidden) of the ancients.
Case/B/974/.06

4th ed. "Daedalus, sive mechanicus" (chap. 19), Sphinx, sive scientia" (chap. 26). Hidden wisdom related to Bacon's ideas for reformation of science? See Rossi, *Bacon*, p. 80.

**Bacon, Francis.**

*Scripta in natvrali et vniversali philosophia.*
Amsterdam: Ludovic Elzevir, 1653.
Subject: Natural history, natural science.
Case/B/245/.048

Contents include Interpretation of nature, description of intellectual globe, "fluxu & refluxu maris."

**Bacon, Francis.**

*Sylva Sylvarum: or a naturall historie, In ten centvries. Written by the right honourable Francis Lo. Verulam Viscount St. Alban. Published after the authors death by William Rawley Doctor of Diuinitie, late his lordships chaplaine.*
London: printed by J. H. for William Lee, 1627.
STC 1169
Subject: Encyclopedias, experimental science, medicine, natural history.
Case/fB/245/.0622

Variant of 1st ed. Has "A table of the experiments" at end. Contains also *New Atlantis, a vvorke unfinished*, and *Magnalia naturae, praecipve qvo ad vsvs hvmanos.*

**Bacon, Francis.**

*Sylva Sylvarum: or a naturall historie. In ten centvries. Written by the right honourable Francis Lo. Verulam Viscount St. Alban. Published after the authors death by William Rawley Doctor of Diuinity, late his lordship's Chaplaine.*
London: printed by J. H. for William Lee, 1628.
STC 1170
Subject: Encyclopedias, experimental science, medicine, natural history.
Case/fB/245/.0623

2d ed.

**Bacon, Francis.**

*Sylva Sylvarum,sive historia natvralis, et novus Atlas.* [*sic*]
Amsterdam: Ludovic Elzevir, 1648.
Subject: Encyclopedias, experimental science, medicine, natural history.
Case/*B/245/.0619

2 pts. in 1 vol. Has "Tabula experimentorum" at beginning.

**Bacon, Francis.**

*Sylva Sylvarum: or, a naturall history. In ten centuries whereunto is newly added the history naturall and experimentall of life and death, or of the prolongation of life.*
London: Printed by J. F. for William Lee, 1651.
STC II B327
Subject: Encyclopedias, experimental science, medicine, natural history.
Case/5A/179

6th ed. Experiments; descriptions of artificial springs, motion of gravity, flame. Cent. I: most nourishing meats and drinks, cure of diseases contrary to predisposition, stanching bleeding, sympathy and antipathy. Cent. II: acoustics. Cent. III: motions by imitation, infectious diseases, exercise. Cent. IV: Epidemic diseases, making gold [alchemy?] pestilential years. Cent. V: plant husbandry, sympathy and antipathy of plants. Cent. VI: plants. Cent. VII: healing of wounds, sneezing. Cent. VIII: medicinal veins of earth, sponges, speed of birds in motion, bathing. Cent. IX: Various kinds of bodies, shellfish, experiments on influence of moon. Cent. X: imagination, receipts for the gout, poultices, baths, plasters.

**Bacon, Francis.**

*Sylva Sylvarum, sive historia natvralis, et nova Atlantis.*
Amsterdam: Elzeviriana, 1661.
Subject: Encyclopedias, experimental science, medicine, natural history.
Case/B/245/.062

**Bacon, Francis.**

*Sylva sylvarum: or a natural history, in ten centuries whereunto is newly added the history naturall and experimentall of life and death, or of the prolongation of life. Published after the authors death by William Rawley . . . whereunto is added articles of enquiry touching metals and minerals.*
London: printed by J. F. & S. G. for William Lee, 1664.
STC II B330
Subject: Encyclopedias, experimental science, medicine, metallurgy, natural history.
Case/fB/245/.049

8th ed. BOUND WITH *Historia vita et mortis, New Atlantis*, and *Novum organvm*, qq.v.

**Bacon, Francis.**

*The tvvo bookes: Of the proficience and aduancement of learning diuine and humane.*
London: printed for Henrie Tomes, 1605.
STC 1164
Subject: Epistemology, experimental science, science and theology.
Case/B/245/.0609

1st ed., 1st issue.

**Bacon, Francis.**

*The tvvo bookes of Francis Bacon. Of the proficience and aduancement of learning, diuine and humane.*
London: printed for Henrie Tomes, 1605.
Subject: Epistemology, experimental science, science and theology.
Ruggles/No. 16

Variant of 1st ed.

**Bacon, Francis.**

*The two bookes of Sir Francis Bacon, of the proficience and advancement of learning, divine and hvmane.*
Oxford: printed by I. L. for Thomas Huggins, 1633. With permission of B. Fisher.
STC 1166
Subject: Epistemology, experimental science, science and theology.
Case/B/245/.061

Universal learning. Object of learning: to use the gift of reason for benefit of men (I, p. 52). To place a value on knowledge in comparison to other things, using "arguments and testimonies divine and humane." Learning aids the digestion and cures diseases of both mind and body. Also teaches art of self-examination and reflection. Bk. II: What have kings and others done to increase advancement of learning? Places, books, and persons of learning. Importance of money to pay educators and scientific experimenters.

## Bacon, Francis.

*Of the advancement and proficience of learning or the partitions of sciences ix bookes written in Latin by the most eminent illustrious & famous Lord Francis Bacon . . . interpreted by Gilbert Wats.*
Oxford: printed by Leon Lichfield for Rob. Young, and Ed. Forrest, 1640.
STC 1167.3
Subject: Epistemology, experimental science, science and theology.
Case/fB/245/.0611

Scientific thought analyzed and systematized. In 6 parts.

## Bacon, Francis.

*Of the advancement and proficience of learning or the partitions of sciences ix bookes written in Latin by the most eminent illustrious & famous Lord Francis Bacon . . . interpreted by Gilbert Wats.*
Oxford: printed by Leon Lichfield for Rob. Young, and Ed. Forrest, 1640.
STC 1167.3
Subject: Epistemology, experimental science, science and theology.
Ruggles/No. 17

1st ed. in English. Large paper copy.

## Bacon, Francis.

*Baconiana . . . or certain genuine remains of Sir Francis Bacon in arguments civil . . . natural, medical.*
London: printed for Richard Chiswell, 1679.
STC II B269
Subject: Epistemology, experimental science, medicine, natural history, philosophy of science.
Case/B/245/.053

"Account of Lord Bacon's works." "Inquisitions touching the compounding of metals." The following two items are continuously paginated: "Baconiana physiologica. Or certain remains of Sir Francis Bacon . . . in arguments appertaining to natural philosophy" and "Baconiana medica, or remains . . . touching medical matters."

## Bacon, Roger, 1214?–1294.

*De l'admirable povvoir et pvissance de l'art et de la nature, ou est traicté de la pierre philosophale, traduict en françois par Iaques Girard de Tournus.*
Lyon: Macé Bonhomme, 1557.
Baudrier, *Bibliographie Lyonnaise* 10:255–56
Subject: Alchemy, philosophers stone.
Case/B/8633/.06

BOUND WITH his *Des choses merueilleuses* and *Miroir d'alquimie*, qq.v.

## Bacon, Roger.

*Des choses merueilleuses en nature, ou est traicté des erreurs des sens, des puissances de l'ame, & des influences des cieux, traduict en françois par Iaques Girard de Tornus.*
Lyon: Macé Bonhomme, 1557.
Subject: Demonology, supernatural, witchcraft.
Case/B/8633/.06

BOUND WITH his *De l'admirable povvoir* and *Miroir d'alquimie*, qq.v.

## Bacon, Roger.

*The cure of old age & preservation of youth by Roger Bacon, a Franciscan Frier translated out of Latin with annotations and an account of his life and writings. By Richard Browne. Also, a physical account of the tree of life by Edward Madeira Arrais, translated likewise out of Latin by the same hand.*
STC II B372
London: printed for Tho. Flesher . . . & Edward Evets . . . 1683.
Subject: Aging, materia medica, medicine, sciences of man.
Case/B/245/.0707

2 pts. in 1 vol. BOUND WITH Madeira Arrais, Duarte, *Arbor vitae*, q.v.

## Bacon, Roger.

*Le miroir d'alquimie de Rogier Bacon philosophe tres-excellent. Traduict de Latin en François par vn gentilhomme du D'aulphiné.*
Lyon: Macé Bonhomme, 1557.
Subject: Alchemy, Cabala, Hermeticism.
Case/B/8633/.06

Of the collection of tracts bound together in this volume, the following are continuously paginated: "Miroir d'alquimie de Rogier Bacon; [T]able d'esmeraude de Hermes trismegiste; Petit commentaire de l'Hortvlain philosophe dict des iardins maritimes, sus la table d'esmeraude d'Hermes trismegiste. Pierre de L'Hortulain; Le livre des secretz d'alquimie compose par Calid filz de Iazic Ivif, translaté d'Hebrieu en Arabic, & d'Arabic en Latin, & de Latin en Francoys; Miroir de Maistre Iean de Mehun." The following two tracts are paginated together (but separately from those preceding): "L'Elixir de philosophes [par Pape Iean XXII]; L'Art transmutatoire de pape Iean XXII de ce nom." BOUND WITH Bacon, R., *De l'admirable puissance* and *Des choses merueilleuses*, qq.v.

## Baerle, Kaspar van, 1584–1648.

*Casparis Baerlaei, rervm per octennivm in Brasilia et alibi nuper gestarum, subpraefectura illustrissimi comitis I. Mavritii, Nassoviae&c.*
Amsterdam: Ioannis Blaeu, 1647.

Subject: Discovery, natural history, New World.
Ayer/*1300.5/B8/B14/1647

Sabin: "This ed. is rare, most copies having been consumed by a fire which destroyed the warehouses of Blaeu, the publisher." Large folio, with maps, landscapes, engravings.

## Baerle, Kaspar van.

*Brasilianische geschichte bey acht jähriger in selbigen landen geführeter regierung seiner fürstlichen gnaden herrn Johann Moritz, fürstens zu Nassau &c. Erstlich in Latein durch Casparem Barlaeum beschreiben.*
Cleve: Tobias Silberling, 1659.
Sabin 3411
Subject: Discovery, natural history, New World.
Ayer/*1300.5/B8/B14/1659

## Baerle, Kaspar van.

*Rervm per octennivm in Brasilia et alibi gestarum . . . cui accesserunt Gulielmi Pisonis medici Amstelaedamensis tractatvs 1. De aeribus, aquis, & locis in Brasilia. 2. De arundine saccharifera. 3. De melle silvestri. 4. De radice altili mandihoca.*
Cleve: Tobias Silberling, 1660.
Subject: Medicine, natural history, New World.
Ayer/*1300.5/B8/B14/1660

2d ed. Work by Piso has running head: "Hist. nat. & medic. lib. I." Includes descriptions and medicinal uses of such new world plants as yucca, cassava.

## Baïf, Lazare de, d. 1547.

*Lazari Bayfi annotationes in L. II. De captivis et postliminio reversis. In qvibvs tractatvr de re navali eivsdem annotationes in tractatum de auro & argento leg. quibus, vestimentorum, & vasculorum genera explicantur. Antonii Thylesii de coloribus, à coloribus vestium non alienus.*
Paris: Robert Stephan, 1536.
Subject: Naval art and science.
Case/F/0231/.07

A kind of encyclopedia on selected subjects. Antonio Telesio (1482–1533), *Libellvs de coloribvs* (relation of colors to humors).

## Baïf, Lazare de.

*Lazari Bayfi annotationes in legem II. De captiuis & postliminio reuersis, in quibus tractatur de re nauali, per autorem recognitae. Eivsdem annotationes in tractatum de auro & argento legato, quibus vestimentorum & vasculorum genera explicantur.*
Basel: H. Frobenius and N. Episcopius, 1537.
Subject: Naval art and science.
Greenlee/5000/B15/1537

BOUND WITH Dolet, Etienne, *De re navali* (Lyon, 1537), q.v.

## Baïf, Lazare de.

*De re navali libellvs, in adolescentulorum bonarum artium studiosorum fauorem, ex Bayfij uigilijs excerptus, & in breuem summulam facilitatis gratia redactus. Addita ubique puerorum, causa uulgari uocabulorum significatione.*
Lyon: Theobald Paganus, 1543.
Subject: Naval art and science.
Case/U/511/.07

Value to boys of studying shipbuilding, types of ships.

## [Bailey, Walter] d. 1592.

*A short discourse of the three kindes of peppers in common vse and certaine special medicines made of the same, tending to the preseruation of health.*
London, 1588.
STC 1199
Subject: Descriptive botany, herbal medicine, natural history.
Case/3A/477

## [Baillet, Adrien] 1649–1706.

*La vie de M. Des-Cartes.*
Paris: Widow Cramoysi, 1693.
Subject: Biography, metaphysics, optics, physics.
E/5/.D/45093

Arranged chronologically: Bk. I, 1596–1619; II, 1619–29; III, 1629–37; IV, 1637–38; V, 1638–41; VI, 1641–44; VII, 1644–50. I, pt.6: dissatisfaction with studies in metaphysics, mathematics; II, pt.2: researches on Rosicrucian brotherhood; III: dioptrics, anatomy, medicine, meteors, condemnation of Galileo.

## [Baillet, Adrien]

*The life of Monsieur Des Cartes containing the history of his philosophy and works.*
London: R. Simpson, 1693.
STC II B451A
Subject: Biography, metaphysics, optics, physics.
Case/E/5/.D/45094

Translated from the French by S. R.

## Baillou, Guillelmus de.

*Gvlielmi Ballonii medici Parisiensis celeberrimi, consiliorvm medicinalivm libri II.*
Paris: Jacobvs Qvesnel, 1635.
Subject: Consilia, medicine.
UNCATALOGUED

Vol. 1: In quo pleraque continentur quae & ad morborum cognitionem, eorundemque curationem propositis exemplis, & obscurorum Hippocratis locorum intelligentiam pertinebunt. Vol. 2 title page: *Epidemiorvm et ephemeridvm libri dvo.* Dated 1640.

## [Baker, Thomas] 1656–1740.

*Reflections upon learning, wherein is shewn the insufficiency thereof, in its several particulars. In order to evince the usefulness and necessity of revelation.*
London: printed for A. Bosvile, 1700.
STC II B521

Subject: Astronomy, calendar (chronology), geography (longitude), natural philosophy.
A/91/.06

3d ed. Latest discoveries in anatomy and botany.

## [Baker, Thomas]

*Reflections upon learning, wherein is shewn the insufficiency thereof, in its several particulars; in order to evince the usefulness and necessity of revelation.*
London: A. Bosvile, 1708.
Subject: Astronomy, calendar, geography, natural philosophy.
A/91/.062

4th ed.

## [Baker, Thomas]

*Traité de l'incertitude des sciences.*
Paris: P. Miquelin, 1714.
Subject: Astronomy, calendar, geography, natural philosophy.
A/91/.064

Translated from English by Nicolas Berger.

## [Baltus, Jean François] 1667–1743.

*Réponse à l'histoire des oracles, de Mr. de Fontenelle, de l'academie Françoise. Dans laquelle on réfute le systéme de Mr. Van-Dale, sur les auteurs des oracles du paganisme, sur la cause & le temps de leur silence; & où l'on établit le sentiment des peres de l'eglise sur le même sujet.*
Strassburg: Jean Renauld Doulssecker, 1707.
Subject: Divination, oracles.
3A/5080

Refutation of Van-Dale's piece on oracles. Effort to treat them rationally. See Thorndike VIII:477n.

## Baranzanus, Redemptus.

*Vranoscopia sev de coelo in qva vniversa coelorum doctrina clarè, dilucidè & breuiter traditur. Peregrines, plurimes de caeli animatione, simplicitate, fluiditate, grauitate, numero, influentiis occultis, lunae maculis, stellarum à terra distantiis, planetarum dignitatibus, coelestis figurae erectione, & ceteris omnibus ad perfectam coelorum cognitionem spectantibus examinantur opiniones.*
Geneva: Petrus & Jacobus Chouët, 1617.
Subject: Astronomy.
Case/B/8635/.072

Part 2: "Vranoscopia sev de caelo . . . in qva singvlarvm sphaerarvm essentia, natvra, proprietas, theoria, praedominium, distantia, magnitudo, motus, & status breuibus exponitur."

## Barbaro, Ermolao, 1454–1493.

[Colophon] *Hermolai Bar. patriarchae Aquileiensis Plinianae castigationes: item aeditio in Plinium secunda: item emendatio in Melam Pomponium: item obscurae cum expositionibus suis uoces in Pliniano codice.*
Rome: Eucharius Argenteus Germanus [Silber], 1493.
H-C *2421^ BMC (XV)IV:113–114^ GW III, 3340^ Pr. 3860^ Still. B88^ Goff B-100.
Subject: Hippocratic medicine, humanism and science, natural history.
Inc./f3860

Manuscript marginalia in early hand (similar to rubrics). Hippocratic medicine, Bk. IX.

## Barbaro, Ermolao.

[Colophon] *Hermolai Barbari Patriarchae Aquileiensis Plinianae castigationes: item aeditio in Plinium secunda: item emendatio in Melam Pomponium: item obscurae cum expositionibus suis uoces in Pliniano codice.*
[Venice, about 1493 or 1494.]
H-C *2420^ BMC (XV)V:587^ GW III, 3341^ Pr. 7422^ Still. B89^ Goff B-101
Subject: Geography, humanism and science, medicine, natural history.
Inc./f5705.5

MS marginalia in early hand throughout.

## Barbaro, Ermolao.

*Hermolai Barbari patricii Veneti. P. Aquileiensis in castigationes Plinianas ad Alexandrum sextum pontificem maximum praefatio.*
Cremona: Carolus de Darleriis, 1495.
H-C *2423^ BMC (XV)VII:959^ GW II, 3342^ Pr. 6929^ Still. B90^ Goff B-102
Subject: Humanism and science, natural history.
Inc./f*6929

Includes material on alchemy, astronomy. Natural science must be part of sound moral and metaphysical framework, of knowledge of both God and man.

## Barbaro, Ermolao.

*Hermolai Barbari P.V. compendivm scientiae natvralis ex Aristotele.*
Venice: Cominus de Tridino Montisferrati, 1545.
Subject: Aristotle, encyclopedias, natural history, natural science.
Case/Y/642/.A8672

In 5 bks. Topics include monsters, formal, material, efficient causes, stars, planets, sun, corruption and generation, 4 elements, comets, winds, vegetative, animal, rational soul, senses, dreams, imagination, memory, contrast between human and divine, demonic intelligences. BOUND WITH his *Compendivm ethicorvm* [Aristotle, *Ethics*].

## [Barbeu du Bourg, Jacques] 1709–1779.

*Lettre d'un garçon barbier à M. l'Abbé des Fontaines, auteur des observations sur les ecrits modernes, au sujet de la Maîtrise-ès-arts.*
Paris, 1743.
Subject: Medical and surgical instruction and training, standards and practice of the medical profession, importance of liberal arts for the physician.
Q/.069

Critique of the state of medicine and the quality of physicians.

## Barcia [Carballido y Zuñiga] Andrés Gonzalez de, 1673–1743.

*Historiadores primitivos de las indios occidentales, que juntò, traduxo en parte, y sacò à luz, illustrados con eruditas notas, y copiosos indices, el illustrissimo señor D. Andres Gonzales Barcia . . . divididos en tres tomos.*
Madrid, 1749.
Subject: Discovery, navigation, New World, voyages.
Ayer/108/B2/1749

3 vols. Vol. 1: Columbus; Cortès; Pedro Alvarado and Diego de Godoy on Cortès; Relacion sumaria de la historia natural de las Indias by Gonzalo Fernandez de Oviedo; Alvar Nuñez Cabeza de Baca, Antonio Ardoino, narrations of voyages and shipwrecks. Vol. 2: Francisco Lopez de Gomara on the Indies and New Spain. Vol. 3: Augustin de Zarate; Hulderico Schmidel, Martin del Barco, Perez de Torres, Manuel de Grova, on Peru, Argentina, and merchant voyages.

## [Barclay, James]

*A treatise on education with reflections on taste, poetry, natural history, and the manner of forming the temper, and teaching youth such moral precepts as are necessary in the conduct of life.*
Edinburgh: printed by Jas. Cochran & Co., 1743.
Subject: Education, natural history, psychology.
I/4/.067

Chap. 12 (pp. 217 ff.): in what manner moral reflections may be instilled into children; with some observations on natural history, musick, and poetry.

## Bardolini, Matteo, 16th cent.

*Coeliplani, sive planispherii, canones.*
[Venice: Ioan., Ant., brothers de Sabio, 1530.]
Subject: Astronomy, planets.
Ayer/8.9/A8/B24/1530

In 3 bks.: 1) nomine coeliplani, de planetae circulo & ordine, de aequatore, de orbe horarium inaequalium siue zodiacium, de soli; 2) de cognitione tabularum coeliplani; 3) table of sun's motion, planets, declination, ascendant.

## Barenghi, Giovanni, 17th cent.

*Considerazioni del signor Giovanni Barenghi sopra il Dialogo de dua [sic] massimi sistemi tolemaico, e copernicano nelle quale si difende il metodo d'Aristotele ne libri del cielo. Le sue dimensioni, per lo moto retto de gl'elementi, e per la quiete della terra nel centro, e per lo moto de gl'orbi celesti, e loro dimensioni, fra' corpi sublunari. Da quanto gl'hà scritto contro il signor Accademico Linceo. Libri tre.*
Pisa: Francesco delle Dote, 1638.
Subject: Astronomy, geocentrism, heliocentrism.
Case/L/9/.07

Ex libris George Sarton. Earth's motion compared to motion of other heavenly bodies.

## Bargagli, Scipione, 1540–1612.

*Delle lodi dell'academie. Oratione di Scipion Bargagli da lui recitata nell' Academia degli accesi in Siena.*
Florence: [Luca Bonetti, Venetiano] 1569.
Subject: Academies and learned societies, quadrivium.
Case/A/91/.071

## Barozzi, Francesco, fl. 1550–1590.

*Cosmographia in quatvor libros distribvta, svmmo ordine, miraqve, facilitate, ac breuitate ad magnam Ptolemaei mathematicam constructionem, ad vniuersamque astrologiam instituens: Francisco Barocio, Iacobi filio, patritio Veneto, autore. Cum praefatione eiusdem authoris, in qua perfecta quidem astrologiae diuisio, & enarratio autorum illustrium, & voluminum ab eis conscriptorum in singulis astrologiae partibus habetur: Ioannis de Sacrobosco verò 84 errores, & alij permulti suorum expositorum, & sectatorum ostenduntur, rationibusque, redarguuntur.*
Venice: Gratiosi Perchacini, 1585.
Subject: Astrology, cosmography.
Ayer/*7/B26/1585

Tables showing relationship of astrology, gnomonics, astronomy, dioptrics. BOUND WITH Orsini, Latino, q.v.

## Barozzi, Francesco.

*Cosmographia in quatvor libros distribvta, svmmo ordine, miraqve, facilitate, ac breuitate ad magnam Ptolemaei mathematicam constructionem, ad vniuersamque astrologiam instituens: Francisco Barocio, Iacobi filio, patritio Veneto, autore. Cum praefatione eiusdem authoris, in qua perfecta quidem astrologiae diuisio, & enarratio autorum illustrium, & voluminum ab eis conscriptorum in singulis astrologiae partibus habetur: Ioannis de Sacrobosco verò 84 errores, & alij permulti suorum expositorum, & sectatorum ostenduntur, rationibusque, redarguuntur.*
Venice: Gratiosus Perchacini, 1598.
Subject: Astrology, cosmography, instrumentation.
Ayer/*7/B26/1598

Continuation of title [also in 1585 ed.]: *Praecesserunt etiam quaedam communia mathematica, necnon arithmetica, & geometria principia,nonnullaeque propositiones, de quibus in toto opere saepe sit mentio: ac demum locupletissimus index eorum, que ipsa cosmographia continentur.* Work contains Errors in Sacro Bosco. Communia mathematica principia [arithmetic and geometry], cosmographia (Sphaera mundi, 4 bks.: stars, their rising, celestial signs, seasonal variance of daylight hours, climate, planets, solar and lunar eclipses).

## Barrère, Pierre, 1690–1755.

*Essai sur l'histoire naturelle de la France equinoxiale ou dénombrement des plantes, des animaux & des mineraux, qui se trouvent dans l'Isle de Cayenne, les isles de Remire, sur les côtes de la mer, & dans le continent de la Guyane. Avec leurs noms differens, Latins, François, & Indiens, & quelques observations sur leur usage dans la médicine & dans les arts.*
Paris: Widow Piget, 1749.
Subject: Materia medica, medicine, natural history.
Ayer/8.9/N15/B27/1749

Pt. I: Alphabetical list of plants, with occasional reference to medicinal use, and to botanical works of others. Pt. II: Animals and minerals, quadrupeds and reptiles, fish, shells, insects, minerals.

### Barros, João de.

*L'Asia del S. Giovanni di Barros. . . . Nella quale oltre le cose appartenenti alla militia, si ha piena cognitione di tutte le città, monti, & fiumi delle parti orientali, con la descrittione de' paesi, & costumi di quei popoli.*
Venice: Vincenzo Valgrisio, 1561.
Subject: Discovery, voyages.
Greenlee/4858/P7/.B27/1561

P. 63 (bk. 4) "instrumento vsato da Portoghesi; per pigliar l'altezza del sole."

### Barros, João de.

*Aanzienelyke scheeps-togt door den grave don Vasco da Gamma.*
Leyden: Pieter Van der Aa, 1707.
Subject: Discovery, voyages.
Greenlee/4850/P7/B27/1707

Plate facing p. 26: natives (of Goa?) fleeing Portuguese ships as they are landing.

### Bartholin, Erasmus, b.1625–1698.

*De aëre Hafniensi.*
Frankfort: Daniel Paulli, 1679.
Subject: Climate.
Case/V/23/.375

BOUND WITH Grube, Hermann, q.v.

### Bartholin, Thomas 1616–1680.

*Anatomia ex Caspari Bartholini parentis institutionibus, omniumque recentiorum & propriis observationibus tertium ad sanguinis circulationem reformata.*
Leyden: Franciscus Hackius, 1651.
Subject: Anatomy, medicine.
Case/QM/21/B37/1651

### Bartholin, Thomas.

*Antiqvitatvm veteris pverperii synopsis a filio Casparo Bartholino commentario illustrata. Cvm Thomae Bartholini ad filium epistola.*
Amsterdam: Henricus Wetstenius, 1676.
Subject: Deafness, medicine.
UNCATALOGUED

BOUND WITH Caspar Bartholin, *De inavribvs vetervm syntagma. Accedit mantissa ex Thomae Bartholini miscellaneis medicis de annvlis narivm* [same imprint].

### Bartholin, Thomas.

*De bibliothecae incendio, dissertatio ad filios.*
Copenhagen: M. Godicchenii for P. Haubold 1670.
Subject: medicine.
Z/491/.B2775

Has list of vols., mostly medical, destroyed in fire. Partly annotated.

### Bartholomaeus Anglicus, fl. 1230–1250.

*De proprietatibus rerum.*
Lyon: Pierre Hongre, 1482.
H-C 2502^ GW III, 3406^ Pr. 8573^ Pell. 1869^ Goff B-134
Subject: Encyclopedias, natural history.
Inc./f*8573

### Bartholomaeus Anglicus.

*De proprietatibus rerum.*
Strassburg, [Printer of the 1483 Jordanus de Quedlinburg (Georg Husner)] 1485.
H *2506^ BMC (XV)I:132^ GW III, 3410^ Pr. 592^ Pell. 1873^ Goff B-138
Subject: Encyclopedias, natural history.
Inc./592

Manuscript notes. Contains a list of authors cited.

### Bartholomaeus Anglicus.

*De proprietatibus rerum.*
[Westminster: Wynkyn de Worde, ca. 1495.]
H-C 2520^ GW III, 3413^ Pr. 9725^ Still. B128^ Goff B-143
Subject: Encyclopedias, natural history.
Inc./f 9698.2

Britwell Court Library copy. 1st ed. of 1st English translation (originally made in 1398 by John of Trevisa). With some woodcuts.

### Bartholomaeus Anglicus.

*De proprietatibus rerum.*
London: Thomas Berthelet, 1535.
STC 1537
Subject: Encyclopedias, natural history.
Case/fA/251/.068

Also contains meteorology. Prologue translator is same as in Batman ed. Berthelet was printer and bookbinder to Henry VIII.

### Bartholomaeus Anglicus.

[*De proprietatibus rerum*]
*Batman vppon Bartholome, his booke De proprietatibus rerum, newly corrected, enlarged and amended: with such additions as are requisite, vnto euery seuerall booke: taken foorth of the most approued authors, the like heretofore not translated in English. Profitable for all estates, as well for the benefite of the mind as the bodie.*
London: imprinted by T. East, 1582.
STC 1538.
Subject: Encyclopedias, natural history.
Case/fA/251/.066

Trans. John Trevisa. "Barthelmew Glantuyle descended of the noble family of the earles of Suffolk. He was a Franciscan frier, and wrote this work in Edward III's time [ca.] 1366. In the year 1397 was this sayd work translated into English and so remained by written coppie, until A.D.

1471, at which time printing began first in England . . . sithence which time this learned and profitable worke was printed by Thomas Berthelet 1535. And last of all augmented and enlarged as appeareath, for the commoditie of the learned and well disposed Christian by me Stephan Batman."

## Bartoli, Cosimo, 16th cent.

*Del modo di misvrare le distantie, le superficie, i corpi, le piante, le prouincie, le prospettiue, & tutte le altre cose terrene, che possono occorre: e a gli huomini. Secondo le uere regole d'Euclide, & de gli altri piu lodati scrittori.*
Venice: Francesco Franceschi, 1564.
Subject: Instrument-making, mensuration, plane and solid geometry, surveying.
Case/4A/3212

The triangle, measurement of, distances by triangulation, cubes, spheres, compass-making, quadrant.

## Bartoli, Daniello, 1608–1685.

*Del svono de' tremori armonici e dell' vdito.*
Bologna: Pietro Bottelli, 1680.
Subject: Acoustics, motion, physics, propagation of sound.
Case/V/22/.076

Wave theory of sound. Sympathetic vibration. "Sperienze, e ragione, che pruouano, nè le vibrationi dell'aria, nè il suono (s'egli non è altro che esse) partir nulla dal vento nè da verum altra dispositione dell'aria. Altre sperienze, e altre ragioni piu valide a dimostrare il contrario" (capo quinto, p. 69). Qualities of sound: speed, intensity, quantity.

## Basilius, St. the Great, abp. of Caesarea, 330 (ca.)–379.

*D. Basili Magni de institvenda stvdiorvm ratione ad nepotes suos oratio paraenetica Graece & Latine. . . .Item Ioanis Pici Mirandule De homine opusculum omnino diuinum.*
Basel: H. Petrum [1537]
Subject: Education, natural philosophy.
Case/I/4/.091

P. 125: "Ioannes Pici Mirandvlae oratio de homine, in qua sacrae & humanae philosophie explicantur."

## Basin, Bernardo, ca. 1445–ca. 1500.

*De magicis artibus et magorum maleficiis.*
Paris [Louis Martineau], 1483.
BMC (XV)VIII:39^ GW III:3719^ Pr. 7923^ Goff B-279
Subject: Magic, witchcraft.
Inc./7923

MS marginalia.

## Bass, Heinrich, 1690–1754.

*Henrici Bassii med. & chirurg. D. und prof. publ. auf der Friedrichs Universität Erläuterter Nuck: oder grundliche anmerckungen uber des berühmten anat. & chirurg. prof. zu Leyden Anthon Nvcks chirurgische hand griffe und experimente worinnen viel neue inventa und instrvmenta vorgestellet werden.*
Halle in Magdeburg: Renger, 1728.
Subject: Experimental medicine, medical instruments.
oRD/32/.B28/1728

Experiment no. 50: Von der transfusion des bluts. Description of experiment is in German; footnotes in double columns, in German and Latin.

## Bate, John, fl. 1606.

*The mysteries of nature and art in foure severall parts. The first of water works. The second of fire works. The third of drawing. The fourth of sundry experiments.*
[London] printed for Ralph Mabb, 1635.
STC 1578
Subject: Chemistry, experimental science and medicine, technology.
Case/A/911/.08

2d ed. Bk. 4: "The book of extravagants: wherein amongst others, is principally contrived divers excellent and approved medicines for severall maladies. . . . Courteous reader, forasmuch as there were divers experiments that I could not conveniently, or rather my occasions would not permit me to dispose in such order as I would have done; I thought it would not be amisse to call them by the names of extravagants, and so to set them downe as I found them, either inserted amongst other my notes, as I put them in practise, or as they came into remembrance."

## [Special material in process]

## Bauderon, Bricius.

*The expert phisician: learnedly treating of all agves and feavers. Whether simple or compound. Shewing their different nature, causes, signes, and cure, viz. the differences of feavers. A continual putrid. An intermitting quartan. Confused erratick feavers.*
London: printed by R. I. for John Hancock, 1657.
STC II B1163
Subject: Medicine, diagnosis and treatment of fevers.
sc/127

14 different kinds of fevers listed on title page. Remedies.

## Baxter, Richard.

*The certainty of the worlds of spirits. Fully evinced by unquestionable histories of apparitions and witchcrafts, operations, voices &c. Proving the immortality of souls, the malice and miseries of the devils and the damned, and the blessedness of the justified. Written for the conviction of saduces and infidels.*
London: printed for T. Parkhurst and J. Salusbury, 1691.
STC II B1215
Subject: How to identify evil spirits, witchcraft.
Case/B/893/.07

Defender of existence of witchcraft. Testimony of witnesses. Classical authorities, e.g., Sebastian Brand, Platerus, Scribonius.

**Baxter, Richard, 1615–91.**

*The signs and causes of melancholy. With directions suited to the case of those who are afflicted by it. Collected out of the works of Mr. Richard Baxter, for the sake of those who are wounded in spirit. By Samuel Clifford, minister of the gospel.*
London: S, Cruttenden & T. Cox, 1716.
Subject: Melancholy, causes and symptoms, possession by devil, psychology.
Case/B/529/.08

**[Bayle, Pierre] 1647–1706.**

*Lettre à M. L. A. D. C. docteur de Sorbonne. Où il est prouvé par plusieurs raisons tirées de la philosophie, & de la theologie, que les cometes ne sont point le presage d'aucun malheur.*
Cologne: Pierre Marteau, 1682.
Subject: Comets, divination, superstition.
B/863/.082

1st ed. of his *Pensées diverses de la comète* [cf. Barbier, *Dictionnaire des ouvrages anonymes*].

**Bekker, Balthasar, 1634–1698.**

*De betoverde weereld, zijnde een grondig ondersoek van 't gemeen gevoelen aangaande de geesten, der selver aart en vermogen, bewinden bedrijf: als ook 't genede menschen door derselver kraght engemeinschop doen.*
Leeuwarden: Hero Nanta, 1691.
Subject: Possession, superstition, witchcraft.
Case/B/88/.08

2 vols. in 1. 1st ed. in Dutch.

**Bekker, Balthasar.**

*Die bezauberte Welt; oder, Eine gründliche untersuchung des allgemeinen aberglaubens betreffend die arth und das vermögen gewalt und wirchtung des satans und der bösen beister über den menschen.*
Amsterdam: D. von Dahlen, 1693.
Subject: Possession, superstition, witchcraft.
Case/B/88/.0833

1st ed. in German. Chap. X: Witchcraft and magic in the Americas (Peru, Brazil, New York, New England, Virginia, and the Carolinas).

**Bekker, Balthasar.**

*Le monde enchanté; ou, examen des communs sentimens touchant les esprits, leur nature, leur pouvoir, leur administration, & leurs operations. Et touchant les éfets que les hommes sont capables de produire par leur communication & leur vertu, divisé en quatre parties par Balthasar Bekker, docteur en théologie, & pasteur à Amsterdam. Traduit du Hollandais.*
Amsterdam, Pierre Rotterdam, 1694.
Subject: Possession, superstition, witchcraft.
Case/B/88/.083

**Bellanti, Lucio, d. 1499.**

*Defensio astrologiae contra Ioannes Picum Mirandulam. Lvcii Bellantii Senensis mathematici ac physici liber de astrologica veritate. Et in dispvtationes Ioannis Pici adversvs astrologos responsiones.*
Venice: Bernardinus Venetus de Vitalibus, 1502.
Subject: Astrology, controversy.
Inc./f5526a

20 "questio de scientia astrologiae" with a table giving abstracts of the questions. In 12 books. BOUND WITH Pico della Mirandola, Giovanni, *Disputationes*, q.v.

**Belidor, Bernard Forest de, 1697?–1761.**

*La science des ingenieurs dans la conduite des travaux de fortification et d'architecture civile.*
Paris: Claude Jombert, 1729.
Subject: Architecture, engineering, military technology.
U/26/.088

**Bellini, Lorenzo, 1643–1704.**

*De urinis et pulsibus, de missione sanguinis, de febribus, de morbis capitis, et pectoris. Opus Laurentii Bellini, dicatum Francisco Redi.*
Frankfurt & Leipzig: printed by Christian Scholvinus for Johannis Gross, 1685.
Subject: Medicine.
Case/R/128.7/B44/1685

**Belon, Pierre, 1517–1564.**

*La nature & diuersité des poissons auec leurs pourtraicts, representez au plus pres du naturel.*
Paris: Charles Estienne, 1555.
Subject: Icthyology.
Case/*P/1/.088

In 2 books. I.: "fish with blood." (vertebrates?) II.: shellfish, anemones, squid, shells. Latin, French, and occasionally Greek and Italian names of fish. Plates with descriptive text.

**Belon, Pierre.**

*Portraits d'oyseavx, animavx, serpens, herbes, arbres, hommes et femmes, d'Arabie & Egypte, obseruez par P. Belon du Mans. Le tout enrichy de quatrains, pour plus facile cognoissance des oyseaux, & autres portraits. Plus y est adiousté la carte du mont Attos, & du mont Sinay, pour l'intelligence de leur religion.*
Paris: Guillaume Cauellat, 1557.
Subject: Biology.
*Wing/ZP/539/.C31

6 categories of birds, e.g., raptors, river birds, birds inhabiting hedges.

**Belot, Jean.**

*Les oeuvres de M. Jean Belot, curé de mil-monts, professeur aux sciences divines & celestes. Contenant la chiromance, physionomie, l'art de memoire de Raymond Lulle: traité des divinations, augures & songes; les sciences steganographique, Paulines, Armadelles & Lullistes; l'art de doctement prêcher & haranguer, &c. Derniere edition revûë, corigée & augmentée de divers traités.*
Lyon: Claude la Rivière, 1654.

Subject: Chiromancy, dream interpretation, mnemonics, physiognomy.
B/864/.0805

## Bembo, Pietro (Cardinal), 1470–1547.

*De Aetna.*
Venice: Aldus Manutius, Romanus, 1495–96.
H-C 2765^ BMC (XV)V:554^ GW III, 3810^ Pell. 2033^ Goff B-304
Subject: Earth sciences, volcanology.
Inc./5550

1st ed. Bembo's 1st work. Dialogue between B. P. and B. F.

## Benedetti, Alessandro, ca. 1450–1512.

*Diaria de bello Carolino.*
[Venice: Aldus Manutius, 1496.]
H *805^ BMC (XV)V:555^ GW I, 863^ Pr. 5552^ Still. A355^ [not in Goff]
Subject: Military science, medicine.
Inc./5552

Observations on type and location of wounds; description of disease, and its prevention in military camp; astrological predictions; treatment of Count Niccolò Pitigliano, an example of early skill in diagnosis; fever and diarrhea in Novara.

## Benjamin ben Jonah, 12th cent.

*Itinerarium Beniamini Tvdelensis, in qvo res memorabiles, qvas ante qvadrigentos annos totum ferrè terrarum orbem notatis itineribus dimensus vel ipse vidit vel à fide dignus suae aetatis hominibus accepit, breuiter atque dilucidè describuntur; ex Hebraico Latinum factum Bened. Aria Montano interprete.*
Antwerp: Christopher Plantin, 1575.
Subject: Natural history, travel.
Case/G/131/.081

Sig. G4 (p. 103): "Est enim aqua illa simul & potus & medicina aduersus huiusmodi repletiones. Fuit autem perpetua quaestio variaque inter homines de Nilo exundatione opinio, sed Aegyptiorum sententia est: eodem tempore quo fluuius hic exundat, vehementer pluere in superioribus regionibus, hoc est in terra Hhabas, quam Hhauilam nominari diximus."

## Benoist, Elie, 1640–1728.

*Mélanges de remarques critiques, historiques, philosophiques theologiques sur les dissertations de M. Toland, intitulées l'une "l'homme sans superstition," et l'autre, "les origines Judaïques."*
Delft: Adrien Beman, 1712.
Subject: Astrology, natural religion, superstition.
Y/672/.L5388

## Bentley, Richard, 1662–1742.

*A confutation of atheism from the origin and frame of the world.*
London: printed for H. Mortlock, 1693.
STC II B1915, 1917–1918
Subject: Cosmography, gravitation, matter, natural theology.
Case/C/52.0886

3 pts. in 2 vols. Boyle Lectures 6–8 (7th in 3 parts). No. 6: reasons for God's existence; no. 7: gravitation; no. 8: earth has to be created by God. Boyle Lectures: a divine or preaching minister to preach 8 sermons per year "for proving the Christian religion against notorious infidels, viz. atheists, deists, pagans, Jews, & Mahometans." For propagation of the Christian religion, not to engage in controversy between Christians.

## Bentley, Richard.

*A confutation of atheism from the structure and origin of humane bodies.*
London: printed for Tho. Parkhurst & H. Mortlock, 1692.
STC II B1919
Subject: Astrology, natural theology, sciences of man.
Case/C/52/.0887

Boyle Lecture no. 3, pt. 1.

## Bentley, Richard.

*A confutation of atheism from the structure and origin of human bodies.* London: printed by J. H. for H. Mortlock, 1693.
STC II B1921
Subject: Astrology, natural theology, sciences of man.
Case/3A/259

Boyle Lecture no. 3. 3rd ed.

## Bentley, Richard.

Eight sermons preach'd at the Honourable Robert Boyle's lecture in the first year MDCXCII by Richard Bentley.
Cambridge: printed for Cornelius Crownfield, 1724.
Subject: Cosmography, natural theology, physics, sciences of man.
C/52/.089

5th ed. With a 9th sermon. Those touching on relevant subjects and not listed separately above: nos. 2, 3, 4 (mechanical philosophy), 5 (chance and fortune).

## Bentley, Richard.

*Matter and motion cannot think: or, a confutation of atheism from the faculties of the soul.*
London: printed for T. Parkhurst and H. Mortlock, 1692.
STC II B 1936
Subject: Natural theolgy, physics.
Case/C/9911/.09

Boyle Lecture no. 2.

## Benzoni, Girolamo, b. 1519.

*De gedenk waardige West-Indise voyagien ge daan door Christoffel Columbus, Americus Vesputius, en Lodewijck Hennepin. Behelzende een naaukeurigeen waarachtige beschrijving de eerste en laaste Americaanse ontdekkingen, door de voornoemde reizigers gedaen, met alle de byzondere voorvallen, hen overgekomen. . . . In t'Italians beschreeven door Hieronymus Benzo, Milanees.*

Leyden: Pieter vander Aa, 1704.
Subject: Discovery, voyages.
Ayer/110/B4/1704

## Berault, Pierre.

*A discourse I. Of the Trinity. . . . II. Of God's existence. . . . III. Of the certainty of the holy Scriptures. . . . IV. Of the immortality of our souls. . . . V. Of physick, metaphysick, and astronomy, for the use of them, who have no time to read the great and intangled volumes of philosophers.*
London: printed by William Redmayne for the author, 1700.
STC II B1950
Subject: Astronomy, metaphysics, natural theology.
Case/C/52/.09

In Latin.

## Berkeley, George, bp. of Cloyne, 1685–1753.

*The medicinal virtues of tar water fully explained by the Right Rev. Dr. George Berkeley, Lord Bishop of Cloyne in Ireland. To which is added the receipt for making it, and instructions to know by the colour and taste of the water when the tar is good, and of the right sort. Together with a plain explanation of the Bishop's physical terms. Designed for the benefit of all who drink it, and those who make it themselves.*
Dublin, printed, London: reprinted, for the proprietors of the Tar-Water Warehouse, 1744.
Subject: Elements, iatrochemistry, materia medica.
Q/.092

BOUND WITH his *Siris*, q.v. Subject said to be philosophy of great chain of being and alchemical theory. See Ritchie in note below. Also BOUND WITH *The miracles of Jesus vindicated.* London: printed for J. Roberts, 1729. Effort to explain transubstantiation, healing of the infirm.

## Berkeley, George, bp. of Cloyne.

*Siris: a chain of philosophical reflections concerning the virtues of tar water, and divers other subjects connected together and arising one from another.*
Dublin, printed, London: reprinted for W. Innys [etc.], 1744.
Subject: Elements, iatrochemistry, materia medica.
Q/.092

2d ed. Philosophy of great chain of being and the alchemical theory, according to Arthur D. Ritchie, *Proceedings of the British Academy* (vol.40). BOUND WITH *Medicinal virtues of tar water*, q.v., & *The miracles of Jesus vindicated.*

## Berkeley, George, bp. of Cloyne.

*Siris: a chain of philosophical reflections and inquiries concerning the virtues of tar water, & divers other subjects connected together and arising one from another.*
Dublin printed, London re-printed for W. Innys & C. Hitch & C. Davis, 1747.
Subject: Elements, iatrochemistry, materia medica.
B/09/.88/v.3

Binder's title: Tracts, philosophical, vol. 3.

## [Berkeley, George] bp. of Cloyne.

*The theory of vision or visual language, shewing the immediate presence and providence of a deity, vindicated and explained.*
London: printed for J. Tonson, 1733.
Subject: Natural theology, optics, senses, vision.
Case/B/542/.092

"Vision is the language of the author of nature" (p. 32).

## Bermingham, Michel M.

*Manière de bien nourrir et soigner les enfans.*
Paris: Barrois, 1750.
Subject: Medicine, pediatrics, infant and child care.
UNCATALOGUED

A manual for mothers(?). Recommends not overdressing infants. Encourages mothers to breast feed their own babies and to avoid the use of wet nurses.

## [Bernard, Jean Frédéric, ed.]

*Recueil de voiages au nord.*
Amsterdam: Jean Frédéric Bernard, 1725–38.
Subject: Navigation, voyages.
Case/G/12/.09

In 10 vols. Title page of vol. 1: *Recueil de voyages au nord, contenant divers mémoires très utiles au commerce & à la navigation.* Vol. 1: "Nouvelle edition, corrigée & mise en meilleur ordre," 1731; vol. 2 (includes description of animals of Spitzbergen and much other natural history), 1732; vols. 3 and 4, 1732; vol. 5, 3d ed., 1734; vol. 6, 3d ed., 1729; vol. 7, 1725; vol. 8, 1727; vol. 9 (the New World, including explorations of Louis Hennepin), 1737; vol. 10, 1738.

## Bernard, Richard, 1567?–1641.

*A guide to grand-ivry men, divided into two bookes: in the first, is the authors best aduice to them what to doe, before they bring in a billa vera in cases of witchcraft, with a Christian direction to such as are too much giuen vpon euery crosse to thinke themselues bewitched. In the second, is a treatise touching witches good and bad, how they may be knowne, euicted, condemned, with many particulars tending thereunto.*
London: printed by Felix Kingston for Ed. Blackmore, 1627.
Subject: Bewitchment and disease, witchcraft.
Case/B/88/.086

A variant of STC 1943. Bk. II: "That strange diseases may happen from onely naturall causes, and neither be wrought by deuils nor witches, and how to bee discerned." Apoplexy, catalepsis.

## Bernardus, Comensis, fl. 1596.

*Lvcerna inqvisitorvm haereticae pravitatis R.P.F. Bernardi Comensis ordinis praedicatorum: et eiusdem tractatus de strigibus . . . additi svnt in hoc impressione dvo tractatus Ioannis Gersoni vnus de protestatione circa materiam fidei, alter de signis pertinaciae haereticae prauitatis.*
Rome: Bartholomaeus Grassi, 1584.
Subject: Heresy, witchcraft.
Case/4A/2268

Pt. I a glossary of terms (and laws) relating to heresy, e.g., ivdaei, mala fides, secta strigiarvm, sortilegio. Pt. II about witches, 14 chaps. including incubus, succubus, fascination, menstruating women, witches interfering with the conjugal act, curses. Pt. III 2 tracts by Gerson. [Colophon: Rome: Vincentius Accoltus, 1584.]

### Bernardus, Comensis.

*Lvcerna inqvisitorvm haereticae pravitatis R.P.F. Bernardi Comensis ordinis praedicatorum: et eiusdem tractatus de strigibus. . . . Additi svnt in hac impressione duo tractatus Ioannis Gersoni, vnus de protestatione circa materiam fidei, alter de signis pertinaciae haereticae prauitatis.*
Venice: Marcus Antonius Zalterius, 1596.
Subject: Heresy, witchcraft
Case/D/49/.095

Contents as in 1584 ed.

### Bernardus, Ioannes Baptista.

*Ioan Baptistae Bernardi Patritii Veneti Seminarium totius philosophiae: opus nouum, & admirabile, & omni hominum generi perquam vtile: quod omnium philosophorum, eorundumque interpretum tam Graecorum, quam Latinorum, ac etiam Arabum quaestiones, conclusiones, sententiasque omnes integras, & absolutas miro ordine digestas conplectitur; vt quiuis vno intuitu, & sine vllo labore, quicquid vnquam à summis sapientiae magistris dictum fuit, perspicere, & eorum opera omnia in unum uelut locum collecta habere possit.*
Venice: Damianus Zenarius, 1582–85.
Subject: Reference sources, medicine, philosophy, science, technology.
Case/fB/05/.092/[1582]

3 vols. Topics listed alphabetically, each with author or authors who discuss or define them, e.g., "Phantasia est sensus quidam imbecillus. Arist. rhet. pri.c.47." Among authors cited in vol. 3 and listed in index: "Mercurij Pimander, trismegisti Asclepius, Apulei, Macrobius in somnio Scipionis, Auerroes in rempublicam Platonis, Mars. Fici. de vita apologia, Ioan Pic heptaplus, Io. Fra Pic. de elementis."

### Bernardus, Ioannes Baptista.

*Seminarivm totivs philosophiae Aristotelicae et Platonicae.*
[Geneva] Iacob Stoer & Franc. Fabri, 1599–1605.
Subject: Reference sources, medicine, philosophy, science, technology.
fB/05/.0922/[1599–1605]

3 vols. Descriptions or definitions of such items as cancer, canes, canum, coitus, tempus, densitas & raritas, dentium, deus. Vol. 1: Aristotle. Vol. 2: Plato (contains material on anatomy, astrology, astronomy). Vol. 3: Stoic philosophy and Seneca's aphorisms.

### [Berners], Juliana, Dame, b. 1388 [supposed author].

*The boke of St. Albans.*
[Westminster: Wynkyn de Worde, 1496]
GW 4933^ Still. B 916^ Goff B-1031
Subject: Natural history, veterinary medicine.
Inc./9704

2d ed. 1st publication of "A treatise on fishing." In black letter beneath woodcut of hunters and falconers: "This present boke sheweth the manere of hawkynge & huntynge: and also of diuysynge of cote of armours, it sheweth a good matetere [sic] belongynge to horses: wyth other commendable treatyses." Medicines for illnesses of hawks, horses. Description of illnesses. Properties of a good horse. Treatise on fishing includes description of various fish.

### Berners, Juliana, Dame.

*The gentlemans academie, or, the booke of St. Albans: containing three most exact and excellent bookes: the first of hawking, the second of all the proper termes of hunting, and the last of armorie: all compiled by Iuliana Barnes, in the yere from the incarnation of Christ 1486. And now reduced into a better method by G.M.*
London: printed by V. Sims for H. Lownes, 1595.
STC 3314 [variant?]
Subject: Natural history, veterinary medicine.
Case/V/112/.09

"Certaine proper termes belonging to all chase": a beuie of roes, a sounder of swine (12 or more wild boars), a rowt of wolues, a trip or heard of goats.

### Beroalde de Verville, 1556–ca. 1612.

*Le cabinet de Minerve. Avqvel sont plvsievrs singularites. Figures. Tableaux. Antiques. Recherches saintes. Remarques serieuses. Obseruations amoureuses. Subtilites agreables. Rencontres joyeuses & quelques histoires meslees es auantures de la sage fenis.*
Tovrs: Sebastien Molin, 1596.
Subject: Cosmography, magic, matter, natural and supernatural phenomena.
Case/Y/762/.B475

A miscellany. In dialogue form. Discusses salamander, magnet. Cabala (p. 221), wind (p. 236), magic (p. 278). Explanation of vision: eye has light within it.

### Beroaldo, Filippo, 1453–1505.

*Philippi Beroaldi opusculum eruditum: quo continentur declamatio philosophi, medici, oratoris de excellentia disceptantium.*
Bologna: Benedictus di Ettore Faelli, 1497. H-C *2963; C II 1005; BMC (XV)VI:844^ GW IV, 4126^ Pr. 6635^ Pell. 2218^ Goff B-473
Subject: Medicine, natural philosophy.
Inc./6635

In 3 parts. I: "Philippi Beroaldi declamatio an orator sit philosopho et medico anteponendvs." Includes sections "Dietetica pharmaceutica, chirurgica, meteorologi pliniana peroratio in medicos." III: Astrological medicine, Hippocrates, old age.

### Beroaldo, Filippo.

*Opusculum eruditum quo continentur declamatio philosophi medici & oratoris de excellentia disceptantium.*

[Paris: Denis Roce, 1501.]
Subject: Medicine, natural philosophy.
Wing/ZP/539/.P532

In 5 parts, separate title pages. Pt. 1: "Orationes" 1499 [Inc. 8402^ GW 4147]. Pt. 2: "Beroaldo de felicitate opusculum" [n.d.] [Inc. 8414.6]^ [GW 4137]. Refers to pleasure, praises rustic life. Obey the body, say Hippocrates and Galen. Pt. 3, n.a. Pt. 4: "Opusculum eruditum quo continentur declamatio philosophi medici" [Inc. 8414.6]. Pt. 5. n.a.

## Beroaldo, Filippo.

*Orationes prelectiones. Praefationes & quaedam mithicae historiae Philippi Beroaldi. Item Plusculae Angeli Politiani. Hermolai Barbari: atque vna Iasonis Maini oratio. Quibus addi possunt varia eiusdem Philippi Beroaldi opuscula cum epigrammatis & eorum commentarijs.*
Paris: Johanne Galthero, [1509].
Subject: Mysticism, Pythagorean symbolism.
Case/Y/682/.B46

## Beroaldo, Filippo.

*Symbola Pythagorae a Philippo Beroaldo moraliter explicata.*
[Bologna? ca. 1500.]
R 705^ Still. P1041^ Goff P-1148
Subject: Mysticism, Pythagorean symbolism.
Inc./6669/.5

Printer unknown. Place, date not given. Not in Hain, Copinger, or Proctor. [Goff: "Not printed after 25 December 1503"] A description of a series of Pythagorean symbols.

## Beroaldo, Filippo.

[*Symbola Pythagorae a Philippo Beroaldo moraliter explicata.*]
[Bologna: B. Hectoris, 1503.]
Subject: Mysticism, Pythagoream symbolism.
Case/B/122/.092

Mentions Galen, music , which soothes passions.

## Besold, Christoph, 1577–1638.

*De natvra populorum, usque pro loci positu, ac temporis decursu variatione: et insimul etiam de linguarum ortv atque immvtatione, philologicus discursus.*
Tübingen: Philbertus Brunni, 1632.
Subject: Anthropology, effect of climate on man, humors, "national types," sciences of man.
Q/17/.093

2d ed. Cap. VII: "De gentium proprietatibus occultis: quae varietati situs, aut terrae apparenti qualitati, minimè adscribi possunt."

## Besold, Christoph.

*Dissertationes singulares.*
Tübingen: Johan-Alexander Cellius, 1619.
Subject: Natural history of the New World.
Case/J/O/.0861

Six miscellaneous tracts. 2: "De novo orbe, coniectanea."

## Besson, Jacques, 16th cent.

*Theatrvm instrvmentorvm et machinarvm Iacobi Bessoni Delphinatis, mathematici ingeniosissimi. Cum Fran. Beroaldi figurarum declaratione demonstratiua.*
Lyon: Barth. Vincentius, 1578.
Subject: Instruments, mechanical engineering, technology.
Wing/fZP/539/.V74

Engraved plates illustrating instruments, machines.

## Bettini, Mario, 1582–1657.

*Apiaria vniversae philosophiae mathematicae, in qvibvs paradoxa, et noua pleraque machinamenta ad vsus eximios traducta . . . opvs non modo philosophis mathematicis, sed & physicis, anatomicis, militaribus viris, machinariae, musicae, poëticae, agrariae, achitecturae, mercaturae professoribus &c. vtilissimum; curiosissimis inuentis refertum, figurarum areis formis cusarum numerosá, & speciosá varietate ornatum.*
Bologna: Io. Baptista Ferroni, 1642.
Subject: Arithmetic, astronomy, encyclopedias, geometry, gnomonics, music, optics.
Case/6A/142

2 vols. Engraved plates, e.g., "aranea geometrizans," demonstrating geometry in nature by means of a spider and its web. To vol. 2 is added *Evclides applicatvs.*

## Beughem, Cornelius a, fl. 1678–1710.

*La France sçavante, id est, Gallia erudita, critica et experimentalis novissima. Seu manuductio ad faciliorem inventionem & cognitionem non tàm scriptorum operumque, quàm experimentorum, observationum, aliarumque rerum notatu dignarum cujusvis facultatis, artis & scientiae, quarum summaria in ephemeridibus eruditorum hujus celeberrimi regni ab ann. 1665 quo coeperunt, usque ad ann. 1687 recensentur.*
Amsterdam: Abraham Wolfgang, 1683.
Subject: Catalogues, weekly lists of publications in sciences.
Case/Z/939/.086

Booksellers catalogues in 6 parts. "Conspectus III, Ephemeridum eruditorum" contains bibliographia medica and physica, medicine, anatomy, physics, chemistry, surgery, botany.

## Beutel, Tobias.

*Geometrischer lust-garten die edele geometria aus dem Euclide gepflantzet. Abgetheilt in 2 bücher, das erste de planis das andere de solidis. . . .Bey dem author in Dressden und Joh. Chr. Tarnovio in Leipzig zu finden.*
[Leipzig] J. Georg, 1672.
Subject: Plane and solid geometry.
UNCATALOGUED

2d ed. (1st ed. 1660). A popular geometry textbook. Instruments, p. 174.

## Beveridge, William bp. of St. Asaph, 1637–1708.

*Institutionum chronologicarum libri II, vna cum totidem arithmetices chronologicae libellis.*

London: Thomas Roycroft, 1669.
STC II B2095
Subject: Calendar, chronology, time.
F/017/.019

"Chronologia est ars tempora rectè distinguendi" (p. 1). Describes Julian and Gregrorian calendars, Arabic, Persian, Judaic, Greek "years."

### Bewerlein, Sixtus.

*Erschröckliche gantz warhafftige geschicht welche sich mit Apolonia Hannsen Geisslbrechts burgers zu Spalt inn dem Eistätter bistumb haussfrawen so den 20. octobris anno 82. von dem bösen feind gar hart besessen und doch den 24 gedachtes monats widerumb durch gottes gnädige hilff auss solcher grossen pein und marter entlediget worden verlauffen hat.*
Ingolstadt: Wolffgang Eder, 1584.
Subject: Account of possession, witchcraft.
Case/B/8847/.1

### Bezançon, Germain de.

*Les médecins à la censure. Ou entretiens sur la medecine.*
Paris: Louis Gontier, 1677.
Subject: Medicine, defense of medical practices.
Case/Q/.095

"Toutes les precautions de la medecine sont inutiles; l'on pourra sans scrupule donner à une malade, brulé d'une fievre chaude, l'hypocras, l'eau de vie, le vin d'Espagne, luy charger l'estomac de viandes grossieres, & luy faire prendre les plus violents purgatifs. On pourra baigner une femme enceinte, saigner abondamment les phthistiques, donner la poudre d'algarot à un foible enfant . . . & presenter de l'opium en telle doze qu'on voudra à un lethargique, s'ils en son tuez, ce ne sera plus la faute du medecin ignorant, mais de la nature du malade qui n'a pas eu l'esprit d'en faire bon usage. . . La nature a bien communiqué aux bestes la connoissance des alimens, & des remedes dont ils ont besoin, comme l'ont remarqué les naturalistes." (p. 107) Some passages on Molière's *Malade imaginaire & Tartuffe*.

### Bèze, Théodore de, 1519–1608.

*Icones, id est verae imagines virorvm doctrina simvl et pietate illvstrivm, qvorvm praecipuè ministerio partim bonarum literarum studia sunt restituta, partim vera religio in variis orbis Christiani regionibus, nostra patrúmque memoria fuit instaurata: additis eorundem vitae & operae descriptionibus, quibus adiectae sunt nonnullae picturae quas Emblemata vocant.*
[Geneva] Ioannes Laonivs, 1580.
Subject: Emblems.
Case/*D/5/.092

Portraits and biographical sketches (of, e.g., Iacobvs Faber Ballvs Stapvlensis, Sorbonicvs theologvs; Conradvs Gesnervs; Sebastianvs Mvnstervs; Joachim Camerarivs), followed by a separate section containing 43 emblems.

* * *

*Bibliotheca anatomica, medica, chirurgica &c. containing a description of the several parts of the body: each done by some one or more eminent physician or chirurgeon; with their diseases and cures.*
London: printed by John Nutt, [1711]–1714.
Subject: Anatomy, medicine, surgery.
UNCATALOGUED

In 3 vols. Title page wanting in vol. 1 (title is taken from vol. 2). Preface to vol. 1: "The compilers of this work took this first design from Bibliotheca anatomica, medica, chirurgica, &c. the second edition whereof was lately published by Daniel Clericus and Jacob Mangetus in two volumes folio." Title page in vol. 2 continues: *Wherein are not only all the tractsof use that are in the second edition of the Bibliotheca Anatomica, . . . but an addition also of near double the number of other curious tracts,which were either omitted in the said Bibliotheca, or have been publish'd since, some of them translated others faithfully abridg'd; very few of which were ever before in English.* Vol. 1: bones, phlebotomy; vol. 2: head, breast, belly; vol. 3: blood and other bodily fluids, disease and cures in the genitals.

### Bilberg, Johan, 1646–1717.

*Disputationem de occultis qualitatibus consensu amplissimae facultates philosophicae in regia academica Upsalensi, sub praesidio clarissimi viri Dn. Joannis Bilberg. . . . publico bonorum examini submittit Gustavus Prosperius G. F.*
Holmiae [Holmstadt?]: printed by Johann. Georg Eberdt, 1687.
Subject: Occult qualities, sympathy and antipathy.
L/0114/.095

### [Binet, Benjamin] 17th cent.

*Idée generale de la theologie payenne, servant de refutation au systeme de Mr. Bekker. Touchant l'existence & l'operation des demons. Ou traitté historique des dieux du paganisme.*
Amsterdam: Jean du Fresne, 1699.
Subject: Demons in Old Testament and mythology.
B/88/.084

Argues in favor of existence of demons, since the belief was historically present in all societies. Parallels between, e.g., Hercules and Joshua, Typhon and Moses, Noah and his sons, and Saturn, Jupiter, Neptune, Pluto.

### Binet, Benjamin.

*Traité historique des dieux et des demons du paganisme. Avec quelques remarques critiques sur le systême de mr. Bekker.*
Delft: André Voorstad, 1696.
Subject: Demons in Old Testament and mythology.
B/88/.085

In epistolary form. Takes issue with B. Bekker on properties and existence of demons past and present.

### [Binet, Etienne] 1569–1639.

*Essay des merveilles de la natvre et des plvs nobles artifices, piece tres necessaire a tovs cevx qvi font profession d'eloqvence. Par Rene Francois, predicateur du Roy.*
Rouen: Romain de Beauuais Iean Osmont, 1622.

Subject: Encyclopedias, mathematics, medicine, natural history.
Case/Wing/Z/4029/.096

61 chapters. Subjects include birds, metals, printing, fish, bees, medicine, mathematics. Chap. 46: "Les devoirs de medicine, de la pharmacie, et chirvrgie."

## Binsfeld, Peter, 1540–1598.

*Tractatus de confessionibus maleficorum & sagarum, an, et qvanta fides ijs adhibenda sit.*
Trier: Henricus Bock, 1589.
Subject: Demonism, possession by devil, witchcraft.
Case/B/88/.089

"Prima conclusio. Malefici, magi, diuinatores, vel quicunque alij pactum cum daemonibus habentes, nulla vera possunt facere miracula" (p. 57). Demonism not transferable to animals.

## Binsfeld, Peter.

*Tractat von bekantnuss der zauberer unnd hexen. Ob und wie viel denselben zu glauben.*
Trier: Heinrich Bock, 1590.
Subject: Demonism, possession by devil, withcraft.
Case/B/88/.09

## Biringucci, Vannuccio, 1480–ca.1539.

*De la pirotechnia. Libri X. Dove ampiamente tratta non solo di ogni sorte & diuersita di miniere, ma anchora quanto si ricerca intorno à la prattica di quelle cose di quel che si appartiene a l'arte de la fusione ouer gitto de metalli come d'ogni altra cosa simile a questa.*
Venice: Venturino Roffinello for Curtio Nauo & brothers, 1540.
Subject: Alchemy, experimental science, metallurgy.
Case/*R/40/.1

1st ed. Lib. IX: Del arte alchimica.

## Biringucci, Vannuccio.

*Of the generation of metalles and their mynes with the manner of fyndinge the same: written in the Italien tounge by Vannuccius Biringuczius in his booke cauled Pyrotechnia.*
London: William Powell, 1555.
Subject: Alchemy, experimental science, metallurgy.
Ayer/*110/.E2/1555

In Eden, Richard. *The decades of the new worlde or West India*, q.v.) From The preface to the booke of metals: "To this booke of the Indies and nauigations I have thought good to adde the booke of metals, for three causes especially me movynge: whereof the fyrst is, that it seemeth to me a thynge undecent to reade so much of golde and syluer, and to know little or nothynge of the naturall generation thereof . . . I will speake of the second cause: which is, that if in trauayling strang and unknowen countreys any mans chaunce shalbe to arriv in such regions where he may knowe by thinformation of thinhabitants or otherwise, that such regions are frutefull of rich metals, he may not bee without sum judgement to make further searche for the same. The thyrde cause is that thowgh this oure realme of Englande be full of metals . . . yet there is fewe or none in Englande that have anye greate skyll thereof" (leaf 325v).

## Blackmore, Sir Richard, d. 1729.

*Creation. A philosophical poem in seven books.*
London: Printed for S. Buckley & J. Tonson, 1712.
Subject: Reconciliation of religion and science.
Y/185/.B5713

## Blackmore, Sir Richard.

*The nature of man. A poem in three books.*
London: printed for Sam. Buckley, 1711.
Subject: Anthropology, natural history.
Case/Y/185/.B5714

Book 1, natural history; bk. 2, descriptive anthropology; bk. 3, politics.

## Blaeu, Willem Janszoon, 1571–1638.

*Gvilielmi Blaev institvtio astronomica de usu globorum & sphaerarum caelestium ac terrestrium: dvabvs partibvs adornata, vna, secundum hypothesin Ptolemaei, per terram qviescentem. Altera, juxtamentem N. Copernici, per terram mobilem. Latinè reddita à M. Hortensio.*
Amsterdam: Ioh. & Cornelius Blaev, 1640.
Subject: Astronomy, Copernican and Ptolemaic systems, celestial and terrestrial navigation.
Ayer/8/B62/1640

Solar quadrants, sundials, clocks, zodiac. Triple movement of the earth.

## Blaeu, Willem Janszoon.

*Institution astronomiqve de l'vsage des globes et spheres celestes et terrestres, comprise en deux parties, l'une suivant l'hypothèse de Ptolemée, qui veut la terre soit immobile; l'autre, selon l'intention de N. Copernicus, qui tient que la terre est mobile.*
Amsterdam: J. Blaeu, 1669.
Subject: Astronomy, Copernican and Ptolemaic systems, celestial and terrestrial navigation.
Ayer/*8/B62/1669

Pt. 1: "suivant l'hypothese impropre de Ptolemée." Longitude and latitude, zodiac, tables of declination. Pt. 2: "suivant la vraye hypothese de N. Copernicus." Finding true north and south, solar quadrants, rising and setting of heavenly bodies.

## Blankaart, Steven, 1650–1702.

*Steph. Blancardi anatomia reformata, sive concinna corporis humani dissectio, ad neotericorum mentem adornata. . . . accessit ejusdem authoris de balsamatione, nova methodus, à nemine antehac hoc modo descripta.*
Leyden: Cornelius Boutesteyn, Jordanus Luchtmans, 1688.
Subject: Anatomy, embalming.
oQM/21/.B55/1688

### Blégny, Étienne de, ca. 1666–1699.

*Les elemens ov premières instructions de la jeunesse.*
Paris: Guillaume Cavelier, 1712.
Subject: Arithmetic, education.
Wing/ZW/739/.B612

"L'aritmétique facile" (p. 213).

### Blégny, Étienne de.

*Nouveaux exemplaires d'escriture d'une beauté singuliar* [*sic*] *ecrits par Estienne de Blegny.*
[Paris: widow of Guillaume Cavelier, 1728.]
Subject: Arithmetic, education.
Wing/ZW/739/.B6125

### Blith, Walter, fl. 1649.

*The English improver improved or the svrvey of hvsbandry svrveyed discovering the improueableness of all lands.*
London: printed for John Wright, 1652.
STC II B3195
Subject: Agricultural instruments, machines, and tools, agronomy, irrigation, soil conservation and improvement.
Case/R/50/.1

3d impression much augmented. Windmills, ploughs, trenching spades, crop rotation.

### [Blome, Richard] d. 1705.

*Britannia: or a geographical description of the kingdoms of England, Scotland, and Ireland, with the isles and territories thereto belonging.*
London: printed by Tho. Roycroft for the undertaker, Richard Blome, 1673.
STC II B3207
Subject: Descriptive geography.
Case/fG/4495/.1

Temperature, produce, domesticated animals, natural resources, topography included as part of description of each shire. Pp. 325–41: Isles and territories belongong to His Majesty in America (including the Caribbean).

### [Blome, Richard]

*The gentlemen's recreation, In two parts. The first being an encyclopedy of the arts and sciences. To wit, an abridgment thereof, which (in a clear method) treats of the doctrine, and general parts of each art, with eliptical tables, comprehending a summary and general division thereof. . The second part treats of horsemanship, hawking, hunting, fowling, fishing, and agriculture.*
London: printed by S. Roycroft for Richard Blome, 1686.
STC II B3213
Subject: Agriculture, encyclopedias.
Case/V/112/.099

"An account of the several arts and sciences treated in this work include metaphysics, arithmetic, natural philosophy, geometry, surveying, gauging, cosmography and astronomy, astrology, geography, navigation, dyalling, architecture, chronology, opticks, perspective."

### Blondel Saint Aubin, Guillaume, fl. 1673.

*Le tresor de la navigation. Divisé en deux parties. La première contient la theorie & pratique de triangles spheriques, enrichie de plusieurs problemes astronomiques & geographiques, tres-vtiles avx navigateurs. La seconde enseigne l'art de naviger par la supuration & démonstration des triangles rectelignes & spheriques; tant par les sinuss que par les logarithmes.*
Havre de Grace: Iacques Gruchet, 1673.
Subject: Astronomy, navigation.
Ayer/8.9*/N2/B64/1673

Woodcut diagrams.

### Blount, Sir Thomas Pope, bart., 1649–1697.

*A natural history: containing many not common observations: extracted out of the best modern writers.*
London: printed for R. Bentley, 1693.
STC II B3351
Subject: Encyclopedias, natural history.
Case/M/O/.11

Subjects include amber, cinnamon, pepper, manna, lignum aloes, tea, opium, diamonds, loadstone, sea compass, petrification, generation of metals, damps in mines, generation of insects, earthquakes.

### Blundeville, Thomas, fl. 1561.

*A briefe description of vniversal mappes and cardes, and of their vse: and also the vse of Ptholemey his tables. Necessarie for those that delight in reading of histories: and also for traueilers by land or sea.*
London: printed by Roger Ward for Thomas Cadman, 1589.
Church 137^ STC 3145
Subject: Cosmography, navigation.
Ayer/*6/P9/B6/1589

1st ed. Glossary contains "Certaine tearmes of cosmographie, brieflie expounded, for those that are not learned in that science, to the intent they may the better vnderstand this treatise." One of earliest English books that made practical use of Ptolemy's geographical tables. Very rare in separate form. Later included in Blundeville's *Exercises*.

### Blundeville, Thomas.

*M. Blvndevile his exercises, containing eight treatises . . . which treatises are verie necessarie to be read and learned of all yoong gentlemen that have not bene exercised in such disciplines, and yet are desirous to haue knowledge as well in cosmographie, astronomie, and geographie, as also in the arte of nauigation. in which arte it is impossible to profite without the helpe of these, or such like instructions.*
London: Iohn Windet, 1597.
STC 3147
Subject: Arithmetic, astronomy, geography, navigation.
Ayer/*6/P9/B6/1597

2d ed. Treatises include (1) Easie arithmeticke. (2) 1st principles of cosmographie. (3) Mercator's terrestrial and celestial globes. (4) Plancius' universal map. (5) Blagraue's astrolabe. (6) Principles of navigation. (7) Vniversall mappes and cards. (8) Tables of sines, tangents, secants. P. 134: "Cosmographie . . . is the description of the whole world, . . . that is to say, of heauen and earth and all that is contained therein. It embraces astronomy, geography, chorography, astrology." P. 376: "A description of Gemma Frizius his instrument called quadratum nauticum."

## Blundeville, Thomas.

*Mr. Blundevil his exercises, contayning eight treatises.*
London: printed by Richard Bishop, 1636.
Subject: Arithmetic, astronomy, geography, navigation.
STC II B3151
Ayer/*6/P9/B6/1636

7th ed. corrected and somewhat enlarged by Ro. Hartwell, Philomathematicus.

## Blundeville, Thomas.

*The fower chiefyst offices belongyng to horsemanshippe. That is to saye, the office of the breeder, of the rider, of the keper, and of the ferrer. In the firste parte wherof is declared the order of breding of horses . . . thirdely howe to dyet them as well when they rest as when they trauell by the way. Fourthly to what diseases they be subiecte, together with the causes of such diseases, the sygnes howe to knowe them, and finally howe to cure the same.*
London: VVyllyam Seres, [1565–66].
STC 3152
Subject: Animal husbandry, veterinary medicine.
Case/4A/822

Each of the 4 sections has a separate title page. Pt. IV: *The order of cvring horses diseases, together with the cavses of svch diseases, the sygnes hovve to knowe them, and finally how to cure the same. With the true art of paring & shooing al manner of houes . . . lately set forth by Thomas Blundeuil of Newton Flotman in Norffolke* (1566). STC 3159^ Contains recipes for ointments, medicines, instructions for bleeding, cauterization, splinting.

## Boate, Gerard, 1604–1650.

*Ireland's natvrall history. Being a true and ample description of its situation, greatness, shape, and nature; of its hills, woods, heaths, bogs; of its fruitfull parts and profitable grounds, with the severall way[sic]of manuring and improving the same: with its heads or promontories, harbours, roades, and bayes; of its springs and fountaines, brookes, rivers, loghs; of its metalls, mineralls, free-stone, marble, sea-coal, turf, and other things that are taken out of the ground. And lastly, of the nature and temperature of its air and season, and vvhat diseases it is free from, or subject unto. Conducing to the advancement of navigation, husbandry, and other profitable arts and professions.*
London: printed for J. Wright, 1652.
STC II B3372
Subject: Descriptive geography, disease, minerals, natural history.
Case/G/42003/.1

Published by Samuel Hartlib. Chap. 16: Of the mines in Ireland, and in particular of the iron-mines. Chap. 24: Of the diseases reigning in Ireland & whereunto that country is peculiarly subject.

## Bobowski, Albert, afterwards Ali Bey, d. 1676.

*Tractatus Alberti Bobovii . . . de Turcarum liturgia peregrinatione Meccana, circumcisione, aegretorum visitatione, &c. . . . subjungitur castigatio in angelum à Sancto Joseph.*
Oxford: Sheldonian theatre, 1690.
Subject: Disease, medicine, social anthropology.
Ayer/7/A2/1691

BOUND WITH Abraham ben Mordecai Farissol, *Itinera mundi*, 1691, q.v. Has separate title page and pagination but continuous signatures.

## Bodin, Jean, 1530–1596.

*De la demonomanie des sorciers.*
Paris: Iacques du Puys, 1580.
Subject: Sorcery, witchcraft.
Case/B/88/.0958

In 4 books. Describes good and evil sorcerers, various methods of divining (via movements of birds, sacrifices, augury using thunder and lightning) magic, bodily transportation by spirits, lycanthropy, witches and natural disasters, ways to prevent becoming possessed, copulation with demons, charms and curses to cause and cure illnesses, necessary proofs to condemn witches.

## Bodin, Jean.

*De magorum demonomania libri IV.*
Basel: Thomas Guarinus, 1581.
Subject: Sorcery, witchcraft.
Case/B/88/.095

Many passages crossed out (in MS, probably censorship). Marginal notes in the preface.

## Bodin, Jean.

*De la demonomanie des sorciers.*
Paris: Iacques du Puys, 1582.
Subject: Sorcery, witchcraft.
Case/B/88/.096

## Bodin, Jean.

*De la demonomanie des sorciers. . . . reueu, corrigé, & augmenté, d'vne grande partie.*
Paris: Iacques dv-Pvys, 1587.
Subject: Sorcery, witchcraft.
Case/B/88/.098

MS marginalia in preface.

## Bodin, Jean.

*De la demonomanie des sorciers. Par I. Bodin, Angevin. Reueüe diligemment, et repurgee de plusieurs fautes qui s'estoyent glissees és precedentes impressions.*
Lyon: Antoine de Harsy, 1598.
Subject: Sorcery, witchcraft.

Case/B/88/.099

4th ed.

## Bodin, Jean.

*Demonomania de gli strigoni, cioè furori, et malie de' demoni, col mezo de gli huomini: diuisa in libri IIII. Di Gio. Bodino francese. Tradotta dal kr. Hercole Cato... Con vna confutatione dell'opinione di Gio. Vuier, per confermare quanto nell'opera si contiene, & contra quelli, i quali niente credono à cosi fatte materie. Di nuouo purgata, & ricorretta.*

Venice: Aldus, 1589.

Subject: Sorcery, witchcraft.

Case/*B/88/.1

This copy does not contain the preface by Niccolo Manassi of Feb. 15, 1587.–cf. Renouard, *Annales des Alde.*

## Bodin, Jean.

*Vniversae natvrae theatrvm. In qvo rervm omnivm effectrices causae, & fines contemplantur, & continuae series quinque libris discutiuntur.*

Hanover: Wechel for Claudius Marnius & Ioann. Aubrius, 1605.

Subject: Encyclopedias, fossils, origins of nature, physics, spirits.

Case/M/O/.115

Dialogue between Theodorus and the master, Mystagogus.

## Boehme, Jakob, 1575–1624.

*The clavis, or key. Or, an exposition of some principall matters, and words in the writings of Jacob Behmen. Very usefull for the better apprehending, and understanding of this booke [Signatura rerum]. Written in the German language in March and April 1624 by Jacob Behmen. Also called Teutonicus Philosophus.*

[n.p.] 1647.

Subject: Alchemy, astrology, doctrine of signatures, mysticism.

Case/C/515/.098

Trans. John Sparrow (1615–1665?). 7 properties of eternal nature: 3 substances, 4 elements. "You understand by science some skill or knowledge: in which you say true, but doe not fully expresse the meaning" (p. 27). BOUND WITH his *Signatura rerum* and *XL questions concerning the soule*, qq.v. For a recent analysis of Boehme's work see Russell H. Hvolbeck, "Seventeenth century dialogues, Jacob Boehme and the new science," (Ph.D diss., University of Chicago, 1984).

## Boehme, Jakob.

*Een gebedt-boeckien, in't welck den waren grondt wordt ghetoont, van't recht bidden. Dorr Jacob Böhme; anders Teutonicus philosophus.*

n.p. 1641

Subject: Alchemy, astrology, doctrine of signatures, mysticism.

Case/C/515/.109

Miscellaneous works and tracts by Boehme, trans. into Dutch 1639–42. Relevant tracts appear to be nos. 4, 5, 8, 11, 13, 14. Three titles of special interest: Aurora, Clavis, Mysterium magnum.

## Boehme, Jakob.

*XL questions concerning the soule. Propounded by Dr. Balthasar Walter. And answered by Jacob Behmen. . . . And in his answer to the first question is the turned eye, or, philosophick globe... Written in the Germane language anno 1620.*

London: printed by Matth. Simmons, 1647.

STC II B3408

Subject: Alchemy, astrology, doctrine of signatures, mysticism.

Case/C/515/.098

Trans. John Sparrow. BOUND WITH *The Clavis* and *Signatura rerum*, qq.v.

## Boehme, Jakob.

*Mysterium magnum. Or, an exposition of the first book of Moses called Genesis. Concerning the manifestation or revelation of the divine word through the 3 principles of the divine essence; also of the originall of the world and the creation. Wherein the kingdome of nature, & the kingdome of grace, are expounded . . . Comprised in three parts: written anno 1623 by Jacob Behm. To which is added the life of the author*[by Durand Hotham, 1653] *and his four tables of divine revelation.*

London: printed by M. Simmons for H. Blunden, 1654.

STC II B3411

Subject: Alchemy, astrology, doctrine of signatures, macrocosm–microcosm, mysticism.

Case/fC/355/.101

Preface signed by John Sparrow. Description of metals (chap. X, p.38): "For the metalls are in themselves nothing else but a water and oyl; which are held by the wrathful properties; viz by the astringent austere desire; that is; by a Saturnine martiall fiery property; in the compaction of sulphur and mercury, to be one body."

## Boehme, Jakob.

*The remainder of books written by Jacob Behme.*

London: printed by M. S. for Giles Calvert, 1661.

Subject: Alchemy, astrology, doctrine of signatures, mysticism.

Case/C/621/.106

Trans. by John Sparrow. 7 pts. in 1 vol. Not in Wing. [1st & 2nd] Apologies to Balthazar Tylcken. Being an answer of the authour, concerning his book the Aurora. . . . Of the 4 complexions. And others.

## Boehme, Jakob.

*The remainder of books written by Jacob Behme.*

London: printed by M. S. for Giles Calvert, 1662.

STC II B3415

Subject: Alchemy, astrology, doctrine of signatures, mysticism.

Case/C/515/.1094

Trans. John Sparrow. "With a catalogue of all the books that are known to be extant written by Jacob Behme and now printed in England."

**Boehme, Jakob.**

*Several treatises of Jacob Behme not printed in English before according to the catalogue here following.*
STC II B3418
London: L. Lloyd, 1661.
Subject: Alchemy, astrology, doctrine of signatures, mysticism.
Case/C/621/.107

I. A book of the great six points: As also a small book of the other six points. II. The 177 theosophick questions. III. Of the earthly and heavenly mystery. IV. [n.a.] V. Of divine vision. To which are annexed the exposition of the table of the three principles: also an epistle of the knowledge of God, and of all things. And of the true and false light. With a table of the revelation of the divine secret mystery.

**Boehme, Jakob.**

*Signatura rerum; or the signature of all things: shewing the sign, and signification of the severall forms and shapes in the creation: and what the beginning, ruin, and cure of every thing is; it proceeds out of eternity into time and again out of time into eternity, and comprizeth all mysteries. Written in high Dutch MDCXXII by Jacob Behmen alius Teutonicus Philosophus.*
London: printed by J. Macock for G. Calvert, 1651.
STC II B3419
Subject: Alchemy, astrology, doctrine of signatures, mysticism.
Case/C/515/.098

Preface of translation signed J. Ellistone. Postscript listing author's works. BOUND WITH his *Clavis*, and *XL questions*, qq.v.

**Boehme, Jakob.**

*Jacob Behmen's theosophick philosophy unfolded; in divers considerations and demonstrations, shewing the verity and utility of the several doctrines or propositions contained in the writings of that divinely instructed author. Also the principal treatises of the said author abridged. And answers given to the remainder of the 177 theosophick questions . . . which were left unanswered by him at the time of his death. By Edward Taylor. With a short account of the life of Jacob Behmen.*
STC II B3421
London: printed for Tho. Salusbury, 1691.
Subject: Alchemy, astrology, doctrine of signatures, mysticism.
Case/C/621/.112

A list of words defined according to Boehme. Includes extracts of several of his works, e.g., *Aurora*, or morning redness, and *The three principles of the divine essence.*

**Boehme, Jakob.**

[*Theosophische werken.*]
Amsterdam, 1682.
Subject: Alchemy, astrology, doctrine of signatures, mysticism, theosophy.
Case/C/515/.105

5 pts. in 1 vol. Each part has engraved frontispiece with mystical symbols.

**Boehme, Jakob.**

*The third booke of the authour being, The high and deepe searching out of the threefold life of man through the three principles.*
London: printed by M.S. for H. Blunden, 1650.
STC II B3422
Subject: Alchemy, astrology, doctrine of signatures, mysticism.
Case/C/621/.128

Trans. John Sparrow. Adam bore the mark of the entire Cross imprinted in the brain pan of his skull. After the Fall from Eden, however, as a further mark of mortal imperfection, only half of the imprint of the Cross occurs in the brain pan of the human skull: half in the skull of man and half in that of woman (chap. 11, p. 165, sidenote).

**Boemus, Joannes, fl. 1500.**

*The manners lawes & cvstomes of all nations. Collected out of the best writers by Joannes Boemus Avbanvs, a Dutchman. With many other things of the same argument.*
London: printed by G. Eld, 1611.
STC 3198.5
Subject: Anthropology, geography, hygiene.
Case/G/117/.092

Making of iron weapons in Spain (p. 378). Washing in urine (p. 379).

**Boemus, Joannes.**

*Omnivm gentivm mores, leges, & ritus ex multis clavissimis rerum scriptoribus, à Ioanne Boëmo Aubano Teutonico nuper collecti, & nouissime recogniti.*
[Colophon] Freiburg im Breisgau: Ioannes Faber Emmevs Ivliacensis, 1540.
Subject: Anthropology, geography, hygiene.
Ayer/*7/H7/1542

Title page dated 1542. BOUND WITH Honter, Johannes, q.v.

**Boerhaave, Hermann, 1668–1738.**

[*Methodus dicendi*]
*A method of studying physick. Containing what a physician ought to know in relation to the nature of bodies, the laws of motion, staticks, hydrostaticks, hydraulicks, and the proprieties [sic] of fluids: chymistry, pharmacy and botany: osteology, myology, splanchnology, angiology and dissection: The theory and practice of physick: physiology, pathology, surgery, diet, &c. And the whole praxis medica interna; with the names and characters of the most excellent authors on all these several subjects in every age: systematicks, observators, &c. their best editions and the method of reading them.*
London: printed by H. P. for C. Rivington, B. Creake, and J. Sackfield, 1719.

Subject: Mechanics, medicine, natural science.
Case/R/128.7/.B66/1719

## Boerhaave, Hermann.

*Praelectiones academicae in proprias institutiones rei medicae edidit; et notas addidit Albertus Haller.*
Gottingen: Abram Vandenhoeck, 1740.
Subject: Medicine.
UNCATALOGUED

2d ed. In 6 vols. Vols. 1 and 2 published 1740; vol. 3 in 1741; vol. 4 in 1743; vols. 5 (in 2 pts.) and 6 in 1744.

## Boguet, Henri, fl. 1603.

*Discovrs des sorciers, avec six advis en faict de sorcelerie. Et vne instrvction pour vn iuge en semblable matiere.*
Lyon: Pierre Rigaud, 1610.
Subject: Demonology, divination, fascination, witchcraft.
Case/B/8839/.113

3d ed. Customs and habits of demons, diviners, witches. Sabbats, lycanthropy, divining by means of animal entrails. The "advis" are sentences appropriate for various types of sorcerers: e.g., users of animal entrails (haruspex, aruspicine) should be burnt. Attendance at a sabbat should be punishable by death.

## Boguet, Henri.

*Discours execrable des sorciers. Ensemble leur procez, fait depuis 2. ans en çà, en diuers endroicts de la France. Auec vne instruction pour vn iuge, en faict de sorcelerie.*
Rouen: Romain de Beavvais, 1606.
Subject: Demonology, witchcraft.
Case/B/8839/.112

"Desquelles maladies les sorciers affligent particulierement les personnes." They can also blight plants and animals (chap. 32, p. 168).

## [Boileau de Bouillon, Gilles] 16th cent.

*La sphere des deux mondes, composée en François par Darinel pasteur des Amadis.*
Antwerp: Iehan Richard, 1555.
Subject: Astronomy, descriptive geography.
Ayer/*7/B6/1555

"Commenté, glosé, & enrichy de plusieurs fables poetiques, par G. B. D. B. C. C. de C. IV.L.OVBLI." In verse with explanatory prose text interpolated. Also a poem, L'autre monde, de Darinel (brief descriptions of various countries?) with reproductions of woodcut maps said to be by Honterus.

## Boissard, Jean Jacques.

*Emblematum liber. Emblems latins de I. I. Boissard, auec l'interpretation françoise du I. Pierre Ioly Messin.*
Metz: Abraham Faber, 1588.
Subject: Emblems.
Case/W/1025/.095

## Boissard, Jean Jacques.

*Emblemata.*
[Frankfurt, 1593?]
Subject: Emblems.
Case/W/1025/.0951

## Boissard, Jean Jacques.

*Tractatus posthumus Jani Jacobi Boissardi Vesvntini de divinatione & magicis praestigiis, quarum veritas ac vanitas solidè exponitur per descriptionem deorum fatidicorum qui olim responsa dederunt; eorundumque prophetarum, sacerdotum, phoebadum, sibyllarum & divinorum, qui priscis temporibus celebres oraculis existerunt. . . . jam modò eleganter aeri incisis per Joh. Theodor, de Bry.*
Oppenheim: H. Galler [1616?].
Subject: Magic, occult, divination.
Case/B/863/.096

1st part of vol. (11 chaps.) De divinatione. Followed by De magia & magicis praestigiis & imposturis. P. 88 begins series of portraits and biographies of famous prophets of antiquity, including Themis, Pythian Apollo, Cumaean Sibyl, Hermes Trismegistus.

## Boissière, Claude de, fl. 1554–1608.

*Nobilissimvs et antiqvissimvs ludus Pythagoreus (qui rythmomachia nominatur) in vtilitatem & relaxationem studiosorum comparatus ad veram & facilem proprietatem & rationem numerorum assequendam, nunc tandem per Claudium Buxerium Delphinatem illustratus.*
Paris: Gulielmus Cauellat, 1556.
Subject: Arithemtic, geometry, numerology, proportion.
Case/V/1652/.107

Illustrated with woodcuts and tables throughout.

## Boissière, Claude de.

*Le tres excellent et ancien ieu Pythagorique, dict rythmomachie, fort propre & tres vtil à la recreation des esprits vertueux, pour obtenir vraye & prompte habitude en tout nombre & proportion: illustré par maistre Claude de Boissiere Daulphinois, & nouuellement amplifié par le mesme autheur.*
Paris: Guillaume Cauellat, 1556.
Subject: Arithemtic, geometry, numerology, proportion.
Case/V/1652/.108

Mentions Jacques le Fèvre d'Etaples.

## Bonet, Theophile, 1620–1689.

*A consiliis medicis. Sepulchretum sive anatomia practica, & cadaveribus morbo denatis, proponens historias et observationes omnivm penè humani corporis affectuum, ipsorumque causas reconditas revelans. Quo nomine pathologiae genuinae, quàm nosocomiae orthodoxae fundatrix, imo medicinae veteris ac novae promptuarium dici meretur.*
Geneva: Leonard Chouët, 1679.
Subject: Disease, medicine, pathology.
UNCATALOGUED

In 4 vols. Vol. 1: Affectus capitis et pectoris; vol. 2: Affectibus medii ventris seu thoracis; vol. 3: Affectibus imi ventris; vol. 4: fevers.

## Bongus, Petrus, d. 1601.

*Numerorum mysteria. Opvs maximarvm rervm doctrina et copia refertvm, in qvo mirus in primis, idemque perpetuus arithmeticae Pythagoricae cum diuinae paginae nvmeris consensus, multiplici ratione probatur.*
Bergamo: C. Ventura, 1599.
Subject: Arithmetic, history and symbolism of numbers, numerology.
Case/B/863/.105

Roman and arabic numerals; scriptural and mystical significance of each number up to 12, after which selected numbers are described and explicated. 144,000 is final number discussed.

## [Bonnefons, Nicolas de]

*The French gardiner instructing how to cultivate all sorts of fruit-trees, and herbs for the garden: together with directions to dry and conserve them in their natural: an accomplished piece, written originally in French, and now translated into English by John Evelyn. Whereunto is annexed, The English vineyard vindicated, by John Rose.*
London: printed by S. S. for Benj. Tooke, 1672.
Subject: Botany, diseases of plants, gardening, husbandry, insect pests, soil chemistry.
Case/R/57/.275

3d ed. Sec. I: Means to recover and meliorate ill ground. Grafting. Curing ills of trees and shrubs. Catalogue of namesof fruits known about Paris (with season when ripe).

## Bonoeil, John.

*His Maiesties graciovs letter to the Earle of Sovth-Hampton, treasurer, and to the councell and company of Virginia heere: commanding the present setting vp of silke works, and planting of vines in Virginia. . . . Also a treatise of the art of making silke: or, directions for the making of lodgings, and the breeding, nourishing, and ordering of silkewormes, and for the planting of mulbery trees, and all other things belonging to the silke art. Together with instructions how to plant and dresse vines, and to make wine, and how to dry raisins, figs, and other fruits, and to set oliues, oranges, lemons, pomegranates, almonds, and many other fruits, &c.*
London: printed by Felix Kyngston, 1622.
Subject: Agriculture, silkworms.
STC 14378
Ayer/150.5/V7/B7/1622

## Bonomi, Giovanni Francesco, b. 1621.

*Chiron Achilles, sive navarchvs humanae vitae, morale emblemata geminato ad felicitatis portem perducens.*
Bologna: H. H. de Duccis, 1661.
Subject: Emblems.
Case/W/1025/.097

"Areana querens curiosus perit," pp. 170–175. Lightning striking scientist. Other illustrations include forge, p. 322, printing press, p. 224.

## Boodt, Anselm Boece de 1550?–1632.

*Symbola diuina & humana pontificvm imperatorvm regvm. Accessit breuis & facilis isagoge Iac. Typotii. Ex mvsaeo Octavii de Strada civis Romani.*
Prague: Egidius Sadeler, 1601.
Subject: Emblems, hierography.
Case/fF/0711/.896

3 vols. in 2. For vols. 2 and 3, see Typotius, Iacobus. Symbols, ecclesiastical, and of aristocracy and royalty, French, English, Italian. Some mystical and occult components.

## Boodt, Anselm Boece de.

*Symbola diuina & humana.*
Frankfurt: Godfridus Schönwetter, 1652.
Subject: Emblems, hierography.
Case/fF/0711/.897

3 vols. in 1. Same engraved title page as 1601 ed., with later imprint superimposed.

## Boodt, Anselm Boece de.

*Symbola varia diversorum principum, archiducum, ducum, comitum & marchionum totius Italiae. Cum facili isagoge D. Anselmi de Boot.*
Amsterdam: Ysbrandum Haring, 1686.
Subject: Emblems, hierography.
Case/W./1025/.1

Coats of arms or symbols of various aristocratic houses. Some magical & mystical symbolism, e.g., snake eating itself: "Cum patientia" (p. 16, no. IX). Work not divinely ordained ("Chimici"), pp. 365–69. Earth viewed from above (over-curious speculation), pp. 321–23.

## Borel, Pierre, 1620–1689.

*Les antiquitez raretez, plantes, minereaux & autres choses considerables de la ville & comté de Castres d'Albigeois, & des leiux qui sont ses enuirons. Auec le roole des principaux cabinets, & autres raretez de l'Europe. Aussi le catalogue des choses rares de maistre Pierre Borel, docteur en medecine.*
Castres: Arnavd Colomiez, 1649.
Subject: Biology, curiosities, natural history, prodigies.
F/399146/.107

2 vols. in 1. Bk. II: topography, catalogue of plants native to Castres, monsters and prodigies, prophesies of Nostradamus. Catalogue of plants, with ca. 200 species listed.

## Borja, Juan de, conde de Mayalde y de Ficallo, b. 1553.

*Emblema moralia scripta quondam Hispanice a Johanne de Boria Latinitate autem sonata a L. C. L. P.*
Berlin: Ulrich Liebpert [for] Johann Michael Rudigeri, 1697.

Subject: Emblems.
Case/W/1025/.107

Emblems include lightning, p. 190, sun, p. 74, volcano, p. 68, earth and heaven, p. 52.

### Borri, Giuseppe Francesco.

*La chiave del gabinetto del cavagliere Gioseppe Francesco Borri, Milanese. Col favor della quale si vedono varie lettere scientifiche, chimiche, e curiosissime.*
Cologne: printed by Pietro del Martello, 1681.
Subject: Alchemy.
B/8633/.108

Letters followed by dialogue. P. 124: 2 letters on transmutation, philosophers stone; p. 382 lists letters by Borri. Topics include formation of metals, congelation of mercury, transformation of mercury to silver,"trarre la semente dall' oro."

### Bosso, Matteo, 1428–1502.

*Dialogus tribus libris seu disputationibus distinctus. Eivsdem de instituendo sapientia animo, siue de vero sapientiae cultu libri octo. Eivsdem de tolerandis aduersis libri duo. Eivsdem de gerendo magistratu iustitiaque colenda opusculum. Eivsdem de immoderato mulierum cultu repraehensoria ad Bessarionem cohortatio.*
Strassburg: Matthias Schürer, 1509.
Subject: Epistemology, neo-Platonism, soul.
Case/C/69/.109

### Bosso, Matteo.

*De institvendo sapientia animo dispvtationes per dies VIII.*
Bologna: (Franciscus) Plato de Benedictis, 1495.
H-C 3675 = 3677*^ Pell. 2781^ Pr. 6609^ BMC (XV)VI:828^ GW 4954^ Goff B-1043
Subject: Epistemology, human wisdom.
Inc./6609

### Bosso, Matteo.

*Recvperationes Fesvlanus lector agnostico.*
Bologna: (Francesco) Plato de Benedictis, 1493.
H-C 3669*^ Pell. 2782^ Pr. 6597^ BMC (XV)VI:826^ GW 4958^ Goff B-1045
Subject: Neo-Platonism.
Wing/Inc./6597

Epistola 86: "Ad Robertum Saluiatum amicorum . . . de Heptaplo Ioannis Mirandula."

### Bosso, Matteo.

*Familiares et secvndae Matthaei Bosso epistolae.*
Mantua: Vincentius Bertochus, 1498.
H-C (+ Add) 3671*^ Pell. 2780^ Pr. 6911^ BMC (XV)VII:934^ GW 4956 (+ var)^ Goff B-1042
Subject: Neo-Platonism.
Inc./f6911

Epistola 34: "Ad inclytum Ioannem Picum Mirandulam ut Hermolaum Barbarum"; epistola 209: "Hieronymum Bentacordum virorum medicina . . . ad Io. Philippum fratrem grauiter aegrotantem."

### Bosso, Matteo.

*De veris ac salvtaribvs animi gavdiis.*
Florence: Franciscus Bonaccursius, 1491.
H-C (+ Add) 3672*^ Pell. 2785^ Pr. 6312^ BMC (XV)VI:674^ GW 4955^ Goff B-1041
Subject: Neo-Platonism, soul.
Inc./6312

### Botelho de Oliveyra, Bernardino.

*Escudo apologetico, physico, optico opposto a varias objecçoens onde se mostra, como, & de que parte se faz, ou se determinaa a sensaçaõ do objecto visivo.*
Lisbon: Mathias Pereyra da Sylva & Joam Antunes Pedrozo, 1720.
Subject: Eyes, imagination, optics, senses, vision.
Greenlee/4504/P855

Describes the way in which the physical and psychological aspects of the senses work together.

### Botero, Giovanni, 1540–1617.

*Geographische landtaffel des gebiets dess grossen Türchen dessen tiranney grosse theyle von Europa, Asia, vnd Africa vnterworssen seindt.*
Cologne: Lambert Andrea, 1596.
Subject: Descriptive geography, ethnography, topography.
Case/5A/547/no./4

Nebenzahl catalogue no. 4 (1961) describes this as the 4th part of the German ed. of Botero's *Theatrvm principvm orbis*, q.v.

### Botero, Giovanni.

*Le relationi vniversali di Giovanni Botero Benese, divise in tre parti.*
Vicenza: Giorgio Greco, 1595.
Subject: Descriptive geography, ethnography, topography.
Case/G/117/.1098

Pt. 1: Le città, i costumi de' popoli, & le conditioni de paesi di tutta la terra: i monti, i laghi, i fiumi, le minere, & opre marauigliose in essa dalla natura prodotte: con l'isole, & penisole dell' oceano e del Mediterraneo; pt. 2: De' maggiore prencipi che siano al mondo; pt. 3: Poi si dà piena contezza de' popoli d'ogni credenza, Catolici, Giudei. Gentili, & Scismatici.

### Botero, Giovanni.

*Le relationi vniversali di Giovanni Botero Benese, divise in qvattro parti.*
Venice: Giorgio Angelieri, 1596.
Subject: Descriptive geography, ethnography, natural history, New World.
Case/G/117/.11

Pt. 1: Europe, Asia, Africa, customs, resources, commerce, New World; pt. 2: major kingdoms; pt. 3: religious beliefs of Catholics, Jews, Gentiles, Schismatics; pt. 4: superstitions of New World natives, difficulty of converting them.

**Botero, Giovanni.**

*Relationi vniversali di Giovanni Botero Benese. Diuise in quatro parti. Novamente reviste, corrette, & ampliate dall' istesso auttore. Et aggiontoui in questa vltima impressione la figurata descrittione intagliata in rame, di tutti i paesi del mondo.*
Brescia: la Compagnia Bresciana, [1598].
Subject: Descriptive geography, ethnography, natural history, New World.
Ayer/7/B7/1598

Pt. 1: Europe. pt. 2: Asia. pt. 3: Africa ("Trogloditi popoli, p. 287). pt. 4: New World, islands.

**Botero, Giovanni.**

*Le relationi vniversali di Giovanni Botero Benese, divise in qvattro parti.*
Venice: Giorgio Angelieri, 1599.
Subject: Descriptive geography, ethnography, natural history, New World.
Ayer/7/B7/1599

In 4 parts (see 1596 eds above).

**Botero, Giovanni.**

*Relaciones vniversales del mundo de Juan Botero Benes, primera y segunda parte, traduzidas a instancia de don Antonio Lopez.*
Valladolid: heirs of Diego Fernandez de Cordoua, 1603.
Subject: Descriptive geography, medicine, natural history, New World.
Ayer/7/B7/1603

P. 32v, col. 2: Aguas medicinales, y baños de varias virtudes. Has section "Nuevo mundo."

**Botero, Giovanni.**

*Le relationi vniversali di Giovanni Botero Benese, divise in sei parti.*
Venice: Alessandro Vecchi, 1612.
Subject: Descriptive geography, ethnography, natural history, New World.
Ayer/7/B7/1612

1st 4 sections titled as in 1599 ed.; pt. V: I capitani illustri nella Europa. Pt. 6: Spain and the church. No plates.

**Botero, Giovanni.**

*Le relationi vniversali di Giovanni Botero Benese, divise in sette parti. Aggivnta alla qvarta parte dell' Indie, . . . Di mostri & vsanze di qvelle parti, e di quei rè con le sue figure al naturale. Raccolte novamente da Alessandro Vecchi.*
Venice: A[lessandro] Vecchi, 1618.
Subject: Descriptive geography, ethnography, natural history, New World.
Ayer/7/B7/1618

6 pts. generally same as 1612, with exception of added section on monsters and natives of New world and other lands. Detailed woodcut full length portraits, with descriptions on facing pp.

**[Botero, Giovanni]**

*Relations of the most famous kingdoms and common-weales thorovgh the world. Discoursing of their scituations, manners, customes, strengthes and pollicies.*
London: printed for Iohn Iaggard, 1608.
STC 3401
Subject: Descriptive geography, ethnography, natural history.
Case/G/117/.112

**[Botero, Giovanni]**

*Relations, of the most famovs kingdoms and commonweales thorough the world. Disoursing of their scituations, manners, customes, strengths, greatnesse, and policies.*
London: printed for Iohn Iaggard, 1616.
STC 3403
Subject: Descriptive geography, ethnography, topography.
Case/G/117/.113

MS marginalia.

**Botero, Giovanni.**

*Relations of the most famovs kingdomes and common-wealths thorowout the world discoursing of their situations, religions, languages, manners, customes, strengths, greatnesse, and policies. Translated out of the best Italian edition by R. I.*
London: printed by Iohn Haviland, 1630.
STC 3404
Subject: Descriptive geography, ethnography, topography.
Ayer/*7/B7/1630

**[Botero, Giovanni]**

*Theatrvm principvm orbis vniversi. In qvo omnes, qvotqvot svnt in orbe terrarvm principes, opibus & viribus conspicui representantur, cum vniuscuiusque regali censu, potentia, regendi forma & principibus ipsis finitimis.*
Cologne: Lambertus Andrea, 1596.
Subject: Descriptive geography, ethnography, topography.
Case/5A/547/no./1

In 4 parts. Pt. 1: Europe; pt. 2: Asia; pt. 3: Africa; pt. 4: Greater Turkish Empire. 1st Latin ed. BOUND WITH 4 additional works. See Metellus, Acosta, Botero, and *Partitio*.

**Bouelles, Charles de, ca. 1470–1553.**

*Aetatvm mundi septem supputatio, per Carolum Bouillum Samarobrinum.*
Paris: Jodoco Badio Ascensio, 1520.
Subject: Universal chronology.
Case/C/274/.112

**[Bouelles, Charles de]**

*In hoc opera contenta. . . Epistola in vitam Raemundi Lullii eremitae.*
[Paris] Ascensianis [1514].
Subject: Alchemy, elements, geometry, mathematics, mysticism, occult, soul.
Case/B/239/.112

Philosophical epistles: Lull, Fols. 41–48. Budé, Fols. 48–51. To Ioannis Labínío, S.J., Fol. 75.

### [Boulton, Richard], fl. 1697–1724.

*A compleat history of magick, sorcery, and witchcraft.*
2 vols. London: printed for E. Curll [and others] 1715–1716.
Subject: Magic, possession, sorcery, witchcraft.
Case/B/88/.106

Vol. I (1715) in 4 pts. Pt. 3: "Account of 1st rise of magicians & witches, shewing the contracts they make with the devil; pt. 4: Confutation of all arguments that have ever been produced against belief of witches, with a judgment concerning spirits by John Locke." Vol. II (1716) in 3 parts. Pt. 1: Salem witch trials; pt. 3: "The survey demoniack. with all the testimonies and informations taken upon oath relating thereunto."

### Boulton, Richard.

*The possibility and reality of magick, sorcery, and witchcraft, demonstrated or, a vindication of a compleat history of magick, sorcery, and witchcraft; in answer to Dr. Hutchinson's Historical essay.*
London: printed for J. Roberts, 1722.
Subject: Magic, possession, sorcery, witchcraft.
B/88/.431

In 2 parts. Pt. 1: "Containing an examination & answer of the positions laid down in [Hutchinson's] book"; pt. 2: "An essay of the nature of material and immaterial substances. How they may affect one another, and alter matter, or work upon human bodies. Proved by reason, philosophy, moral proof, and the testimony of Scripture." A refutation of Bp. Francis Hutchinson's *Historical essay concerning witchcraft*, q.v.

### Boulton, Richard.

*The possibility and reality of magick, sorcery, and witchcraft, demonstrated* [another copy].
Case/B/88/.107

### Bovet, Richard.

*Pandaemonium, or, the devil's cloyster. Being a further blow to modern sadduceism, proving the existence of witches and spirits in a discourse deduced from the fall of the angels, the propagation of Satan's kingdom before the Flood: the idotlatry of the ages after, greatly advancing diabolical confederacies. With an account of the lives and transactions of several notorious witches, some whereof have been popes. Also a collection of several authentick relations of strange apparitions of daemons and spectres and fascinations of witches, never before printed.*
London: printed for Tho. Malthus, 1684.
STC II B3864
Subject: Divination, sorcery, witchcraft.
Case/B/88/.11

### Bowen, Emanuel, fl. 1752.

*A complete system of geography. Being a description of all the countries, islands, cities, chief towns, harbours, lakes, and rivers, mountains, mines, &c. of the known world. Shewing the situation, extent and boundaries, of the several empires, . . . their climate, soil, and produce; their principal buildings, manufactures, and trade; . . To which is prefixed, an introduction to geography, as a science: an explanation of the maps; the doctrine of the sphere: the system of the world: and a philosophical treatise of the earth, sea, air, and meteors.*
London: printed for William Innys [etc.] 1747.
Subject: Geography.

2 vols. With 70 maps "for the use of gentlemen, merchants, mariners, and others, who delight in history and geography." Beginning of vol. I contains an introduction to the system of geography, a natural history of the elements in 2 books, and a translation of the Scientia naturalis of Mr. Le Clerc, pp. i–xxviii.

### Boyceau de la Baraudière, Jacques.

*Traité dv jardinage, selon les raisons de la natvre e de l'art. Divisé en trois livres.*
Paris: Michel Vanlochom, 1638.
Subject: Agriculture, gardening.
fR/57/.114

I: Elements, earth in general, water, sun, wind, sea, moon, fertilizer. II: Trees, grafts, transplanting, diseases of trees. III: Gardens, design.

### Boye, Henricus.

*Dissertatio historico-physica de miris qvorundam humani corporis defectuum supplementis, quam supremi numinis gratiâ & nobil. facul. philosophicae suffragiô munitus amicis antagonistis impugnada exhibet Mag. Sixtus Anspach unà cum ingeniosissimo & ornatissimo defendente Henrico Boye medic. stud.: in auditorio.*
[Copenhagen:] J. P. Bockenhoffer, 1690.
Subject: Medicine.
UNCATALOGUED

### Boyle, Robert, 1627–1691.

*The Christian virtuoso: showing that by being addicted to experimental philosophy, a man is rather assisted than indisposed, to be a good Christian. To which are subjoyn'd I. A discourse about the distinction, that represents some things as above reason, but not contrary to reason: II. The first chapters of a discourse, entituled, greatness of mind promoted by Christianity.*
London: printed by Edw. Jones for John Taylor, 1690.
STC II B3931A (?)
Subject: Corpuscular philosophy, experimental science.
C/O/.109

Half title: The Christian virutoso: in two parts [part 1 only]. BOUND WITH his *A disquisition about the final causes of natural things*, q.v.

### [Boyle, Robert.]

*A discourse of things above reason. Inquiring whether a philosopher should admit there are any such. By a fellow of the*

*Royal Society. To which are annexed by the publisher . . . some advices about judging of things said to transcend reason.*
London: printed by E. T. & R. H. for J. Robinson, 1681.
STC II B3944
Subject: Geometry, natural philosophy, physics, reason vs. revelation, science.
Case/B/49/.1124

Experience, authentic testimony, and mathematical demonstration; 3 ways to distinguish what is within realm of reason. 4 speakers in discourse. *Advices* is separately paginated but does not have a separate title page.

## Boyle, Robert.

*A disquisition about the final causes of natvral things: wherein it is inquir'd, whether, and (if at all) with what cautions, a naturalist should admit them? To which are subjoyn'd, by way of appendix, some vncommon observations about vitiated sight.*
London: printed by H.C. for J. Taylor, 1688.
STC II B3946 or II B3946A (2d ed.)
Subject: Biology, Natural science, ophthalmology, pathology.
C/O/.109

14 "observations" in "Vitiated sight"; a series of case studies. BOUND WITH The Christian virtuoso, q.v.

## Boyle, Robert.

*A free enquiry made into the vulgarly receivd notion of nature.*
London: printed by H. Clark for John Taylor, 1686.
STC II B3979
Subject: Biology, disease, comets, macrocosm–microcosm, corpuscular philosophy, pneumatics, reason.
L/0/.11

Nature functions above and beyond man's will or consciousness, Is nature an immaterial or corporeal entity?

## Boyle, Robert.

*General heads for the natural history of a country, great or small drawn out for the use of travellers & navigators. To which is added, other directions for navigators, &c. with particular observations of the most noted countries in the world: by another hand.*
London: printed for John Taylor, 1692.
STC II B3980
Subject: Botany, climate, descriptive geography, disease, mining, metals, minerals, navigation and navigational instruments.
Case/L/0114/.11

## Boyle, Robert.

*Nova experimenta physico-mechanica de vi aëris elastica & ejusdem effectibus, facta maximam partem in nova machina pneumatica, . . . Editio postrema.*
Rotterdam: Arnold Leers, 1669.
Subject: Experimental science, hydraulics, magnetism, mechanics, natural science, physics, pneumatics.
Case/-B/245/.0967

43 experiments. BOUND WITH his *Tractatus de cosmicis*, q.v. Binder's title: *Robert Boile: Experimenta mechanica.*

## Boyle, Robert.

*Occasional reflections upon several svbiects. Whereunto is premis'd a discourse about such kind of thoughts.*
London: printed by W. Wilson for Henry Herringman, 1665.
STC II B4005
Subject: Epistemology, medicine, natural history, optics, physics.
Case/Y/145/.B71

1st ed. In 6 sections. Meditations include: sec. 2: Occasional reflections upon the accidents of an ague (p. 187); Upon being in danger of death (p. 221); Upon the apprehensions of a relapse (p. 235); sec. 3: Looking through a prismaticall or triangular glass; sec. 4: Angling improv'd to spiritual uses. Sec. 6: Upon the sight of a branch of corral among a great prince's collection of curiosities. [Analogous to man's life on earth: coral, drab and colorless while alive, gains its glory after death.]

## Boyle, Robert.

*The origine of formes and qualities, (according to the corpuscular philosophy) illustrated by considerations and experiments (written formerly by way of notes upon an essay about nitres).*
STC II B4015
Oxford: printed by H. Hall for Ric. Davis, 1667.
Subject: Chemistry, corpuscular philosophy, experimental science, generation and corruption, local motion, medicine, physics.
Case/B/42/.116

2d ed.

## Boyle, Robert.

*The philosophical works of the honourable Robert Boyle, Esq., abridged, methodized and disposed under the general heads of physics, statics, pneumatics, natural history, chymistry, and medicine. The whole illustrated with notes, containing the improvements made in the several parts of natural and experimental knowledge since his time. By Peter Shaw, M.D.*
London:printed for W. Innys & R. Manby & T. Longman, 1738.
Subject: Chemistry, experimental science, medicine, natural history,physics, pneumatics, statics.
B/245/.098

2d ed. corrected. 3 vols. Includes a chronological catalogue of Boyle's writings, giving date of 1st publication and "best" edition.

## Boyle, Robert.

*Some considerations touching the vsefvlness of experimental natural philosophy. Propos'd in a familiar discourse to a friend, by way of invitation to the study of it.*
Oxford: printed by Hen. Hall for Ri. Davis, 1664–71.
STC (vol. 1) II B4030^ (vol. 2) II B4031

Subject: Anatomy, biology, experimental science, iatrochemistry, mathematics, mechanical philosophy, medicine, pathology.
Case/B/42/.117

2d ed. Vol. 1, 1664; vol. 2, 1671.

### Boyle, Robert.

*Tractatus de cosmicis rerum qualitatibus.*
Amsterdam: Johannes Janssonius à VVaesberg & Hamburg, [*sic*] & GottfriedSchultz, 1671.
Subject: Alchemy, chemistry, corpuscular philosophy, natural science.
Case/-B/245/.0967

"De cosmicis rerum qualitatibus. De cosmicis suspicionibus. De temperie subterranearum regionum. De temperie submarinum regionum. De fundo maris. Quibus praemittitur introductio ad historiam qualitatum particularium."

### Boyle, Robert.

*Tracts written by the Hon. Robert Boyle.*
Oxford: printed by W. H. for Ric. Davis, 1671.
STC II B4057
Subject: Alchemy, chemistry, corpuscular philosophy, natural science.
Case/B/245/.097

"The cosmicall qualities of things. The cosmicall superstitions. The temperature of the subterraneall regions. The temperature of the submarine regions. The bottom of the sea. To which is praefixt, an introduction to the history of particular qualities."

### Brack, Wenceslaus.

*Vocabularius rerum.*
[Strassburg: Johann Reinhard Gruninger, ca. 1486.]
H-C *3697^ BMC (XV)I:104–5^ GW IV, 4986^ Pr. 448^ Pell. 2804^ Goff B-1060
Subject: Dictionary encyclopedias, philosophy, quadrivium.
Inc./f448

In 6 books, preceded by vocabulary list with Latin terminology translated into German. Last chap. of bk. VI is on magic art.

### Brack, Wenceslaus.

*Vocabularius rerum.*
Strassburg: Johann Prüss, 1489.
H-C *3705^ BMC (XV)I:122^ GW IV, 4990^ Pr. 541^ Still. B948^ Goff B-1064
Subject: Dictionary encyclopedias, philosophy, quadrivium.
Inc./541

### Bradley, Richard, 1688–1732.

*A general treatise of husbandry & gardening, containing such observations as are new & useful for the improvement of land; with an account of such extraordinary inventions & natural productions as may help the ingenious in their study & promote universal learning.*
London: Woodward, 1724.
Subject: Agriculture, animal husbandry, gardening, water power.
R/5/.12

3 vols. Beekeeping, construction of cold-frames and hot-beds, grafting, water-clocks. Vol. 3 contains an alphabetical list of locations in world with degrees of latitude. Folding plate (p. 173) shows course of sun relative to earth for each month of the year.

### Bradley, Richard.

*The gentleman and farmer's guide for the increase and improvement of cattle, viz: lambs, sheep, hogs, calves, cows, oxen. Also the best manner of breeding & breaking horses both for sport and burden: with an account of their respective distempers, and the most approved medicines for the cure of them.*
London: printed for G. S.: 1732.
Subject: Animal husbandry, veterinary medicine.
H/3145/.117

2d ed. Bradley was a professor of botany, Cambridge.

### Brand, Adam.

*A journal of an embassy from their majesties John and Peter Alexowits, . . &c, into China. . . . Written by Adam Brand, . . . with some curious observations concerning the products of Russia. by H[enry] W[illiam] Ludolf.*
London: printed for D. Brown, 1698.
STC II B4246
Subject: Fauna, flora, minerals, travel.
G/60/.115

The Journal is primarily anthropological. Ludolf's work has separate title page, but continuous pagination. Descriptions of the flora and fauna. Full title: "Some observations concerning the products of Russia, which may serve as a supplement to the preceding treatise. Written originally in Latin by Henry William Ludolf."

### Braun, Ernst.

*Novissimum fundamentum & praxis artilleriae, oder nachitziger besten mannier neu vermehrter und gantz gründlicher unterricht was diese höchst-nützliche kunst vor fundamenta habe und erfordere denn auch was vor neue urthen canonen feuer-mörser und haubitzen heutiges tages im rechten gebrauch sind und zu felde geführet auch wie selbige gegossen detzgleichen wie die neu erfundene gransten brand-kugeln und ernst feuer-wercke nebst denen darzu gebörigen materiálien müssen laboriret, versextiget und in die weite geworssen werder; ferner eine wolkommene beschreibung der lust-sachen und haupt-feuer wercke auff dem lande und wasser wie solche zu arbeiten und vor könige und fursten darzustellen sind. Alles nicht allein sonst vor dem gebräuchlichen sondern auch mit allerhand neuen inventionen vermehret und mit aussführlichen und zu diesem werck dienlichen und gehörig en kupffer-stücken versehen.*
Danzig: Johann Friedrich Gräfen, 1682.

Subject: Geometry, metallurgy, military science, pyrotechnics, trajectory.
fU/444/.117

## Brereton, John, fl. 1603.

*A briefe and true relation of the discouerie of the north part of Virginia; being a most pleasant, fruitfull and commodious soile: made this present yeere 1602, by Captaine Bartholomew Gosnold, and diuers other gentlemen their associates, by the permission of the honourable knight, Sir Walter Ralegh, &c. Written by Mr. Iohn Brereton one of the voyage. Whereunto is annexed a treatise conteining the important inducements for the planting in those parts. and finding a passage that way to the South Sea, and China. Written by M. Edward Hayes, a gentleman long since imploied in the like action.*
STC 3610
London: Geor. Bishop, 1602.
Subject: Discovery, natural history.
Ayer/*116/G6/B7/1602a

1st ed. 1st impression. P. 12 "A briefe note of such commodities as we saw in the countrey notwithstanding our small time of stay." [Arranged alphabetically under headings: trees, fowles, beastes, fruits, plants and herbs, fishes, mettals and stones.]

## Brereton, John.

*A briefe and true relation of the discouerie of the north part of Virginia, . . . with diuers instructions of speciall moment newly added in this second impression.*
London: Geor. Bishop, 1602.
STC 3611
Subject: Discovery, natural history.
Ayer/*116/G6/B7/1602

Added material includes: "Inducements to the liking of the voyage intended towards Virginia in 1585 by R. Hakluyt." [Ends of the voyage: 1. to implant the Christian religion, 2. to traffic, 3. to conquer.] Description of flora and fauna in Florida and southern Virginia taken from works by De Soto, Laudonniere, and Thomas Hariot.

## Brickell, John.

*The natural history of North Carolina. With an account of the trade, manners, and customs of the Christian and Indian inhabitants.*
Dublin: printed by James Carson for the author, 1737.
Subject: Natural history, travel.
Case/G/865/.12

Has a section on beasts, birds, and fishes of North Carolina.

## Bricot, Thomas, ed., 15th cent.

*Cursus optimarum quaestionum super totam logicam. Cum interpretatione textus: secundum viam modernorum: ac secundum cursum magistri Georgii per Magistrum Thomam Bricot sacre theologiae professorum emendate.*
[Freiburg im Breisgau: Kilian Fischer, about 1495]
H 3969*^ Pell. 2989^ Goff G-148
Subject: Natural science, philosophy, scientific logic.
Inc./f3215.4

Predicabilium, predicamentorum, peryarminias, priorum, posteriorum, L. thopicorum, elenchorum. BOUND WITH another ed., q.v.

## Bricot, Thomas, ed.

*Cursus optimarum quaestionum super philosophiam Aristotelis. Cum interpretatione textus secundum viam modernorum: ac secundum cursum magistri Georgii: per magistrum Thomam Bricot sacre theologiae professorum emendate.*
[Freiburg im Breisgau, Kilian Fischer, not after 1496]
H *3975^ BMC (XV)III:696, 784^ Pr. 7609^ Pell. 2990^ Goff G-147
Subject: Aristotle, natural history, philosophy.
Inc./f3215.5

Liber phisicorum 8 books, followed by De celo & mundo, 4 bks. De generatione & corruptione, 2 bks. De metheorum, 3 bks. De anima, 3 bks. De sensu & sensato. Liber de sompno et vigilia. De memoria & reminiscentia. De longitudine & breuitate vite. Explicit states: "Et sic est finis pauorum naturalium et totius philosophie naturalis." Also includes Liber metaphisice, 6 bks. This section and the one on logic follow the form "queritur, oppositum, dubitatem." Bound in oak boards, covered in blind-stamped leather. Pastedowns are 13th cent. MS of same text. BOUND WITH Bricot, preceding work.

## Briggs, William, 1642–1704.

*Guilielmi Briggs ophthalmo-graphia: sive oculi ejusque partium descriptio anatomica, nec non, ejusdem nova visionis theoria; Regiae Societati Londinensi proposita.*
Leyden: Peter vander Aa, 1686.
Subject: Anatomy of the eye, comparative anatomy, medicine, ophthalmology, optics.
Case/oQM/511/.B75/1686

## Bright, Timothy, 1551?–1615.

*A treatise of melancholie. Containing the cavses thereof, & reasons of the strange effects it worketh in our minds and bodies: with the phisicke cure, and spirituall consolation for such as haue thereto adioyned an afflicted conscience. The difference betwixt it, and melancholie with diuerse philosophicall discourses touching actions, and affections of soule, spirit, and body: the particulars whereof are to be seene before the booke.*
London: Thomas Vautrollier, 1586.
STC 3747
Subject: Body and soul, emotions, melancholic disorders, diagnosis and cures, passions of the mind.
Case/B/529/.12

1st. ed. With errata leaf. This ed. was one of Shakespeare's sources. [Bright also invented modern shorthand.]

## Brinley, John.

*A discovery of the impostures of witches and astrologers.*
London: printed for John Wright, 1680.
STC II B4698

Subject: Astrology, witchcraft.
Case/B/88/.122

In 2 pts. Pt. 1 is about witches. Pt. 2 is a discourse of the impostures practiced in judicial astrology. Only God can foretell future. Witches exist, but devils cannot cause diseases.

* * *

*The British Apollo.*
London: printed for the authors by J. Mayo [etc.], 1708–1711.
Subject: Popular medicine and science.
Case/A/51/.1611

Index for each year is bound in immediately following the collected issues for that year.

* * *

*The British Apollo: in three volumes. Containing two thousand answers to curious questions in most arts and sciences, serious, comical, and humorous; approved of by many of the most learned and ingenious of both universities, and of the Royal Society. Perform'd by a society of gentlemen.*
London: printed by James Bettenham for Charles Hitch, 1740.
Subject: Health, hygiene, medicine, natural history, navigation, questions on miscellaneous topics.
Y/O/.112

4th ed. 3 vols. Index at end of vol. 3. Pharmacological questions (is "sage of virtue" more wholesome than Indian tea?) 1:19. Astronomy 1:36. Is there such a creature as a salamander? 2:451. [There are a number of periodical miscellanies of this type, but I include only this one. They tend to contain a lot of nonscientific and popular material. Though they may be worth listing separately (as they reflect the interest of the public in scientific questions) their bulk, derivativeness, lack of organization, and superficiality diminish their scientific validity for the purposes of this checklist.]

## [Brocklesby, Richard] 1722–1792.

*Reflections on antient and modern musick, with the application to the cure of diseases.*
London: printed for M. Cooper, 1749.
Subject: Music and medicine.
V/23/.12

Chap. 1: Origin of music and how it affects the mind; chap. 2: operation of music on bodily organs; chap. 3: power of music in disorders of the mind; chap. 4: difference between ancient and modern music; chap. 5: old age delayed by application of music. Tarantula bite cured by music (p. 59). Quotes classical and Elizabethan authorities on music's curative properties, e.g., melancholy. Brocklesby was Samuel Johnson's physician, and a friend of Edmund Burke.

## Brome, James, d. 1719.

*Travels over England, Scotland, & Wales . . . caves & wells.*
London: printed by Abel Rogers [etc.], 1700.
STC II B4861
Subject: Geography, travel.
Case/G/4459/.12

P. 274: description of stones, fossils, etc. in Scotland. Refers reader to John Ray.

## Browne, Edward, 1644–1708.

*An account of several travels through a great part of Germany . . . Wherein the mines, baths & other curiosities of those parts are treated of.*
London: B. Tooke, 1677.
STC II B5109
Subject: Balneology, descriptive geography, metallurgy, mining, natural history, travel.
Case/G/47/.116

Plates illustrating miners of Hungary, etc.

## Browne, Edward.

*A brief account of some travels in divers parts of Europe, viz. Hungaria, Servia, Bulgaria, Macedonia, Thessaly, Avstria, Styria, Carinthia, Carniola, and Frivli. Through a great part of Germany and the low countries. Through Marca Trevisana, and Lombardy on both sides the Po. With some observations on the gold, silver, copper, quick-silver mines, and the baths and mineral waters in those parts. As also the description of many antiquities, habits, fortifications and remarkable places.*
London: printed for Benj. Tooke, 1685.
STC II B5111
Subject: Balenology, descriptive geography, metallurgy, mining, natural history, travel.
Case/G/307/.126

2d ed. with many additions.

## Browne, Edward.

*A brief account of some travels in Hungaria, Servia, Bulgaria, Macedonia, Thessaly, Austria, Styria, Carinthia, Carniola, & Friuli. As also some observations on gold, silver, copper, quick-silver mines, baths, mineral waters in those parts: with figures of some habits and remarkable places.*
London: printed by T. R. for B. Tooke, 1673.
STC II B5110
Subject: Balneology, descriptive geography, metallurgy, mining, natural history, travel.
G/307/.125

1st ed.

## Browne, John, 1642–ca. 1702.

*Myographia nova sive musculorum omnium (in corpore humano hactenus repertorum) accuratissima descriptio, in sex prae lectiones distributa.*
Amsterdam: Joannes Wolters, 1694.
Subject: Anatomy.
Case folio/QM/151/.B7/1694

## Browne, Sir Thomas, 1605–1682.

*Hydriotaphia, urne-buriall, or, a discourse of the sepulchrall urnes lately found in Norfolk. Together with the Garden of Cyrvs, or the quincunciall, lozenge, or net-work plantations of the ancients, artificially, naturally, mystically considered.*

London: printed for Hen. Brome, 1658.
STC II B5154
Subject: Anthropology, burial customs, life and death.
Ruggles/Coll./no. 42

1st ed. *Garden of Cyrus* (continuous pagination) subject: comparison of plants and animals, numerology.

**Browne, Sir Thomas.**

*Pseudodoxia epidemica: or, enquiries into very many received tenents, and commonly presumed truths.*
London: printed by T. H. for Edward Dod, 1646.
STC II B5159
Subject: Natural history, rational skepticism vs. superstition in science, religion.
Case/Y/145/.B838

1st edition of the "Vulgar errors." Causes and kinds of popular errors, including received opinions (e.g., elephant has no joints), some deduced from Holy Scripture.

**Browne, Sir Thomas.**

*Pseudodoxia epidemica: or, enquiries into very many received tenents, and commonly presumed truths. Together with some marginall observations.*
London: printed by A. Miller for Edw. Dod and Nath. Elkins, 1650.
STC II B5160
Subject: Natural history, rational skepticism vs. superstition in science, religion.
Case/fY/145/.B 83805

2d ed. corrected and much enlarged by the author.

**Browne, Sir Thomas.**

*Pseudodoxia epidemica: or, enquiries into very many received tenents, and commonly presumed truths.... With marginal observations and a table alphabetical. Whereunto are now added two discourses the one of urn-burial, or sepulchrall urns, lately found in Norfolk. The other of the garden of Cyrus, or network plantations of the antients.*
London: printed for E. Dod, 1658.
STC II B5162
Subject: Natural history, rational skepticism vs. superstition in science, religion.
Case/Y/145/.B 8381

4th ed. *Pseudodoxia* lacking title page. BOUND WITH *Hydriotaphia*, 1658. *Garden of Cyrus* has continuous pagination, separate title page (imprint also 1658) but no publisher.

**Browne, Sir Thomas.**

*Pseudodoxia epidemica: or, enquiries into very many received tenents and commonly presumed truths, together with the Religio medici.*
London: printed by J. R. for N. Elkins, 1672.
STC II B5165
Subject: Natural history, rational skepticism vs. superstition in science, religion.
Case/4A/1842

Pt. II (*Religio medici*) wanting in this copy. "The sixth and last edition, corrected and enlarged by the author, with many explanations, additions and alterations throughout."

**[Browne, Sir Thomas]**

[*Religio Medici*]
[London] 1643.
STC II B5169
Subject: Material and immaterial world, miracles, occult, rational skepticism.
Case/*C/53/.1197

"A true and full coppy of that which was most imperfectly and surreptitiously printed before under the name of Religio medici." 1st authorized ed. "All places, all ayres make unto me one country; I am in England, everywhere, and under any meridian; I have been shipwrackt, yet am not enemy with the sea or winds" (p. 134). "'Tis not only the mischief of diseases, and the villanie of poysons that make an end of us, we vainly accuse the fury of gunnes, and the new inventions of death; 'tis in the power of every hand to destroy us, and we are beholding unto every one wee meete hee doth not kill us" (p. 99).

**Browne, Sir Thomas.**

*Religio medici.*
London: [Andrew Crooke] 1643.
STC II B5169
Subject: Material and immaterial world, miracles, occult, rational skepticism.
Ruggles/Coll./no. 41

**[Browne, Sir Thomas]**

*Religio medici.*
Paris, 1644.
Subject: Material and immaterial world, miracles, occult, rational skepticism.
Case/C/53/.1198

Trans. into Latin by J. Merryweather.

**[Browne, Sir Thomas]**

*Religio medici. With annotations never before published, upon all the obscure passages therein. Also, observations by Sir Kenelm Digby now newly added.*
London: printed by Tho. Milbourn for Andrew Crook, 1659.
STC II B5174
Subject: Material and immaterial world, miracles, occult, rational skepticism.
Case/C/53/.1199

5th edition, corrected and amended. BOUND WITH Digby's "Observations," q.v.

**Brucioli, Antonio, d. 1556.**

*Dialogi. Della naturale philosophia.*
[Venice: Alessandro Brucioli & brothers, 1544–55.]
Subject: Natural philosophy, moral, human, metaphysical.
Case/B/235/.1043

5 vols. in 1. Each volume has separate title page. 1: Moral philosophy (1549); 2: natural philosophy, human (1544); 3: natural philosophy, the world (1545); 4: metaphysics (1545); 5: wisdom and folly (1538).

## Bruno, Giordano, 1548?–1600.

*Le ciel reformé. Essai de traduction de partie du livre italien, Spaccio della bestia trionfante.*
[n.p., n.p.] 1750.
Subject: Constellations, Hermeticism, natural religion, Pythagorean philosophy.
Y/712/.B842

Translated by Louis Valentin de Vougny. Has translation and transcription of original title page, 1584.

## Bruno, Giordano.

*De imaginvm, signorvm, & idearum compositione. Ad omnia inuentionum, dispositionum, & memoriae genera.*
Frankfurt: Ioan. VVechel and Petrus Fischer, 1591.
Salvestrini, 2d ed., 207 (variant)
Subject: Epistemology, mnemonics, mysticism, synthetic philosophy.
Wing/ZP/547/.W386

On the composition of images, signs, and ideas for all sorts of discoveries, dispositions, and recollections. Written for John Henry Heinz. In 3 books. Bk. 1: Generalities dealing with diverse kinds of meaning; bk. 2: the 12 figures of princes; bk. 3: Figures of the 30 seals. Partial contents: De luce radio & speculo. De imaginum momento physico, mathematico, logico iuxta uniuersas modorum istorum acceptiones. De distinctione imaginum & signorum.

## Bry, Johann Theodor de, 1561–1623.

[*America, India Occidentalis*]
[Frankfurt, 1590–1634.]
Church 140–175
Subject: Discovery, exploration, geography, navigation, New World, voyages.
Ayer/*110/B9/1590

13 vols. (For full description of this and following editions consult Church, Elihu Dwight, *Catalogue of books relating to the discovery and early history of North and South America.* 5 vols. New York: Dodd, 1907.)

## Bry, Theodor de, 1528–1598.

*America India Orientalis.*
Frankfurt(?) de Bry & sons(?) 1590.
Church 146–168
Subject: Discovery, exploration, geography, navigation, New World, voyages.
Ayer/*110/B9/1590a

9 vols. in 4.

## [Bry, Theodor de]

*India occidentalis.*
[Frankfurt, 1590–1602]
Church 140–168
Subject: Discovery, exploration, geography, navigation, New World, voyages.
Ayer/*110/B9/1590b

9 vols. in 5.

## Bry, Theodor de.

*India Occidentalis.*
[Frankfurt, 1591–94]
Sabin III: 30–32
Subject: Discovery, exploration, geography, navigation, New World, voyages.
Ayer/*110/B9/1590c

2 pts. in 1 vol. Binders title: *De Morgues narratio in Florida Americae. Bezono Americae historiae.*

## [Bry, Theodore de]

[*Indiae ocidentalis*]
[Frankfurt: Ioannis Wechel for Theodor de Bry, 1591]
Sabin III: 30–32
Subject: Discovery, exploraton, geography, navigation, New World, vogages.
Ayer/*110/B9/1590d

Vol. 2 only (1st ed., 1st impression).

## [Bry, Theodor de]

[*Indiae occidentalis*]
*Merveillevx et estrange rapport tovtes fois fidele, des commoditez qvi se trovvent en Virginia, des facon des natvrels habitans d'icelle, laquelle a esté novvellement descovverte par les Anglois qve Messire Richard Greinvile . . . a la charge principale de Messire Walter Raleigh chevalier svrintendant des mines d'estain. Par Thomas Hariot servitevr dv svsdit Messire Walter l'vn de cevx de la-dite colonie, et qvi y a esté employé a descovvrir.*
[Frankfurt, 1590]
Church 203
Subject: Discovery, exploration, geography, navigation, New World.
Ayer/*110/B9/1590e

Vol. 1 only. Text trans. from English. Plates have Latin captions. Hand-colored title page.

## [Bry, Johann Theodor de]

[*Indiae occidentalis. Germanicè*]
[Frankfurt, Oppenheim, Hanau, 1597–1630]
Church 177–202
Subject: Discovery, exploration, geography, navigation, New World, voyages.
Ayer/*110/B9/1597c

14 pts. in 3 vols.

## [Bry, Johann Theodor de, ed. & comp.]

[*Indiae orientalis*]
[Frankfurt, 1598–1628]
Church 205–224
Subject: Discovery, exploration, geography, navigation, voyages.

Ayer/*110/B9/1598

12 vols.

## [Bry, Johann Theodor de, ed. & comp.]

[*Indiae orientalis*]
[Frankfurt, 1598–1613]
Church 205–222
Subject: Discovery, exploration, geography, navigation, voyages.
Ayer/*110/B9/1598a

10 pts. in 5 vols. Except for pt. 7, W. Richter, 1606 (Church 216), all pts. conform to 1598 ed. above.

## Bry, Johann Theodor de.

*Regnum Congo hoc est vera descriptio regni Africani, qvod tam ab incolis quam Lusitanis Congus appellatur.*
Frankfurt: Erasmus Kempffer for the heirs of Theodor de Bry, 1624.
Church 206
Subject: Discovery, exploration, geography, navigation, voyages.
Ayer/*110/B9/1598b

"Small voyages." BOUND WITH *Appendix regni Congo. Qva continentur navigationes quinque Samuelis Brunonis, cuius & chirurgi Basileensis, quas recenti admodum memoria animosè suscepit & feliciter perfecit.* Frankfurt: printed by Caspar Rötelius for heirs of Theodor de Bry, 1625. Church 225.

## Bry, Johann Theodor de.

*Newe welt und Americanische historien. Warhafftige und vollkommene beschreibungen aller West Indianischen landschafften, insulen, königreichen und provintzien, seecusten, fliessenden und stehenden wassern, port vnd anländungen, gebürgen, thälern, stätt, flecken und wohnplätzen, zusampt der natur vnd eygenschafft dess erdrichs der lufft der mineren vnd metallen der brennenden vulcanen oder schwefelbergen, der siedenden vnd anderer heilsamen quellen, wie auch der thier vögel, fisch, vnd gewürm in denselben, sampt andern wunderbaren creaturen vnd miraculn der natur, in diesem halben theil dess erdkreyses.*
Frankfurt: Mattheus Merian, 1655.
Subject: Subject: Discovery, exploration, geography, navigation, New World, voyages.
Ayer/*108/G68/1655

Abridged by Joh. Ludwig Gottfried. Folding plate of N. America. Bk. I: West Indies. Peru, Quito. Engraved plates throughout, including Florida, goldsmithing, Virginia.

## Bry, Johann Theodor de.

*Emblemata secvlaria varietate secvli hvivs mores ita experimenta, vt sodalitatum symbolis insigniisque conscribendis & depingendis peraccomodata sint. Versibvs Latinis, rhythmisque Germanicis, Gallicis, Belgicis specialitem declamatione literarum studiis exornata.*
Oppenheim: Hieronymus Galler for J. T. de Bry, 1611.
Subject: Emblems.
Case/W/1025/.1268

In Latin and German. See no. 53, alchemist.

## Bry, Johann Theodor de.

*Florilegivm renovatvm et avctvm: das ist: vernewertes und vermehrtes blumenbuch: von mancherley gewächsen, blumen unn pflantzen welche uns deren schönheit lieblicher geruch, gebrauch, und manigfaltiger unterschied angenehme machet, die nicht allein auss der von uns befandter sondern auch den alten unbefandter welt fruchtbaren schoss uns herfür gegeben werden.*
Frankfurt: Mattheus Merian, 1641.
Subject: Herbals.
Case/W/765/.13

First illustration in the vol. has a Latin inscription dated 1647. 14 pp. of German text, followed by engraved plates of flowers and plants (used in T. de Bry's *Indiae occidentalis*).

## Bry, Johann Theodor de.

*Proscenium vitae humanae siue emblematvm secvlarivm ivcvndissima, & artificiosissima varietate vitae humanae & seculi huius deprauati mores, ac studia peruersissima adumbrantium: et latinis versibvs explicatorum decades septem. . . . Sculptore Ioan Theodoro de Bry.*
Frankfurt: William Fitzer, 1627.
Subject: Alchemy, emblems, medicine.
Case/W/1025/.127

XLVI, medicus circumforaneus; LIII, alchemist; LXV, stultorum medicus. Partly derived from Breughel's *Proverbia*.

## Bueno, Diego.

[*Arte de leer con elegancia: escrivir, y contar &c.*
Zaragoça (Zaragoza): 1700.]
Subject: Arithmetic.
Wing/fZW/640/.B862

Title page wanting (author and publication information taken from privilege page).

## Buenting, Heinrich, 1545–1606.

*Chronologia hoc est omnivm temporvm et annorvm series ex sacris bibliis, aliisqve fide dignis scriptoribvs, ab initio mvndi ad nostra vsqve; tempora fideliter collecta, & calculo astronomico exactissimè demonstrata & confirmata, cui insertae sunt ipsae picturae ac imagines eclipsum solis & lunae, aliorumque motuum coelestium obseruationes, ex tabulis Prutenicis retro supputatae ad initium vsque mundi, adscripto ipso calculo, additis etiam compluribus in locis integris calendarijs, Hebraicis, Graecis, Aegyptijs & Latinis. Continentvr in hoc libro res gestae omnivm temporvm. . Inserta est etiam huic operi irreprehensibilis defensio anni Juliani, & certissima refutatio calendarij Gregoriani per theses & antitheses.*
Zerbst: Bonaventura Faber for Ambrosius Kirchner, 1590.
Subject: Astronomy, calendar, eclipses, universal chronology.
Case/folio/D/11/.B8/no. 1

Has list of authors cited. Chronological introduction, differentiating between calendars of different civilizations, e.g., Hebrew, Egyptian, Greek. Catalogue of eclipses. 1st year of world: 3967 B.C. BOUND WITH *Vandaliae & Saxoniae Alberti Cranzii continvatio.* Wittenberg: Johannis Cratonis, 1586.

## Buenting, Heinrich.

*Chronologia.* [Another copy]
Case/fF/017/.128

Binding contemporary blind-stamped vellum, gold stamped with initials and date 1591. Last page of index wanting and supplied in photocopy.

## [Bulkeley, John] fl. 1743.

*A voyage to the South Seas, and to many other parts of the world, performed from the month of September in the year 1740 to June 1744, by Commodore Anson, in his Majesty's ship the Centurion, having under his command the Gloucester, Pearl, Severn, Wager, Trial, and two store-ships. By an officer of the squadron.*
London: R. Walker, 1745.
Subject: Voyages.
Ayer/118/A 62/V 97/1744

Description of game and hunting wild bulls, pp. 772–75. Local fauna in Mexico; plate depicting sea lions.

## Bulwer, John, fl. 1648.

*Anthropometamorphosis: man transform'd: or, the artificiall changeling historically presented, in the mad and cruell gallantry, foolish bravery, ridiculous beauty, filthy fineness, and loathesome loveliness of most nations, fashioning and altering their bodies from the mould intended by nature; with figures of those transfigurations. To which artificiall and affected deformations are added, all the native and nationall monstrosities that have appeared to disfigure the humane fabrick. With a vindication of the regular beauty and honesty of nature.*
London: printed by William Hunt, 1653.
STC II B5461
Subject: Deformities, created and natural, scarification, tattooing, use and abuse of the body.
Case/F/03/.13

MS portrait and frontispiece inserted to replace missing (probably engraved) leaves.

## Bulwer, John.

*Chirologia: or the natvrall langvage of the hand. Composed of the speaking motions. and discoursing gestures therof. Whereunto is added Chironomia: or, the art of manvall rhetoricke. Consisting of the naturall expressions, digested by art in the hand, as the chiefest instrument of eloquence, by historicall manifesto's, exemplified out of the authentique registers of common life, and civill conversation. With types, or chirograms: a long wished-for illustration of this argument.*
London: printed by Tho. Harper [1644].
STC II B5462A
Subject: Chirology, sign language.
Case/Y/996/.13

Engraved title page has representations of uses of "hand language": Natura loquens, elocutio manualis, arithmetica naturalis. "With our hands we sue, intreat, beseech, sollicite, call, allure, intice, dismisse, threaten, bewaile, forbid, ask mercy, chafe, fume, rage, revenge, weepe, relieve, perswade, resolve, speake to." Chirograms show natural gestures of the fingers.

## Buoni, Tommaso.

*Academiche lettioni di tutte le specie de gli amori humani . . . in cui con stile graue si tratta dell' amor naturale, sociabile, humano, dell' amor de giouani, de' maritati, de' progenitori, de' figliuoli, di se medesimo, de gli amici, della sapienzia, della patria, dell'oro, dell' intemperato, & del diuino.*
Venice: Gio. Battista Colosini, 1605.
Subject: Emotions, psychology.
Case/B/529/.125

## Buoni, Tommaso.

*Discorsi academici de' mondi.*
Venice: Gio. Battista Colosini, 1605.
Subject: Macrocosm–microcosm.
B/42/.131

In 2 pts. Each part has separate pagination and separate title page. Pt. 1 macrocosm, pt. 2, microcosm. Title page, pt. 2: *Discorsi academici delle grandezze del microcosmo parte seconda de' mondi di Tomaso Bvoni.*

## Buoni, Tommaso.

*Problemes of beavtie and all humane affections.*
London: printed by G. Eld, for Edward Blount and William Aspley [1606].
STC 4103
Subject: Emotions, psychology.
Case/B/529/.127

Translated into English by S. L., gent. Among subjects treated in table of problems: Love, hatred, desire, sorrow, hope, despair, why is love so potent, compassion.

## Buonincontro, Lorenzo, ca. 1411–1502.

*Lavrentii Bonincontri Miniatensis, de rebus coelestibus, aureum opusculum, ab L. Gaurico Neapolitano.*
[Venice: Ioannes & brothers de Sabio, 1526.]
Subject: Astronomy.
Ayer/438/B94/1526

In 3 books. 1: rerum naturalium & diuinarum; 2: moon, Mercury, Venus; 3: sun, Mars, Jupiter, Saturn.

## Bureau d'adresse et de rencontre, Paris.

*Premiere [quatrieme] centvrie des questions traictées ez dv Bvreav d'adresse.*
Paris: Bureau d'adresse, 1635–41.
Subject: Health and hygiene, medicine, natural history, questions on miscellaneous topics, witchcraft.
Case/A/9/.641

5 vols. Vol. 5: *Recveil general des questions traitees és conferences du Bureau d'addresse.* Paris: I. le Gras, 1655. The 1st 4 vols. brought together by Eusèbe Renaudot and augmented, corrected and enriched by a 5th. Subjects include elements, monsters, universal spirit, sympathy and antipathy, cabala, artificial memory, magnetic cure of illnesses, the 3 suns, eclipses, sleep, rainbows, incubi and succubi, spontaneous generation, Rosicrucians, R. Lull, natural magic.

## Buridanus, Joannes, fl. 1317–1358.

*Acutissimi philosophi reuerendi magistri Johannis Buridani subtilissime questiones super octo phisicorum libros Aristotelis diligenter recognite et reuise a magistro Johanne Dullaert de Gandauo antea uusorum impresse.*
[Paris: Petrus Le Dru, for Denis Roce, 1509.]
Subject: Physics.
Case/*Y/642/.A/8203

Binder's title: Opera Buridani. BOUND WITH Buridanus, *Metaphysicales*, q.v.

## Buridanus, Joannes.

*[Metaphysicales quaestiones breues & vtiles super libros metaphysices Aristotelis quem ab excellentissimo magistro Ioanne Buridano diligentissima cura & correctione fuerunt in vltima preelectione ipsius recognitem.*
[Paris: Jodocus Badius Ascensius, 1518.]
Subject: Aristotle, metaphysics.
Case/*Y/642/.A/8203

BOUND WITH Buridanus, *Phisicorum libros Aristotelis*, q.v.

## Burley, Walter, 1275–1345?

*Expositio . . . in libros octo de physico auditu. Aristo. stragerite* [sic] *emendata diligentissime.*
Venice: Bonetus Locatellus for Octavianus Scotus, 1491.
H-C *4139^ GW V, 5777^ Pr. 5028^ Pell. 3078^ Goff B-1305
Subject: Physics.
Inc./f5028

Incipit: Clarissimi philosophi gualterij de burleis anglici fidelissimisque interpretis Aristotelis & sui commentatoris Auerrois expositio in libros octo de physico auditu feliciter incipit. MS marginalia in bks. 1, 2, and 5.

## Burley, Walter.

*Preclarissimi viri Gualtery Burlei Anglici sacre pagine professoris excellentissimi super artem veterum Porphyry & Aristotelis expositio.*
Venice: Octavianus Scotus for Bonetus Locatellus, 1488.
H-R 4131^ GW 5769^ Goff B-1310
Subject: Philosophy.
Inc./f5017.5/Inc./5110.5/Inc./4785.5/Inc./716B

MS marginalia passim. Includes "Liber sex principiorum Gilberti Porectani" and Burley's commentary on the Perihermenias of Aristotle. BOUND WITH tracts and commentaries on Aristotle's works by Thomas Aquinas, Gratia Dei de Esculano, Gaetano de Thienis, and Joannes de Janduno, qq.v.

## Burnet, Thomas, 1635?–1715.

*Archaeologiae philosophicae sive doctrina antiqua de rervm originibvs libri duo.*
London: R. N. for Gualt. Kettilby, 1692.
STC II B5943
Subject: Anthropology, archaeology, cosmogony, doctrines of origins and philosophy considered by nationality.
B/42/.134

Bk. 1: ethnic and national groups (e.g., Arabs, Scythians, Hebrews and their cabala, etc.) Bk. 2: creation of earth, Mosaic prognostications. Note on front flyleaf: "Burnet—suppressed." See *DNB*.

## Burnet, Thomas.

*Telluris theoria sacra: orbis nostri.*
London: R. N. for Gualt. Kettilby, 1681.
STC II B5948
Subject: Earth science reconciled with scripture.
Case/C/257/.128

In 2 books (separate title pages). Bk. 1: Diluvio & paradiso. Bk. 2: Tellure primigeniâ & de paradiso.

## Burnet, Thomas.

*Telluris theoria sacra: orbis nostri originem & mutationes generales, quas aut jam subiit aut olim subiturus est, complectens. Libri duo priores de diluvio & paradiso.*
London: R. N. for Gualt. Kettilby, 1689.
STC II B5949
Subject: Earth science reconciled with scripture.
Case/C/257/.129

2d ed. In 4 books.

## Burnet, Thomas.

*The theory of the earth: containing an account of the original of the earth, and of all the general changes which it hath already undergone, or is to undergo, till the consummation of all things. The two first books concerning the deluge, and concerning paradise.*
London: printed by R. Norton for Walter Kettilby, 1684.
STC II B5950
Subject: Earth science reconciled with scripture.
Case/C/257/.13

Preface: "This theory of the earth may be called *sacred* because it is not the common physiology of the earth . . . but respects only the great turns of fate, and the revolutions of our natural world; such as are taken notice of in the sacred writings."

## [Burnet, Thomas]

The sacred theory of the earth: containing an account of the original of the earth, and of all the several changes which it hath already undergone, or is to undergo, till the consummation of all things. In two volumes. The two first books concerning the deluge, and concerning paradise. The two last books concerning the burning of the world, and concerning the new heavens and new earth.

London: J. Hooke, 1726.
Subject: Earth science reconciled with scripture.
C/257/.131

6th ed. "We are not to suppose that any truth concerning the natural world can be an enemy to religion" (preface). "It is reasonable to believe that the bottom of the sea is much more rugged, broken and irregular than the face of the land" (p. 178).

## Burthogge, Richard, 1638?–ca.1700.

*An essay upon reason, and the nature of spirits.*
London: printed for John Dunton, 1694.
STC II B6150
Subject: Epistemology, metaphysics, supernatural.
Case/B/48/.13

Mentions Fludd, Boehme, among others. Discusses mind/matter, motion and energy, existence of spirits, witchcraft, magic, perception, reason, nature of knowledge.

## [Burton, Robert] 1577–1640.

*The anatomy of melancholy, what it is. With all the kindes, cavses, symptomes, prognostickes, and severall cvres of it. In three maine partitions with their seuerall sections, members, and svbsections. Philosophically, medicinally, historically, opened and cvt vp. By Democritvs, Iunior. With a satyricall preface, conducing to the following discourse.*
Oxford: printed by Iohn Lichfield and Iames Short for Henry Cripps, 1621.
STC 4159
Subject: Diseases, melancholy, psychology.
Case/B/529/.128

1st ed. Pt. I: causes of melancholy; pt. II: cure of melancholy; pt. III: love melancholy, religious melancholy.

## [Burton, Robert]

*The anatomy of melancholy.*
Oxford: printed by Iohn Lichfield & James Short for Henry Cripps, 1624.
STC 4160
Case/fB/.529/.129

2d ed., corrected and augmented by the author. Among recommended cures for "head melancholy" is bloodletting.

## Burton, Robert.

*The anatomy of melancholy.*
Oxford: printed for Henry Cripps, 1628.
STC 4161
Case/fB/529/.1294

3d ed. augmented and corrected by the author. 1st appearance of engraved title page.

## Burton, Robert.

*The anatomy of melancholy.*
Oxford: printed for Henry Cripps, 1632.
STC 4162
Case/fB/529/.1295

4th ed, corrected and augmented by the author.

## Burton, Robert.

*The anatomy of melancholy.*
Oxford: printed for Henry Cripps, 1638.
STC 4163
Case/fB/529/.1296

5th ed., corrected and augmented by the author.

## Burton, Robert.

*The anatomy of melancholy.*
London: printed for H. Cripps, 1660.
STC II B6183
Case/fB529/.13

7th ed. Engraved title page with "Argument of the frontispiece."

## Busche, Hermann von dem, 1468–1534.

*Aureum reminiscendi memoriandia que per breue opusculum mirum in modum naturali prestans memorie vberrimum suffragium litteris quoque alphabeticis ac figuris varie dispositionis ornatum quarum occasione quelibet res memoranda facilius ac citius ad memoriam reduci potest.*
Zuolis: Arnoldus Kempen, 1502.
Subject: Mnemonics.
Case/*W/9102/.13

Begins with "Tractatus de arte memoratiua."

## Butler, Charles, d. 1647.

*The feminine monarchie; or, the historie of bees. Shewing their admirable nature and properties, their generation, and colonies, their gouernment, loyaltie, art, industrie, enemies, warres, magnaminitie, &c. Together with the right ordering of them from time to time: and the sweet profit arising thereof.*
London: printed by Iohn Haviland for Roger Iackson, 1623.
STC 4193
Subject: Apiculture, natural history.
Case/O/97/.128

2d. ed. Includes recipes for wax, remedies employing wax, honey.

## Butler, Charles.

*The feminine monarchie; or, the history of bees. Shewing their admirable nature and properties; their generation, and colonies; their government, loyalty, art, industry, enemies, vvars, magnaminitie, &c. Written out of experience by Charles Butler, Magd.* Oxford: printed by W. Turner for de [*sic*] author, 1634.
STC 4194
Subject: Apiculture, natural history.
Case/O/97/.13

3d ed. Printed according to the author's orthographic system, proposed in his English grammar (1633). In 10 chapters, including nature and properties of bees, breeding of bees.

# C

## [Caesarius, Johannes] ed., 1468?–1550?

*Introductio Jacobi Fabri Stapulensis in arithmecam [sic] diui Seuerini Boetii pariter & Jordani.*
[Cologne, 1507?]
Subject: Arithmetic, geometry, squaring the circle.
Case/L/130/.142

Ms marginalia. Other tracts included in this vol.: "Ars supputandi tam per calculos qui per notas arithmeticas suis quidem regulis eleganter expressa Judoci Clichtouei Neoportunensis"; "Questio haud indigna de numerorum et per digitos & per articulos finita progressione ex Aurelio Augustino"; "Epitome rerum geometricarum ex geometrico introdutorio Caroli Bouilli"; "De quadratura circuli demonstratio ex Campano."

## Calandri, Filippo, 15th cent.

*Philippi Calandri . . . de arimethrica [sic] opusculum.*
Florence: Lorenzo de Morgiani [and Giovanni Thedesco da Maganza] 1491/92. H-C-R 4234^ BMC (XV)VI:681^ Pr. 6352^ Still. C30^ GW 5884^ Goff C-34
Subject: Arithmetic.
Inc./6352

## Calcagnini, Celio, 1479–1541.

*Catalogum operum post praefationem inuenies, & in calce elenchum. In dicanda enim erant retrusiora quaedam ex utriusque lingue thesauris, quae passim inferciuntur, & ad ueterum scripta intelligenda pernecessaria sunt.*
Basel Froben, 1544.
Subject: Astronomy, encyclopedias, magic.
Case/fY/682/.C108

Contains treatise "That Heaven is motionless and the Earth moves perpetually." Contents include "De rebus Aegyptiacis," and "De re nautica." "Paraphrasis trium librorum meteorum Aristotelis."

## Calcagnini, Celio.

*Dissertationvm lvdicrarvm et amoenitatvm. Scriptores varij.*
Leyden: Franciscus Heger & Hack, 1638.
Subject: Humorous accounts of animals, gout, insect pests.
Case/Y/6894/.23

Engraved title page. Contents include praise of gout by Bilibald Pirckheimer, encomium on the flea by Calcagnini, on the ant by Philipp Melanchthon, on the louse by Daniel Hensius.

* * *

*Il calendario gregoriano perpetvo . . . tradotto dal Latino nell' Italiano idioma dal Reuerendo M. Bartholomeo Dionigi da Fano.*
Venice: Gio. Bapt. Sessa & brothers, 1582.
Subject: Calendar reform, golden number.
Case/3A/952

Six canons; no. 6, movable feasts.

## Camdem, William, 1551–1623.

*Britannia sive florentissimorvm regnorvm, Angliae, Scotiae, Hiberniae, et insvlarvm adiacentium ex intima antiquitate chorographica descriptio.*
London: Ralph Newbery, 1586.
STC 4503
Subject: Descriptive and historical geography.
Case/G/4495/.133

## Camden, William.

*Britannia sive florentissimorvm regnorvm, Angliae, Scotiae, Hiberniae, et insvlarvm adiacentium ex intima antiquitate chorographica descriptio.*
London: Ralph Newbery, 1587.
STC 4504
Subject: Descriptive and historical geography.
Case/G/4495/.134

## Camden, William.

*Britannia sive florentissimorvm regnorvm, Angliae, Scotiae, Hiberniae, et insvlarvm adiacentium ex intima antiquitate chorographica descriptio.*
Frankfurt: printed by Ioannes Wechel for Peter Fischer & heirs of Henricus Tackius, 1590.
Subject: Descriptive and historical geography.
Case/G/4495/.135

In 2 parts.

## Camden, William.

*Britannia sive florentissimorvm regnorvm, Angliae, Scotiae, Hiberniae, et insularum adiacentium ex intima antiquitate chorographica descriptio.*
London: George Bishop, 1594.
STC 4506
Subject: Descriptive and historical geography.
Case/G/4495/.136

On title page: "Nunc quartò recognita, & magna accessione post Germanicum editionem adaucta."

## Camden, William.

*Britannia sive florentissimorvm regnorvm, Angliae, Scotiae, Hiberniae, et insularum adiacentium ex intima antiquitate chorographica descriptio.*
London: George Bishop, 1600.
STC 4507
Subject: Descriptive and historical geography.
Case/G/4495/.1365

"Nunc postremò recognita, & magna accessione post Germanicum aeditionem adaucta."

## Camden, William.

*Britannia sive florentissimorvm regnorvm, Angliae, Scotiae, Hiberniae, et insularum adiacentium ex intima antiquitate chorographica descriptio.*
London: George Bishop & Ioannis Norton, 1607.
STC 4508
Subject: Descriptive and historical geography.
Case/fG/4495/.137

## Camden, William.

*Britain, or a chorographicall description of the most flourishing kingdomes, England, Scotland, and Ireland, and the ilands adioyning, out of the depth of antiqvitie.*
London: George Bishop and Joannis Norton, 1610.
STC 4509
Subject: Descriptive and historical geography.
Case/fG/4495/.1379

"Translated newly into English by Philemon Holland.... Finally, revised, amended, and enlarged with sundry additions by the said author."

## Camden, William.

*Britain, or a chorographicall description of the most flourishing kingdomes, England, Scotland, and Ireland, and the islands adjoyning, out of the depth of antiqvitie.*
London: printed by F. Kingston R. Young and I. Legat for Andrew Hee, 1637.
STC 4510
Subject: Descriptive and historical geography.
Case/fG/4495/.138

Has an index, "The names of severall nations, cities, and great townes, rivers, promontories or capes, &c. of Britaine in old time . . . together with the later and more moderne names."

## Camden, William.

*Camden's Britannia, newly translated into English: with large additions and improvements.*
London: printed by F. Collins, for A. Swalle . . . and A. & J. Churchil, 1695.
STC II C359
Subject: Descriptive and historical geography.
Case/G/4495/.139

Published by Edmund Gibson of Queens College in Oxford. Preface to the reader: "The catalogues of plants at the end of each county were communicated by the great botanist of our age, Mr. Ray."

## Camerarius, Georgius, n.d.

*Emblemata amatoria.*
[Venice: Typographia Sarcinea, 1627]
Subject: Emblems.
Case/–W/1025/.137

Printed for P. Tozzi. Emblems include Zodiacus, p. 124, cupids at forge, p. 38. 400 symbola & emblemata. 1st long and unrivaled emblematic work aiming at a systematic study of natural phenomena. Each of Camerarius's 4 Centuria deals with birds, quadrupeds, air, water, respectively. Practical (kitchens and pharmacies) and didactic (moral, religious instruction). Uses of emblem literature, a descriptive and interpretive concern with nature. Fluid boundaries between natural history and emblematics: Aldrovandi quotes Camerarius who quotes Gesner who quotes Alciati. Description of nature merely a 1st step in the task of interpretation for purposes of understanding. See "The natural sciences and the arts: aspects of interaction from the Renaissance to the 20th century," *Acta Universitatis Upsaliensis, Figura Nova,* ser. 22 (Stockholm, 1985).

## Camerarius, Joachim, 1534–1598.

*Opuscula aliquot elegantissima, nempe: erratvm Aeolia phaenomena, prognostica, planetae disticha.*
Basel [Balthasar Lasius & Thomas Platter], 1536.
Subject: Astronomy.
Case/Y/682/.V826

BOUND WITH Gaza, Theodor, *Liber de mensibus,* q.v.

## Camerarius, Joachim.

*Comentarii vtriusque linguae, in quibus est*[in Greek]*hoc est diligens exqvisitio nominvm, qvibvs partes corporis humani appellari solent. Additis et fvnctionvm nomenclatvris, & aliis his accedentibus: positis ferè contra se Graecis ac Latinis uocabulis.*
Basel: Ioannes Heruagius, 1551.
Subject: Anatomy.
Case/fX/642/.144

## Camerarius, Joachim.

*Symbolorvm emblematvm ex animalibvs qvadrupedibvs desvmtorvm centvria altera collecta a Ioachimo Camerario.*
Nuremberg: Paulus Kavfmann, 1595.
Subject: Animal emblems, natural history.
Case/W/0149/.14

## Camerarius, Joachim.

*Symbolorvm & emblematvm ex re herbaria desvmtorvm centvria vna collecta.*
Frankfurt: Johannes Ammony [1654] 1661.
Subject: Emblems.
Case/W/1025/.139

In 4 parts: 1) flora; 2) quadrupeds; 3) flying creatures and insects; 4) water creatures and reptiles.

## Camerarius, Ludwig, 1573–1651.

*Symbolorum et emblematum centuriae tres . . . accessit noviter centuria.*
Nuremberg: Voegeliniana, 1605.
Subject: Emblems, natural history.
Case/*W/1025/.138

2d ed. Engraved title page refers to Joachim Camerarius. Letterpress title lists the following sections: 1) Ex herbis &

stirpibus;2) Ex animalibus quadrupedibus; 3) Ex volatilibus & insectis; 4) Ex aquatilibus & reptilibus.

## Camilli, Giovanni.

*Enthosiasmo di Gio. Camilla filosofo e medico Genovese. De'misterii, e maravigliose cause della compositione del mondo.*
Venice: Gabriel' Giolito de'Ferrari, 1564.
Subject: Elements, geometry, malign spirits, natural history, planets.
Case/L/O/.14

## Campanella, Tommaso, 1568–1639.

*Lvdovico Ivsto XIII . . . dedicat Fr. Thomas Campanella . . . tres hosce libellos, videlicet: Atheismus triumphatus, seu contra antichristianismum, &c. De gentilismo non retinendo. De praedestinatione & auxiliis diuinae gratiae cento Thomisticus.*
Paris: Tvssanvs Dubray, 1636.
Subject: Astrology, experimental philosophy, reconciliation of Christian doctrines and naturalistic and observational science.
C/52/.144

## Campanella, Tommaso.

*De libris propriis & recta ratione studendi. Syntagma.*
Paris: widow of Gvilielmus Pelé, 1642.
Subject: Mathematics.
Case/I/4112/.142

Chap. 2, article 6: "De ordine legendi mathematicos." Chap. 4, article 5: "De mathematicis." BOUND WITH Bruni, Leonardo (n.a.).

## Campanella, Tommaso.

*De sensu rerum et magia. Libros quatuor, in qvibvs mvndvm esse vivam dei statvam, omnésque illius partes, partiumque particulas sensv donatas esse, alias clariori, alias obscuriori, quantus ipsarum sufficit conseruationi, ac totius in quo consentiunt, probatur. Ac arcanorum naturalium rationes aperiuntur.*
Paris: Dionysius Bechet, 1637.
Subject: Magic, mysticism, sympathy and antipathy.
Case/4A/1096

*Correctos et defensos à stupidorum incolarum mundi calumniis per argumenta& testimonia diuinorum codicum, naturae, sc. ac scripturae, eorundémque interpretum, scilicet, theologorum & philosophorum, exceptis atheis.*

## Canepari, Pietro Maria.

*De atramentis cvivscvnqve generis opvs sanè nouum hactenus à nemine promulgatum in sex descriptiones digestum.*
Venice: Evangelista Deuchinus, 1619.
Subject: Alchemy, iatrochemistry, ink making, spagyric medicine.
Case/Wing/Z/3055/.144

1. De lapide pyrite metallorum. . . . 2. De atramento metallico. . . . 3. De atramento sutorio, vulgo vitriolo. . . . ac eius praeparatione & vsu aduersus pestem & alios morbos. 4. De atramento scriptorio. 5. De indico Dioscoridis, cum aliis pigmentis diuersi coloris, scripturis atque medicinae aptis. 6. De varijs operationibus ex vitriolo gerendis, ac de multiplici modo eliciendi oleum à vitriolo. . . . His sparsim multa penitiora tum medicinae, arcana, cum alijs facultatibus commoda leguntur.

## Canepari, Pietro Maria.

*De atramentis cujuscunque generis. Opus sanè novum hactenus à nemine promulgatum in sex descriptiones digestum.*
London: J. M. for Jo. Martin [etc.], 1660.
STC II C425B
Subject: Alchemy, iatrochemistry, ink-making, spagyric medicine.
Wing/Case/Z/3055/.145

## Canobbio, Alessandro.

*Breve trattato . . . sopra le Academie.*
Venice: A. Bòchino & brothers, 1571.
Subject: Geometry, mathematics in service of music.
Case/A/9/.943

Praising the Academia Filarmonica di Verona.

## Capaccio, Giulio Cesare, 1550–1631.

*Delle imprese trattato di Giulio Cesare Capaccio. In tre libri diuiso.*
Naples: Gio. Giacomo Carlino, & Antonio Pace, 1592.
Subject: Emblems, symbolism.
Case/W/1025/.145

Each part has separate title page, separate pagination. I: Del modo di far l'impresa da qualsiuoglia oggetto, o naturale, o artificioso con nuoue maniere si ragiona. II: Tvtti ieroglifici, simboli, e cose mistiche in lettere sacre, o profane si scuoprono; e come da quegli causar si ponno l'imprese. III: Nel figvrar degli emblemi di molte cose naturali per l'imprese si tratta.

## Capo Bianco, Alessandro.

*Corona e palma militare di artiglieria. Nelle quale si tratta dell' inuentione di essa, e dell' operare nelle fattioni da terra, e mare, suochi artificiati da giuoco, e guerra; & d'un nuouo instrumento per misurare distanze.*
Venice: Francesco Bariletti, 1602.
Subject: Instrument-making, military technology, surveying.
Case/fU/O/.032

Lf. 51 verso: "Sopra il fabricar l'instrumento." BOUND WITH Altoni, Giovanni, q.v.

## Cardano, Girolamo, 1501–1576.

*Hieronymi Cardani Mediolanensis philosophiae ac medici celeberrimi opera omnia: tam hactenvs excvsa; hîc tamen aucta & emendata; quàm nunquam aliàs visa; ac primùm ex auctoris ipsius autographus eruta: curâ Caroli Sponii.*
Lyon: Ioannis Antonius Hvgvetan & Marcus Antonius Ravavd, 1663.
Subject: Natural science, medicine.
fY/682/.C176

10 vols. Vol. 1: n.a.; vol. 2: Moralia et physica. Includes de natura, de principiis rerum, de secretis, de gemmis & coloribus, de aqua & aethere, de fulgure; vol. 3: Physica; vol. 4: De subtilitate; vol. 5: Astronomia, astrologia; vols. 6–9: Medicinalivm; vol. 10: Opvscvla miscellanea.

## Cardano, Girolamo.

*Arcana politica, sive De prudentia civili liber singularis.*
Lyon: Elzeviriana, 1635.
Subject: Astrology, natural science.
Case/B/692/.1452

Chap. 26: De natura. Chap. 128: De geniis, fato, astrologia.

## Cardano, Girolamo.

*Hieronymi Cardani mediolanensis medici & philosophi praestantissimi, in Cl. Ptolemaei Pelvsiensis IIII de astrorum iudicijs, aut, ut uulgò uocant, quadripartitae constructionis, libros commentaria, quae non solum astronomis & astrologis, sed etiam omnibus philosophiae studiosis plurimum adiumenti adferre poterunt. Praeterea, eiusdem Hier. Cardani geniturarum XII et avditv mirabilia et notatu digna, & ad hanc scientiam recte exercendam obseruatu utilia, exempla. . . . Ac denique eclipseos, quam grauissima pestis subsecuta est, exemplum.*
Basel: [Henrichvs Petri, 1554].
Subject: Astrology.
Case/fL/900/.712

Original ed. of Cardanus on astrological works of Ptolemy. BOUND WITH Firmicus Maternus, Julius, *Ad mavortium*, 1551, q.v.

## Cardano, Girolamo.

*Hieronymi Cardani in Cl. Ptolemaei de astrorum ivdicijs avt (vt vvlgò appellant) qvadripartitae constructionis lib. IIII, commentaria, ab avtore postremùm castigata, & locupletata. His accesservnt, eivsdem Cardani, de septem erraticarvm stellarvm qualitatibus atque uiribus liber posthumus, antè non uisus. Genitvrarvm item XII ad hanc scientiam rectè exercendam obseruatu utilium, exempla. Item, Cvnradi Dasypodii, mathematici Argent. scholia et resolvtiones sev tabvlae in lib. IIII. Apotelesmaticos Cl. Ptolemae 1: Vnà cum aphorismis eorundem librorum. Denique breuis explicatio astronomici horologii Argentoratensis, ad ueri & exacti temporis inuestigationem extructi.*
Basel: Henricpetrina [1578].
Subject: Astrology, astronomy.
Case/fL/900/.714

Bk. 1 of Ptolemy's De astrorum ivdiciis has MS marginalia.

## Cardano, Girolamo.

*De consolatione libri tres.*
[Milan, 1543.]
Subject: Arithmetic, natural philosophy.
Case/B/235/.151

## Cardano, Girolamo.

*Contradicentium medicorum liber primvs. Continens contradictiones centum & octo. Liber secvndvs, continens contradictiones centum & octo. Addita praeterea eivsdem avtoris, de sarzaparilia consilium pro dolore uago, quaedam aliae non inutiles. De cina radice, disputationes etiam.*
Lyon: Seb. Gryphivs, 1548.
Subject: Medicine.
UNCATALOGUED

2 vols. in 1. Some scattered MS marginalia.

## Cardano, Girolamo.

*Hieronymi Cardani medici mediolanensis, contradicentivm medicorvm libri duo. . . . Addita praeterea eiusdem autoris de sarzaparilia, de cina radice, eiúsque vsu, consilium pro doloro vago, disputationes etiam quedam aliae non inutiles. Accesserunt praeterea Iacobi Peltarij contradictiones ex lacuna desumptae, cum eiusdem axiomatibus.*
Paris: Iacobus Macaeus, 1564.
NLM 837
Subject: Medicine.
UNCATALOGUED

Contradictio IIII (leaf 142v): "Absinthium an nocent ventriculo"; leaf 300v: "Historia morbi validissimi ad exemplar trivm librorvm epidemiorvm qvos olim agens in saccensi oppido conscripsi." Peltarius's work, *De conciliatione locorum Galeni*, has same imprint.

## Cardano, Girolamo.

*Metoposcopia. H. Cardani medici mediolanensis metoposcopia libris tredecim et octingentis faciei hvmanae eiconibvs complexa.*
Paris: Thomas Iolly, 1658.
Subject: Astrology, physiognomy (facial lines related to planets).
fB/56/.146

Title page lacking, supplied by photocopy. Bk. 1 is photocopy; bks. 2–13 letterpress original.

## Cardano, Girolamo.

*De propria vita liber. Ex bibliotheca Gab. Navdaei. Adjecto hac secunda editione de praeciptis ad filios libello.*
Amsterdam: Joannes Ravestein, 1654.
Subject: Autobiography, medicine.
Case/E/5/.C/17935

Chap. 40: Felicitas in curando.

## Cardano, Girolamo.

*De rerum varietate.*
Basel [Heinrich Petri], 1557.
Subject: Encyclopedias, natural science, occult.
Case/A/911/.15

In 17 books, among which are metals, plants, motion, chemistry and alchemy, navigation, supernatural, witches, natural magic.

## Cardano, Girolamo.

*De sapientia libri qvinqve, in qvibvs omnis hvmanae vitae cursus uiuiendique ratio explicatur. Eiusdem de consolatione libri tres. His propter similitudinem argumenti & ipsus Cardani commendationem adiecti sunt.*
Orléans: Petrus & Iacobus Chouët, 1624.
Subject: Arithmetic, natural philosophy, soul.
Y/682/.C177

BOUND WITH Petrus Alcyonius *De exilio.*

## Cardano, Girolamo.

*Somniorvm synesiorvm omnis generis insomnia explicantes, libri IIII. Qvibvs accedvnt eivsdem haec etiam: de libris proprijs.*
Basel, S. Henricpetri [1585].
Subject: Encyclopedias, natural science, occult, sleep.
Case/B/578/.146

Among topics are De secretis, Actio in Thessalicum medicum, De uno.

## Cardano, Girolamo.

*De subtilitate libri XXI . . . Addita insuper apologia aduersus calumniatorem, qua uis horum librorum aperitur.*
Basel: Sebastian Henricpetri [1582].
Subject: Encyclopedias, natural science, occult.
Case/A/911/.152

Includes a tabular comparison of *Rerum varietate* with this vol. Among topics are alchemical mixtures, biology, metaphysics.

## Cardano, Girolamo.

*[De subtilitate]*
*Les livres de Hiersome Cardanvs medecin Milannois intitulez de la subtilité, & subtiles inuentions, ensembles les causes occultes, & raisons d'icelles.*
Paris: Abel l'Angelier, 1584.
Subject: Encyclopedias, natural science, occult.
Case/A/911/.153

In 21 books. Trans. Richard le Blanc.

## Carena, Cesare, fl. 1645.

*Tractatvs de Officio Sanctissimae Inqvisitionis, et modo procedendi in causis fidei. In tres partes divisvs.*
Cremona: Io. Baptista Belpierus, 1655.
Subject: Demonology, witchcraft.
Case/folio/oBX/710/.C3/1655

Pt. 1: De Romanae pontificis potestate; pt. 2: De haeresi; Pt. 3: Theorica & practica criminalis. Titulus XII: De sortilegiis. Includes material on incantation of demons, divination by the stars.

## Carpentarius, Jacobus [Jacques Charpentier] 1524–1574.

*Artis analiticae, sive judicandi, descriptio.*
Paris: Gabriel Buon, 1561.
Subject: Logic.
UNCATALOGUED

Association copy of rare work on logic with author Charpentier's marginal notes, additions, etc. Defense of scholasticism, Aristotelianism against attacks of Ramus.

## Carpentarius, Jacobus.

*Platonis cvm Aristotele in universa philosophia, comparatio. Quae hoc commentario, in Alcinoi institutionem ad eiusdem Platonis doctrinam, explicatur.*
Paris: Iacob du Puys, 1573.
Subject: Natural philosophy, natural science.
Case/Y/642/.P5302

2 vols. in 1. Sig. YY (p. 321): De coeli natura qualis esset apud Platonem, antea disputatum est. Apud Aristotelem autem totius coelestis regionis vna est quinta essentia. Chap. 7: De tribvs vniversae natvrae. Sig. GGG4 (p. 378): Hominis origo explicatur. Chap. 16: Respirationis modus & vsus in homine explicatur: deinde morborum causae, potissimùm febrium, petutur ex elementis, quibus nostra corpora constant, & ex humoribus qui illis proportione respondent. Vol. 2 *Pars posterior Platonicae et Aristotelicae comparationis in vniversa philosophia.* [same imprint as above]

## Carpenter, Nathanael.

*Geography delineated forth in two bookes. Containing the sphaericall and topicall parts thereof.*
Oxford: printed by Iohn Lichfield and William Tvrner for Henry Cripps, 1624.
STC 4670
Subject: Geography, spherical and topographical.
Ayer/7/C 29/1625

Bk. 1: Terrestrial globe, magnetism in earth's sphere, proportion etc. of earth relative to the heavens, circles of the terrestrial sphere (equator, tropics) artificial globes, planispheres, measure of the earth, zones, climates, parallels, longitude and latitude. Bk. 2: Of topography, measuring land, "Of the finding out of the angle of position by some dioptrick instrument at two or more stations", hydrography, motion, "depth of traffic of the sea", lakes, rivers, topographical features, floods, earthquakes.

## Carpi, Ugo da, ca.1455–ca.1523

*Thesavro de scrittori. Opera artificiosa la quale con grandissima arte si per pratica come per geometria insegna a scriuere diuerse sorte littere.*
Rome, 1532.
Subject: Geometry, proportion.
Wing/ZW/535/.C222

## Carpi, Ugo da.

*Thesavro de scrittori. Opera artificiosa laquale con grandissima arte si per pratica come per geometria insegna a scriuere diuerse sorte littere.* [N.p.] 1535.
Subject: Arithmetic, geometry, proportion.
Wing/ZW/535/.C223a

"Intagliato per Ugo da Carpi." With *Ragione d'abbaco* by Angelis Mutinensis.Penciled on front flyleaf: Roma: Antonio Blado, 1535.

### Carpi, Ugo da.

*Thesauro de scrittori.* [Another copy]
[N.p.] 1535.
Wing/ZW/535/.C223

Inked in on title page: "1553."

### Carpi, Ugo da.

*Thesauro de scrittori.* [Another copy]
Wing/ZW/535/.C2228

### Casaubon, Meric, 1599–1671.

*Of credulity and incredulity; in things divine and spiritual: wherein (among other things) a true and faithful account is given of the Platonick philosophy, as it hath reference to Christianity: as also the business of witches and witchcraft.*
London: printed by T. N. for Samuel Lownds, 1670.
STC II C806
Subject: Occult, witchcraft.
Case/B/86/.149

P. 171: Supernatural operations.

### Casaubon, Meric.

*Of credulity and incredulity, in things natural, civil and divine. Wherein, among other things, the sadducism of these times, in denying spirits, witches, and supernatural operations, by pregnant instances, and evidences, is fully confuted.*
London: T. Garthwaite, 1668.
STC II C807
Subject: Occult, witchcraft.
Case/B/86/.15

2 parts: natural and civil.

### Casaubon, Meric.

*A treatise concerning enthvsiasme, as it is an effect of nature: but is mistaken by many for either divine inspiration, or diabolical possession.*
London: printed by R. D., 1655.
STC II C812
Subject: Passions of the mind, revelations, visions.
Case/B/529/.15

Supernatural enthusiasm = possession, natural enthusiasm = fervency of soul, spirits, or brain. 8 species of natural enthusiasm: contemplative, rhetorical, poetical, precatory or supplicatory, musical, martial, erotic, mechanical. Are natural enthusiasms (that come from natural causes) contagious? Enthusiasm as "incidental to corporall diseases" (p. 16).

### Casaubon, Meric.

*A treatise proving spirits, witches and supernatural operations, by pregnant instances and evidences: together with other things worthy of note.*
London: printed for Brabazon Aylmer, 1672.
STC II C812
Subject: Occult, witchcraft.
Case/B/86/.151

First published under the title of *Credulity and Incredulity* (see above). This copy has a 2d letterpress title page, of same date, pasted on front flyleaf, on verso of which the title page of the 1668 ed. is copied in MS. 2 MS leaves are added at end to supply missing errata leaves.

### Casaubon, Meric.

*A true & faithful relation of what passed for many yeers between Dr. John Dee (a mathematician of great fame in Q. Eliz. and King James their reignes) and some spirits: tending (had it succeeded) to a general alteration of most states and kingdomes in the world. His private conferences with Rodolphe Emperor of Germany, Stephen, K. of Poland, and divers other princes about it. The particulars of his cause, as it was agitated in the emperors court; by the Pope's intervention: his banishment, and restoration in part. As also the letters of sundry great men and princes (some whereof were present at some of these conferences and apparitions of spirits) to the said D. D. Out of the original copy written with Dr. Dees own hand.... With a preface confirming the reality (as to the points of spirits) of this relation: and shewing the several good uses that a sober Christian may make of all. By Meric Casaubon, D. D.*
London: printed by D. Maxwell for T. Garthwait, 1659.
STC II D811
Subject: Occult, spirits.
Case/fB/894/.22

Contains a list of printed and unprinted (some unfinished) books by Dee.

### Cassebohm, Ioanne Friderico, d. 1743.

*Tractatus quatuor anatomici de aure humana, tribus figurarum tabulis illustrati.*
Halle: printed by Orphana Trophei, 1734.
Subject: Anatomy (and deformities) of the ear, medicine.
UNCATALOGUED

### Casserio, Giulio, d. 1616.

*Ivlii Casserii Placentini philosophi atque medici Patavii vtranqve medicinam exercentis de vocis avditvsqve organis historia anatomica singvlari fide methodo ac indvstria concinnata tractatibvs dvobvs explicata ac variis iconibvs aere excvsis illustrata.*
[Ferrara: Victorius Baldinus, 1601.]
Subject: Anatomy of the ear, medicine.
UNCATALOGUED

### Cassini de Thury, César François, 1714–1784.

*La meridienne de l'observatoire royal de Paris, vérifiée dans toute l'étendue du royaume par de nouvelles observations; pour en déduire la vraye grandeur des degrés de la terre, tant en longitude qu'en latitude, & pour y assujettir toutes les opérations géométriques faites par ordre du roy, pour lever une carte générale de la France. Avec des observations d' histoire naturelle, faites dans les provinces traversées par la méridienne, par M. le Monnier, ... docteur en medecine.*
Paris: Hippolyte-Louis Guerin, & Jacques Guerin, 1744.
Subject: Geography, surveying.
Case/5A/543

In 3 parts. 1. Verifying meridional line of Paris. 2. Details of operation & result of calculations. 3. Observations made in different locations. Chap. 3: "Observations des etoiles pour déterminer les differences des latitudes." Separate title page: "Observations d'histoire naturelle, faites dans les provinces méridionales de la France pendant l'année 1739." Describes primarily plants, mines, and minerals.

### Castello Branco, Anselmo Caetano Muhoz de Abreu Gusmão e, 1759–?

*Oraculo prophetico, prolegomeno da teratologia, ou historia prodigiosa em que se dà completa noticia de todos os monstros composto, para confuzao de pessoas ignorantes, satisfaçaõ des homens sabias, exterminio de prophecias falsas, e explicaçaõ de verdadeiras prophecias.*
Lisbon: Mauricio Vicente de Almeida, 1733.
Subject: Prodigies, prognostication.
Greenlee/4504/P855

Pt. 1.

### Castello Branco, Francisco Correa do Amaral e, b. 1683.

*Noticia de him caso raro, e extra ordinario succedido neste pezente anno de 1733. Em Villa Franca de Xira.*
Lisbon: Pedro Ferreyra, 1733.
Subject: Medicine, monsters, unusual births.
Greenlee/4504/P855

### Castro, Rodrigo de, 1546–1628.

*Medicus-politicus: sive de officiis medico-politicis tractatus, quatuor distinctus libris: in quibus no solvm bonorvm medicorum mores ac virtutes exprimuntur, malorum verò fraudes & imposturae deteguntur: verum etiam pleraque alia circa novum hoc argumentum utilia atque jucunda exactissimè proponuntur. Opvs admodvm vtile medicis aegrotis, aegrotorum assistentibus, & cunctis aliis litterarum, atque adeo politicae disciplinae cultoribus.*
Hamburg: Frobenius, 1614.
Subject: Disease, iatrochemistry, medical ethics, practice of medicine, supernatural, surgery.
Q/156

In 4 books. Bk. 4: On fascination, philtres, "De triplici mundi et quod uniuscujusque in homine pulcherrime reperiatur imago."

### Cataneo, Girolamo, 16th cent.

*Dell' arte militare libri cinqve, ne' qvali si tratta il modo di fortificare . . . vna fortezza.*
Brescia: Thomaso Bozzola, 1584.
Subject: Geometry, mensuration, military technology.
Case/U/O/.154

Bk. I, chap. 1: "Di alcvne operationi geometriche pertinenti al fabricare fortezze." BOUND WITH Cataneo, *Dell' arte misvrare*, q.v.

### Cataneo, Girolamo.

*Dell' arte misvrare libri dve, nel primo de' qvali s'insegna a misvrare, et partir' i campi. Nel secondo a misvrar le mvraglie' imbottar grani, vini, fieni, & strami; col liuellar dell' acque, & altre cose necessarie à gli agrimensori.*
Brescia: T. Bozzola, 1584.
Subject: Arithmetic, mensuration, surveying.
Case/U/O/.154

BOUND WITH Cataneo, *Dell' arte militare*, q.v.

### Cataneo, Pietro.

*La pratiche delle dve prime matematiche di Pietro Cataneo con la aggionta, Libro d'albaco e geometria con il pratico e vero modo di misurar la terra, non piv mostro da altri.*
Venice: Giouanni Grissio, 1559.
Subject: Mathematics, geometry, mensuration.
Case/L/10/.15

* * *

*Catastrophe mundi: or, Merlin reviv'd, in a discourse of prophecies & predictions, and their remarkable accomplishment. With Mr. Lilly's hieroglyphicks exactly cut, and notes and observations theron. As also a collection of all the antient (reputed) prophecies that are extant, touching upon the grand revolutions like to happen in these latter ages.*
London: printed by John How etc., 1683.
STC II L2214
Subject: Hieroglyphics, predicitions, prophecies, secret knowledge.
Case/B/863/.154

Woodcuts, pp. 23–55.

### Catesby, Mark, 1679?–1749.

*The natural history of North Carolina, Florida, and the Bahama Islands: containing the figures of birds, beasts, fishes, serpents, insects and plants: particularly, the forest-trees, shrubs, and other plants, not hitherto described, or very incorrectly figured by authors. Together with their descriptions in English and French. To which, are added observations on the air, soil, and waters: with remarks upon agriculture, grain, pulse, roots, &c. To the whole is prefixed a new and correct map of the countries treated of.*
London: printed at the expense of the author and sold by W. Innys [etc.], 1731–43.
Subject: Natural history, North Carolina etc.
Case/+W/764/.15

In 2 vols. Vol. 1: birds; vol. 2: fish. Vol. 1 has engraved slip pasted on front flyleaf: "This volume was coloured by George Edwards, F.R.S." Map atbeginning of vol. 2 of Virginia., the Carolinas, Florida. the Bahamas, etc.

### Catholic Church. Pope, 1621–1623 (Gregory XV)

*Serenissimi D. N. D. Gregorii Papae XV. constitvtio aduersus maleficia, seu sortilegia committentes.*
Rome: Apostolic printing office, 1623.

Subject: Witchcraft.
Case/6A/265/no./9

Papal bull on witchcraft.

## Cats, Jacob, 1577–1660.

*Silenus Alcibiadis; sive Proteus, humanae vitae ideam emblemate trifariàm variato, oculis subjiciens.*
Amsterdam: Guiljelmi Ianssonij, 1619–20.
Subject: Emblems.
Case/*Y/972/.C28

In 7 vols. Vols. 1, 4, and 6 have emblems depicting ships, medicine, dentistry, winds.

## Cats, Jacob.

*Proteus ofte minne-beelden verandert in sinne-beelden.*
Rotterdam: Pieter van Waesberge, 1627.
Subject: Emblems.
Case/Y/972/.C2886

## Cattaneo, Cristoforo.

*La geomance du seigneur Christofe de Cattan, gentilhomme Geneuois. Liure non moins plaisant & recreatif, que d'ingenieuse inuention, pour sçauoir toutes choses, presentes, passees, & à aduenir.*
Paris: Gilles Gilles, 1567.
Subject: Astrology, divination.
Case/B/863/.156

In 3 books: bk. 1: Nature and quality of geomancy; bk. 2: Zodiac; bk. 3: Practice of the art.

## Caus, Salomon de, b. 1576?

*Institution harmoniqve, diuisée en deux parties. En la premiere sont monstrées les proportions des interualles harmoniques et en la deuxiesme les compositions dicelles.*
Frankfurt: Jan Norton, 1615.
Subject: Harmonics, mathematics of proportion.
Case/6A/243

BOUND WITH *La pratique . . . des horloges solaires*, q.v.

## Caus, Salomon de.

*La perspective avec la raison des ombres.*
London: Jan Norton, 1612.
STC 4869
Subject: Optics, perspective, plane and solid geometry.
Wing/+ZP/6465/.M74

Engraved title page has incised in base of architectural frame: "Frankfort ches la vesue de Hulsius." Jan Norton, "imprimeur du roy de la Grande Bretaigne, aus langues estrangeres."

## Caus, Salomon de.

*La pratique et demonstration des horloges solaires, avec vn discovrs svr les proportions, tiré de la raison de la 35 proposition du premier liure d'Euclide, & autres raisons & proportions, & l'vsage de la sphere plate.*
Paris: Hyerosme Droüart, 1624.
Subject: Clocks, mathematics, proportion, solid geometry.
Case/6A/243

BOUND WITH his *Institution harmonique*, q.v.

## Cella, Anselmus.

*Evropae descriptio . . . per Ancelmum atque Christophorum Cellae. Prognosticon Antonii Torquati, medicinae doctoris Ferrarieñ. Clarissimique astrologi, ab anno M.CCCC.LXXX, vsque ad annum M.D. XXXVIII Matthiae Vngariae regi dicatum. De fide et moribus Aethiopum. libellus Christianis lectu planè dignus.*
Antwerp: Ioannes Steelsius, 1535.
Subject: Anthropology, geography, prognostications.
Case/G/30/.158

In 3 parts. Pt. 1: Geographical description of Africa, Egypt, Asia, Assyria, Persia, India, etc., and Europe, Spain, France, Italy, Sweden, Poland, Russia, etc.; pt. 2: Prognosticon; pt. 3: Ethiopian faith, religion, customs, and ceremonies.

## Cellini, Benvenuto, 1500–1571.

*Dve trattati vno intorno alle otto principali arti dell' oreficeria. L'altro in materia dell'arte della scultura; doue si veggono infiniti segreti nel lauorar le figure di marmo, & nel gettarle di bronzo.*
Florence: Valente Panizzi and Marco Peri, 1568.
Subject: Alloys, chiseling, enameling, metallurgy, metalworking, soldering, welding.
Case/W/979/.149

Technological as well as fine arts approach to handling stone and metal.

## Celsus, Aulus Cornelius, 1st Cent. A.D.

*De medicina.*
Milan: Leonhard Pachel and Ulrich Scinzenzeler, 1481.
H-C 4836^ BMC (XV)VI:750^ GW VI, 6457^ Pr. 5940^ Still. C326^ Goff C-365
Subject: Medicine.
Inc./5940

2d ed. In 8 books.Bk. 1: Medicinae intentio & diuiso; bk. 2: Quae anni tempora: quae tempestatum genera; bk. 3: De morborum generibus; bk. 4: De humani corporis interioribus sedibus; bk. 5: De simplicibus facultatibus; bk. 6: De uiciis singularum corporis partium de sycosi & eius curatione; bk. 7: De chirurgica; bk. 8: De positu & figura totius humani corporis. MS marginalia passim, esp. index.

## Celsus, Aulus Cornelius.

*De re medica libri octo, inter Latinos eius professionis autores facilè principis: ad ueterum & recentium exemplarium fidem, necnon doctorum hominum iudicium, summa diligentia excusi. Accessit huic thesauris ueriis, quàm liber Scribonii Largi titulo compositionum medicamentorum, nunc primum tineis & blattis, ereptus industriae Ioannis Rvellii doctoris disertissimi.*
Paris: C. Vuechel, 1529.
Subject: Medicine.

Case/5A/470/no. 1

BOUND WITH Galen, *Liber de plenitudine*, q.v.

## Celsus, Aulus Cornelius.

*Aurelii Cor. Celsi de re medica libri octo. Item Q. Sereni liber de medicina. Q. Rhemnii Fanij Palaemonis de ponderibusque opera mensuris liber.*
Lyon: S. Gryphivm, 1542.
Subject: Medicine.
Case/3A/290

In 8 books. BOUND WITH Apicius, *De re culinaria*, 1541, q.v.

## Celsus, Aulus Cornelius.

*Die acht bücher des hoch berumpten Aurelii Cornelii Celsi von beyderley medicine: das ist von der leib und mund artznei: zu errettung menschlichs lebens: inn allen kranckheyten seer dienstlich und behütsam newlich jetzo verdeutscht durch I. Johanem Rhüffner von Ratemberg am Yne.*
[Mainz: J. Schöffer, 1531.]
Subject: Medicine.
Case/f*R/50/.202

Bound with Crescenzi, Pietro de. *Vom ackerbaw*, q.v.

## Celtes, Conradus [Konrad Pickel], 1459–1508.

*Conradi Celtis protucij, libri odarum quatuor, cum epodo.*
Strassburg, 1513.
Subject: Astrology, nature mysticism, neo-Platonism.
Case/Y/682/.C34

Ode no.21: De solario per norī cum astrologvm in vento. Bk. 3, no.23: Ad Bernhardvm Valerum Barbatvm, mathematicum astronomvm et philosophvm; no. 24: Ad Ioannem Reuchlin; bk. 2, no. 3: Ad Benedictvm Tyctelivm medicvm et philosophvm viennensem; bk. I no. 17: Ad Albertvm Brvtvm astronomvm.

## Celtes, Conradus.

*De situ et moribus Germaniae.*
Wittenberg: Ioannes Lufft, 1557.
Subject: Topography
Case/F/4709/.565

BOUND WITH Melanchthon, Philipp. *Germania Cornelii Taciti.* Celtes' tract begins sig. K2. Topics include "De sideribvs verticalibus Germaniae. De qvatvor lateribvs Germaniae. De tribvs ivgis et montibus Germaniae. De qvalitate tellvris per Germaniam."

## Chales, Claude François Milliet de, 1621–1678.

*Cursus seu mundus mathematicus. Tomus quartus complectens mvsicam, pyrotechniam, astrolabivm.*
Lyon: Anissonios, Joan. Posuel & Claud. Rigaud, 1690.
Subject: Acoustics, mathematics.
Case/6A/145

This vol. contains only the "Musica" portion of the text (49 pp.). Includes "Qualis sit motus, qui sonus est." "De echone seu reflexione soni."

## [Chales, Claude, Franois Milliet de]

*L'art de naviger demontré par principes, & confirmé par plusieurs observations tirées de l'experience.*
Paris: Estienne Michallet, 1677.
Subject: Navigation.
Case/L/995/.158

In 7 books. 1. Common terms; 2. Principles of the sphere; navigation by stars, using simple instruments; 3. The compass; 4. Loxodromic lines, geometry for navigation; 5. Hydrographic maps; 6. Estimating latitude; problem of the longitude; 7. Practical methods at sea: tides, winds, currents, computing time.

## Chambers, Ephraim, c. 1680–1740.

*Cyclopaedia: or, An universal dictionary of arts and sciences, containing the definitions of the terms, and accounts of the things signify'd thereby, in the several arts, both liberal and mechanical, and the several sciences, human and divine: the figures, kinds, properties, productions, preparations, and uses of things natural and artificial; the rise, progress, and state of things ecclesiastical, civil, military, and commercial: with the several systems, sects, opinions, &c. among philosophers, divines, mathematicians, physicians, antiquaries, criticks, &c.*
London: J. & J. Knapton [etc.] 1728.
Subject: Encyclopedias.
fA/21/.157

In 2 vols. Has diagram of knowledge, natural and artificial. Definitions of various sciences (note in Preface, pp. iii ff.).

## Chambers, Ephraim.

*Cyclopaedia: or, an universal dictionary of arts and sciences; containing an explication of the terms, and an account of the things signified thereby.*
London: printed for D. Midwinter [etc.], 1738.
Subject: Encyclopedias.
fA/21/.158

2nd ed. "Corrected and amended; with some additions."

## Chambers, Ephraim.

*Dizionario universale delle arti e delle scienze, che contiene la spiegazione de' termini, e la descrizion delle cose significate per essi, nelle arti liberali e meccaniche, e nelle scienze umane e divine: le figure, le spezie, le proprietà, le produzioni, le preparazioni, e gli usi delle cose si naturali, come artifiziali.*
Venice: G. Pasquali, 1748–1749.
Subject: Encyclopedias.
A/21/.1585

9 vols. Engraved plates (e.g., istoria natural, navigazione, sezioni coniche) bound in at end of each vol.

## Champier, Symphorien, d. 1537.

*Index librorum.*
[Lyon: J. de Campis, 1506.]
Subject: Medicine.
Case/*Y/682/.C36

In 12 parts, including "De medicine claris scriptoribus" in 5 tracts; Galen; aphorisms of Champier; "Opera parua Hippocratis." See Paul Allut (*Etude bibliographique sur Symphorien Champier* [Lyon, 1859]) for attribution of date and place printed.

## Champier, Symphorien.

*Index librorum.*
[Lyon: Iacques Marechal, 1517.]
Subject: Natural history.
Case/Y/682/.C361

In 4 parts, including (pt. 4) "De mirabilibus mundi. Primus de morbis & mirabilibus gentium & prouinciarum orbis tabulas ptolomei."

## Champier, Symphorien.

*Liber de quadruplici vita. Theologia Ascelpij Hermetis Trismegisti discipuli cum commentarijs eiusdem domini Simphoriani. Sixti philosophi Pythagorici enchiridion Isocratis ad demonicum oratio preceptiua. Silue medicinales de simplicibus: cum nonnullis in medice facultatis praxim introductorijs. Quedam ex Plinij iunioris practicae. Tropheum gallorum quadruplicem eorundem complectens historiam.*
Lyon: S. Gueynardi; J. Huguetani, arte vero J. de Campis, 1507.
Subject: Astrology, demons, Hermeticism, materia medica, medicine.
Case/Y/682/.C362

"Opus de quadruplici vita videlicet de vita sana: longa: Celtus comparanda et supracelesti. Theologia Asclepii Hermetis discipuli ad ammonem regem: cum commentarijs. Silue medicinales eiusdem domini Simphoriani ad complementum librorum de vita sana et longa spectantes. Introductiones eiusdem in practicam medicine. Quedam in praxim medicine ex Plinio iuniore excerpta."

## Champier, Symphorien.

*Libri VII de dialectica, rhetorica, geometria, arithmetica, astronomia, musica, philosophia naturali, medicina & theologia: et de legibus & repub. eaquae parte philosophiae quae de moribus tractat. Atque haec omnia sunt tractata ex Aristotelis & Platonis sententia.*
Basel: H. Petrus [1537].
Allut, p. 260
Subject: Medicine, natural philosophy, quadrivium.
Case/Y/672/.C79

BOUND WITH *Enarrationes Bartholomaei Latomi*, 1539. [n.a.]

## Champier, Symphorien.

*Rosa gallica aggregatoris Lugdunensis domini Symphoriani Champerii omnibus sanitatem affectantibus vtilis & necessaria. Quae in se continent praecepta, auctoritates, atque sententias memoratu dignas, ex Hippocratis, Galeni, Erasistrati, Asclepiadis, Diascoridis, Rasis, Haliabatis, Isaac, Auicennae, multorumque aliorum clarorum virorum libris in vnum collectas: quae ad medicam artem rectaque viuendi formam plurimum conducunt. Vna cum sua preciosa margarita: de medici atque egri officio.*
Paris: Jodocus Badius Ascensius, 1514.
Allut, pp. 167–71^ Brunet I 1766–7
Subject: Disease, herbal medicine, materia medica.
Case/Q/.162

## Champier, Symphorien.

*De triplici disciplina cuius partes sunt philosophia naturalis, medicina, theologia, moralis philosophia integrantes quadriuium.*
[Lyon, 1508.]
Allut VIII (p. 153)
Subject: Medicine, natural philosophy, quadrivium.
Case/Y/682/.C356

Has "Vocabulorum medicinalium epitoma" [running title], with MS marginalia.

## Champlain, Samuel de, 1567–1635.

*Des savvages, ov voyage de Samvel de Champlain de Brovage fait en la France nouuelle, l'an mil six cens trois. Contenant les moeurs, façons de viure, mariages, guerres, & habitations de sauuages de Canadas. De la descouuerture de plus de quatre cens cinquante lieuës dans le pays des sauuages, quels peuples y habitent, des animaux qui s'y trouuent, des riuieres, lacs, isles, & terres, & quels arbres & fruicts elles produisent. De la coste d'Arcadie, des terres que l'on y a descouuertes, & de plusieurs mines qui y sont, selon le rapport des sauuages.*
Paris: Clavde de Monstr'oeil, 1604.
Church 327 A
Subject: Geography, topography, voyages.
Ayer/*121/C6/1604

## Champlain, Samuel de.

*Les voyages dv Sievr de Champlain Xaintongeois, capitaine ordinaire pour le roy, en la marine. Divisez en deux livres. Ou, iovrnal tres-fidele des observations faites és descouuertures de la nouuelle France: tant en la description des terres, costes, riuieres, ports, haures, leurs hauteurs, & plusieurs declinaisons de la guide-aymant; qu'en la créance des peuples, leur superstition, façon de viure & de guerroyer: enrichi de quantité de figures. Ensemble deux cartes geografiques, la premiere seruant à la nauigation, dressée selon les compas qui nordestent, sur lesquels les mariniers nauigent: l'autre en son vray meridien, auec ses longitudes & latitudes: à laquelle est adiousté le voyage du destroict qu'ont troué les Anglois, au dessus de Laborador, depuis le 53e degré de latitude, iusques au 63e en l'an 1612, cherchans vn chemin par le nord, pour aller à la Chine.*
Church 360
Paris: Iean Berjon, 1613.
Subject: Discovery, natural history, navigation, voyages.
Ayer/*121/C6/1613

Engraved folding maps and plates.

### Champlain, Samuel de.

*Voyages et descovvertvres faites en la Novvelle France depuis l'année 1615 jusques à la fin de l'année 1618 par le sieur de Champlain, cappitaine ordinaire pour le roy en la mer du Ponant. Où sont descrits les moeurs, coustumes, habits, façons de guerroyer, chasses, dances, festins, & enterrements de diuers peuples sauuages, & de plusieurs choses remarquables qui luy sont arriuées audit païs, auec vne description de la beauté, fertilité, & temperature d'iceluy.*
Paris: Clavde Collet, 1619.
Church 375
Subject: Discovery, natural history, voyages.
Ayer/*121/C6/1619

### Champlain, Samuel de.

*Voyages et descovvertvres faites en la novvelle France depuis l'année 1615 jusques à la fin de l'année 1618.*
Paris: Clavde Collet, 1620.
Church 378
Subject: Discovery, natural history, voyages.
Ayer/*121/C6/1620

Engraved title page dated 1619. Letterpress title page same as 1619 except for date.

### Champlain, Samuel de.

*Voyages et descovvertvres faites en la novvelle France depuis l'année 1615 jusques à la fin de l'année 1618.*
Paris: Clavde Collet, 1627.
Subject: Discovery, natural history, voyages.
Ayer/*121/C6/1627

2d ed. Engraved title page, 1619. Not in Church.

### Champlain, Samuel de.

*Les voyages de la Novvelle France occidentale, dicte Canada, faits par le Sr. de Champlain. . . .toutes les descouuertes qu'il a faites en ce païs depuis l'an 1603 iusques en l'an 1629. . . Avec vn traitté des qualitez & conditions requises à vn bon & parfaict navigateur pour cognoistre la diuersité des estimes qui se font en la nauigation. . . Et la maniere de bien dresser cartes marines auec leurs ports, rades, isles, sondes & autre chose necessaire à la nauigation. Ensemble vne carte generalle de la description dudit pays faicte en son meridien selon la declinaison de la guide aymant, & vn catechisme ou instruction traduicte du François au langage des peuples sauuages de quelque contrée; auec ce qui s'est passé en ladite nouuelle France en l'année 1631.*
Paris: chez Lovis Sevestre, 1632.
Church 420
Subject: Discovery, natural history, navigation, voyages.
Ayer/*121/C6/1632

Text as originally printed, without cancel.

### Champlain, Samuel de.

*Les voyages de la Novvelle France occidentale.*
Paris: chez Lovis Sevestre, 1632.
Church 420
Case/G/82/.1611

2 leaves canceled (pp. 27–30) because 1 paragraph (p. 27) was construed as a reflection on Cardinal Richelieu. (Replaced with cancellans.)

### Champlain, Samuel de.

*Les voyages de la Novvelle France occidentale.*
Paris: [chez Clavde] Collet, 1632.
Church 420
Graff/642

[Another copy with same cancel and replacement as above.]

### Champlain, Samuel de.

*Les voyages de la Novvelle France occidentale.*
Paris: chez Clavde Collet, 1632.
Church 420
Ruggles/Coll./No. 54

[Another copy with same cancel and replacement as above.]

### Champlain, Samuel de.

*Les voyages de la Novvelle France occidentale.*
Paris: C. Collet, 1640.
Church 446
Subject: Discovery, natural history, navigation, voyages.
Ayer/*121/C6/1640

[Full title as in 1632 ed.] Has both cancellanda (leaves to be canceled) and cancellans (replacement leaves).

* * *

*The Character of a quack-astrologer: or, the spurious prognosticator* [i.e., John Gadbury] *anatomiz'd.*
London, 1673.
STC II S49
Subject: Astrology.
Case/B/8635/.162

Attributed to J. S.

### Charleton, Walter 1619–1707.

*Natvral history of the passions.*
[London] printed by T. N. for James Magnes, 1674.
STC II C3684A
Subject: Passions of the mind.
Case/B/529/.16

In 6 sections. 1: Introduction; 2: Sensitive soul; 3: Rational soul;4: Passions of the mind in general; 5: In particular [lists various passions and describes them: love, hate, pride, humility, joy grief, remorse, etc.]; 6: Conclusion. There are "simple" and "mixed" passions. Generosity, divine providence, deliberation, and virtue are remedies against failings caused by desires.

### Charleton, Walter.

*Onomasticon zoicon, plerorumque animalium differentias & nomina propria pluribus linguis exponens. Cui accedunt mantissa anatomica; et quaedam de variis fossilivm generibus.*
London: Jacobus Allestry, 1668.

STC II C3688
Subject: Biology, fossils, taxonomy.
Case/O/02/.162

Binding gold-stamped "Robert Harley," with his bookplate. Quadrupeds, birds, insects, fish (includes hippopotamus, p. 171).

## Charleton, Walter.

*Physiologia Epicuro-Gassendo-Charltoniana: or a fabrick of science natural, upon the hypothesis of atoms, founded by Epicurus, repaired by Petrus Gassendus, augmented by Walter Charleton.*
London: printed by Tho. Newcomb, for Thomas Heath, 1654.
STC II C3691
Subject: Atomism, natural philosophy, natural science.
UNCATALOGUED

"The first part." In 3 books. Subjects include Of time and eternity, corporeity and inanity (the presence of bodies is confirmed by sense), properties, motions, etc. of atoms, origins of qualities, occult qualities made manifest (chap. XV). Conclusion promises a second part, apparently never published.

## [Charleton, Walter]

*Two discourses. I. Concerning the different wits of men. II. Of the mysterie of vintners.*
London: printed by R. W. for William Whitwood, 1669.
Subject: Enology, intelligence, plant diseases, psychology.
STC II C3694
Case/B/5/.162

Title page to part I: *A brief discourse concerning the different wits of men . . . in the year 1664*. Title page to part II: *The mysterie of vintners. or, A brief discourse concerning the various sicknesses of wines, and their respective remedies, at this day commonly used. . . .*

## Charlevoix, Pierre François Xavier de, 1682–1761.

*An account of the French settlements in North America: shewing from the latest authors, the towns, ports, islands, lakes, rivers, &c. of Canada.*
Boston: printed by Rogers & Fowle, 1746.
Evans 5725^ Howes A 29^ Sabin 95
Subject: Geography.
Case/4A/1153

## Charlevoix, Pierre François Xavier de.

*Histoire et description generale de la nouvelle France, avec le journal historique d'un voyage fait par ordre du roi dans l'Amérique septentrionnale.*
Paris: Pierre-François Giffart, 1744.
Subject: Descriptive geography, voyages.
Graff/650

In 3 vols. Vol. 1 contains "fastes chronologiques," with 1248 given as year of 1st navigation to Greenland, said by author to have been known to the Norwegians since the 9th cent. Vol. 3: "Journal d'un voyage . . . dans l'Amerique septentrionnale."

## Charlevoix, Pierre François Xavier de.

*Histoire et description generale de la nouvelle France, avec le journal historique d'un voyage fait par ordre du roi dans l'Amérique septentrionnale.*
Paris: Nyon sons, 1744.
Subject: Botany, descriptive geography, voyages.
Ayer/169.1/C 47/1744

In 3 vols. 1st 35 pp. of vol. 2: "Description des plantes principales de l'Amerique septentrionnale." 98 plants described. Vol. 3: "Journal" as above.

## Charlevoix, Pierre François Xavier de.

*Histoire et description generale de la nouvelle France, avec le journal historique d'un voyage fait par ordre du roi dans l'Amérique septentrionnale.*
Paris: Rolin sons, 1744.
Subject: Botany, descriptive geography, voyages.
F/82/.16

In 3 vols. Vols. II & III bear imprint: the widow Ganeau. 55 pp. at back of vol. III: "Description des plantes principales."

## Charlevoix, Pierre François Xavier de.

*Histoire et description generale de la nouvelle France, avec le journal historique d'un voyage fait par ordre du roi dans l'Amérique septentrionnale.*
Paris: Pierre-François Giffart, 1744.
Subject: Botany, descriptive geography, voyages.
Hermon/Dunlap/Smith/Coll./Nos. 9/& 10

In 6 vols. 1st 4 vols. [Smith no. 9]: history and general description; vols. 5 and 6 [Smith no. 10]: the journal.

## Chaucer, Geoffrey, 1340?–1400.

*Workes of Geffray Chaucer newly printed with dyuers workes whiche were neuer in print before.*
[London: Thomas Godfray, 1532.]
STC 5068
Subject: Astrolabe
Case/6A/112

1st complete ed. of Chaucer, 1st folio. "The conclusions of the astrolabie," fols. 299–307.

## Chaucer, Geoffrey.

*The workes of Geffray Chaucer.* [Another copy]
Case/6A/114

## Chaucer, Geoffrey.

*The workes of Geffray Chaucer newlye printed, wyth dyuers workes whych were neuer in print before.*
[London]: prynted by John Reynes, 1542.
STC 5070
Subject: Astrolabe.
Case/Y/185/.C4055

"The conclusions of the astrolabie," fols. 291–300.

## Chaucer, Geoffrey.

*The workes of Geffray Chaucer newly printed with dyuers workes whiche were neuer in print before.*
[London: W. Bonham, ca. 1551.]
STC 5071
Subject: Astrolabe.
Case/6A/113

"The conclusyons of the astrolaybe," fols. 278v–287v.

## Chaucer, Geoffrey.

*The woorkes of Geffrey Chaucer, newly printed, with diuers addicions, whiche were neuer in printe before.*
[London: John Kyngston for Jhon Wight, 1561.]
STC 5076
Subject: Astrolabe.
Wing/ZP/545/.K6116

4th ed., 3d issue. "The conclucions of the astrolabie," fols. 261–269v.

## Chaucer, Geoffrey.

*The workes of Geffrey Chaucer, newlie printed, with diuers addicions, whiche were neuer in print before.*
[London: imprinted by Jhon Kyngston for Jhon Wight] 1561.
STC 5075
Subject: Astrolabe.
Safe/2/Y/185/.C4057

4th ed., 1st issue. "The conclucions of the astrolabie," fols. 261–269v.

## Chaucer, Geoffrey.

*The workes of our antient and learned English poet, Geffrey Chavcer, newly printed.*
London: printed by Adam Islip for Bonham Norton, 1598.
STC 5078
Subject: Astrolabe.
Case/fY/185/.C4059

"The conclusion of the astrolabie," pp.261–269.

## Chaucer, Geoffrey.

*The workes of ovr ancient and learned English poet, Geffrey Chavcer, newly printed.*
London: printed by Adam Islip, 1602.
STC 5080
Subject: Astrolabe.
Case/fY/185/.C406

"The conclusions of the astrolaby," fols. 249–257v.

## Chaucer, Geoffrey.

*The works of our ancient, learned, & excellent English poet Jeffrey Chaucer: As they have lately been compar'd with the best manuscripts; and several things added, never before in print.*
London, 1687.
STC II C3736
Subject: Astrolabe.
Case/Y/185/.C4068

"The conclusions of the astrolabie," pp. 445–460.

## Chauvet, Jacques, 16th cent.

*Methodiqves institvtions de la vraye et parfaicte arithmetique de Iacqves Chavvet, diuisée en six parties. Reveve, corrigee et amplifiee d'exemples geometriques, extraction des racines quarrées & cubes, & autres choses appartenantes à la geometrie, auec les figures & pratique d'icelles.*
Rouen: Nicolas Loyselet, 1648.
Subject: Arithmetic, geometry.
L/130/.164

Parts include whole numbers, fractions, proportions, astronomical fractions.

## Chauvin, Etienne, 1640–1725.

*Nouveau Journal des Sçavans dressé à Rotterdam par le Sieur* C**** [Etienne Chauvin].
Rotterdam: Pierre vander Slaart, 1694.
Subject: Alchemy, medicine.
A/54/.668

Jan./Feb. 1694–Nov./Dec. 1694. Some relevant pieces: "Idée générale des maladies & des passions de l'homme," p. 131; "Mélanges curieux, ou ephémérides de médicine & de physique," Art. VI, p. 48; "Six exercitations de Martin Lister sur quelques maladies chroniques," Art. VII, p. 558; "Le grand-oeuvre," Art. VIII, p. 570.

## Chaves, Jeronimo de, 1523–1574.

*Chronographia o reportorio de los tiempos, el mas copioso y precisso que hasta ahora ha salido a luz.*
Seville: Fernando Diaz for Juan Francisco de Cisneros, 1580.
Subject: Astrology, astronomy, calendar, chronology.
Ayer/47/C49/1580

In 4 tracts. 1. Time and its division. 2. Elements, planets, zodiac. 3. Calendar, movable feasts, eclipses. 4. Judicial astrology: critical days for purging, bleeding, other medical treatments, weather forecasting.

## Cheyne, George, 1671–1743.

*Philosophical principles of natural religion: containing the elements of natural philosophy, and the proofs for natural religion arising from them.*
London: printed for George Strahan, 1705.
Subject: Natural philosophy, natural religion, natural science.
B/72/.165

In 4 chaps. 1: Astronomical, biological, physical, laws, origin of man; 2: Origin of world; 3: Existence of a deity; 4: Finiteness and infiniteness, limits of human knowledge.

## Cheyne, George.

*Principij filosofici di religione naturale, ovvero elementi della filosofia, e della religione da essi derivanti . . . tradotta dall' idioma inglese dal cavaliere Tommaso Dereham.*
Naples: Moscheni & Co., 1729.

Subject: Natural philosophy, natural religion, natural science.
B/72/.166

In 6 parts: 1: Delle leggi fisiche della natura; 2: Dell' attrazione, o gravitazione ne' corpi; 3: Della origine, e del presente stato delle cose; 4: Della eterna produzione; 5: Della esistenza di un dietà; 6: Prove della essenza di uno iddio. . . . della struttura umana.

## Cheyne, James, 1545–1602.

*De geographia libri dvo. Accessit Gemmae Phrysii medici ad mathematici de orbis diuisione & insulis, rebusque nuper inuentis opusculum longè, quàm antehac, castigatius.*
Douai: Lodouic de VVinde, 1576.
Subject: Geography, geometry, hydrography, latitude and longitude.
Ayer/*7/C395/1576

Contains Gemma Reinerus, Frisius. *De orbis*, q.v.

## Chifflet, Jean Jacques, 1588–1660.

*Lilivm francicvm, veritate historica, botanica, et heraldica illvstratvm.*
Antwerp: Plantiniana, Balthasar Moreti, 1658.
Subject: Botany, emblems, heraldry.
Case/F/0739/.16

Botanical description, chap. 17, p. 116.

## Childrey, Joshua, 1623–1670.

*Britannica Baconica: or, the natural rarities of England, Scotland, & Wales. According as they are found in every shire. Historically related, according to the precepts of the Lord Bacon; methodically digested; and the causes of many of them philosophically attempted, with observations upon them, and deductions from them, whereby divers secrets in nature are discovered, and some things hitherto reckoned prodigies, are fain to confess the cause whence they proceed. Usefull for all ingenious men of what profession or quality soever.*
London: printed for the author, 1661.
STC II C3871
Subject: Astrology, fossils, natural wonders, rational science.
Case/M/045/.16

From "Preface to the reader": "That I have one or two reflections on astrology, I hope the reader will pardon me. . . . I cannot but profess, that I have an affection for the study."

## Cholieres, Nicolas de, 1509–1592.

*Les neuf matinees du seigneur de Cholieres.*
Paris: I. Richter, 1585.
Subject: Metallurgy, numerology.
Case/Y/762/.C4515

Medicine more significant than law.

## Chomel, Noel, 1632–1712.

*Dictionnaire oeconomique, contenant divers moyens d'augmenter son bien, et de conserver sa santé. Avec plusieurs remedes assurez et éprouvez, pour un tres-grand nombre de maladies, & de beaux secrets pour parvenir à un longue & herueuse vieillesse. Quantité de moyens pour élever, nourrir, guérir & faire profiter toutes sortes d'animaux domestiques, comme brébis, moutons, boeufs, chevaux, mulets, poules, abeilles, & vers à soye. Differens filets pour la pêche de toutes sortes de poissons, & pour la chasse de toutes sortes d'oiseaux & animaux &c. Une infinité de secrets découverts dans le jardinage, la botanique, l'agriculture, les terres, les vignes, les arbres; comme aussi la connoissance des plantes des païs étrangers, & leurs qualitez specifiques, &c. . . .*
Paris: Etienne Ganeau & Jacques Etienne, 1718.
Subject: Agricultural encyclopedias, animal husbandry, home medicine.
fA/24/.168

In 2 vols. Entries arranged alphbetically.

## [Cingularis, Hieronymus] 1464 [or 65]–1558.

*Totius philosophiae humanae in tres partes, rationalem, naturalem, & moralem digestio, earundumque partium luculentiss. descriptio, tribus libris consummatam noticiam complectens.*
Basel, Ioannes Oporinus [1555].
Subject: Natural philosophy, natural science, soul.
Case/B/0/.162

Dedicatory letter and Epitome both written by Hieronymus VVildenberg. Table or chart of divisions of philosophy. "Totius philosophiae naturalis in octo libros digestio," pp. 115 ff. "In Aristotelis de anima libros," p. 188, has colophon: "Datae Turuni ex aedib. nostris anno . . . 1551. Domino Nicolao à Lynde, & Domino Ioanne Strobando, magistratum gerentibus."

## Ciruelo, Pedro, 1470 ca.–1560.

*Cursus quattuor mathematicarum artium liberalium quas recollegit atque correxit magister Petrus Ciruelis Darocensis theologus simul & philosophus.*
[Alcalà: A. G. de Brocar, 1516.]
Subject: Geometry, mathematics, numerology, perspective.
Case/fL/10/.172

Proportion in music.

## Citolini, Alessandro, 16th cent.

*La tipocosmia di Alessandro Citolini, de Serraualle.*
Venice: Vincenzo Valgrisi, 1561.
Subject: Cosmography, encyclopedias, epistemology.
Ayer/7/C4/1561

Divided into 7 days. 1: Inteligible world; 2: Celestial world; 3: Mixtures, stones, plants, animals, man; 4: Knowledge, religion, sciences, mathematics, metaphysics, cosmography, physics; 5: Actions of man, technology; 6: Arts, cosmetics, acting spectacles, perfumery, grammar, voyages, funerals, end of the world; 7: Summary, day of rest.

## Clarke, Samuel, 1675–1729.

*A collection of papers, which passed between the late learned Mr. Leibnitz and Dr. Clarke, in the years 1715 and 1716. Relating to the principles of natural philosophy and religion.*
London: printed for James Knapton, 1717.
Subject: Dynamics, gravity, natural theology, space, time.
B/72/.17

Controversy between Clarke, defender of Newtonianism, and Leibniz, who attacked the Newtonian position, ended in a draw with death of Leibniz in 1716.

## Clarke, Samuel.

*A mirrour or looking-glass both for saints and sinners.... Whereunto are added a geographical description of all the countries in the known world: as also the wonders of God in nature, and the rare, stupendioüs, and costly works made by the art and industry of man... The fourth edition very much enlarged: especially in the geographical-part, wherein all the counties in England and Wales are alphabetically described: together with the cities, and most remarkable things in them: As also the four chief English plantations in America.*
London: printed by Tho. Milbourn for Robert Clavel [etc.] 1671.
STC II C4552
Subject: Geographical encyclopedia, natural history.
Case/fA/15/.17

2 vols. Vol. I in 4 parts, with separate title pages. Pt. 1: Alphabetically arranged examples of God's mercies and judgments, includes plagues, astrologers. List of all the universities in Europe. Pt. 2: "A geographical description of all the countries in the known world.... Together with the rarest beasts, fowls, birds, fishes, and serpents which are least known amongst us." Pt. 3: "A true and faithful account of the four chiefest plantations of the English in America... with the temperature of the air: the nature of the soil: the rivers, mountains, beasts, fowls, birds, fishes, trees, plants. &c. London: printed for Robert Clavel, 1670." Included in this part, pp. 84–85: "A remedy against the stone" (STC II C4558). Pt. 4: "Examples of the wonderful works of God in the creatures." Includes "of strange stones, of strange beasts and serpents," as well as descriptions of trees, herbs, plants, gums. Vol II is a supplementary list of God's mercies and judgments. Has table at end: "Of the most remarkable histories in this book." Includes such items as a boy possessed by the devil, of strange ants at Lichfield, of the dreadful plague in London.

## Clavius, Christophe, 1538–1612.

*Gnomonices libri octo, in qvibvs non solum horologiorum solarum, sed aliarum quoque rerum, quae ex gnomonis umbra cognosci possunt, descriptiones geometricê demonstrantur.*
Rome: Franciscvs Zanettvs, 1581.
Subject: Astronomy, gnomonics, plane and spherical geometry.
Case/fR/25/.176

5 types of clocks or sundials: horizontal, vertical, meridian, polar, equinoctial.

## Clavius, Christophe.

*Horologiorvm nova descriptio.*
Rome: Aloysius Zanettus, 1599.
Subject: Astronomy, geometry, gnomonics, horology.
Case/R/25/.178

## Clavius, Christophe.

*Christophori Clavii Bambergensis ex societate Iesv, in sphaeram Ioannis de Sacro Bosco commentarivs. Nunc quartò ab ipso auctore recognitus, & plérisque in locis locupletatus.*
Lyon: printed by Gvichardvs Ivllieron for Ioannes & David de Gabiano [brothers], 1593.
Subject: Commentary on Sacro Bosco's Sphere.
Ayer/*6/S2 C61/1593

In 4 books. Bk. 1: Compositionem sphaerae, quid sit polus mundi, quod sint sphaerae, quae sit forma mundi; bk. 2: De circulis, ex quibus sphaera materialis componitur, & illa super coelestis, quae per istam repraesentatur, componi intelligitur; bk. 3: De ortu, & occasu signorum, & de diuersitate dierum, & noctium, & diuisione climatum; bk. 4: De circulis, & motibus planetarum, & de causis eclipsium solis & lunae.

## [Cleonides]

*Le liure de la musique d'Euclide, traduit par P. Forcadel lecteur du roy es mathematiques.*
Paris: Charles Perier, 1566.
Subject: Mathematical harmonics.
Case/3A/734

Harmonics attributed to Cleonides, not Euclid. BOUND WITH C. *Ptolemaei Mathematicae constructionis*, 1557, q.v.

## Clermont, sieur de.

*L'arithmetique militaire, ou l'arithmetique pratique de l'ingenieur et de l'officier, divisée en trois parties, ouvrage également necessaire aux officiers, aux ingenieurs & aux commerçans.*
Strassburg: Jean Renauld Doulsseker, 1707.
Subject: Commercial and military arithmetic.
U/26/.178

2d ed.

## Clichtove, Josse [with Kaspar Schatzgeyer].

*De mystica numerorum significatione opusculum: eorum praesertim qui in sacris litteris vsitati habentur, spiritualem ipsorum designationem succincte elucidans.*
Paris, Henricus Stephanus [1513].
Subject: Numerology.
Case/4A/1807/no. 3

BOUND WITH *Examen novarvm doctrinarivm* and *De bello et pace* (1523), n.a.

## Cockburn, William, 1669–1739.

*The present uncertainty in the knowledge of medicines, in a letter to the physicians in the commission for sick and wounded seamen. With a postscript to physicians, shewing the necessity of a true theory of diseases.*
London: printed by R.J. for Benj. Barker, 1703.

Subject: Diagnosis, disease, materia medica, medicine.
Case/F/4557/.655

No. 23 in a series of misc. tracts, with binder's title: *Pamphlets.*

### Cocker, Edward, 1631–1675.

*The pen's transcendency: or fair writings store-hovse. Furnished with examples of all the curious hands practised in England, and the nations adjacent. . . . Also, so many and plain directions for all hands, and such rare discoveries of the secrets of art, as will enable all that are ingenious to write any hand compleatly without a teacher. With the explication of a geometrical scheme, which comprehends the whole body and mysterie of fair writing. Likewise a choice receipt for ink, and to write with gold.*
London: Are to be sold by R. Walton, 1660.
STC II C4850(?)
Subject: Geometry of lettering, ink making.
Wing/+ZW/645/.C64

"The excellency of a curious hand consisteth in the geometricall proportion of each letter, with true distances and a cleare cariage thereof" (example no. 30).

### Cocker, Edward.

*Art's glory, or the pen-man's treasury . . . with directions, theorems, and rare principles of art . . . also a receipt for ink, and to write with gold.*
London: John Overton, 1674.
STC II C4831A
Subject: Ink making.
Wing/ZW/645/.C633

### Cocker, Edward.

*Penna volans, or the young mans accomplishment being the quintessence of those curious arts writing & arithmetick whereby ingenious youths may soone be made for cleark-ship fit, or management of trade.*
London: printed for John Ruddiard, 1661.
STC II C4849
Subject: Arithmetic, pen making.
Wing/ZW/645/.C 638

"On the excellencies of that exquisite art of arithmetick" (p. 15).

### Cocles [i.e. Bartolommeo Della Rocca], (called), 1467–1504.

*Chiromantia. Expositione del Tricasso Mantuano sopra il Cocle.*
[Venice: Helisabetta de Rusconi, 1525.]
Subject: Astrology, chiromancy, physiognomy, prognostication.
Wing/ZP/535/.R902

In 3 books.

### Cocles [i.e., Bartolommeo Della Rocca], (called).

*Physiognomiae Barptolomeo Coclite Bononiensi conscriptum, cum absolutissima chiromantiae ratione Andrea Corui Mirandulani, nunc uero recens, summa diligentia, cum pluribus distinctum & complectum capitibus, tum à foedis repurgatum mendis prodit in lucem per I. Multagrum.*
Strassburg: M. Iacobus Camerlander [1541].
Subject: Chiromancy, humors, physiognomy, physiology.
Case/*B/56/.18

### Codronchi, Giovanni Battista, fl. 1597–1620.

*De vitiis vocis, libri dvo. In quibus non solum vocis definitio traditur, & explicatur, sed illus differentiae, instrumenta, & cause aperiuntur. Vltimo de vocis conseruatione, praeseruatione, ac vitiorum eius curatione tractatur. . . .Cui accedit consilium de Raucedine, ac methodus testificandi, in quibusuis casibus medicis oblatis, postquam formulae quaedam testationum proponantur. Opusculum non modo neotericis medicis, sed & iurisperitis, ac iudicibus vtilissimum.*
Frankfurt: heirs of A. Wechel (Claudius Marinus & Ioannes Aubrius), 1597.
Subject: Disease, consilia, voice.
Case/3A/47

Bk. 1: Anatomy, kinds of voices; bk. 2: Diseases of the voice and their cure. Consilium of Ravcedine: remedies, forensic medicine(?)

### [Codrus, Antonio Urseo, called] 1446–1500.

*Orationes.*
Bologna: Ioannes Antonius Platonides [1502]
Subject: Medicine, supernatural, veterinary medicine.
Case/fY/682/.C638

Ed. by Philippo Beroaldo. MS marginalia. "In hoc primo Codri sermone de his rebus mentio sit. De metamorphosi humana in beluas. . . De medicis, clinicis, chirurgis & ueterinariis."

### Codrus, Antonio Urseo, called.

*Orationes, seu sermones vt ipse appellabat.*
[Venice: P. Liechtenstein, 1506]
Subject: Medicine, supernatural, veterinary medicine.
Case/fY/642/.A8479

Dedication signed "Philippus Beroaldus." MS marginalia. Contents as in 1502 ed. BOUND WITH Aristotle, *Habentvr hoc volvmine Theodoro Gaza interprete* and Codrus, qq.v.

### Coeffeteau, Nicolas, bp. 1574–1623.

*A table of humane passions. With their causes and effects.*
London: printed by Nicholas Okes, 1621.
STC 5473
Subject: Passions of the mind, psychology.
Case/B/529/.18

Trans. Edward Grimeston. Love, hatred, choler cupidity, pleasure, etc.

### Coeffeteau, Nicolas, bp.

*Tableav des passions hvmaines, de levrs cavses et de levrs effets.*
Paris: Sebastien Cramoisy, 1626.
Subject: Passions, psychology.

B/529/.179

Edition nouuelle, reueuë, & augmentée. What the passions are, how many there are, what kind, good and evil passions, and passions appropriate to age of individual. (1st ed. published 1619).

## Cogan, Thomas, 1545?–1607.

*The hauen of health, chiefly made for the comfort of students, and consequently for all those that haue a care of their health, amplified vpon fiue words of Hippocrates, written Epid. 6. labour, meat, drink, sleepe, Venus. . . . Hereunto is added a preseruation from the pestilence with a short censure of the late sicknesse at Oxford.*
London: Melch. Bradvvood for Iohn Norton, 1612.
STC 5483
Subject: Diet, hygiene.
Case/Q/.183

A "Table containing the effect of the whole booke," in which the 5 key subjects are subdivided, e.g., Labor into body, mind, Sleep into time, place, lying of body, amount of sleep.

## Coler, Johann, d. 1639.

*Oeconomia ruralis et domestica. Darinn das ganz ampt allertrewer hauss-vättern und hauss-mutter beständiges und allgemeines hauss-buch, vom hausshalten, wein-acker, gärten, blumen, und feld, bau begriffen auch wild un vögelfang, weid werck, fischereyen, vie hezucht, holtzfällung, und sonsten von allern was zu bestellung und regirung wolbestellen mäyer hofs, länderey, gemeinen feld und hausswessens nützlich und vonnöhten seyn möchte.*
Frankfurt: Johann Baptista Schönwetters, 1680.
Subject: Agriculture, encyclopedias, medicine.
Case/fR/50/.184

In 3 books. Bk. 1 includes distilling, fishing, hunting; bk. 2: "Haus artzney"; bk. 3: plague, purging, bloodletting, interpretation of dreams.

* * *

*A collection of scarce and valuable treatises, upon metals, mines, and minerals.*
London: printed by C. Jephson for Olive Payne, 1738.
Subject: Metallurgy, mines and mining.
Ayer/108/C6/1738

In 4 parts. Pts. 1 and 2: Containing the art of metals written originally in Spanish by Albaro Alonso Barba, director of the mines at Potosí in the Spanish West Indies. Trans. earl of Sandwich, 1669 (includes chaps. on generation of stones and metals); pt. 3: G. Plattes, discovery of mines (from gold to coal); pt. 4: Houghton's compleat miner.

* * *

*A collection of voyages and travels, some now first printed from original manuscripts.*
London: printed for Awnsham and John Churchill, 1704.
Subject: Discovery, travel.
Bonaparte/Collection/No. 10,920

8 vols. Vol. 1 contains a catalogue and character of most books of travels (a bibliographic essay, with the books mentioned being divided by language) and a miscellany of tracts on, e.g., fishery, history of navigation; vol. 2: travels by, e.g., John Nieuhoff, John Greaves, Christopher Borri; vol. 3 includes Alonso de Ovalle on Chile, naval tracts by Sir William Monson; vol. 4 works by, e.g., Pelham, William Ten Rhyne; vols. 5 and 6 dated 1732; vols. 7 and 8 dated 1747; vol. 8 published by Thomas Osborne.

* * *

*A collection of voyages and travels, consisting of authentic writers in our own tongue, which have not before been collected in English. . . . And continued with others of note, that have published histories, voyages, travels, journals or discoveries . . . relating to any part of the continent of Asia, Africa, America, Europe, or the islands thereof, from the earliest account to the present time . . . compiled from the . . . library of the earl of Oxford.*
London: printed for Thomas Osborne, 1745.
Subject: Discovery, travel.
Case/fG/12/.385

2 vols. With an introductory discourse concerning geography (vol. 1).

## Colonna, Francesco, 1432–ca.1527.

*Hypnerotomachia Poliphili, vbi hvmana omnia non nisi somnivm esse docet. Atqve obiter plvrima scitv sane qvam digna commemorat.*
[Venice: Aldus Manutius, 1499.]
H-C *5501^ BMC (XV)V:561–62^ GW VI, 7223^ Pr. 5574^ Still. C669^ Goff C-767
Subject: Alchemy, dream symbolism.
Inc./f*5574

In Italian, with some Latin headings. Has a second title page: "Poliphili Hypnerotomachia, vbi hvmani omnia non nisi somnivm esse ostendit, atque obiter plvrima scitv saneqvam digna commemorat."

## [Colonna, Francesco]

*La Hypnerotomachia di Poliphilo cioé pvgna d'amore in sogno. Dov' egli mostra, che tvtte le cose hvmane non sono altro che sogno: & doue narra molt' altre cose degne di cognitione.*
Venice: Aldus sons, 1545.
Subject: Alchemy, dream symbolism.
Wing/fZP/535/.A363

"Ristampato di novo, et ricorretto."

## [Colonna, Francesco]

*La Hypnerotomachia di Poliphilo cioé pvgna d'amore in sogno. Dov' egli mostra, che tvtte le cose hvmane non sono altro che sogno: & doue narra molt' altre cose degne di cognitione.*
Venice: Aldus sons, 1545.
Subject: Alchemy, dream symbolism.
Case/6A/74

"Ristampato di novo, et ricorretto."

## Colonna, Francesco.

*Hypnerotomachie, ov discours du songe de Poliphile, deduisant comme l'amour le combat a l'occasion de Polia. Soubz la*

*fiction de quoy l'aucteur monstrant que toutes choses terrestres ne sont que vanité, traicte de plusieurs matieres profitables & dignes de memoire. Nouuellement traduict de langage italien en francois.*
Paris: Iacques Keruer, 1554.
Subject: Alchemy, dream symbolism.
Wing/fZP/539/.M39

2d French ed. Ed. Jacques Gohory. Revised by Jean Martin.

## [Colonna, Francesco]

*Le tableav des riches inventions couuertes du voile des feintes amoureuses, qui sont representées dans le songe de Poliphile desvoilées des ombres du songe, & subtilement exposées par Beroalde* [de Verville, François].
Paris: Matthiev Gvillemot, 1600.
Subject: Dream interpretation, mysticism.
Case/Y/712/.C712

Has "Recveil steganographiqve, contenant l'intelligence du frontispiece de ce livre."

## [Colonna, Francesco Maria Pompeo] d. 1726.

*Les principes de la nature, suivant les opinions des anciens philosophes, avec un abregé de leurs sentimens sur la composition des corps: ou l'on fait voir que toutes leurs opinions sur ces principes, peuvent se réduire aux deux sectes, des atomistes, et des academiciens.*
Paris: André Cailleau, 1725.
Subject: Atomism, matter and spirit, natural philosophy, numerology.
B/1/.178

2 vols. Vol. 2, pp. 164–66: "Opinions des anciens philosophes, raportées par Macrobes, Songe de Scipion, ch. 14."

## [Colonna, Francesco Maria Pompeo]

*Les secrets les plus cachés de la philosophie des anciens, découverts et expliqués, a la suite d'une histoire des plus curieuses. Par M. Croisset de la Haumerie* [pseud.].
Paris: d'Houry son, 1722.
Subject: Alchemy, spagyric medicine.
Case/B/8633/.18

Tracts in "natural order": 1. Metallic seeds in the earth; 2. Extraction methods from 3 kingdoms, animal, vegetable, mineral, with rules for preserving health; 3. Effects of poorly compounded medicines; 4. Pure essence of gold essential for Great Work; 5. Obscure language used in Great Work explained; 6. Description of vessels used by alchemists; 7. Explanation of different kinds of fires, natural, unnatural, against nature; 8. Necessity of following nature, using the metallic mode; 9. Necessity of having material to work on, and also an object on which to meditate.

## Columna, Pietro, fl. 1480–1539.

*Petri Galatini opus de arcanis Catholicae veritatis, hoc est, in omnia difficilia loca Veteris Testamenti, ex Talmud, alijsque Hebraicis libris . . . absolutissimus commentarius. Ad haec Ioannis Reuchlini Phorcensis de arte cabalistica libri tres, omnigena eruditione pleni.*
Basel: Ioannes Hervagivs, 1550.
Subject: Cabala, mysticism.
Case/C/54/.185

Reuchlin has half title only, begins p. 719, sig. ff4.

## Comenius, Johann Amos, 1592–1670.

*Naturall philosophie reformed by divine light: or, a synopsis of physicks: by J. A. Comenius: exposed to the censure of those that are lovers of learning, and desire to be taught of God. Being a view of the world in generall, and of the particular creatures therein conteined: grounded upon scripture principles. With a brief appendix touching diseases of the body, mind and soul, with their generall remedies.*
London: printed by Robert and William Leybourn for Thomas Pierrepont, 1651.
STC II C5522
Subject: Mysticism, natural history, natural science, natural theology.
Case/B/76/.18

In 12 books, including, bk. 4: the qualities of things (with alchemical implications); bk. 8: stars, meteors, minerals; bk. 9: plants; bk. 10: living creatures; bk. 11: man.

## Comenius, Johann Amos.

*Orbis sensvualivm pictus: hoc est, omnium fundamentalium in mundo rerum, & in vitâ actionum, pictura & nomenclatura. John Amos Comenius's visible world, or a picture and nomenclature of all the chief things that are in the world; and of mens employments therein. A work newly written by the author in Latine, and high-Dutch (being one of his last essays, and the most suitable to childrens capacities of any that he hath hitherto made) and translated into English by Charles Hoole, M.A. for the use of young Latine-scholars.*
London: printed for Charles Mearne, 1685.
STC II C5525
Subject: Dictionary encyclopedias.
Wing/ZP/645/.M46

Latin-English illustrated dictionary with definitions of sciences, arts, anatomy, technical terms.

## Comenius, Johann Amos.

*Orbis sensvalivm pictus trilinguis . . . Latina, Germanica, Hungarica.*
Nuremberg: Martin Endter, 1708.
Subject: Dictionary encyclopedias.
Case/X/675/.188

## Comenius, Johann Amos.

*Orbis sensualium picti. Hoc est: omnium fundamentalium in mundo rerum, & in vita actionum, pictura et nomenclatura. Der sichtbaren welt.*
Nuremberg: Martin Endter, 1729–30.
Subject: Dictionary encyclopedias.
Wing/ZP/747/.E565

Latin-German. In 2 pts., with separate title pages. Pt. 2 title page: "Orbis sensualium picti denuò aucti pars secunda, CL. figuris instructa & illustrata, in qua tyronibus facillimâ methodo & summâ voluptate ingens elegantium phrasium ac rarismorum terminorum artium in prima parte non extantium, copia instillari potest" (1730).

## Comenius, Johann Amos.

*A reformation of schooles, designed in two excellent treatises: the first whereof summarily sheweth, the great necessity of a generall reformation of common learning. What grounds of hope there are for such a reformation. How it may be brought to passe. The second answers certaine objections ordinarily made against such undertakings, and describes the severall parts and titles of workes which are shortly to follow. Written many yeares agoe in Latin. . . . And now upon the request of many translated into English, and published by Samuel Hartlib.*
London: printed for Michael Sparke senior, 1642.
STC II C5529
Subject: Educational reform.
Case/I/4012/.19

## Comiers, Claude, d. 1693.

*Pratique curieuse, ou les oracles des sibylles, sur chaque question proposée. Tirée des manuscrits de la bibliotheque de Mr. Comiers.*
The Hague: Abraham Troyel, 1694.
Subject: Astrology, fortune-telling, numerology, prognostication.
Case/B/863/.19/no. 1

* * *

*The compleat book of knowledge. Treating of the wisdom of the antients; and shewing the various and wonderful operations of the signs and planets, and other celestial constellations, on the bodies of men, women and children; and the mighty influences they have upon those that are born under them. Compiled by the learned Albubetes, Benesaphan, Erra Pater, and other of the antients. To which is added, the country man's kalendar; with his daily practice, and perpetual prognostication for weather, according to Albumazar, Ptolomy, and others. Together with a catalogue of all the market-towns, fairs, and roads in England and Wales.*
London: W. Onley, 1698.
STC II C5629
Subject: Almanacs, astrology.
Case/A/15/.18

## Cooper, Thomas, fl. 1626.

*The mystery of witchcraft. Discouering, the truth, nature, occasions, growth and power thereof. Together with the detection and punishment of the same. As also, the seuerall stratagems of Sathan, ensnaring the poore soule by this desperate practize of annoying the bodie. . . . Very necessary for the redeeming of these atheisticall and secure* [sic] times.
London: printed by Nicholas Okes, 1617.
STC 5701
Subject: Witchcraft.
Case/B/88/.196

In 3 books. Bk. 1: Nature, causes, effects of witchcraft (charms, etc.); bk. 2: Detection and conviction of witches; bk. 3: Evils of the present day.

## Cordo, Simon A. Genuensis, fl. 1288.

*Clavis sanationis.*
Venice: Guglielmus Anima mia, Tridinensis, 1486.
H-C *14749^ BMC (XV)V:410^ Pr. 5109^ Still. S477^ Goff S-528
Subject: Drugs, herbal medicines, materia medica.
Inc./f5109

In alphabetical order, dictionary style. Coloph: Guielmus de Tridino ex Monteferato, 1486.

## Cordovero, Moses Ben Jacob, 1522–1570.

*'Or negerav.*
[Furth: Model of Ansbach, 1701]
Subject: Cabala.
Case/C/13/.19

BOUND WITH Mordecai Ben Judah Löb Ashkenazi, fl. 1701, q.v.

## Cornaro, Luigi, 1467?–1566.

*Discorsi della vita temperate sobria del sig. Lvigi Cornaro. Ne' quali con l'essempio di se stesso dimostra con quai mezzi possa l'huomo conseruarsi sano insin' all' vltima vecchiezza.*
Rome: Giacomo Mascardi, 1616.
Subject: Health, hygiene, temperance.
Case/Q/.187

## Corneille, Thomas, 1625–1709.

*Le dictionnaire des arts et des sciences par M. D. C. de l'Académie françoise.*
Paris: Jean Baptiste & widow Coignard, 1694.
Subject: Dictionary encyclopedias.
Case/+A/24/.194

2 vols.

## Cornut, Jacques Philippe, 1606–1651.

*Iac. Cornvti doctoris medici parisiensis Canadensivm plantarvm, aliarúmque nondum editarum historia. Cui adiectum est ad calcem enchiridion botanicvm parisiense, continens indicem plantarum, quae in pagis, siluis, pratis, & montosis iuxta Parisios locis nascuntur.*
Paris: Simon Le Moyne, 1635.
Subject: Botany, New World.
Ayer/*109.9/B6/C81/1635

Plates accompany descriptions of plants of Canada's various districts.

## Coronelli, Marco Vincenzo, 1650–1718.

*Epitome cosmografica, o compendiosa introduttione all' astronomia, geografia, & idrografia, per l'usq, dilucidatione, e fabbrica delle sfere, globi, planisferj, astrolabj, e tavole geografiche, e particolarmente degli stampati, e spiegati nelle*

*publiche lettioni dal P. Maestro Vincenzo Coronelli M. C. cosmografo della . . . republica di Venetia, e lettore di geografia in quella università, per l'Accademia Cosmografica degli Argonauti.*
Cologne, 1693.
Subject: Cosmography, geography, instrumentation.
Ayer/7/C 78/1693

In 3 books. P. 376: "Della fabbrica, & vso dell' astrolabio armillare." With volvelle.

## Corrozet, Gilles, 1510–1568.

*Hecatomgraphie. C'est à dire les descriptions de cent figures & hystoires, contenans plusieurs appophtegmes prouerbes, sentences & dictz tant des anciens que des modernes.*
Paris: Denys Ianot, 1541.
Subject: Emblems.
Case/3A/533

Emblems include "Contre les astrologues," sig. K.vi (verso)–K.vii. "Contre les magiciens," H.iv (verso).

## Corte, Claudio.

*Il cavallerizzo di Clavdio Corte da Pavia, nel qual si tratta della natura de' caualli, delle razze, del modo di gouernarli, domarli, & frenarli. Et di tvtto qvello, che à caualli, & à buon caualerizzo s'appartiene.*
Venice: Giordano Ziletti, 1573.
Subject: Animal husbandry, veterinary medicine.
Case/V/137/.194

## Cortés, Hernando, 1485–1547.

*De insvlis nvper inventis Ferinandi Cortesii ad Carolum V. Rom. imperatorem narrationes, cum alio quodam Petri Martyris ad Clementem pontificem maximum consimilis argumenti libello.*
[Cologne: Arnold Birckman], 1532.
Church, no. 63^ Harrisse 168^ John Carter Brown, II:103
Subject: Discovery, exploration, natural history, New World.
Ayer/*655.51/C8/1532

"De insvlis nvper inventis . . . lucidissima Petri Martyris narratio. Ferdinandi Cortesii de nova maris oceani Hispania narratio secunda. Tertia Ferdinandi Cortesii narratio. De fratrvm minorvm regvlaris observantae profectv & animarum lucro in Huketan siue Noua Hispania, epistola venerandi Martini de Valentia."

## Cortés, Hernando.

*Ferdinandi Cortesii. Von dem Newen Hispanien so im meer gegem nidergang zwo gantz lustige vnnd fruchtreiche historien an den grossmächtigisten vnüberwindtlichisten herren Carolvm. v. Römischen kaiser &c. könig in Hispanien &c. Die erst im M.D.XX. jar zügeschriben in wellicher grundtlich vnd glaubwirdig erzelt wirt der abendtländern . . . Die andere im 1524. jar wie Temixtitan so abgefallen wider erobert nachmals andere herrliche syg sampt der erfindung des meers svr so man für das indianisch meer achtet.*
Augsburg: S. Vlrich for Philipp Vlhart, 1550.
Subject: Discovery, natural history, New World, West Indies.
Ayer/*655.51/C8/1550

"Die andere," is taken from Peter Martyr *De orbe novo*, 4th decade. Volume also contains second and third narrations of Cortez, and letter by Oviedo.

## Cortès, Jeronimo, fl. 1600.

*Libro y tratado de los animales terrestres y volatiles, con la historia, y propriedades deilos; alabando de cada vno de los terrestres la virtud en que mas se auentajò, y señalò: con autoridad de doctos, y santos.*
Valencia: Iu[a]n Chrysostomo Garriz, 1615.
Subject: Zoology.
Case/O/500/.195

In 2 parts: terrestrial animals and flying creatures (including insects).

## Cortés, Martin, fl. 1551.

*Breue compendio de la sphera y de la arte de nauegar con nueuos instrumentos y reglas exemplificado con muy subtiles demonstraciones: compuesto por Martin Cortes natural de burjalaroz en el reyno de Aragon y de presente vezino de la ciudad de Cadiz.*
[Seville: Anton Aluarez, 1551]
Subject: Astronomy, latitude and longitude, navigation, tides.
Ayer/*7/C8/1551

In 3 parts. Pt. 1: La composicion del mundo y de los principios vniuersales que para el arte de la nauigacion se requieren; pt. 2: De los mouimientos del sol y de la luna y delos effectos que de sus mouimientos se causan; pt. 3: De la composicion y vso de instrumentos del arte de la nauegacion.

## Corvinus, Laurentius, 1465 ca.–1527.

*Dominici marii nigri veneti geographiae libri commentariorum XI, nunc primum in lvcem magno stvdio editi, qvibus non solum orbis totius habitabilis loca, regiones, prouinciae, urbes, montes, insulae, maria, flumina, & caetera, ut nostro tempore sunt sita & denominata, uerum etiam omnium ferè populorum, & uariarum gentium mores . . . ita ut uel ipso Strabone utilior nostris temporibus, autor hic doctorum quorundam iudicio meritò habeatur. Vna cvm Lavrentii Corvini Nouoforensis geographia. Et Strabonis epitome per D. Hieronymvm Gemvsaevm translata, quam adiecimus ut quo cum marium hunc nostrum lector conferat, habent.*
Basel [Henrichvs Petri, 1557].
Subject: Astronomy, geography.
Case/fG/117/.62

4 tracts: "Dominici mari" (11 bks.); "Geographia ostendens"; "Strabonis geographicorum epitome, Hieronymo Gemusaeo interprete" (17 bks.); "Rvdolfvs Agricola iunior Rhetus & Iachimo Vadiano. De nonnullis locis dissertatio."

## Cosmographia, Aethici.

*Aethici Cosmographia: Antonii Avgvsti itinerarivm provinciarvm: ex bibliotheca P. Pithoei, cum scholiis Iosiae Simleri.*
Basel [T. Guarinus], 1575.
Subject: Cosmography.
Case/–G/112/.195

Contents include "Itinerarij Antonini Pii fragmentum. Rutilij Claudij Numatiani ... Liber cui titulus itinerarium. Vbi sequestri liber de fluminibus, fontibus, lacubus, montibus, nemoribus." P. 32: "Aethici alia totivs orbis descriptio."

## Costa, Manuel Goncalves da, 1606–1688.

*Noticias astrologicas & vniuersal influencia das estrellas.... Em particvlar prognostico deste reyno do anno de 1660 ... & eleger os meyos da conseruaçaõ da saude fructos da terra, tratos do mar.*
Lisbon: Antonio Craesbeeck, 1659.
Subject: Astrology.
Greenlee/4504/P855

## Cotta, John, 1575?–1650?

*Infa[ll]ible true and assured vvitch; or the second edition, of the tryall of witchcraft.*
London: printed by I. L[egatt] for R. Higgenbotham, 1624.
STC 5837
Subject: Disease, sorcery, witchcraft.
Case/B/8845/.19

"Whether the diseased are bewitched. The marks of witches ... how to be tried and known from all naturall diseases. The necessitie of consulting with the physition ... in all diseases supposed to be inflicted by the diuell" (from chap. 10, "How men may by reason and nature be satisfied, concerning such sicke persons as are indeed and truly bewitched").

## Cousin, Jean, 16th cent.

*Livre de perspective de Iehan Cousin senonois, maistre painctre à Paris.*
Paris: Iehan le Royer, 1560.
Subject: Plane and solid geometry, perspective.
Wing/fZP/539/.L565

## [Covarrubius Horozco, Sebastian de] fl. 1611.

[*Emblemas morales.*]
[Madrid: L. Sanchez, 1610]
Subject: Emblems.
Case/W/1025/.253

Includes emblem against alchemy (fol. 11), forqe, archipendula, and various depictions of machines and levers.

## Cramer, Daniel, 1568–1637.

*Octoginta emblemata moralia nova, e sacris literis petita, formandis ad veram pietatem accommodata, & elegantibus picturis aeri incisis repraesentata.*
Frankfurt: Lvca Jennisius, 1630.
Subject: Emblems.
Case/W/1025/.2

Emblems depiciting monster or devil (p.181), music and medicine (p.73), earth and heaven's mystery (p.41).

## Crescenzi, Pietro de, 1230–1321.

*Il libro della agricultura.*
Florence: Nicolaus Laurentius, Alamanus, 1478.
H-C-R 5837^ BMC (XV)VI:627^ GW VII, 7826^ Pr. 6115^ Still. C864^ Goff C-973
Subject: Agriculture, animal husbandry, botany, veterinary medicine.
Inc./f6115

In 12 books. Bk. 6: Herbal, kitchen garden; bk. 9: Animal husbandry; bk. 10: Falconry, veterinary medicine; bk. 12: Calendar listing activities appropriate to each month.

## Crescenzi, Pietro de.

*Il libro dell agricultura.*
Vicenza: Leonhardus Achates, 1490.
H-C 5838^ BMC (XV)VII:1033^ GW VII, 7827^ Pr. 7128^ Still. C865^ Goff C-974
Subject: Agriculture, animal husbandry, botany, veterinary medicine.
Inc./f7128

## Crescenzi, Pietro de.

[*Liber ruralium commodorum*]
[Augsburg: Johann Schüssler, 1471.]
H-C 5828^ BMC (XV)II:328^ Pr. 1590^ Goff II C-965
Subject: Agriculture, animal husbandry, botany, veterinary medicine.
NL/Inc./1590/(+)

## Crescenzi, Pietro de.

*Pietro Crescentio tradotto novamente per M. Francesco Sansovino nel quale si trattano le cose della villa. Con le figvre delle herbe poste nel fine. Con vn vocabolario delle voci difficili che sono in questa opera, & con i dissegni de gli stromenti co quali si cultiua & si lauora la terra.*
Venice: Francesco Rampazetto, 1564.
Subject: Agriculture, animal husbandry, botany, veterinary medicine.
Case/R/50/.204

In 12 books.

## Crescenzi, Pietro de.

*Vom ackerbaw erdtwücher und bawleüte. Von natur, art, gebrauch und nutzbarkeit aller gewechsz früchten, thyeren sampt allem dem so dem menschen dyenstlich in speysz und artzenyung.*
Strassburg: Hans Knoblouch the younger, 1531.
Subject: Agriculture, animal husbandry, botany, veterinary medicine.
Case/f*R/50/.202

BOUND WITH Celsus, *Die acht bücher ...*, q.v.

## Crescenzi, Pietro de.

*Von dem nutz der ding die in äckeren geburt werden. Von nutz & buwleut. Von natur, art, gebruch und nutzbarkeit aller gewächsz früchten, thyereren, und alles des der mensch geleben, oder in dienstlicher übung haben soll.*
[Strassburg: Joannes Schott for Johannes Knoblouch & Paul Götz] 1518.
Subject: Agriculture, animal husbandry, botany, veterinary medicine.
Case/f*/R/50/.2

In 12 books.

## [Crouch, Nathaniel] 1637?–1725?

*Admirable curiosities, rarities, and wonders, in Great-Britain and Ireland. Being an account of many remarkable persons and places; and likewise of the battles, sieges, earthquakes, inundations, thunders, lightenings, fires, murthers, and other considerable occurrences, and accidents for several hundred years past. With the natural and artificial rarities in every country, and many other observable passages, as they are recorded by credible historians of former and latter ages. By Robert Burton*
London: Printed for A. Bettesworth, 1728.
Subject: Geography, natural phenomena.
Case/G/4496/.2/[pseud.]

9th ed.

## Crusius, Thomas Theodorus, 1648–1728.

*De philologia studiis liberalis doctrinae, informatione & educatione litteraria generosorum adolescentum, comparanda prudentia juxta & eloqventia civili, libris & scriptoribus ad eam rem maximè aptis, qvôqve ordine scriptorum historiae Romanae monumenta sint legenda, tractatus Gvilielmi Bvdaei, Thomae Campanellae, Joachim Pastorii, Joh. Andreae Bosii, Joh. Schefferi, & Petri Angelii Bargaei. Quos Thomas Crenius collegit, recensuit, emendavit, in incisà sive commata distinxit, & notis suis.*
[Leyden]: David Severini, 1696.
Subject: Arithmetic, humanistic studies, medical education.
I/89/.201

Includes Campanella, "De libris propriis," pt. 3.

## Crusius, Thomas Theodorus.

*De philologia studiis liberalis doctrinae.* [Another copy]
Case/4A/.714

## Cubero Sebastián, Pedro, b. 1645.

*Breve relacion de la peregrinacion qve ha hecho de la mayor parte del mvndo Don Pedro Cvbero Sebastian, predicado apostolico del Asia, natural del reyno de Aragon; con las cosas mas singulares que le han sucedido, y visto, entre tan barbaras naciones, su religion, ritos, ceremonias, y otras cosas memorables, y curiosas que hà podido inquirir; con el viage por tierra, desde España, hasta las Indias Orientales.*
Madrid: printed by Iuan Garcia Infançon, 1680.
Subject: Anthropology, voyages.
Ayer/*2062/C96/1680

## Cubero Sebastián, Pedro.

*Peregrinacion qve ha hecho de la mayor parte del mundo Don Pedro Cvbero Sebastian.*
Zaragoza: printed by Pascal Bueno, 1688.
Subject: Anthropology, voyages.
Ayer/*2062/C96/1688

2d impression. P. 243, cap. 39: "Cuenta el autor un terrible terremoto, que huvo en las Islas Filipinas."

## Cudworth, Ralph, 1617–1688.

*An abridgment of Dr. Cudworth's true intellectual system of the universe. In which all the arguments for and against atheism are clearly stated and examined. To which is prefix'd, an examination of what that learned person advanc'd touching the doctrine of a trinity in unity, and the resurrection of the body.*
London, 1732.
Subject: Atomism, Hermeticism, laws of nature, natural theology.
B/245/.2

2 vols. In 14 chapters. Vol. I, chap. 4: Hermeticism and theology of Greeks and Egyptians.

## Cudworth, Ralph.

*Systema intellectvale hvivs vniversi sev de veris natvrae rervm originibvs comentarii qvibvs omnis eorvm philosophia, qvi devm esse negant, fvnditvs evertitvr. Accedvnt reliqva eivs opvscvla. Ioannes Lavrentivs Moshemivs, Theol. D. seren. Dvcis Brvnsv. A consiliis rervm sanctiorvm, abbas coenobiorvm vallis S. Mariae et lapidis S. Michaelis, reliqva omnia ex anglico Latine vertit, recensvit, variisqve observationibvs et dissertationibvs illvstravit at avxit.*
Jena: widow Meyer, 1733.
Subject: Atomism, Hermeticism, laws of nature, natural theology.
B/245/.207

2 vols in 1.

## Cudworth, Ralph.

*The true intellectual system of the universe; the first part; wherein all the reason and philosophy of atheism is confuted; and its impossibility demonstrated.*
London: printed for Richard Royston, 1678.
STC II C7471
Subject: Atomism, Hermeticism, laws of nature, natural theology.
Case/fB/245/.198

From Preface, sig. ***: "Knowledge, or science, added to this faith (according to the Scripture advice) will make it more firm and steadfast." Attempted reconciliation of God and science, refutation of the necessity of atheism. Discussion of Hermeticism, Casaubon (pp. 320 ff.).

## Cudworth, Ralph.

*The true intellectual system of the universe: The first part; wherein all the reason and philosophy of atheism is confuted, and its impossibility demonstrated.*

London: printed for J. Walthoe [etc.] 1743.
Subject: Atomism, Hermeticism, laws of nature, natural theology.
B/245/.202

2d ed. With an account of the life and writings of the author, by Thomas Birch.

## Culpeper, Nicholas, 1616–1654.

*Culpeper's last legacy: left and bequeathed to his dearest wife, for the publick good, being the choycest and most profitable of those secrets which while he lived were lockt up in his breast, and resolved never to be publisht till after his death. Continuing sundry admirable experiences in several sciences, more especially in chyrurgery and physick: viz. compounding of medicines; making of waters, syrups, oyles, electuaries, conserves, salts, pills, purges, and trochischs. With two particular treatises; the one of feavers, the other of pestilence: as also rare and choyce aphorisms and receipts fitted to the understanding of the meanest capacities.*
London: printed by Tho. Ratcliffe for Nath. Brooke, 1668.
STC II C7520
Subject: Materia medica, medicine.
Case/Q/.19

4th impression.

## Culpeper, Nicholas.

*Culpeper's school of physick: or the experimental practice of the whole art. Wherein are contained all the inward diseases from the head to the foot, with their proper and effectual cures; such diet set down as ought to be observed in sickness or in health. With other safe waies for preserving of life, in excellent aphorisms, and approved medicines, so plainly and easily treated of, that the free-born student rightly understanding this method, may judge of the practice of physick, so far as it concerns himself, or the care of others, &c.*
London: printed for Obadiah Blagrave, 1678.
STC II C7545
Subject: Diet, disease, iatrochemistry, herbal medicine, materia medica.
Case/Q/.2

## Cuningham, William, b. 1531.

*The cosmographical glasse, conteyning the pleasant principles of cosmographie, geographie, hydrographie, or nauigation.*
London: John Day, 1559.
STC 6119
Subject: Cosmography, geography, hydrography, navigation.
Wing/fZP/545/.D27

In 5 books. Bk. 1: Definition of geography, cosmography; parts of the world, heavenly, elementary; sphere, geometrical principles; zodiac, equinoctial, tropics; bk. 2: Climate, longitude, latitude; bk. 3: "The making and protracture of the face of the earth"; bk. 4: Hydrography, navigation; bk. 5: Cosmographical descriptions of regions, provinces, islands, etc.

## Cureau de La Chambre, Marin, 1594–1669.

*L'Art de connoistre les hommes.*
Amsterdam: Iacques le Jeune, 1660.
Subject: Physiognomy, psychology.
Case/B/56/.198

In 2 books. Astrological signs (p. 199). Definition of metoposcopy (study of signs imprinted on forehead by stars). Chiromancy (study of signs imprinted on hands by stars).

## Cureau de La Chambre, Marin.

*The art how to know men. Originally written by the Sieur de la Chambre, counsellour to his majesty of France, and physician in ordinary. Rendered into English by John Davies of Kidwelly.*
London: printed for Thomas Basset, 1670.
STC II L128A
Subject: Physiognomy, psychology.
Case/B/56/.2

Has same engraved title page as 1660 French ed. but reversed and with title in English. Comparison between male and female anatomy: man's neck thicker, body more muscular. "The figure of the parts denotes the inclinations," i.e., largeness of shoulders and breast, openness of nostrils, and wideness of mouth denote courage.

## Cureau de La Chambre, Marin.

*A discovrse of the knowledg of beasts, wherein all that hath been said for and against their ratiocination, is examined.*
London: printed by Tho. Newcomb for Humphrey Mosele, 1657.
STC II L131.
Subject: Comparative psychology, imagination, reason in beasts.
Case/B/58/.2

In 4 parts.

## Cureau de La Chambre, Marin.

*Les characteres des passions par Sr. de la Chambre, medecin de Monseigneur le Chancelier.*
Paris: P. Rocolet & P. Blaise, 1640–1659.
Subject: Passions of the mind, psychology.
B/529/.193

3 vols. Vol. 1: 5 passions: love, joy, laughter, desire, hope. 6 simple passions: love, desire, pleasure, hate, aversion, sorrow (dated 1640); vol. 2: Character of the passions, treating of the nature and effects of the courageous passions (dated 1645); vol. 3: "Où il est traitté de la nature & des effets de la haine & de la dovlevr. De la haine des animavx qui est fondée sur les qualitez occultes" (dated 1659).

## Cureau de La Chambre, Marin.

*Les characteres des passions par Sieur de la Chambre, medecin de Monseigneur le Chancelier.*
Paris: P. Rocolet & P. Blaise, 1643.
Subject: Passions of the mind, psychology.
B/529/.194

In 6 chaps. Each chap. devoted to one of the passions: 1: character of the passions in general; 2: love; 3: joy; 4: laughter; 5: desire; 6: hope; from "Advis av lectevr": "Je veux examiner les passions, les vertus & les vices, les moeurs & les costumes des peuples, les diuerses inclinations des hommes, les temperamens, les traicts de leur visages; en vn mot, où je pretens mettre ce que la medecine, la morale, & la politique ont de plus rare & de plus excellent."

## Cureau de La Chambre, Marin.

*Les characteres des passions.*
Amsterdam: Antoine Michel, 1658.
Subject: Passions of the mind, psychology.
Case/B/529/.195

In 2 vols. Vol. 1 in 7 parts: 1. character of the passions; 2. Nature of animals who contribute to this scientific knowledge; 3. Beauty of men and women, their inclinations; 4. Difference between bodies and customs among people; 5. Temperaments and their effects on body and soul; 6. Connection between passions and habits; 7. Ordering all the signs in order to know men. Vol. 2: nature and effects of the courageous passions.

## Cureau de La Chambre, Marin.

*The characters of the passions. Written in French by the Sieur de la Chambre, physitian to the Lord Chancellor of France.*
London: printed by T. Newcomb for J. Holden, 1650.
STC II L129
Subject: Passions of the mind, psychology.
Case/B/529/.2

In 7 parts.

## Curvo Semmedo, João, 1635–1719.

*Atalaya da vida contra as hostilidades da morte; fortificada, e guarnecida com tantos desenssores, quantos saõ os remedios, que no discurço de sinocenta, & oyto annos experimentou Joaõ Curvo Semmedo, cavalleyro professo da ordem de Christo, familiar do santo officio, & medico da caza real.*
Lisbon: Ferreyrenciana, 1720.
Subject: Medical dictionary.
Greenlee/4532.1/M48/C98/1720

Alphabetical list of diseases, medicines, treatments.

## Cyrano de Bergerac, Savinien 1619–1655.

*Les Novvelles oevvres de Monsieur de Cyrano Bergerac, contenant l'histoire comiqve des etats & empires du soleil, plvsievrs lettres, et avtres pieces diuertissantes.*
Paris: Charles de Sercy, 1662.
Subject: Anatomy of the eye, cosmography, geometry, motion, physics, senses.
Y/762/.C99

Contains "fragment de physique ou la science des choses naturelles" (title pt. 2, p. 261). Espouses Copernican world view, based on Giordano Bruno. In 2 pts.; pt. 1: Physics; pt. 2: Cosmography.

# D

## Dalechamps Jacques, 1513–1588.

*Historia generalis plantarvm in libros XVIII per certas classes artificiose digesta, haec plusquam mille imaginibus plantarum locupletior superioribus, omnes propemodum quae ab antiquis scriptoribus, Graecis, Latinis, Arabibus, nominantur: necnon eas quae in orientis atque occidentis partibus, ante seculum nostrum incognitis, repertae fuerunt, tibi exhibet. Habes etiam earundum plantarum peculiaria diuersis nationibus nomina: habes amplas descriptiones, è quibus singularum genus, formam vbi crescant & quo tempore vigeant, natiuum temperamentum, vires denique in medicina proprias cognosces.*
Lyon: Gvlielmvm Rovilivm, 1586–87.
Pritzel 2035^ Arber, Herbals 98–99
Subject: Botany, herbals.
Case/*N/.405/21

2 vols. in 1. Vol. 1 imprint, 1587; vol. 2, 1586. Describes trees, grains and legumes, vines, shade plants, poisonous and medicinal plants, among others.

## Dampier, William, 1652–1715.

*A new voyage round the world. Describing particularly, the isthmus of America, several coasts and islands in the West Indies, the isles of Cape Verd, the passage by Terra del Fuego, the south sea coasts of Chili, Peru, and Mexico; the isle of Guam one of the Ladrones, Mindanao, and other Philippine and East-India islands near Cambodia, China, Formosa, Luconia, Celebes, &c. New Holland, Sumatra, Nicobar isles; the Cape of Good Hope, and Santa Helena. Their soil, rivers, harbours, plants, fruits, animals, and inhabitants. Their customs, religion, government, trade, &c.*
STC II D163
London: printed for James Knapton, 1698.
Subject: Anthropology, voyages.
Ayer/*118/D2/1698/vol. 1

3d ed. corrected. Illustrated with "maps and draughts."

## Dampier, William.

*Voyages and descriptions vol. II. In three parts, viz; 1. A supplement of the voyage round the world, describing the countreys of Tonquin, Achin, Malacca, &c. their product, inhabitants, manners, trade, policy, &c. 2. Two voyages to Campeachy; with a description of the coasts, product, inhabitants, logwood cutting, trade, &c. of Jucatan, Campeachy, New-Spain, &c. 3. A discourse of trade-winds, breezes, storms, seasons of the year, tides and currents of the torrid zone throughout the world: with an account of Natal in Africk, its product Negro's, &c.*
London: printed for James Knapton, 1700.
Subject: Anthropology, descriptive geography, voyages.
Ayer/118/D2/1700/vol. 2

2d ed. 2nd voyage to Campeachy describes the east coast of Campeche, "its vegetables, weather, animals, &c."

## Dampier, William.

*Voyages and descriptions vol. II.*
London: printed for James Knapton, 1705.
Subject: Anthropology, descriptive geography, voyages.
Ayer/3A/126

3d ed. of vol. 2.

## Daneau, Lambert, ca 1530–1595?

*Devx traitez novveavx, tres-vtiles pour ce temps. Le premier tovchant les sorciers, auquel ce qui se dispute auiourd'huy sur cette matiere, est bien amplement resolu & augmenté de deux proces extraicts des gresses pour l'eclaircissement & confirmation de cet argument. Le second contient vne breue remonstrance sur les ieux de cartes & de dez. Reueu & augmenté par l'auteur.*
[Geneva?] Jacques Bavmet, 1579.
Subject: Witchcraft.
Case/B/88/.214

## Daneau, Lambert.

*Les sorciers. Dialogve tres-vtile et necessaire povr ce temps. Avqvel ce qvi se dispute auiourdhui des sorciers & eriges, est traité bien amplement, & resolu.*
[Geneva] Iaques Bourgeois, 1574.
Subject: Witchcraft.
Case/B/88/.215

7 points discussed: 1) Meaning of the word,"sorcerer"; 2) Are there sorcerers? What kinds of people are they? 3) Over what things sorcerers have power; 4) How do they do their evil work and corrupt [people]? 5) What authority it is that condemns sorcerers, and to what punishment they are entitled; 6) Is it legal to help sorcerers overcome their malady? 7) How to protect oneslf from sorcerers.

## Danet, Pierre, d. 1709.

*A complete dictionary of the Greek & Roman antiquities; . . . also an account of their navigations, arts & sciences, & the inventors of them. . . . Compiled originally in French . . . by Monsieur Danet. Made English with the addition of very vseful mapps.*
London: printed for John Nicholson, 1700.
STC II D171
Subject: Archaeology, dictionary encyclopedias.
Case/F/0231/.214

## Danfrie, Phillippe, fl. 1558–1589.

*Declaration de l'usage du graphometre par la pratique duquel l'on peut mesurer toutes distances des choses. . . . A la fin de ceste declaration est adiousté par le dict Danfrie vn traicté de l'vsage du trigometre.*
Paris: Danfrie, 1597.

Subject: Instrumentation, mensuration, surveying.
Wing/ZP/539/.D183

### [Daniel, Gabriel] 1649–1728.

*A voyage to the world of Cartesius. Written originally in French, and now translated into English.*
London: printed by T. Bennet, 1692.
STC II D201
Subject: Motion, physics.
Case/Y/1565/.D2

Satirical treatment of Cartesian philosophy.

### [Daniel, Gabriel]

*A voyage to the world of Cartesius.*
London: Thomas Bennet, 1694.
STC II D202
Subject: Motion, physics.
Case/Y/1565/.D21

2d ed. In 4 pts. Trans. T. Taylor.

### Dannewaldt, Matthias.

*Cometologia oder historischer discurs, was von vielen seculis herauff cometische erscheinungen sich begeben ingleichen derselben kürtzliche betrachtung und was etwa der im Decembr. dieses 1664sten jahrs enstandene comet vor muthmatzliche bedeutung nach sich ziehen möchte. Mit beygefügten abrissen wie er zu Augspurg/Nürnberg/Hamburg und allhier zu Leipzig gesehen worden. Darbey auch der annoch als eine göttliche zorn ruthe am himmel stehende anderwertige comét fürzlich berühret.*
Leipzig: Christian Kirchner [1664].
Subject: Comets, history and description.
Case/B/863/.217

Mentions that comet of 1618 was accompanied by plague. Lists comet sightings from biblical times onward.

### Danti, Ignazio, 1537–1586.

*Dell' vso et fabbrica dell' astrolabio, et del planisferio. Di maestro Egnatio Danti publico lettore delle mathematiche nello studio di Bologna. Nuouamente ristampato, & accresciuto in molti luoghi, con l'aggivnta dell' vso, & fabbrica di noue altri istromenti astronomici, come nella faccia seguente si contiene.*
Florence: Giunta, 1578.
Subject: Astronomy, instrumentation.
Ayer/7/D19/1578

"Istromenti astronomici di nvovo aggivnti" includes Sfera armillare [of Ptolemy], Torquetto astronomico, Astrolabio armillare, gran regola astronomica, Quadrante astronomico, Armilla equinoziale, Diottra d'Hipparco, Gnomone astronomico, & geometrico, Anemoscopio verticale.

### [Daugis, Antoine Louis]

*Traité sur la magie, le sortilege, les possessions, obsessions et malefices, où l'on démontre la verité & la réalité; avec une methode sûre & facile pour les discerner, & les reglemens contre les devins, sorciers, magiciens, &c. ouvrage très-utile aux ecclesiastiques, aux medecins, & aux juges.*
Paris: Pierre Prault, 1732.
Subject: Astrology, magic, supernatural, witchcraft.
B/88/.219

In 2 books. Bk. 1: De la réalité de la magie; titre 7: Des magiciens, des sorciers, des devineurs, & l'astrologie judiciaire, & la division de ces arts en leurs differentes especes; bk. 2: Des possessions, obsessions, & malefices.

### Davies, Sir John, 1569–1626.

*Nosce teipsum. This oracle expounded in two elegies.*
London: printed by Richard Field for Iohn Standish, 1599.
STC 6355^ Pforzheimer 266.
Subject: Epistemology, human development, local motion, macrocosm–microcosm.
Case/3A/524

2d ed. 2 elegies: 1: Of humane knowledge; 2: Of the soule of man, and the immortalitie therof. 1st section deals with validity of scientific knowledge, and with epistemology of "humane knowledge." See Robert Schuler, *English magical and scientific poems to 1700: An annotated bibliography* (New York & London: Garland, 1979).

### [Davies, Sir John]

*Nosce teipsum.*
London: printed by Richard Field for Iohn Standish, 1602.
STC 6356
Subject: Epistemology, human development, local motion, macrocosm–microcosm.
Case/Y/185/.D318

"Newly corrected and amended."

### Davies, Sir John.

*Nosce teipsum.*
London: printed by Augustine Matthews for Richard Hawkins, 1622.
STC 6359
Subject: Epistemology, human development, local motion, macrocosm–microcosm.
Case/Y/185/.D32

### Davies, Sir John.

*The original nature and immortality of the soul.*
London: printed for W. Rogers, 1697.
STC II D405
Subject: Senses, soul.
Case/Y/185/.D322

### Davys, John, 1550?–1605.

*A rutter or briefe direction for readie sayling.*
In *Purchas, Samuel, his pilgrimes*, vol. I, bk. 4, pp. 444–55, q.v.

### Davys, John.

*The worldes hydrographical description. Wherein is proued not onely by aucthoritie of writers, but also by late experience of trauellers and reasons of substantiall probabilitie, that the*

***worlde in all his zones clymats and places, is habitable and inhabited, and the seas likewise vniuersally nauigable without any naturall annoyance to hinder the same whereby appeares that from England there is a short and speedie passage into the South Seas, to China, Molucca, Phillipina, and India, by northerly nauigation, to the renowne, honour and benefit of her Maiesties state and communalty.***
London: Thomas Dawson, 1595.
Church 249^ STC 6372
Subject: Discovery, navigation, northwest passage to the Orient.
Ayer/*7/D2/1595

"By late experience to proue that America is an iland, and may be sayled round about contrary to the former obiection" [sig. A7]; "To prooue by experience that the sea fryseth not" [sig. B7v].

## Deacon, John, and Walker, Iohn.

*Dialogicall discourses of spirits and divels. Declaring their proper essence, natures, dispositions, and operations: their possessions and dispossessions: with other the appendantes, peculiarly appertaining to those speciall points. Verie conducent, and pertinent to the timely procuring of some Christian conformitie in iudgement: for the peaceable compounding of the late sprong controuersies concerning all such intricate and difficult doubts.*
London: Geor. Bishop, 1601.
STC 6439
Subject: Occult, witchcraft.
Case/C/559/.22

BOUND WITH Deacon and Walker, *A svmmarie ansvvere*, q.v.

## Deacon, John, and Walker, Iohn.

*A svmmarie ansvvere to al the material in any of Master Darel his bookes. More especiallie to that one booke of his, intitluled, The doctrine of the possession and dispossession of demoniaks out of the word of God.*
London Geor. Bishop, 1601.
STC 6440
Subject: Occult, witchcraft.
Case/C/559/.22

BOUND WITH Deacon, John, and Walker, Iohn, *Dialogicall discourses*, q.v.

## Decembrio, Angelo, 1415 (ca.)–1462.

*Politiae literariae Angeli Decembrii ad summum pontificum Pium II. libri septem, multijuga eruditione refertissimi, ante annos octoginta plus minus scripti, & Rhomes in bibliotheca pontificis thesauri loco reconditi, clade uero Romana Carolo Borbonio & Georgio Fronspergio ducibus clarissimis, anno 1527, eruti, & per nos magnos labore & diligentia in lucem aediti, quod omnibus studiosis faustum, foelixque sit.*
Augsburg: Henricus Steyner, 1540.
Subject: Botany.
Case/fX/0163/.221

1st printed ed. I am indebted for this citation to the late Jon Pearson Perry, "Practical and ceremonial uses of plants materials as 'literary refinements' in the libraries of Leonello d'Este and his courtly literary circle. Angelo Decembrio's *De politia literaria*, book 1, part 3, and book 2, part 21," *La Bibliofilia* 91, no. 2 (July 1989). Decembrio describes the use of certain aromatic plants and flowers to discourage insect infestation in his library. Decembrio, younger brother of Per Candido Decembrio, modeled this work after *Noctes Atticae* of Aulus Gellius, q.v.

## Decembrio, Angelo.

*De politia literaria libri septem, multa & uaria eruditione referti: ante annos centum scripti, & nunc tandem ab infinitis mendis repurgati, atque omnino rediuiui.*
Basel: Ioannes Heruagius, 1562.
Subject: Botany.
Case/X/0167/.222

2d printed ed. (MS completed 1463.)

## Dee, John, 1527–1608.

*Monas hieroglyphica.*
Antwerp: Guliel. Silvius, 1564.
Subject: Anagogical reasoning, astrology, astronomy, cabala, magic, hieroglyphics, mathematics.
Wing/ZP/5465/.S5863

## Defoe, Daniel, 1661(?)–1731.

*A system of magick or a history of the black art. Being an historical account of mankind's most early dealing with the devil; and how the acquaintance on both sides first began.*
London: printed and sold by J. Roberts, 1727.
Subject: Conjuring, devil, magic, sorcery.
Case/B/88/.22

1st ed.

## Dekkers, Frederic, 1644–1720.

*Frederici Deckers medicinae doctoris exercitationes medicae practicae circa medendi methodum observationibus illustrate cum capitum, rerum, verborum ac medicamentorum.*
Leyden: printed by Daniel, Abraham, & Adrian Gaesbeek, 1673.
Subject: Diagnosis, materia medica, medicine.
oR/128.7/.D4/1673

## Del Rio, Martin Antoine, 1551–1608.

*Disquisitionvm magicarvm libri sex: quibus continentur accurata curiosarum artium, & vanarum superstitionum confutatio, vtilis theologis, jurisconsultis, medicis, philologis.*
Lyon: Horatio Cardon, 1612.
Caillet 2967
Subject: Magic, sorcery, witchcraft.
fB/88/.224

Bk. 1: Kinds of magic, superstition; definition and division of magic into natural magic, artificial magic, alchemy; bk. 2: Demonic magic; bk. 3: Witches (maleficium); bk. 4: Divination; bks. 5 and 6, respectively: Offices of judge, confessor.

## Del Rio, Martin Antoine.

*Disqvisitionvm magicarvm libri sex, qvibvs continetvr [sic] accvrata cvriosarvm artivm et vanarum superstitionum confutatio, vtilis theologis jurisconsultis, medicis, philologis.*
Cologne: [printed for] Hermann Demen, 1679.
Subject: Magic, sorcery, witchcraft.
Case/B/88/.225

## Derham, William, 1657–1735.

*Astro-theology or a demonstration of the being and attributes of God, from a survey of the heavens.*
London: W. & J. Innys, 1721.
Subject: Astronomy, natural theology, religion and science.
B/72/.213

4th ed. In 8 books, including 1: Magnitude of the heavens; 2: Number of heavenly bodies; 3: Situation of heavenly bodies; 4: Motions of the heavens; 5: Phases of the moon and Venus, Jupiter's belts; 6: Gravity; 7: Light and heat.

## Derham, William.

*Physico-theology. Or, a demonstration of the being and attributes of God from his works of creation. Sermons at Mr. Boyle's lectures.*
London: printed for W. Innys, 1714.
Subject: Astronomy, natural theology, religion and science.
Case/B/72/.214

2d ed. In 11 books, including 1): Terraqueous globe, clouds, wind, rain, light; 4): Animals in general, the senses; 5): Tribes of animals, man; 6): Quadrupeds; 7): Birds; 8): Insects and reptiles; 9): Inhabitants of the waters; 10): Vegetables.

## Derham, William.

*Physico-theology: or, a demonstration of the being and attributes of God, from his works of creation. Being the substance of 16 sermons preached in St. Mary le Bow-Church, London, at the honourable Mr. Boyle's lectures in the year 1711 and 1712. With large notes, and many curious observations.*
London: W. Innys, 1716.
Subject: Astronomy, natural theology, religion and science.
B/72/.2142

4th ed. corrected.

## Descartes, René, 1596–1658.

*Discours de la methode pour bien conduire sa raison, & chercher la verité dans les sciences. Plus la dioptriqve. Les meteores. Et la geometrie. Qui sont des essais de cete methode.*
Leyden: Ian Maire, 1637.
Subject: Experimental science, geometry, optics.
Case/4A/877

1st ed.

## Descartes, René.

*Discovrs de la methode povr bien condvire sa raison, & chercher la verité dans les sciences. Plvs la dioptriqve, les meteores, la mechaniqve, et la mvsiqve, qui sont des essais de cette methode. Par René Descartes. Auec des remarques & des éclaircissemens necessaires.*
Paris: Charles Angot, 1668.
Subject: Experimental science, geometry, optics.
Case/B/239/.22

In 5 parts. *La dioptrique* consists of 10 parts, including refraction, how to make glasses, anatomy of the eye, light, vision. Section on meteors in 10 parts, including weather, rain, snow, thunderstorms, rainbows, clouds, nature of earthly bodies, ocean. BOUND WITH *Traité de la mechanique*, q.v.

## Descartes, René.

[*Musicae compendium*]
*Renatus Des-Cartes excellent compendium of musick: with necessary and judicious animadversions thereupon.*
London: printed by Thomas Harper for Humphrey Moseley, 1653.
STC II D1132
Subject: Consonance and dissonance, mathematics of music, proportion.
Case/V/5/.227

## Descartes, René.

*Renati Des-Cartes Musicae compendium.*
Amsterdam: Joannes Jansonius, Jr., 1656.
Subject: Consonance and dissonance, mathematics of music, proportion.
Case/V/22/.226

## Descartes, René.

*Traité de la mechaniqve, composé par monsieur Descartes. De plvs l'abregé de mvsiqve dv mesme autheur mis en François. Avec des eclaircissemens necessaires.*
Paris: Charles Angot, 1668.
Subject: Mechanics.
Case/B/239/.22

Description of engines and machines, with the aid of which force can be exerted to lift heavy burdens. Examples of the vise, the lever, the wedge, pulley, gear, inclined plane. BOUND WITH *Discovrs de la methode*, q.v.

## [Diacetius] Cattani da Diacetto, Francesco, bp. of Fiesole, 1446–1522.

*Discorso . . . sopra la svperstizzione dell' arte magica.*
Florence: Valente Panizzi & Marco Peri C., 1567.
Subject: Demons, incantation, natural magic.
Case/B/88/.155

## Díaz de Isla, Ruy, 16th cent.

*Tractado contra el mal serpentino: que vulgarmente en España es llamado bubas que fue ordenado en el ospital de todos los santos de Lisbona: fecho por Ruy Diaz de Ysla.*
[Seville: Dominico de Robertis, 1539]
Subject: Antidotes, medicine, snakebite.
Ayer/*109.9/D4 D5/1539

* * *

*Dictionnaire géographique portatif, ou description de tous les royaumes, . . . dans lequel on indique en quels royaumes . . . ces lieux se trouvent, les rivières, bayes, mers, montagnes, &c.*
Paris: Didot, 1749.
Subject: Geographical dictionary.
3A/3534

Trans. (Monsieur Vosgien) from the 13th ed. of Laurence Eachard's *Gazetteer's or newsman's interpreter* (library has 11th and 14th eds., qq.v.).

## Diemerbroeck, Ijsbrand van, 1609–1674.

*Isbrandi de Diemerbrock in academia Vltraiectina medicines & anatomes professoris opera omnia anatomica et medica, partim jam antea excusa, sed plurimis locis ab ibso auctore emendata, & aucta, partim nondum edita. Nunc simul collecta, & diligenter recognita per Timannvm de Diemerbroeck Isb. fil. medicinae doctorem & rei publicae Trajectinae poliatrum.*
Geneva: Samvel de Tovrnes, 1687.
Subject: Anatomy, medicine.
UNCATALOGUED

In 2 vols.; library has vol. 1 only, consisting of human anatomy.

## Digby, Sir Kenelm, 1603–1665.

*Observations upon* Religio medici *occasionally written by Sir Kenelm Digby knight.*
London: printed by A. M. for L. C., 1659.
STC II D1444
Subject: Material and immaterial world, miracles, occult.
Case/C/53/.1199

BOUND WITH Browne, Sir Thomas, *Religio medici*, q.v.

## Digby, Sir Kenelm.

*Of bodies, and of mans soul. To discover the immortality of reasonable sovls. With two discourses of the powder of sympathy, and of the vegetation of plants.*
London: printed by S. G. and B. G. for John Williams, 1669.
STC II D1445
Subject: Botany, experimental science, sympathetic medicine.
Case/B/4/.23

Treatise 1: Of bodies; treatise 2: The nature and operations of man's soul. Of the sympathetic powder. A discourse in solemn assembly at Montpellier. Made in French by Sir Kenelm Digby, Knight 1657. A Discourse concerning the vegetation of plants. Spoken by Sir Kenelm Digby, at Gresham Colledge . . . 1660. At a meeting of the society for promoting philosophical knowledge by experiments.

## [Digby, Sir Kenelm]

*Two treatises: in the one of which the natvre of bodies; in the other the nature of mans soule, is looked into: in way of discovery of the immortality of reasonable sovles.*
London: printed for Iohn Williams, 1645.
STC II D1449
Subject: Botany, experimental science, soul, sympathetic medicine.
Case/B/79/.242

Light, magnetism, natural motion, plants and animals.

## [Digby, Sir Kenelm]

*Two treatises: in one of which the nature of bodies; in the other the nature of man's soule is looked into: in way of discovery of the immortality of reasonable sovles.*
London: printed for Iohn Williams, 1665.
STC II D1451
Subject: Botany, experimental science, soul, sympathetic medicine.
Case/B/79/.243

2 vols. in 1. Binders title: "Immortality." BOUND WITH *The philosophical touch-stone*, by Alexander Ross, q.v.

## Digges, Leonard, ca. 1530–1570.

*A booke named tectonicon. Briefly shewing the exact measuring, and speedie reckoning all manner of land, squares, timber, stone, steeples, pillers, globes &c. Further, declaring the perfect making and large use of the carpenters ruler, containing a quadrant geometricall: comprehending also the rare vse of the square. And in the end a little treatise adioyning, opening the composition and appliancy of an instrument called the profitable staffe. With other things pleasant and necessary, most conductible for surveyors, land meaters, ioyners, carpenters, and masons. Published by Leonard Digges, gentleman, in 1556.*
London: Felix Kingston, 1637.
STC 6857
Subject: Instrumentation, mensuration, plane and solid geometry, surveying.
Case/L/20/.23

## Diodorus, Siculus, 1st cent. B.C.

[*Bilbiothece historiae libri: In quibus prisce res: fabulae: & multa ac uaria de situ locorum ac moribus gentium continentur.*]
Venice: Andreas de Paltasichis, 1476–77.
H-C 6189^ BMC (XV)V:251^ Pr. 4421^ Pell. 4267^ Goff D-211
Subject: Geography, origins of world.
NL/Inc./4421(+)

MS marginalia. Describes creation of world, first man, Nile floods, diseases.

## Diogenes, the Cynic, 412?–323 B.C.

*Epistolae Diogenis, Bruti, et Hippocratis.*
Florence: Antonius Francisci, Venetus, 1487.
H-C 6194^ BMC (XV)VI:677 Pr. 4272^ Still. D173^ Goff D-217
Subject: Disease, medicine.
Inc./6329A

"In epistolas Hippocratis medici praestantissimi e Graeco in Latinvm per renvtivm tradvctas ad Nicolavm V. Pon. Max." MS marginalia, esp. in sig. L.

## Diogenes, the Cynic

*Epistolae Diogenis, Bruti, et Hippocratis.*

Florence: Antonius Francisci Venetum. [i.e., Venice: Thomas di Piasiis, 1492.]
H-R6195^ H-C Add 6193^ BMC (XV) V:475^ GW VII:8397^ Pr. 6329^ Still. D174^ Goff D-218
Subject: Disease, medicine.
Inc./6329

Variant of reprint of the geniune Florentine edition of 1487, according to Goff. (From a sale catalogue entry: "Diogenes Cynicus [pseudo-Diogenes Sinopensis] Fictitious letters of Diogenes of Sinope, founder of a philosophical sect of cynics—written more than 300 years after his death in the time of Augustus—and translated from the Greek by the humanist Francesco Griffolini of Arezzo. Followed by likewise fictitious letters of Brutus [Caesar's murderer] and Hippocrates. The Hippocratic letters, translated by Rinuccio of Castiglione, were composed in the late Hellenistic period. The colophon was taken over from the 1487 edition. It was identified by Proctor as the product of a Venetian, by the British Museum Catalogue [Pollard] as Christophorus de Pensis, and by Gesamtkatalog as Thomas de Piasiis.")

## Dionysius Periegetes, fl. 100 A.D.

[*Orbis descriptio.*]
Basel: Ioannes Bebel, 1523.
Subject: Cosmography, geography.
Wing/ZP/538/.B387/no. 1

Title and text in Greek and Latin. Pages numbered in Greek portion, leaves signed (beginning with A) in Latin portion. Tracts include: "Dionysij orbis descriptio, Arati astronomicon, Procli sphaera. Cum scholijs Ceporini."

## Dionysius Periegetes.

*Dionysii Alex. et Pomp. Melae situs orbis descriptio. Aethici Cosmographia. C. I. Solini polyistor. In Dionysii poematium commentarij Evstathii: interpretatio eiusdem poematij ad verbum, ab Henr. Stephano scripta: necnon annotationes eius in idem, & quorundam aliorum. In Melam annotationes Ioannis Oliuarii: in Aethicvm scholia Iosiae Simleri: in Solinvm emendationes Martini Antonii Delrio.*
Geneva: Henricus Stephanus, 1577.
Subject: Cosmography, geography.
Case/Y/6309/.23

In 4 tracts: 1: Dionysius's poem [De situ orbis] in Greek, with annotations in Greek and Latin surrounding the text. Commentary of Stephan and Eustathius follow; 2: Pomponius Mela De situ orbis libri III; 3: C. Ivlii Solini polyhistor, appellatus à quibusdam rervm toto orbe memorabilivm thesavrvs: ab ipso autem Solino priùs inscriptus, collectanea rervm memorabilivm. A Martino Antonio Delrio; 4: Aethici Cosmographia.

## Dionysius Periegetes.

*Dionysii Alex. et Pomp. Melae situs orbis descriptio.*
Geneva: Henricus Stephanus, 1577.
Subject: Cosmography, geography.
Case/G/112/.23

Another ed.

## Dionysius Periegetes.

*Dionysii Alexandrini de sitv orbis liber interprete Andrea Papio Gandensi.*
Antwerp: Christophorus Plantin, 1575.
Subject: Cosmography, geography.
Wing/ZP/5465/.P70175

Text in Greek and Latin on facing pp. In hexameter verse.

## Dionysius Periegetes.

[*De situ orbis*]
Venice: Franz Renner, 1478
H 6227^ Pell. 4294^ Pr. 4173^ BMC (XV) V:195^ Goff D-254
Subject: Cosmography, geography.
Wing/Inc./4173

"Eloquentissimi uiri Domini Antonii Becharies ueronensis prooemium in Dionysii traductionem de situ orbis habitabilis ad clarissimum physicum magistrum Hieronymum de Leonardis."

## Dionysius [de Rubertis] de Burgo, d. 1342.

*Commentarius super Valerium Maximum.*
[Strassburg: Adolf Rusch, ca. 1475.]
H-C *4103^ BMC (XV)I:63^ GW VII:8411^ Pr. 237^ Pell. 3059^ Still. D198^ Goff D-242
Subject: Prodigies.
Inc./f237

Bk 1, leaves 26v–34v: De prodigiis; leaves 35–49: De sompniis; leaves 49–63: De miraculis.

## Dioscorides (Pedacius), Pedanius of Anazarbos, 1st cent.

*Libri octo graece et latine. Castigationes in eosdem libros.*
Paris: Petrus Haultinus, 1549.
Subject: Materia medica, medicine.
Wing/ZP/539/.H293

In 8 books, in Greek and Latin. "Pedanii Dioscoridis Anazarbei de medicinali materia libri primi praefatio. Io. Ruellio Suessionensi interprete." Herbal medicine, aromatics, unguents, venoms, wines and metals. Bk. 7: poisonous plants and venomous animals.

## Dioscorides (Pedacius), Pedanius of Anazarbos.

[*De materia medica.*]
[Venice: Aldus Manutius, 1499.]
H-C 6257^ BMC (XV) V:560–61^ Pr. 5571^ Pell. 4338^ Goff D-260
Subject: Materia medica, medicine.
Wing/Inc./5571(+)

1st ed. Title and text in Greek. MS marginalia in Greek.

ALL/ Materia medica/ Medicine

* * *

[*Discours*] *Le plaisant discovrs d'un medecin savoyart, emprisonné, pour avoir donné aduis au Duc de Sauoye de ne croire son deuin.*

[N.p.] 1600.
Subject: Astrology.
Bonaparte Coll./No./3973

In verse.

* * *

*A dissertation upon earthquakes, their causes and consequences; comprehending an explanation of the nature and composition of subterraneous vapours, their amazing force, and the manner in which they operate; ... Together with a distinct account of, and some remarks upon, the shock of an earthquake, felt in the cities of London and Westminster, on Thursday, February 8, 1749–50.*
London: printed for James Roberts, 1750.
Subject: Earthquakes, earth sciences, geology.
H/5305/.38

Bound with several tracts having a general title page on vails giving (n.a.). Binders title *Vails giving*.

## Dodoens, Rembert, 1516–1585.

*Cosmographica in astronomiam et geographiam isagoge, per Rembertum Dodonaeum Malinatem, medicum & mathematicum.*
Antwerp: Jan van der Loe, 1548.
Subject: Astronomy, cosmography, climate, zodiac.
Case/L/900/.234

In 4 books: 1): De mundo; 2): De coelo; 3): De terra; 4): De motu.

## Dodoens, Rembert.

*A nievve herball, or historie of plantes: wherein is contayned the vvhole discourse and perfect description of all sortes of herbes and plantes: their diuers & sundry kindes: their straunge figures, fashions, and shapes: their names, natures, operations, and vertues: and that not onely of those which are here growying in this our countrie of Englande, but of all others also of forrayne realmes, commonly vsed in physicke.*
London: Gerard Dewes, 1578.
STC 6984
Subject: Botany, herbals.
Case/N/405/.23

Trans. by Henry Lyte. Many woodcut illustrations. For each plant is given the physical description, place it grows, name (Greek, Latin, and others, e.g., English, French), its "vertues," season it grows, dangers, and remedies.

## Döpler, Jacob, fl. 1693.

*Theatrum poenarum, suppliciorum et executionum criminalium. Oder schaü-blatz derer leibes und lebens-straffen, welche nicht allein vor alters bey allerhand nationen und völckern in gebrauch gewesen, sondern auch noch heut zu lage in allen vier welt-theilen üblich sind. Darinnen zugleich der gantze inquisitions-process, captur, examination, confrontation, tortur, bekantnis und ratification derselben; item die abstraffung der verbrecher auch endliche hinrichtung der malefiz-personen, und wie bey jedweden legaliter und gewissenschaft zu verfarhren, enthalten.*
Sondershausen: Ludwig Heinrich Schönermarck for the author, 1693.
Subject: Witchcraft.
Case/I/2/.235

In 2 vols. (library has vol. 1 only).

## Dolce, Lodovico, 1508–1568.

*Dialogo di M. Lodovico Dolce, nel quale si ragiona del modo di accrescere e conseruar la memoria.*
[Venice: Gio. Battista, et Marchio Sessa brothers, 1562.]
Subject: Mnemonics, psychology.
Wing/ZP/535/.S505

Woodcut diagram of cranium (leaf 5) illustrates physical location of, e.g., memory.

## Dolet, Stephan.

*De re navali liber ad Lazarvm Bayfium.*
Lyon: Seb. Gryphivs, 1537.
Subject: Dictionary encyclopedia of naval science.
Greenlee/5000/B15/1537

A series of definitions of nautical and related terms, e.g., aqua, fluctus, navigatio, ostivm, stagnvm, lacvs, anchora, etc. [Possibly a response to criticisms by Baïf.] BOUND WITH Baïf, Lazar de, *Annotationes*, q.v.

## Doni, Antonio Francesco, 1513–1574.

*Mondi celesti, terrestri, et infernali, de gli academici pellegrini. Composti dal Doni; mondo piccolo, grande, misto, risibile, imaginato, de pazzi, & massimo, inferno, de gli scolari, de mal maritati, delle puttane, & ruffiani, soldati, & capita ni poltroni, dottor cattiui, legisti, artisti, de gli vsurai, de poeti & compositori ignoranti.*
Venice: Domenico Farri, 1567.
Subject: Macrocosm–Microcosm, natural philosophy.
Case/Y/712/.D7234

In "tavola del mondo piccolo: Del corpo dell' huomo & del corpo del mondo. Autori di nuoue openioni [*sic*] delle sfere . Supplica de gli hortolani. Autori di nuoue openioni grande. Legge eterna, legge, naturale, legge mosaica, legge euangelica, & legge humana." In "tavola del mondo de pazzi: Spiriti, maligni & amorosi."

## Doni, Giovanni Battista 1593–1647.

*Compendio del trattato de' generi e de' modi della mvsica. Con vn discorso sopra la perfettione de' concenti. Et vn saggio à due voci di mutationi di genere, e di tuono in tre maniere d'intauolatura: e d'vn principio di madrigale del principe, ridotto nella medesima intauolatura.*
Rome: Andrea Fei, 1635.
Subject: Harmonics, music.
Case/V/22/.23

"Diuisione harmonica & aritmetica."

## Dornau, Caspar, 1577–1632.

*Amphitheatrum sapientiae Socraticae joco-seriae, hoc est, encomia et commentaria avtorvm, qva vetervm, qva recentiorvm prope omnivm: qvibvs res, avt pro vilibvs vvlgo aut damnosis*

*habitae, styli patrocinio vindicantur, exornantur, opvs ad mysteria natvrae discenda, ad omnem amoenitatem, sapientiam, virtutem, publice priuatimque vtilissimum; in dvos tomos.*
Hanover: Wechel for Daniel & David Aubri & Clement Schleichius, 1619.
Subject: Disease, fauna, medicine.
Case/fY/6809/.24

2 vols. in 1. An encyclopedia of "encomia" on drinking, gout, quartan fever, various animals (horse, dog, peacock, ass, crow, etc., by Aldrovandi). Sig. Bbb: "Dess adels dess esels."

## Dornau, Caspar.

*Menenivs Agrippa, hoc est, Corporis hvmani cvm repvblica perpetva comparatio: observationibvs historicis, ethicis, oeconomicis, politicis, physicis, medicis, illustrata.*
Hanover: Wechel, for heirs of Ioannes Aubri, 1615.
Subject: Medicine, philosophy.
J/15/.239

BOUND WITH Jacobi Bornitius, *De nummis* (n.a.).

## Drelincourt, Charles, 1633–1697.

*Opuscula medica, quae reperiri potuere omnia.*
The Hague: Gosse & Neaulme, 1727.
Subject: Medicine.
UNCATALOGUED

## Dresser, Matthaeus, 1536–1607.

*Historien und bericht von dem newlicher zeit erfundenen königreich China. . . . Item von dem auch new erfundenen lande Virginia.*
Leipzig: printed by Beyer for Frantz Schnelboltz, 1598.
Subject: Exploration, New World, voyages.
Ayer/150.5/V7/D7/1598

BOUND WITH Hariot, Thomas, q.v. Dresser's work may be out of scope. Harriot's work has continuous pagination but separate title page as follows: *Wunderbarlicher doch warhafftiger bericht. Von der landschafft Virginia inn der newen welt, welche newlich im jahr Christi 1585. Von den Engelendern erfunden ist. Erstlichen in Engelendischer sprach beschrieben durch Thomam Hariot vnd hernach in Teutsch gebracht durch Christophorvm P.* [N.p. n.d.] Includes "Von schwartz kunstern oder zauberern. Von vogeln. Von cederbaum. Von früchten. Von fischen. Von einem holtz genant sassafras."

## Du Bartas, Guillaume de Salluste, seigneur, 1544–1590.

*Bartas his deuine weekes and workes.*
London: Humfrey Lownes, 1605.
STC 21649
Subject: Arithemtic, astrology, astronomy, creation, religion and science.
Case/Y/762/.D8308

Trans. Joshua Sylvester. The first week or birth of the world, the second week or childhood of the world. *Encyclopaedia Britannica:* "[Du Bartas] is filled with the indiscriminate information that passed under the name of science in the sixteenth century."

## Du Bartas, Guillaume de Salluste.

*His deuine weekes and works.*
London [printed by Humfrey Lownes, 1613]
STC 21652
Subject: Arithemtic, astrology, astronomy, creation, religion and science.
Case/Y/762/.D 831

Trans. Joshua Sylvester. "4thly corrected." The first week or birth of the world where-in in seaven dayes the glorious work of the creation is diuinely handled: 1. The chaos; 2. The elements; 3. The sea and earth; 4. The heavens, svn, moon, &c.; 5. The fishes & fovles; 6. The beasts and man; 7. The sabaoth. Du Bartas his second week, disposed (after the proportion of his first) into seaven dayes: (viz.) 1. Adam; 2. Noah; 3. Abraham; 4. David; 5. Zedechias; 6. Messias; 7. Th' eternal sabbath.

## Du Bartas, Guillaume de Salluste.

*Divine weeks and works.*
London: printed by Robert Young, 1633.
STC 21654
Subject: Arithemtic, astrology, astronomy, creation, religion and science.
Case/Y/762/.D833

Engraved title page: *Du Bartas his diuine weekes and workes with a compleat collection of all the other most delightfull workes translated and written by [that] famous philomusus, Iosvah Sylvester gent.* Included in this collection (beginning on p. 563) is a tract by Sylvester, "Tobacco battered and the pipes shattered."

## Du Bartas, Guillaume de Salluste.

*Divine weeks and works.*
London: printed by Robert Young, 1641.
STC II D2405
Subject: Arithemtic, astrology, astronomy, creation, religion and science.
Case/*Y/762/.D834

Title page: *Du Bartas his divine weekes and workes: VVith a complete collection of all the other most delightfull vvorkes, translated and vvritten by that famous philomusus Josuah Sylvester, Gent. VVith additions.*
Includes Sylvester's tract (p. 569).

## Du Bartas, Guillaume de Salluste.

*Les oeuvres de G. de Salvste Sr. dv Bartas. Reueües corrigees, augmentees de nouueaux. Commentaires, annotations en marge, et embellie de figures sur tous les jours de la sepmaine. Plus a esté adiouste la premiere et seconde partie de la suitte auecq l'argument general, et amples sommaires au commencement de chacun liure.*
Paris, 1611 '10.
Subject: Arithemtic, astrology, astronomy, creation, religion and science

Wing/fZP/639/.B645

"Derniere edition."

### Du Bartas, Guillaume de Salluste.

*Les oeuvres poetiqves & Chrestiennes de G. de Salvste Sr du Bartas, prince des poëtes François.*
Geneva: Pierre Chouët, 1632.
Subject: Arithemtic, astrology, astronomy, creation, religion and science.
Case/Y/762/.D83

"Ceste edition contient plus que les precedentes, par l'addition de quelques vers & fragmens: & le tous agencé suyuant la disposition ordonée par l'autheur." In addition to the 1st week, it has the 1st 4 days of the second week: "les autres trois iours restez à faire par le decez de l'auteur."

### Du Bartas, Guillaume de Salluste.

*Commentaires et annotations svr la sepmaine, de la creation dv monde.*
Paris: Abel l'Angelier, 1583.
Subject: Arithemtic, astrology, astronomy, creation, religion and science.
Case/Y/762/.D83582

### Du Bartas, Guillaume de Salluste.

*Ioannis Edoardi Du Monin Bvrgvdionis Gyani Beresithias sive mundi creatio, ex Gallico G. Salusti du Bartas heptamero expressa.*
Paris: Ioannes Parant, 1579.
Subject: Arithemtic, astrology, astronomy, creation, religion and science.
Case/Y/762/.D8357

### Du Bartas, Guillaume de Salluste.

*La sepmaine, ov creation dv monde, de G. de Salvste seignevr dv Bartas.*
Paris: for Michel Gadoulleau, 1580.
Subject: Arithemtic, astrology, astronomy, creation, religion and science.
Case/Y/762/.D8358

### Du Bosc, Jacques, d. 1660.

*The compleat vvoman. VVritten in French by Monsieur Du-Bosq, and by him after severall editions reviewed, corrected, and amended.*
London: printed by Thomas Harper and Richard Hodgkinson, 1639.
STC 7266
Subject: Cosmetics, psychology.
Case/B/68/.24

### [Du Bosc, Jacques]

*L'honneste Femme.*
Paris: Iean Iost, 1633–34.
Subject: Cosmetics, psychology.
B/68/.2388

Seconde edition, reveve, corrigée & augmentée par l'autheur. 2 vols. in 1. Pt. 1: De l'inclination à la vertu. Pt. 2: De la naissance & de l'education. Each vol. has separate title page & separate pagination.

### [Du Bosc, Jacques]

*L'honneste femme. divisée en trois parties.*
Rouen: Iacqves Besongne, 1650.
Subject: Cosmetics, psychology of women.
Case/B/68/.239

3 vols. in 1. Derniere edition, reueuë, corrigée & augmentée par l'autheur.
Pt. 1 includes "De l'humeur gaye & de la melancolique"; pt. 2: Self love;pt. 3: "Il me semble qu'entre les sciences humaines il n'y en a de plus importantes, que la medecine, la jurisprudence, & la morale."

### [Du Breuil, Jean], 1602–1670.

*La perspective pratiqve, necessaire a tovs peintres, gravevrs, scvlptevrs, architectes, orfevres, brodevrs, tapissiers, & autres se seruans du dessin.*
Paris: Melchior Tavernier & François L'Anglois, dit Chartres, 1642.
Subject: Perspective, proportion.
Case/Wing/ZA/3/.244

Includes section on "mesvres, et proportions des figvres aux perspectives, tableavx, et ovvrages de bosse," and on "pratiqves povr trovver les ombres natvrelles, tant au soleil, & au flambeau, qu'à la chandelle & à la lampe."

### Ducret, Toussaint, 16th cent.

*De arthritide vera assertio, eivsqve cvrandae methodvs, adversvs Paracelsistas, avthore Tvssano Dvcreto Cabilunense, medico.*
Lyon Bartholomaeus Vincentius, 1575.
Subject: Anti-Paracelsian medicine, diseases of the joints, materia medica.
Case/oRC/927/.D83/1575

In 2 parts. Pt. 1: Definitions of etymology and terms; pt. 2: causes of diseases of the joints.

### Dudley, Sir Robert, 1532–1588.

*Dell' arcano del mare.*
Florence: Francesco Onofri, 1646–47.
Subject: Military and naval technology, navigation, navigational instruments.
Ayer/*135/D8/1646

3 vols. Vol. 1 (bks. 1 and 2): latitude and longitude, shipbuilding; vol. 2 (bks. 3 and 4): science of navigation, "cioè spirale, ò di grancircoli." These 2 vols. dated 1646. Vol. 3 (dated 1647) consists of bks. 5 and 6 in 2 elephant folios. Engraved title page vignette of book 6, "Orsa minore," a sort of beaverish-looking representation of the constellation, with placement of the stars indicated. Bk. 5 includes engravings of instruments, and volvelles.

### Dürer, Albrecht, 1471–1528.

*Etliche Vnderricht zu befestigung der Stett, schloss, vnd flecken.*
Nuremberg: Hieronymus Andreae, 1527.

Subject: Geometry, military technology.
Case/Wing/fZ/4035/.2459/no. 2

Dürer's principal work on architecture. BOUND WITH his *Vier bücher*, q.v.

## Dürer, Albrecht.

*Hierinn sind begriffen vier bücher von menschlicher proporcion.*
Nuremberg: Ieronymus Formschneyder [for Dürer's widow], 1528.
Subject: Anatomy, geometry, perspective, proportion.
Case/Wing/fZ/4035/.2459/No. 3

## Dürer, Albrecht.

[*Institutionum geometricarum*]
*Albertvs Dvrervs Nvrembergensis pictor hvivs etatis celeberrimvs versus è germanica lingua in latinam, pictoribus, fabris, erariis ac lignariis, lapicidis, statuariis, & vniuersis demum qui circino, gnomone, libella, aut alioqui certa mensura opera sua examinant propè necessarius, adèo exacte quatuor his suarum institutionum geometricarum libris lineas, superficies & solida corpora tractauit, adhibitis designationibus ad eam rem accommodissimus.*
Paris: Christian Wechel, 1532.
Subject: Anatomy, geometry, perspective, proportion.
Case/Wing/fZ/4035/.247

MS note on flyleaf and bookplate of Gilbert Redgrave, "This is the first ed. of the translation into Latin."

## Dürer, Albrecht.

[*Institutionum geometricarum*]
*Albertvs Dvrervs Nvrembergensis pictor hvivs aetatis celeberrimus, versus è Germanica lingua in Latinam, pictoribus, fabris aerariis ac lignariis, lapicidis, statuariis, & vniuersis demum qui circino, gnomone, libella, aut alioqui certa mensura opera sua examinant, propè necessarius, adeò exactè quatuor his suarum institutionum geometricarum libris, lineas, superficies & solida corpora tractauit, adhibitis designationibus ad eam rem accommodatissimis.*
Paris: Christian Wechel, 1535.
Subject: Anatomy, geometry, perspective, proportion.
Case/fY/642/E615/No. 2

## Dürer, Albrecht.

[*Institutionum geometricarum*]
*Di Alberto Durero pittore e geometra chiarissimo. Della simmetria de i corpi humani, libri quattro. Nuouamente tradotti dalla lingua Latina nella Italiana, da M. Gio. Paolo Gallvcci Salodiano. Et accresciuti del quinto libro, nel quale si tratta, con quai modi possano i pittori, & scoltori mostrare la diuersità della natura de gli huomini, & donne, & con quali le passioni, che sentono per li diuersi accidenti, che li occorrono. Hora di nuouo stampati. Opera a i pittori, e scoltori non solo vtile, ma necessaria, & ad ogn' altra, che di tal materia desidera acquistarsi perfetto giudicio.*
Venice: Domenico Nicolini, 1591.
Subject: Anatomy, geometry, perspective, proportion.
Case/W/51/.23

## Dürer, Albrecht.

*Les quatres livres d'Albert Dvrer, peinctre et geometrien tres excellent, de la proportion des parties & pourtraicts des corps humains.*
Paris: Charles Perier, 1557.
Subject: Anatomy, geometry, perspective, proportion.
Wing/fZP/539/.P412

Trans. Loys Meigret.

## Dürer, Albrecht.

*Underweysung der messung mit dem zirckel vnd richtscheyt in linien ebnen vnnd gantzen corporen. Durch Albrect Dürer züsamen gessogen, und zü nuss allen kunstlieb habenden mit zü gehörigen figuren in truckt gebracht.*
Nuremberg: Hieronymus Andreae [for Dürer], 1525.
Subject: Anatomy, geometry, perspective.
Case/Wing/fZ/4035/.2459/No. 1

1st ed. of Dürer's work on perspective for the use of sculptors, painters, and craftsmen. Typed note, "In it he explains the various instruments he had developed, among them a type of camera obscura, for drawing in perspective." BOUND WITH *Vier bücher* and *Befestigung*, qq.v.

## Dürer, Albrecht.

*Underweysung der messung mit dem zirckel vnd richtscheyt in linien ebnen unn gantzen corporen, durch Albrecht Dürer zusamen gesogen unn durch in selbs (als er noch auff erden war) an vil orten gebessert in sonderheyt mit xxij figuren gemert die selbigen auch mit eygner handt auffgerissen wie es dann eyn yder werckmann erkennen wirdt aun aberzu nutz allen kunst liebhabenden in truck geben.*
Nuremberg: Hieronymus Formschneyder, 1538.
Subject: Anatomy, geometry, perspective.
Case/Wing/Z/4035/.246

Another ed. of the handbook for draughtsmen.

## Dürer, Albrecht.

*Alberti Dvreri pictoris et architecti praestantissimi de vrbibvs, arcibvs, castellísque condendis, ac muniendis rationes aliquot, praesenti bellorum necessitati accommodatissimae: nunc recens è lingua Germanica in Latinam traductae.*
Paris: Christian Wechel, 1535.
Subject: Architecture, engineering, military technology.
Case/fY/642/.E615/No. 3

## Du Laurens, André, 1558–1609.

*Toutes les oevvres de Me. André dv Lavrens sieur de Ferrieres, con[seill]er & premier medecin du tres-chrestien roy de France & de Nauarre, Henry le Grand, & son chancelier en l'vniuersité de Montpellier.*
Roven: Iacqves Besongne, 1661.
Subject: Anatomy, diagnosis, medicine.
UNCATALOGUED

In 2 parts. Pt. 1: anatomy; pt. 2: medicine, diagnosis, diseases. "Recveilles et tradvittes en Francois par Me. Theophile Gelée medecin ordinaire de la ville de Dieppe."

## Du Laurens, André.

*Il tesoro della vecchiezza di Andrea Lavrenti, diuiso in quattro discorsi. Della vista, Della malinconia. De' catarri, e della vecchiezza. Tradotto dalla lingua Francese.*
Venice: Marco Ginammi, 1637.
Subject: Gerontology, medicine, melancholy.
Q/.247

Preservation of health, saving teeth.

## Du Laurens, François.

*Solutiones aliquot quaestionum, quibus vera primae philosophiae principia astruuntur. Opus metaphysicum Francisci Dulaurens.*
The Hague: Adrian Vlaq, 1693.
Subject: Epistemology, metaphysics, psychology.
Ayer/6/P9/1663

BOUND WITH Ptolemy, *Tractatus de judicandi facultate*, 1663, q.v.

## Dumas, Louis, 1676–1744.

*La biblioteque des enfans ou les premiers elemens des lettres.*
Paris: Pierre Simon, 1733.
Subject: Education, science.
Wing/ZP/739/.S 596

4 vols. in 1. "Supplement sur l'aritmetique, sur le calendrier, sur l'ecriture," etc. (pp. 237 ff.). Has a table of precious stones, distances of planets, and other astronomical miscellany.

## Dumolinet, Claude, 1620–1687.

*Le cabinet de la bibliothèque de Sainte Genevieve divisé en deux parties. Contenant les antiquitez de la réligion des Chrétiens, des Egyptiens, & des Romains; des tombeaux, des poids & des médailles; des monnoyes, des pierres antiques gravées, & des mineraux; des talismans, des lampes antiques, des animaux les plus rares & les plus singuliers, des coquilles les plus considérables, des fruits étrangers, & quelques plantes exquises.*
Paris: Antoine Dezallier, 1692.
Subject: Curiosities, natural history.
Case/+F/0236/.246

Part 2: natural history. Many engravings, including a series depicting the interior of the bibliothèque. Beneath Dumolinet's portrait: "Vous qui de cet autheur contemplés le visage, / Portés sur ses vertus vos regards curieux. / Il fut humble, sçavant, officieux, et sage, / Et ce que son portrait n'offre point à vos yeux, / Vous le découvrirés en lisant cet ouvrage."

## Du Moulin, Antoine, b. ca. 1520.

*De diversa hominvm natvra, provt a veteribus philosophis ex corporum speciebus reperta est, cognoscenda liber, Antonii Molinii Matisconensis diligentia nunc primum in lucem emergens.*
Lyon: Ioan Tornaesius, 1549.
Subject: Comparative anatomy, physiognomy.
Case/B/56/.246

At top of 1st text page: "De physiognomiae ratione libellus ex veterum philosophorum monumentis summo compendio collectus." Subtitles include: "Quòd homines sint animalibus similes."

## Dupleix, Scipion, 1569–1661.

[Binder's title] *Philosophie morale.*
Paris: Iacques Bessin, 1631.
Subject: Natural science, dreams, psychology.
B/239/.244

3 separately paginated tracts, with separate title pages. 1: "L'etiqve ov philosophie morale"; 2: "La cvriosité natvrelle redigée en questions selon l'ordre alphabetique"; 3: "Les cavses de la veille et dv sommeil des songes & de la vie & de la mort." In discourses. What are the causes of dreams? Can one discover the state of health through dreams? Definitions in pt. 2 include "étoiles, oiseav, sang, venin."

## Duport, François, 1540?–1624?

*Francisci Porti Crespeiensis valesii mediciqve parisiensis medica decas, eivsdem avthoris in singvla librorum capita commentarijs illustrata. Opvs scitv facillimvm ob metrum, & ad praxim vtilissimum.*
Paris: Abraham Savgrain, 1613.
Subject: Disease: etiology, semiotics, therapeutics.
4A/2137

In 10 books. Signs of perfect health: blood, bile. Signs of melancholy, of phlegmatic condition: watery humor, flatulence. Imminent signs of disease or illness: fever, illnesses in chest. Bk. 2: Illnesses in the head: frenzy, lethargy, apoplexy, epilepsy. Bk. 3: esophagus, stomach, kidneys. Bk. 7: nose, tonsils. Bk. 10: women, hysteria, menstrual disorders, sterility.

## [Durand, David], 1680–1763.

*The life of Lucilio (alias Julius Caesar) Vanini, burnt for atheism at Thoulouse. With an abstract of his writings. being the sum of the atheistical doctrine taken from Plato, Aristotle, Averroes, Cardanus, and Pomponatius's philosophy. With a confutation of the same; and Mr. Bayle's arguments in behalf of Vanini compleatly answered.*
London: printed for W. Meadows, 1730.
Subject: Astrology, religion and science, supernatural.
E/5/.V318

With a catalogue of Vanini's writings.

## Durante, Castore, 1529–1590.

*Herbario novo di Castore Dvrante medico, et cittadino Romano. Con figure, che rappresentano le viue piante, che nascono in tutta Europa & nell' Indie orientali & occidentali. Con versi Latini, che comprendono le facoltà de i semplici medicamenti. Con discorsi, che dimonstrano i nomi, le spetie, la forma, il loco, il tempo, le qualità, & le virtu mirabili dell' herbe, insieme col peso, & ordine da vsarle, scoprendosi rari secreti, & singolari rimedij da sanar le più difficili infirmità del corpo humano. Con dve tavole copiosissime, l'vna delle herbe, & l'altra dell infirmità, & di tutto quello che nell' opera si*

*contiene. Con aggionta in quest' vltima impressione de i discorsi à quelle figure, che erano nell' appendice, fatti di Gio. Maria Ferro, spetiale alla sanità.*
Venice: Gio. Giacomo Hertz, 1667.
Subject: Botany, herbals, materia medica.
Ayer/*8.9/B7/D8/1667

## Durastanti, Giano Matteo, fl. 16C.

*L'aceto scillino dell' eccellentiss. Messer Giano Matteo Durastanti da san Giusto. Cioè le sue; trè compositioni; & mirabbili forze nell' allungar l'humana, sanità, & vita; & li conuenuoli modi dell' vsarlo.*
[Macerata: Sebbastiano Martellini, 1576]
Subject: Medicine.
Wing/ZP/535/.M36

## Dutertre, Jean Baptiste, 1610–1687.

*Histoire generale, des isles de Christophe, de la Gvadelovpe, de la Martiniqve, et avtres dans l'Ameriqve. Où l'on verra l'establissement des colonies Françoises, dans ces isles; leurs guerres ciuiles & etrangeres, & tout ce qui se passe dans les voyages & retours des Indes. Comme aussi plusieurs belles particularitez des Antisles de l'Amerique: Vne description generale de l'Isle de Guadeloupe: de tous ses mineraux, de ses pierreries, de ses riuieres, fontaines, & estangs: & de toutes ses plantes. De plus, la description de tous les animaux de la mer, de l'air, & de la terre.*
Paris: Iacques Langlois & Emmanuel Langlois, 1654.
Subject: Natural history, New World, voyages.
Ayer/*1000/D8/1654

## Du Verney, M. [Joseph Guichard], 1648–1730.

*Tractatus de organo auditus continens structuram, usum et morbos omnium auris partium.*
Nuremberg: printed by Joannes Michael Spörlini for Johannes Zieger, 1684.
Subject: Anatomy of the ear, diseases, hearing.
Case/RF/120/.D816/1684

## Du Verney, M.

*Tractatus de organo auditus; oder abhandlung vom gehör worinnen nicht allein die structur des ohres, desgleichen auch das hören an sich selbst deutlich erkläret wird; sondern auch alle kranckheiten welche diesem theile zustossen können, grundlich gezeiget, und die dawieder dienende hülst-mittel angewiesen werden.*
Berlin: Johann Andreas Rudiger [1732].
Subject: Anatomy of the ear, diseases, hearing.
Case/RF/120/.S815/1732

# E

## Eachard, John, 1636?–1697.

*Mr. Hobbs's state of nature considered.*
London, 1672.
STC II E57a
Subject: Natural philosophy.
Case/B/245/.412

2d ed. "State of nature is a state of war." (p. 68) *And some opinions of Mr. Hobbs considered in a second dialogue.* London: printed by J. Macock for Walter Kettilby, 1673. STC II E64.

## Eachard, John.

*Dr. Eachard's works. III. Mr. Hobbs's state of nature considered: in a dialogue between Philautus, & Timothy.*
London: printed for J. Phillips et al., 1705.
Subject: Natural philosophy.
C/7826/.252

4 pts. in 1 vol. 5th ed. corrected by the author.

## Eachard, Laurence, 1670?–1730.

*An exact description of Ireland: Chorographically surveying all its provinces & counties.*
London: printed for Tho. Salusbury, 1691.
Subject: Chorography, geography.
Case/G/4/2003/.25

Attire, way of living, superstitions, rivers, products.

## Eachard, Laurence.

*The gazetteer's: or newsman's interpreter: being a geographical index of all the considerable cities, patriarchships, . . . towns, ports . . . &c. in Europe. Shewn in what kingdoms . . . they are . . . their distances (in English miles) from several other places of note; with their longitude and latitude, according to the best approved maps.*
London: printed for Tho. Salusbury, 1693.
STC II E145
Subject: Geography.
Case/G/005/.258

2d ed. corrected.

## Eachard, Laurence.

*The gazetteer's; or newsman's interpreter. Being a geographical index. . . of special use for the true understanding of all modern histories of Europe. . . and for the conveniency of cheapness and pocket-carriage, explained by abbreviations and figures.*
London: printed for John Nicholson . . . & Samuel Ballard, 1716.
Subject: Geography.
3A/160

11th ed. corrected and very much enlarged.

## Eachard, Laurence.

*The gazetter's or, newsman's interpreter. Being a geographical index of all the considerable provinces, cities, patriarchships . . . in Europe. . . . Of special use for the true understanding of all modern histories of Europe; . . . and for the conveniency of cheapness and pocket-carriage, explained by abbreviations and figures.*
London: printed for R. Robinson, etc., 1738.
Subject: Geography.
G/005/.26

2 vols. in 1. Pt. 1, 14th ed. Pt. 2, . . . *Being a geographical index of all the empires . . . in Asia, Africa and America,* [same imprint] 7th ed.

## Eachard, Laurence.

*A most compleat compendium of geography, general and special: describing all the empires, kingdoms, and dominions, in the whole world. Shewing their bounds, situation, dimensions, ancient and modern names, . . . commodities, divisions, . . . mountains, lakes. . . . Together with an appendix of general rules for making a large geography, with the great uses of that science.*
London: printed for J[ohn] Salusbury, 1697.
STC II E150
Subject: Geography.
Case/G/11/.254

4th ed. corrected and much improved.

* * *

*The eating of blood vindicated: in a brief answer to a late pamphlet intituled, A blood tenent confuted.*
London: printed for H. Shepheard . . . & W. Ley, 1646.
STC II E111
Subject: Dietary strictures, ethics, medicine.
Case/C/726/.25

## Eden, Richard.

*The history of trauayle in the vvest and east Indies and other countreys lying eyther way, towardes the fruitfull and ryche Moluccaes. As Moscouia Persia, Arabia, Syria, Aegypte, Ethiopia, Guinea, China in Cathayo, and Giapan: VVith a discourse of the northwest passage. Gathered in parte, and done into Englyshe by Richarde Eden. Newly set in order, augmented, and finished by Richard VVilles.*
London: Richard Iugge, 1577.
Church 119^ John Carter Brown no. 312^
Subject: Discovery, herbal medicine, metallurgy, natural history, navigation, occult religion, travel.
Ayer/*110/E2/1577

In 4 parts: 1. New World, Spanish discovery, by P. Martyr, Gonzalo Ferdinando Oviedo, Vasquez Nunez, John Grisalua. Includes Peru, Yucatan, Florida (Cortez),

Cuba, West Indies. Describes gold mining, navigation, metals, customs, animals and birds, fish, topography, plants and trees, and"divers other thinges that are engendred there both on land and in the water." 2. Northwest passage. 3. North eastern frosty seas "Moscouia, Schondia or Denmarke, Groenlande, Laponia, Norway" (Jacob Ziegler, pp. 275 ff.). 4. Voyages into Guinea or south east Africa, Egypte, Ethiopia, Arabia, Syria, Persia, and East Indies. With P. Martyr's last 4 decades on the conquest of Mexico.

## Eden, Richard, ed.

*The decades of the new worlde or west India, conteyning the nauigations and conquestes of the Spanyardes, with the particular description of the moste ryche and large landes and ilandes lately founde in the west ocean perteyning to the inheritance of the Kinges of Spayne. In the which the diligent reader may not only consyder what commoditie may hereby chaunce to the hole Christian world in tyme to come, but also learne many secreates touchynge the lande, the sea, and the starres, very necessarie to be knowen to such as shal attempte any nauigations, or otherwise haue delite to beholde the strange and woonderfull woorkes of God and nature. Written in the Latine tounge by Peter Martyr of Angleria, and translated into Englysshe by Rycharde Eden.*
London: Richard Iugge, 1555.
Church 102^ Sabin 1561
Subject: Discovery, natural history, voyages.
Ayer/*110/E2/1555

Includes Biringucci, *Generation of metals*, preface (p. 326) and text (p. 327). Describes animals, insects, birds, trees, plants, fish, and Indians' manner of fishing, also the "familiaritie that the Indians have with the deuyl." Tract by Jacob Ziegler, *Of the north regions*.

## Edrisi, 1100–1166.

*De geographia vniversali. Hortvlvs cultissimus mirè orbis regiones, prouincias, insulas, vrbes, earumque dimensiones & orizonta describens.*
Rome: in typographia Medicea, 1592.
Subject: Geography.
Wing/ZP/535/.M468

Latin and Arabic titles, with text in Arabic only.

* * *

*Eeuwig duerende almanach, beginnende je. 1681.*
Amsterdam: Iacobus Robyn [1681].
Subject: Almanacs.
Wing/ZP/646/.R57

Pocket-sized, bound in vellum. With volvelles showing phases of moon for each of 4 seasons as well as zodiac and names of saints' days.

* * *

*Eine grawsame erschreckliche vnd wunderbarliche geschicht oder newe zeitung, welche warhafftig geschehen is in diesem 1559 jahr zur platten zwo meyl weges vom Joachims thal allda hat ein schmiedt ein tochter die ist vom bösen feindt dem Teuffel eingenommen vnd besessen worden. Der hat so wunderbarlich und seltzam ding aus ir geredt mit den priestern die teglich bey ihr gewest sind. Und wie er leizlich von ihr ausgetrieben worden ist durch der priester und viel frommer Christen des gemeinen volcks gebet und seufftzen welchs sie teglich für sie zu Gott gethan haben. Den frommen gottfürchtigen Christen etwas tröstlich. Aber den gottlosen und uns bussfertigen etwas erschrecklicher sie zur busse zu vermanen. VVie denn solches der böse geist selbs wider seinen willen hat reden und anzeigen müssen.*
Erffurdt: Georgius Bawman, 1559.
Subject: Possession, witchcraft.
Case/B/8847/.365

## Elyot, Sir Thomas, d. 1546.

*The castell of helth corrected and in some places augmented by the first author thereof.*
[London: Thomas Berthelet, 1549.]
STC 7647
Subject: Health, medicine.
Case/B/692/.26

In 4 books. Bk. 1: humors, anatomy; bk. 2: diet. There are things natural (powers, complexions, elements, humors, members, operations, spirits); things not natural (food and drink); and things against nature (sickness, cause of sickness, and accidents that follow sickness).

## Engel, Johann, 1463–1512.

*Astrolabium planum in tabulis ascendens: continens qualibet hora atque minuto equationes domorum celi: morarum nati in vtero matri cum quodarum tractatum natiuitatum vtili ac ornato: necnon horas in equales pro qualibet climate mundi.*
Venice: Lucas Antonius de Giunta, 1502.
Subject: Judicial astrology.
Wing/ZP/535/.G43

In 3 parts. Pt. 1: "Tabulas pro quolibet climate duas in se pertinet"; pt. 2: "Tabula cuiuslibet climatis huius prime partis equatoris domorum celi pertinet"; pt. 3: Tabulam more infantis in vtero matris includit per quam quis scire poterit & tempus conceptionis & tempus natiuitatis hominis alcuius."

* * *

English political cartoons and miscellaneous illustrations from the Narcissus Luttrell collection, 1721– (items on medical or scientific subjects [or using medicine or science as a metaphor but illustrating contemporary practices]). Pieces are numbered within folders, and the entire group of folders is boxed.
Case/W/778/.186

"Women examining and preparing the [silkworm] eggs in order to put them into boxes" (no. 195, folder, "not in BM catalogue," in undated box); the following pieces are all in folder BM 1715–2455, dated 1721–40: [Hogarth] "Cunicularii or the wise men of Godliman in consultation" (no. NL 2, BM 1779, dated 1726); "The doctors in labour" (NL 3, BM 1781, dated 1726); "Prenez les pilules" (NL 5, BM 1987, dated 1739).

### Erastus, Thomas, 1523–1583.

*Repetitio dispvtationis de lamiis; sev, strigibvs: in qua plenè, solidè, & perspicuè, de arte earum, potestate, itemque poena disceptatur.*
Basel: Petrus Pernam, [1578].
Subject: Witchcraft.
Case/B/88/.28

Cf. Caillet v. 2, no. 3647.

### Ercker, Lazarus, ca. 1530–1594.

*Beschreibung aller furnemisten mineralischen erzt und bergkwercks arten wie die selbigen und eine jede in sonderheit irer natur und eigenschafft nach auff alle metaln probirt und im kleinen fewer sollen versucht werden mit erklärung ettlicher fürnemer nützlicher schmelzwerck im grossen fewer auch scheidung goldts, silbers und anderer metaln sampt einem bericht dess kupffersaigerns messing brennens und salpeter siedens auch aller salzigen minerischen proben und was denen allen anhengig in fünff bücher verfast dess gleichen zuuorn niemals in druck kommen.*
Frankfurt: Lazarus Ercker, 1580.
Subject: Alchemy, metallurgy.
Case/*R/40/.27

Illustrations of smelting, etc.

### Erizzo, Sebastiano, 1525–1585.

*Trattato di Messer Sebastiano Erizzo, dell' istrvmento et via inventrice de gli antichi.*
Venice: Plinio Pietrasanta, 1554.
Subject: Instruments and inventions.
Case/B/49/.272

BOUND WITH [Plato] *Il dialogo di Platone*, q.v. (Mentions universal principles of Pythagoras: sole, lvce, lvme, splendore, calore, generatione.)

### Errard, Jean, 1554–1610.

*La fortification redvicte en art et demonstree. Par J. Errard de Bar-le-dvc ingenieur du tres-Chrestien roy de France et de Nauarre. Premierement imprimee à Paris. Maintenant mis en lumiere par la vesue, & les deux fils de Theodore de Bry.*
Frankfurt: Wolfg. Richter, 1604.
Subject: Engineering, fortification, geometry.
Case/f/U/26/.268

In 4 books. Chap. 1, bk. 1: Dv canon, de sa longvevr, du calibre, de la poudre, & des proportions necessaires.

### Eschuid, Joannes [John Eastwood], d. ca. 1379.

[*Summa astrologiae judicialis*]
[Venice: Joannes Lucilius Santritter, 1489.]
H-C 6685^ BMC (XV)V:462^ Pr. 5184^ Pell. 4626^ Goff E-109
Subject: Judicial astrology.
Wing/Inc./5184(+)

MS marginalia, astronomical diagrams, map, p. 44, copy of Macrobius.

### Espinosa, Juan de 1540–1595.

*Dialogo en lavde de las mvgeres. Intitulado gin aecepaenos.*
Milan: Michel Tini, 1580.
Subject: Psychology, study of types of women, attitudes, emotions.
Case/K/7/.26

In 5 parts.

* * *

*An Essay concerning adepts: or, A resolution of this inquiry, how it cometh to pass that adepts, if there are any in the world, are no more beneficial to mankind than they have been known hitherto to be and whether there could be no way to encourage them to communicate themselves. With some resolutions concerning the principles of the adepts; and a model, practicable, and easy, of living in community. By a philadept.*
London: printed by J. Mayos [etc.], 1698.
STC II E3279.
Subject: Alchemy, magic, philosophers' stone.
Case/B/8633/.272

Among sectons in pt. 1: That it cannot reasonably be denied that there is such a thing as the philosophers stone; pt. 2: What great men might reasonably do in order to obtain the elixir. "When I give myself the title of Philadept, the signification of that is not that I am actually acquainted with any adepts, but that I am a well-wisher to them" (p. 51).

### Estienne, Charles, 1504–1564.

*Agricoltvra nvova, et casa di villa, di Carlo Stefano Francese, tradotta dal cavalier Hercole Cato. Nella quale si contiene tutto quel che può esser necessario per fabricare vna casa di villa; preudere le mutationi, & diuersità de' tempi & stagioni; medicare i lauoratori ammalati; notrire, & medicare caualli, buoi vacche, & animali; & volatili di tutte le sorti; far horti, ordinare giardini da fiori, & herbe odorifere, & da semplici, & herbe medicinali; gouernare l'api, preparare il mele, & la cera, fare composte, confettare i frutti, fiori, radici, & scoize, piantare, innestare, & medicare ogni sorte di fruttari, & conseruare i frutti; fare ogli per vso del viuere, & della medicina; distillare l'acque si medicinali come da odore, & da lisci; mantanere i pratti; fare & conseruare le peschiere con tutti i modi, & vie di pescare; lauorare le terre da grani, mietere, battere, & conseruare le biaue; acconciare le vigne, fare i vini si d'vna, come di frutti, fare aceti, & agreste; piantare i boschi per i legnami da opera, & per la legna da fuoco. Fabricare la garenna per con vn discorso della caccia del cerno, del cinghiale, della lepre, della volpe de' tassi, del coniglio, & del lupo; & vn trattato di tutti i cani da caccia, segni della bellezza, & bontà loro, & de' modi d'auuezarli & degli vcelli da rapina, cioè falconi, sparuieri & altri, che possono vsarsi à far preda d'vcelli, con le maniere d'ammaestrarli, gouernarli, & medicarli de' mali loro. Tvtta di nvovo corretta et megliorata.*
Venice: Mattio Valentin, 1606.
Subject: Agriculture, animal and plant husbandry, materia medica, medicine.
R/5/.272

In 6 books, with treatise on hunting.

### Estienne, Charles.

*L'agricvltvre et maison rvstiqve, de MM. Charles Estienne, et Iean Liebavlt, doctevrs en medicine. Reueuë & augmentee de beaucoup en ceste derniere edition. Plus un bref recueil des chasses du cerf, du sanglier, du lievre, du renard, du blereau, du connil, du loup, des oiseaux, & de la fauconnerie. Item la fabrique & vsage de la iauge, ou diapason.*
Paris: Iamet & Pierre Mettayer, 1602.
Subject: Agriculture, animal and plant husbandry, materia medica, medicine.
R/5/.271

"Derniere edition." Chap. 52: "Herbes de bonnes sentevr."

### Estienne, Charles.

*Dictionarivm historicvm geographicvm, poeticvm.*
Ursellis: C. Sutorius, 1601.
Subject: Geographical and historical dictionary.
A/251/.269

### Estienne, Charles.

*Dictionarium historicvm, geographicvm, poeticvm, authore Carlo Stephano, Editio novissima; cui praeter collium, silvarum, desertorum, insularum, populorum, pagorum, tribuum, fontium, lacuum, torrentium, paludumque ingentem recentium, veterumque nominum acervum ex Ferrario, aliisque libris typis excusis, calamo exaratis, chartis geographicis, marmoribus vetustis, nummis atque tabulis antiquis diligenter & fideliter excerptum, magna historiarum insignium, ac rerum copia adjicitur.*
Oxford: G[uilielmi] H[all] & G[uileilmi] D[owning] by Gul. Wells & Rob. Scott, 1671.
STC II E3347
Subject: Geographical and historical dictionary.
Case/fA/251/.27

2 title pages (1): Oxford: [printed by] Guilielmi Hall; [for himself] & Guilielmi Downing, 1670; (2): Oxford: tipis G. H. & G. D. apud Gul. Wells & Rob. Scott, 1671.

### Estienne, Charles.

*Dictionarium historicum, geographicum, poeticum: gentium, hominum . . . complectens & illustrans . . . ad incudem verò revocatum, innumerisque penè locus auctum & emaculatum per Nicolaum Lloydium . . . editio novissima. In qua historico-poetica & geographica seorsim sunt alphabeticè digesta.*
London: B. Tooke [etc.], 1686.
STC II E3349
Subject: Geographical and historical dictionary.
Case/fA/251/.271

In 2 parts.

### Estienne, Charles.

*De latinis et graecis nominibus arborum, fruticum, herbarum, piscium, & auium liber: ex Aristotele, Theophrasto, Dioscoride, Galeno, Aetio, Paulo Aegineta, Actuario, Nicandro, Athenaeo, Oppiano, Aeliano, Plinio, Hermolao Barbaro, & Iohanne Ruellio: cum Gallica eorum nominum appellatione.*
Paris: Rob. Stephan, 1547.
Subject: Botany, dictionary encyclopedias.
Case/X/682/.27

3d ed.

### Estienne, Charles.

*Maison rustique, or, the country farme. Compyled in the French tongue by Charles Stephens, and Iohn Liebavlt, doctors of physicke. . . . Newly rieuiewed, corrected, and augmented, with diuers large additions, out of the works of Serres his agriculture, Vinet his maison champestre, Albylerio, Grilli, and other authors. . . . the husbandrie of France, Italie, and Spaine, reconciled and made to agree with ours here in England: by Gervase Markham.*
London: printed by Adam Islip for John Bill, 1616.
STC 10549
Subject: Agriculture, animal and plant husbandry, materia medica, medicine.
Case/R/5/.275

In 7 books. Bk. 1, chap. 12: "The remedies which a good huswife must be acquainted withall, for to help her people when they be sicke."

### [Estienne], Charles.

*Praedium rusticum in quo cuiusuis soli vel culti vel inculti plantarum vocabula ac descriptiones earumque conserendarum atque excolendarum instrumenta suo ordine describuntur.*
Paris: Carolus Stephanus, 1554.
Subject: Botany, dictionary encyclopedias, herbals.
Case/R/5/.274

With "Index plantarvm."

### [Estienne, Charles]

*De re hortensi libellvs, vvlgaria herbarum, florum, ac fruticum, qui in hortis conseri solent nomina Latinis vocibus efferre docens ex probatis authoribus.*
Paris: Robert Stephan, 1535.
Subject: Botany, dictionary encyclopedias, herbals.
Case/R/570/.274

### Estienne, Henry, 1528–1598.

*The Art of making devises: treating of hieroglyphicks, symboles, emblemes, aenigma's, sentences, parables, reverses of medalls, armes, blazons, cimiers, cyphres and rebus.*
London: printed by W. E. & J. G. and are to be sold by R. Marriot, 1646.
STC II E3350B
Subject: Secret writing.
Case/F/0711/.27

Trans. by Tho. Blount. Includes chapter on hieroglyphics, symbols, etymology and definition of devises, a "mysticall medley of pictures and words, representing . . . some secret meaning."

### Euclid, fl. 3d cent. B.C.

*Euclidis Megarensis philosophi acutissimi mathematicorumque omnium sine controuersia principis opera a Campano*

*interprete fidissimo tralata que cum antea librariorum detestanda culpa mendissedissimis adeo deformis erent: vt vix Euclidem ipsum agnosceremus. Lucas Paciolus theologus insignis: altissima mathematicarum disciplinarum scientia rarissimus iudicio castigatissimo detersit: emendauit. Figuras centum & vndetriginta que in alijs codicibus inuerse & deformate erant: ad rectam symmetriam concinnauit: & multas necessarias addidit. Eundem quoque plurimis locis intellectu difficilem commentariolis sane luculentis & eruditis. aperuit: enarrauit: illustrauit. Adhec vt elimatior exiret Scipio Vegius mediol. vir vtraque lingua: arte medica: sublimioribusque studijs clarissimus diligentiam: & censuram suam prestitit.*
[Venice] A. Paganinio Paganinus, [1509].
Subject: Geometry, mathematics.
Wing/fZP/535/.P123

Probably refers to the mathematician Euclid. Euclides of Megara (450?–374 B.C.), a philosopher, was a disciple of Socrates. Only the titles of his works survive. See *Encyclopaedia Britannica*, s.v. Euclid.

## Euclid.

[*Works*]
*Euclidis Megarensis mathematici clarissimi elementorum geometricorum lib. XV. Cum expositione Theonis in priores XIII àBartholomaeo Veneto latinitate donata, Campani in omnes, & Hypsiclis Alexandrini in duos postremos. His adiecta sunt phaenomena, catoptrica & optica, deinde protheoria Marini & Data, postremum uero, opusculum de Leui & Ponderoso, hactenus non uisam, eiusdem autoris.*
Basel: Iohannes Hervagius, 1537.
Subject: Geometry, mathematics.
Case/fY/642/.E614

Euclid the mathematician, confused with Euclid of Megara. See note in entry above. Copious contemporary MS notes. Blind-stamped vellum, with owner's name, Ivstvs Ionas [a friend of Martin Luther], stamped in cover.

## Euclid.

*Die sechs erste bücher Euclidis, vom anfang oder grund der geometrj. In welchen der rechte grund nitt allain der geometrj (versteh alles kunstlichen gwissen und vortailigen gebrauchs des zirckels, linials oder richtscheittes und andrer werckzeüge so zü allerlai abmessen dienstlich) sonder auch der fürnemsten stuck und vortail der rechenkhunst furgeschriben und dargethon ist. Auss Griechischer sprach in die Teütsch gebracht aigenttlich erklärt auch mit verstentlichen exempeln gründlichen figurn und allerlaj den nutz fürangen steilenden anhängen geziert dermassen vormals in Teütscher sprach nie gesehen worden.*
[Basel: Jacob Kündig for Joannes Oporinus, 1562].
Subject: Geometry, mathematics.
Case/Wing/fZA/3/.5

BOUND WITH Hans Lencker: *Perspective*, 1571, q.v.

## Euclid.

*Euclides Megarensis mathematici clarissimi elementorum geometricorum. Lib. XV. Cum expositione Theonis in priores XIII à Bartholomaeo Veneto latinate donata, Campani in omnes, & Hypsiclis Alexandrini in duos postremos. His adiecta phaenomena catoptrica & optica, deinde protheoria Marini & Data, postremum uero, opusculum de Leui & Ponderoso hactenus non uisum, eiusdem autoris.*
Basel: Iohannes Hervagius, 1537.
Subject: Geometry, mathematics.
Case/fY/642/.E615/no. 1

Selections from Euclid's works. Not Euclid of Megara. See above. Preface by Melanchthon wanting in this copy. BOUND WITH Dürer, Albrecht, *Institutionum geometricarum* and *Pictoris et architecti*, qq.v.

## Euclid.

*Elementa geometria.*
Venice: Erhard Ratdolt, 1482.
H-C 6693^ BMC (XV)V:285^ Pr. 4383^ Pell. 4630^ Goff E-113.
Subject: Geometry, mathematics.
Wing/Inc./4383

MS marginalia. First printed Latin version of Campana of Novara. (First Latin trans. ascribed to Adelard of Bath.) Said to be 1st book printed with mathematical diagrams.

## Euclid.

[*Elementa geometriae.*]
[Vicenza: Leonardus Achates, 1491.]
HC 6694^ Pell. 4631^ Goff E-114.
Subject: Geometry, mathematics.
Wing/Inc./7130

[Colophon] Impressus Vincentiae per Magistrum Leonardus de Basilea & Gulielmum de Papia Socios.

## Euclid.

*The elements of geometrie of the most auncient philosopher Evclide of Megara. Faithfully (now first) translated into the Englishe toung, by H. Billingsley, citizen of London. Whereunto are annexed certaine scholies, annotations, and inuentions, of the best mathematiciens, both of time past, and in this our age. With a very fruitfull praeface made by M. I. Dee, specifying the chiefe mathematicall sciences, what they are and whereunto commodious: where, also, are disclosed certaine new secrets mathematicall and mechanicall, vntill these our daies greatly missed.*
London, Iohn Daye [1570].
Subject: Geometry, mathematics.
Case/fY/642/.E62

Euclid the mathematician, not Euclid of Megara (see note above). According to Casaubon (*A true and faithful relation*), q.v., Dee's mathematical preface to Euclid was first published in this year. Dee describes archemastry, or experimental science. From the preface: "All thinges (which from the very first originall being of thinges, haue bene framed and made) do appeare to be formed by the reason of numbers. For this was the principall example or patterne in the mind of the creator."

## Euclid.

*Evclidis posteriores lib IX. Accessit XVI de solidorvm. Regularium cuiuslibet intra quodlibet comparatione: omnes perspicvis demonstrationibus accuratisque scholiis illustrati: nunc iterum editi ac multarum rerum accessione locupletati: avctore Christophoro Clavio Bambergensi.*
Rome: Bartholomaeus Grassius, 1589.
Subject: Geometry, mathematics.
Case/oQA/31/.E83/1589

Includes books 14–15 by Hypsicles and book 16 by François de Foix, Conte de Candale. Has index "Problematvm qvae hisce comment. continentvr," and index "Theorematvm qvae hisce comment. continentvr."

## Euler, Leonhard, 1707–1783.

*Scientia navalis sev tractatvs de constrvendis ac dirigendis navibvs.*
Petropoli [St. Petersburg]: Academiae Scientiarum, 1749.
Subject: Military technology, naval engineering, physics.
Case/U/9/.276

In 2 vols. Vol. 1 includes chapters on stability of bodies in water, resistance in water of certain shapes. Vol. 2 contains chapters on the motion of ships, action of wind on sails.

## Euler, Leonhard.

*Tentamen novae theoriae mvsicae ex certissimis harmoniae principiis dvlcide expositae.*
Petropoli [St. Petersburg]: Academiae Scientiarum, 1739.
Subject: Acoustics, harmonics, mathematical proportions of music.
Case/V/5/.277

## Eustachi, Bartolomeo, d. 1574.

*Tabulae anatomicae.*
Amsterdam: R[vdolphvs] & G[erardvs] Wetstenivs, 1722.
Subject: Human anatomy.
Case oversize/QM/21/.E92/1722

## Evelyn, John, 1620–1706.

*Fumifugium: or, the inconveniencie of the aer and smoak of London dissipated. Together with some remedies humbly proposed by J. E. Esq.*
London: printed by W. Godbid for Gabriel Bedel & Thomas Collins, 1661.
STC II E3488
Subject: Air pollution, botany.
Case/F/4595/.885

Binders title: *Tracts on London, 1643–78.* From dedicatory epistle: "I am able to enumerate a catalogue of native plants, and such as are familiar to our country and clime, whose redolent and agreeable emissions would . . . perfectly improve and meliorate the aer about London." Evelyn suggests a law to prohibit burning of certain substances within the city. Would encourage the planting of fragrant shrubs: sweetbriar, periclymenas, woodbines, jasmine, syringa, guelder-rose, musk, bays, juniper, lignum vitae, lavender, and esp. rosemary. Also, balsam, mint, marjoram, pinks.

## Evelyn, John.

*Numismata: A discourse of medals, antient and modern. . . . To which is added a digression concerning physiognomy.*
London: printed for Benj. Tooke, 1697.
STC II E3505
Subject: Physiognomy.
W/56/.274

Digression concerning physiognomy, pp. 292–342.

* * *

*Experimental philosophy asserted and defended against some late attempts to undermine it.*
London: printed by J. Bettenham, 1740.
Subject: Atomism, experimental philosophy.
Case/B/42/.275

5 "positions" whose truth will be asserted and supported by experiments and observations, e.g., (1) matter was created in atoms (small and indivisible). (4) Air is a mixture of atoms and grains. P. 71: "There are mechanical natural agents by which the operations of nature are performed . . . which have determined me in favour of experimental philosophy, in opposition to occult qualities, sypmpathy and antipathy, or attraction and repulsion." The author refutes the notion that "there are powers inherent in matter acting without means."

# F

## Fabricius, Johann Albert, 1668–1736.

*Hydrotheologie oder versuch, durch aufmerksame betrachtung der eigenschaften, haften, reichen austheilung und bewegung der wasser, die menschen zur liebe und bewunderung ihres gütigsten, we eisesten, mächtigsten schöpfers zu ermuntern. Nebst einem verzeichniss von alten und neuen see-und wasserrechten, wie auch materien und schriften, die dahin gehoren unter XL titul gebracht.*
Hamburg: König and Richter, 1734.
Subject: Hydrography, water power.
Case/B/72/.281

## Fabricius, Johann Albert.

*Mathematische remonstratione, dass Herr Leonhard Christoph Sturm in seinen diese tage herausgegebenen so genannten mathematischen beweiss von dem Heil. Abendmahl, seine neue erklärung der worte der einsetzung nicht bündig demonstriret habe sondern die von ihm her fürgebrachte ubersetzung das wörtleins.*
Hamburg and Leipzig, 1714.
Subject: Magic, mathematics.
3A/180

BOUND WITH *Promotoris edlen ritters von orthoptera theosophische bedancken von der macht der finsternis oder von der gewalt des teufels in der lufft.* 1709.

## Fabricius ab Aquapendente, Hieronymus, 1537–1619.

*Opera chirvrgica in duas partes diuisa. Qvarum prior pars operationes chirvrgicas, in totvm corpvs humanum à vertice capitis adimos pedes vsq. peragi solitas, plurimis raris obseruationibus, & nouis inuentis, chirurgie dexteritatem, & incunditatem spectantibus, refertas comprehendit. . . . Altera pars libros qvinqve chirvrgiae, privs in Germania impressos, & sub nomine pentatevchi chirvrgici diuulgatos; nunc vero ab ipsomet auctore in multis locis emendatos, & duabus ad librum I & II appendicibvs locupletatos continet.*
Venice: Robertvs Megliettvs, 1619.
Subject: Medicine, surgery.
UNCATALOGUED

## Fabricius ab Aquapendente, Hieronymus.

*De visione, de voce, de avditv.*
Venice: Franciscus Balzetta, 1600.
Subject: Anatomy, medicine.
UNCATALOGUED

## Fabricius Hildanus, Guilelmus, 1560–1634.

*Gvilelmi Fabricii Hildani Paterniacensis chirurgi ordinarij observationvm & curationum chirurgicarum centvriae. in qua inclusae sunt viginti & quinque, antea seorsim aeditae: reliquae nunc cum nonnullis instrumentorum, ab autore inventorum delineationibus, in gratiam & vtilitatem artis chirurgicae in lucem prodeunt.*
Basel: Lvdovici Regis, 1606.
Subject: Medicine, obstetrics, surgery.
UNCATALOGUED

In 2 vols. Also in vol. 1: *Centvria secvnda. Epistolis nonnullis virorum doctissimorum, nec non instrumentis cheirurgicis, ab authore inuentis illustrata.* Geneva: Petrus & Iacobus Chouët, 1611. Vol. 2: [centuria 3] *Epistolis nonnvllis. . . . Accessit epistola, de nova, rara, & admiranda herniae uterinae, & partus Caesarei historia.* Oppenheim: printed by Hieronymus Galler for heirs of Johan-Theod. de Bry, 1614.

## Fabricius Hildanus, Guilelmus.

*Guilhelmi Fabrici Hildani . . . observationum et curationum chirurgicarum centuria quarta. Epistolis nonnullis virorum doctissimorum, simul & instrumentis ab authore inventis illustrata. Accessit eivsdem avthoris epistolarvm ad amicos, eorundumque ad ipsum centuria prima. In qua passim medica, chirurgica, aliaque lectione digna continentur.*
Oppenheim: printed by Hieron. Galler for Johannes Theodor de Bry, 1619.
Subject: Medicine, surgery.
UNCATALOGUED

BOUND WITH *Observationum & curationum chirurgicarum centuria V.*
Frankfurt: Mattheus Merianus, 1627.

## Fabricius Hildanus, Guilelmus.

*Observationum & curationum chirurgicarum centvriae, nunc primum simul in vnum opus congesta, ac in duo volumina distributae.*
Lyon: Joan. Antonii Hvgvetan, 1641.
Subject: Medicine, obstetrics, surgery.
UNCATALOGUED

2 vols. in 1. Vol. 1: centuriae 1–3; vol. 2: centuriae 4–5 (with history and description of Caesarean births).

* * *

*FACTUMS, et arrest du Parlement du Paris, contre des bergers sorciers executez depuis peu dans la province de Brie.*
Paris: Rebuffé, 1695.
Subject: Witchcraft.
Case/B/8839/.282

* * *

*A faire in Spittle Fields, where all the knick knacks of astrology are exposed to open sale, to all that will see for their love, and buy for their money. First, Mr. William Lilley presents you with his pack. The introduction. Nativities calculated. The great ephemeridies. Monarchy or no monarchy. The caracter of K. Charles. Annus tenebrosus. Second, Nicholas Culpeper*

*brings under his velvet jacket His chalinges against the doctors of phuisick. A pocket medicine. An almanack & conjuring circle. Third Mr. Bowker unlocked his pack, the 12 signes of the Zodiack. The 12 houses. The 7 planets. The yeares predictions, and the starry globe. Written by J. B. gent.*
London: printed by J. C., 1652.
STC II B101
Subject: Astrology.
Case/Y/195/.F16

In verse.

## Fairfax, Nathaniel, 1637–1690.

*A Treatise of the bulk and selvedge of the world. Wherein the greatness, littleness and lastingness of bodies are freely handled.*
STC II F131
London: printed for Robert Boulter, 1674.
Subject: Metaphysics, time.
Case/B/45/.283

## Fallopius, Gabriel, 1523–1562.

*Observationes anatomicae.*
Venice: Marus Antonius Vlmus, 1562.
Subject: Anatomy.
UNCATALOGUED

* * *

*Fama fraternitatis des löblichen ordens des Rosencreutzes. Einfältigs antwort schreiben An die hocherleuchte Frat. dess löblichen ordens vom Rosencreutz. Auff ihre an die gelehrten Europae, aussgesande faman & confess. &c.*
[Leipzig, 1617?]
Subject: Rosicrucians.
K/975/.257

* * *

*Fama fraternitatis oder entdeckung der brüderschafft des löblichen ordens dess rosen creutzes. Beneben der confession. Oder bekanntnuss derselben fraternitet an alle gelehrte und häupter in Europa geschrieben. Jetzo von mehrern erraten als hiebevorn geschehen entlediget . . ., und zum andern mal in druck verfertiget. Sampt dem sendtschreiben Iuliani de Campis, und Georgii Moltheri Med. C. und ordinarii zu Wetzlar relation von einer diss ordens gewissen person.*
Frankfurt am Main: J. Bringern & J. Berwern, 1617.
Subject: Rosicrucians.
K/975/.036

P. 57: "An all welche von der newen bruderschafft dess ordens vom rosen creuz genannt etwas gelesen oder von andern per modum discursus der sachen beschaffenheit vernommen." p. 34: "An der weissheit begierigen leser der confession." p. 72: "Im aussschreiben der brüder vom rosen creuz wirt gesazt dass sey bey j[e]nen ein parergon." BOUND WITH Libavius, q.v.

* * *

*Fama remissa ad fratres Roseae Crucis. Antwort auff die Famam und confessionem der löblichen brüderschafft vom Rosencreutz.*
[N.p.] 1616.
Subject: Rosicrucians.
K/975/.284

Dedication by H. Ar. No:R.

## [Farnworth, Richard] d. 1666.

*VVitchcraft cast out from the religious seed and Israel of God. And the black art, or nicromancery inchantments, sorcerers, wizards, lying divination, conjuration, and witchcraft, discovered, with the ground fruits and effects thereof; as it is proved to be acted in the mistery of iniquity, by the power of darknesse, and witnessed against by Scripture, and declared against also, from, and by them that the world scornfully calleth Quakers.*
London: printed for Giles Calvert, 1655.
STC II F513
Subject: Witchcraft.
Case/B/8845/.285

## Favolius, Hugo, 1523–1595.

*Theatri orbis terrarvm enchiridion, minoribvs tabvlis per Philippvm Gallaevm exaratvm: et carmine heroico ex variis geographis & poëtis collecto.*
Antwerp: Philippus Gallaeus Christophorus Plantin, 1585.
Subject: Cosmography, geography.
Ayer/*7/F27/1585

In verse.

## Fazio degli Uberti, 1310–ca.1370.

*Opera chiamato ditto mundi. Vuolgare.*
[Venice: printed by Christofaro di Pensa da Mandelo, 1501.]
Subject: Astrology, geography.
Case/Y/712/.F283

Treatise in verse popularizing new geographical knowledge.

## Felgenhauer, Paul, fl. 1623.

*Apologeticus contra invectivas aeruginosas Rostij. Kurtze verantwortung auff das heldenbuch vom rosengarten oder gründtlichen apologetischen bericht von dem newen himmlischen propheten Rosencreutzern, chiliasten, enthusiasten.*
[N.p.] 1622.
Subject: Rosicrucians.
K/975/.291

## Felgenhauer, Paul.

*Speculum temporis. Zeit spiegel darinnen neuen vermahnung aller welt vird vor augen gestellet, was für eine zeit jetzt sey unter allerley ständen, besonders unter den meisten geistlich genandten und gelehrten. Hierinnen ist auch ein kurze doch deutliche erweisung des geheimnüss der drey letzten gemeinen, in der offenbahrung Johannis.*
[N.p.] 1620.
Subject: Rosicrucians
K/975/.292

## Fernandes, Jacome.

*Consequencias do fenomeno que appareceu em sinco de Agosto do anno prezento sobre Constantinopla.*
Lisbon: Mauricio Vicente de Almeyda, 1732.
Subject: Apparitions, astrology, supernatural.
Greenlee/4504/P855

## Fernel, Jean, 1497–1558.

*De abditis rerum causis libri duo.*
Venice [Petri & I. de Nicolini de Sabio for A. Arriuabeni], 1550.
Subject: Medicine, metaphysics.
Case/Q/.292

Cf. Bibliothèque Nationale Cat., Graesse. "Animam nostram non ab elementis ortam Galeni decreto" (bk. 2, chap. 4).

## Fernel, Jean.

*Ioannis Fernelli Ambianatis, de naturali parte medicinae libri septem.*
Lyon: Joan. Tornaesius, & Gulielmus Gazeius, 1551.
Subject: Medicine, physiology.
Case/oR/128.6/.F47/1551

## Fernel, Jean.

*Ioannis Fernelii neotericorvm principis, et Franciaè archiatri vniversa medicina: A doctissimo et experientissimo medico diligenter recognita, & ab innumeris mendis & erroribus, quibus priores scatebant editiones repurgata, collatis inuicem vetustissimis & optimis exemplaribus. . . . Addita sunt Fernelij consilia: & Guliel. Planti scholia in pharmacopoeam seu librum therapevtices septimum.*
Geneva: Petrus Chouët, 1638.
Subject: Anatomy, materia medica, pathology.
UNCATALOGUED

## Ferrand, Jean, 1497–1558.

*[Erotomania] or a treatise discoursing of the essence, causes, symptomes, prognosticks, and cure of love or erotique melancholy.*
Oxford: L. Lichfield, 1640.
STC 10829
Subject: Love-melancholy, medicine, melancholy, psychology.
Case/K/73/.294

## Ferrari, Giacomo.

*Democrito et Eraclito. Dialogo del riso, delle lagrime & della malinconia.*
Mantua: Aurelio, & Lodovico Osanna, brothers, 1627.
Subject: Passions, imagination, memory.
B/5293/.294

Democrito (Del riso); Eraclito (Delle lagrime); Discorso della malinconia del Andrea Lavrentio.

## Ferrariis, Theophilus de.

*Propositiones ex omnibus Aristotelis libris excerptae.*
Venice: Joannes and Gregorius de Gregoriis, 1493.
H-C *6997^ BMC (XV)V:344^ Pr. 4531^ Still. F93^ Goff F-117
Subject: Metaphysics, physics, zoology.
Inc./4531

"Incipivnt propositiones copiosissime ac fidissime ex omnibvs Aristotelis libris collectae per Fratrem Theophilvm de Ferrariis Cremonensem vitae regularis sacri ordinis praedicatorvm: et primo ex libris metaphysicae: annotatioque lectionvm vt exposito divi Thomae Aquinatis requirit posita est." Includes commentaries or propositions from the Metaphysics, Physics, De animalium, De generatione animalium, Peri hermenios, Posterior analytics, and Prior analytics. In this vol. also "Tabula per alphabetum in omnium operum Aristotelis auctoritates atque sententias."

## Ferrer Maldonado, Lorenzo, d. 1625.

*Imagen del mundo, sobre la esfera, cosmografia, y geografia, teorica de planetas, y arte de nauegar.*
Alcalá: Iuan Garcia and Antonio Duplastre, 1626.
Subject: Astronomy, cosmography, geography, navigation, zodiac.
Ayer/8.9/N2/F385/1626

In 8 parts. 1: Composition and order of heavens; 2: Four elements, parts of the world; 3: Form, figure, and quantity of the universe (circular motion); 4: Artificial sphere; 5: Rising and setting of signs, artificial days; 6: Planets, their spheres; 7: Eclipses and aspects of the lights; 8: Hydrography, art of navigation, use of astrolabe.

## Ferrer de Valdecebro, Andres, 1620–1680.

*Govierno general, moral, y politico. Hallado en las fieras, y animales sylvestres. Sacado de svs natvrales virtvdes, y propriedades. Con particvlar tabla para sermones varios de tiempo, y de santos.*
Barcelona: Thomàs Loriente, 1696.
Subject: Animals as role models, animal characterization.
B/692/.294

Morals drawn from characterization of various animals. Analogues for human behavior based on virtues of animals, e.g., lion, elephant, rhinoceros, unicorn, tiger, lynx, leopard, hyena, wolf, bee, boar, deer, bull, camel, horse, dog, baboon.

## Ferrier, Auger, 1513–1588.

*Des ivgemens astronomiqves svr les nativitez.*
Lyon: Iean de Tovrnes, 1550.
Subject: Astronomy, judicial astrology.
Case/B/8635/.261

In 3 books. 1: Casting nativities, latitudes and aspects of planets; Concerning family (brothers, parents, infants, servants, illnesses, marriage); 2: Significance of planets, 12 signs of zodiac; 3: Directions and revolutions—lords of tripleness can be good or evil omens.

## Ferrier, Auger.

*A learned astronomical discourse of the iudgement of natiuities. Deuided into three bookes, and dedicated first to Katherin the French Queene, by Oger Ferrier, her physition.*
London: printed at the Widdow Charlewoods house for Edward White, 1593.
STC 10833a
Subject: Astronomy, judicial astrology.
Case/B/863/.886

Translated by Thomas Kelway. Binders title: *Tracts on Astrology*. Bk. 1: Of the gyuer of life called of the Arabians Hyleg. Of the yeeres, called of the Arabians Alcocoden. Of the lord of the natiuity; bk. 2: Significations of planets. Aspects of planets between them. Of the lords of the tryplicities of the [12] houses; bk. 3: Of directions, of the eclipses. BOUND WITH *A true coppie of a prophesie*; Iohn Melton, *Astrologaster or, figvre-caster*; and *Catastrophe mundi*, qq.v.

## Feuillée, Louis, 1660–1732.

*Journal des observations physiques.*
Paris: Pierre Giffart, 1714–25.
Subject: Astronomy,botany, topography.
Ayer/1263/F42/1714

3 vols. Vol. 1 (1714): "Journal des observations physiques, mathematiques et botaniques, faites par l'ordre du roy sur les côtes orientales de l'Amerique meridionale, & dans les Indes Occidentales, depuis l'année 1707 jusques en 1712"; vol. 2 (1714); vol. 3 (1725): "Journal des observations . . . dans un autre voyage fait par le même ordre à la Novelle Espagne & aux isles de l'Amerique." Following vol. 3: "Histoire des plantes medicinales qui sont le plus en usage aux royaumes du Perou & du Chily dans l'Amerique meridional," with "Tables des declinaisons du soleil pour tous les degrez et minutes de l'ecliptique."

## Ficino, Marsilio, 1433–1499.

*Opera & quae hactenus extitêre, & quae in lucem nunc primùm prodiêre omnia, omnium artium & scientiarum, maiorumque facultatum multipharia cognitione refertissima, in duos tomos digesta, quorum seriem uersa pagella reperies. Vna cvm gnomologia, hoc est, sententiarvm ex iisdem operibus collectarum farragine copiosissima, in calce totius voluminis adiecta.*
Basel: Henricus Petri, 1561.
Subject: Astrology, astronomy, Hermeticism, medicine, mysticism, neo-Platonism.
Case/B/235/.304

2 vols. in 1. Contents include: bk. 1 (chap. 12): De creatione rerum; (chap. 16) Epidemiarum antidotus, tutelam bonae ualetudinis continens; (chap. 28) Oratio de laudibus medicinae; (chap. 32) Disputatio contra iudicia astrologorum; bk. 2 (chap. 5): Mercurij Trismegisti Pymander, de potestate ac sapientia dei item Asclepius de uoluntate dei; (chap. 7) Jamblichus de mysteriis Aegytiorum, Caldeorum, atque Assiriorum; (chap. 9) Proclus de sacrificio & magia; (chap. 12) Psellus de daemonibus; (chap. 16) Pithagore aurea uerba & symbola. Original blind-stamped vellum binding.

## Ficino, Marsilio.

*Opera, et quae hactenus extitêre & quae in lucem nunc primum prodiêre omnia, omnium artium & scientiarum, maiorumque facultatum multifaria cognitione refertissima, in duos tomos digesta, & ab inumeris mendis hac postrema editione castigata quorum seriem versa pagella reperies. Vna cum gnomologia.*
Paris: Gvillemvs Pelé, 1641.
Subject: Astrology, astronomy, Hermeticism, medicine, mysticism, neo-Platonism.
Case/B/235/.307

In 2 vols.

## Ficino, Marsilio.

*Il comento di Marsilio Ficino sopra il conuito di Platone & esso conuito, tradotti in lingua toscana per Hercole Barbarasa da Terni.*
Rome, 1544.
Subject: Love, neo-Platonism, passions.
Case/Y/642/.P5723

Cap. 3C 55: Dell'anime delle sfere, & de demoni. With Plato, Il fedro.

## Ficino, Marsilio.

[*Delle divine lettere*]
*Tomo primo. Delle divine lettere del gran Marsilio Ficino, tradotte in lingua toscana per M. Felice Figliucci.*
Venice: Gabriel Giolito de Ferrari, 1546.
Subject: Health, medicine, neo-Platonism.
Case/B/235/.3077

In 5 books.

## Ficino, Marsilio.

[*Delle divine lettere*]
*Tomo primo delle divine lettere del gran Marsilio Ficino tradotti in lingva thoscana per M. Felice Figlivcci senese.*
Venice: Gabriel Giolito di Ferrarii, 1549.
Subject: Health, medicine, neo-Platonism.
Case/B/235/.3078

## Ficino, Marsilio.

*Delle divine lettere del gran Marsilio Ficino tradotte per M. Felice Figlivcci Senese. Novamente ristampate con due tauole.*
Venice: Gabriel Giolito de' Ferrari, 1563.
Subject: Health, medicine, neo-Platonism.
Case/B/235/.308

## Ficino, Marsilio.

*Epistolae.*
Nuremberg: Anton Koberger, 1497.
H-C*7062^ BMC (XV)II:443^ Pr. 2113^ Pell. 4792^ Goff F-155
Subject: Astrology, Hermeticism, medicine.
Inc./2113

In 10 books. Among topics: Contra Aueroem, Medicine corpus. Musica spiritum theologia animum curat. MS marginalia.

**Ficino, Marsilio.**

*Epistole Marsilii Ficini Florentini.*
Venice: Matteo Capcasa for Hieronymus Blondus, 1495.
H-C 7059^ Pell 4791^ BMC (XV)V:486^ Pr. 5001^ Goff F-154
Subject: Astrology, Hermeticism, medicine.
Wing/Inc./5001

In 20 books. Colophon: Epistoles familiares. A few MS marginalia.

**Ficino, Marsilio.**

*Index eorvm qvae hoc in libro habentvr. Iamblichus de mysteriis Aegyptiorum, Chaldaeorum, Assyriorum.*
Venice: Aldus & Andreas Socerus, 1516.
Subject: Hermeticism, natural magic, neo-Platonism.
Case/Y/6409/.288

Proclus in Plationicum Alcibiadem de anima, atque daemone. Proclus de sacrificio, & magia. Porphyrius de diuinis, atque daemonibus. Synesius Platonicus de somniis. Psellus de daemonibus. Expositio Prisciani, & Marsilii in Theophrastum de sensu, phantasia, & intellectu. Alcinoi Platonici philosophi, liber de doctrina Platonis. Speusippi Platonis discipuli, liber de Platonis definitionibus. Pythagorae philosophi aurea uerba. Symbola Pythagorae philosophi. Xenocratis philosophi Platonici, liber de morte. Mercurii Trismegisti Pimander. Eiusdem Asclepius. Marsilii Ficini de triplici uita lib. II. Eiusdem liber de uoluptate. Eiusdem de sole & lumine libri II. Apologia eiusdem in librum suum de lumine. Eiusdem libellus de magis. Quod necessaria sit securitas, & tranquillitas animi. Praeclarissimarum sententiarum huius operis breuis annotatio.

**Ficino, Marsilio.**

*De le tre vite, cioé a qual guisa si possono le persone letterate mantenere in sanità. Per qual guisa si possa l'huomo prolungare la uita. Con che arte, e mezzici possiamo questa sana, e lunga uita prolungare per uia del cielo. Recato tvtto di Latino in buona lingua uolgare.*
[Venice: Michel Tramezzino, 1548.]
Subject: Health, hygiene, medicine.
Case/Q/.295

Bk. 1, chap. 7: "Che cinque sono i principali nemici de letterati, la pituita, la atrabile, il coito, la repletione, & il dormire di matina."

**Ficino, Marsilio.**

*De triplici vita.*
[Basel: Johann Amerbach, 1498?]
H-C *7063^ BMC (XV)III:759^ Pr. 7650^ Pell. 4798^ Goff F-160
Subject: Health, hygiene, medicine.
Inc./7650

*De vita sana,* chap. 11: De cura stomachi; chap. 19: De sirupis; chap. 20: De pilulis; chap 21: De medicina liquida. *De vita longa,* chap. 2: Quo uitalis calor nutriant humore: quo deficiente sit resolutio: quo excendte sit suffocatio; chap. 7: De dieta: uictu & medicina sensum. *De vita coelitus comparanda,* chap. 2: De concordia mundi: de natura hominis cum stellas: quo fiat attractam ab una quaque stella. Followed by: Apologia quaedam: In qua de medicina astrologia uita mundi: item de magis quae Christum statim natum salutauerunt.

**Ficino, Marsilio.**

[*De vita sana*]
*Le premier liure de Marsille Fiscine de la vie saine traduict de Latin en Francoys par Maistre Jehan Beaufilz aduocat ou chastelet de Paris.*
Paris: Denys Janot, 1541.
Subject: Health, hygiene, medicine.
Wing/ZP/539/.J263

1st edition of 1st French translation. *Le second liure de Marsille Fiscine pour viure longuement.*

**Ficino, Marsilio.**

*De vita libri tres, recens iam à mendis situque uindicati. Quorvm primus, de studiosorum sanitate tuenda. Secundus, de vita producenda. Tertius, de vita coelitus comparanda. Eiusdem apologia. His accessit epidemiarum antidotus tutelam quoque bonae valetudinis continens, eodem autore.*
Basel: [Andreas Cratander & Io. Bebelivs] 1532.
Subject: Health, hygiene, medicine.
Case/B/235/.3048

**Ficino, Marsilio.**

*De vita libri tres.*
[n.p.] Joannes Le Preux, 1595
Subject: Health, hygiene, medicine.
Case/B/235/.305

Bks. 1 and 2: De studiosorum sanitate tuenda, siue eorum, qui literis operam nauant, bona valetudine conseruanda; bk. 3: De vita coelitus comparanda; De vita; Apologia in qva de medicina, astrologia, vita mundi, item de magis qui Christum statim natum salutauerunt, agitur; Quod necessaria sit ad vitam securitas & tranquilitas animi; Epidemiarum antidotus.

**Fieschi, Maurizio, conte di.**

*Decas de fato annisque; fatalibus tam hominibus, quam regnis mundi.*
Frankfurt: Joann. Baptista Schönwetter, 1665.
Subject: Astrology, numerology, prognostication.
B/8635/.27

10 spheres, described, beginning with no. 10, God, creation. 8th sphere: judicial astrology; 7th, physiognomy; 3d, medical astrology; 2d, critical days; 1st, Platonic numbers.

**Filelfo, Francesco, 1398–1481.**

*Epistolarum familiarum libri xxxvij. Ex eius exemplari transumpti: ex quibus vltimi xxj nouissime reperti fuere: & impressorie traditi officine.*
Venice: Ioannes & Gregorius de Gregoriis, brothers, 1502.

Subject: Calendar.
Case/fE/5/.F4748

Leaf 131, calendar, volvelle.

## Filmer, Sir Robert, d. 1653.

*An advertisement to the jury-men of England, touching witches; together with a difference bewteen an English and Hebrew witch.*
N.p., n.d.
Subject: Witchcraft
Case/J/O/.298

In his *The free-holders grand inquest* (n.a.).

## Finé, Oronce, 1494–1555.

*Novvs orbis regionvm ac insvlarvm veteribus incognitarvm, unà cum tabula cosmographica, & aliquot aliis consimilis argumenti libellis, quorum omnium catalogus sequenti patebit pagina.*
Paris: Antonius Augerellus for Ioannes Paruus & Galeoto à Prato [1532].
Subject: Cosmography, navigation.
Ayer/*110/N8/1532

With double-page circular map, recto looking "down" on earth from above north pole, verso looking "down" from perspective of south pole. Includes works by Grynaeus, Münster, Aloysius Cadamustus, Columbus, P. Alonso, Pinzoni, Vespucci, and Peter Martyr, among others.

## Finé, Oronce.

*Protomathesis: opus uarium, ac scitu non minus utile quàm iucundum, nunc primùm in lucem foeliciter emissum.*
Paris: Gerardus Morrhius & Ioannis Petri, 1532.
Subject: Cosmography, geometry, instrument making, mathematics.
Wing/fZP/539/.M83

1st ed. Includes De arithmetica practica lib. IIII; De geometria lib. II; De cosmographia sive mundi sphaera lib. V, propriis ipsius authoris commentariis elucidati; De solaribus horologiis et quadrantibus lib. IIII. Diagrams and woodcuts by Finé.

## Finé, Oronce.

*De re & praxi geometrica, libri tres, figuris & demonstrationibus illustrati. Vbi de quadrato geometrico, & virgis seu baculis mensoriis, necnon aliis, cum mathematicis, tum mechanicis.*
Paris: Gilles, Aegidius Gourbin, 1556.
Subject: Plane and solid geometry, surveying.
Wing/ZP/539/.G738

## Finé, Oronce.

*De rebus mathematicis, hactenus desideratis, libri IIII. Quibus inter caetera, circuli quadratura centum modis, & suprà, per eundem Orontium recenter excogitatis, demonstratur.*
Paris: Michael Vascosanus, 1556.
Subject: Geometry, mathematics.
Wing/fZP/539/.V472

In 4 books. Bk. 1: Inventionem dvarvm rectarum inter datas extremas continuè proportionalium pluribus & hactenus inauditis modis exponit;bk. 2: Rationem circvnferentiae ad circuli diametrum exprimit; bk. 3: Inventionem lateris cuiuslibet polygoni regularis in dato circulo descripti; bk. 4: Omnimodam solidorum transmutationem, cum ipsa sphaerae cubicatione & uersione cubi in sphaeram aequalem.

## Finé, Oronce.

*Sphaera mundi, sive cosmographia qvinque libris recèns auctis & emendatis absoluta: in qua tum prima astronomiae pars tum geographie, ac hydrographie rudimenta pertractantur.*
Paris: Michael Vascosanus, 1552.
Subject: Astronomy, cosmography, geography, geometry, hydrography, zodiac.
Wing/ZP/539/.V461

Latin trans. of French ed. of 1551, q.v.

## Finé, Oronce.

*Le sphere dv monde, proprement ditte cosmographie, composee nouuellement en françois, & diuisee en cinq liures, comprenans la premiere partie de l'astronomie, & les principes uniuersels de la geographie & hydrographie. Auec une epistre, touchant la dignité, perfection & utilité de sciences mathematiques. Par Oronce Finé, natif dv Daulphiné, lecteur mathematicien du treschrestien Roy de France.*
Paris: Michel de Vascosan, 1551.
Subject: Astronomy, cosmography, geography, geometry, hydrography, zodiac.
Ayer/*7/F495/1551

*Sphere du monde* in 5 bks.; *Sensvit vne epistre, touchant la dignité* in verse.

## Fioravanti, Leonardo.

*Dello specchio di scientia vniversale libri tre. Nel primo de' quali, si tratta di tutte l'arti liberali, & mecanice, & si mostrano tutti i secreti piu importanti che sono in esse. Nel secondo si tratta di diuerse scientie, & di molte belle contemplationi de filosofi antichi. Nel terzo si contengono alcune inuentioni notabili, vtilissime & necessarie da sapersi.*
Venice: heirs of M. Sessa, 1583.
Subject: Encyclopedia of arts, sciences, and crafts.
Case/L/O/.297

Includes agriculture, instrument making, printing, medicine and surgery, alchemy, weaving, gardening, fishing and hunting, politics and government, superstition, lead and its effects, various remedies.

## Fioravanti, Leonardo.

*Mirroir universel, des arts et sciences en general, . . . diuisé en trois liures. Au premier est traité de tous les arts liberaux & mecaniques, & se monstrent tous les secretz qui sont en iceux de plus grande importance. Au second, de diuerses sciences, histoires, & belles contemplations des philosophes anciens. Au troisiesme, sont contenus plusieurs secretz & notables inuentions tres-vtiles & necessaires à sçauoir.*
Paris: chez P. Cavellat, 1584.

Subject: Encyclopedia of arts, sciences, and crafts.
Case/L/O/.298

Translated by Gabriel Chapuys.

## Firmicus Maternus, Julius, 4th cent., A.D.

*Ivlii Firmici Materni Ivnioris Sicvli V. C. ad mavortivm Lollianvm matheseos librorum octo generalis elenchos.*
Venice: Aldus Manutius Romanus, 1499.
H-C 14599*^ BMC (XV)V:560^ Pr. 5570^ Goff F-191
Subject: Astrology, astronomy, horoscope, judicial astrology,zodiac.
Inc./f5570

Binders title: *Astronomi veteres.* Ed. Franciscus Niger. Bound into this vol. (with coninuous pagination) are the following: (1) "Ivlii Firmici astronomicorum libri octo integri, & emendati, ex Scythicis oris ad nos nuper allati"; (2) "Marci Manilii astronomicorum libri quinque" [in verse]; (3) "Arati phaenomena Germanico Caesare interprete cum commentarius & imaginibus"; (4) "Arati eiusdem phaenomenon fragmentum Marco T. C. interprete" [with woodcuts of the constellations]; (5) "Arati eiusdem phaenomena Ruffo Festo Auienio paraphraste"; (6) "Arati eiusdem phaenomena Graece"; (7) "Theonis commentaria copiosissima in Arati phaenomena Graece" [Theon's Greek commentary surrounds Aratus's text in no. 6]; (8) "Procli Diadochi sphaera Graece"; (9) "Procli eiusdem sphaera, Thoma Linacro Britanno interprete." Table of contents is bound in following no. 1.

## Firmicus Maternus, Julius.

*Ad Mavortium Lollianum, astronomicon libri VIII, per Nicolavm Prvcknervm astrologum nuper ab innumeris mendis uindicati. His accesservnt, Clavdii Ptolemaei Pheludiensis Alexandrini [aptelesmaton?] quod quadripartitum uocant, De inerrantium stellarum significationibus, Centiloquium eiusdem. Ex Arabibvs et Chaldaeis: Hermetis uetustissimi astrologi centum aphoris.; Bethem centiloquium, Eivsdem de horis planetarum; Almanzoris astrologi propositiones; Zahelis Arabis de electionibus; Messahalah de ratione circuli & stellarum; Omar de natiuitatibus; Marci Manilii poetae disertissimi astronomicon; Othonis Brvnfelsii de diffinitionibus & terminis astrologiae libellus isagogicus.*
Basel: Ioannes Hervagius, 1551.
Subject: Astrology.
Case/fL/900/.29

## Firmicus Maternus, Julius.

*Ad Mavortium Lollianum, astronomicon libri VIII.* [Another copy]
Case/fL/900/.712

BOUND WITH *Hieronymi Cardani . . . in Cl. Ptolemaei* (1554), q.v.

## Fisher, George [pseud.?]

*The American instructor: or, young man's companion. Containing, spelling, reading, writing, and arithmetick, in an easier way than any yet published. . . . Also merchants accompts, and a short and easy method of shop and book-keeping; with a description of the several American colonies. Together with the carpenter's plain and exact rule: shewing how to measure carpenters, joyners, sawyers, bricklayers, plaisterers, plumbers, masons, glasiers and painters work. With Gunter's line, and Coggeshal's description of the sliding-rule. Likewise the practical gauger made easy; the art of dialling, and how to erect and fix any dial; with instructions for dying colouring, and making colours. To which is added, the poor planters physician. With instructions for marking on linnen; how to pickle and preserve; to make divers sorts of wine; and many excellent plaisters and medicines, necessary in all families.*
Phildelphia: B. Franklin & D. Hall, 1748.
Subject: Household encyclopedia, practical arithmetic.
Wing/ZP/783/.F8528

9th ed.

## Fletcher, Phineas, 1582–1650

*The purple island, or the isle of man: together with piscatorie eclogs and other poeticall miscellanies. By P. F.*
Cambridge: Printed by the printers to the University of Cambridge,1633.
Subject: Anatomy.
Case/Y/185/.F65
1st ed. An elaborate allegorical poem describing the human body. See Robert Schuler, *English magical and scientific poems to 1700* (New York: Garland, 1979), no. 205. (Piscatory eclogues have separate pagination.)

## Fludd, Robert, 1574–1637.

*Katholicon medicorvm katoptron* [title transliterated from Greek] *in quo, qvasi specvlo politissimo morbi praesentes more demonstratiuo clarissime indicantur, & futuri ratione prognostica aperte cernuntur, atque prospicuntur, siue, tomi primi, tractatvs secvndi, sectio secvnda, de morborvm signis.*
[N.p.] 1631.
Bibliotheca Osleriana 2627^ De La Vallière Catalogue 1:518–20, no. 13^Ebert 2:587, no. 14
Subject: Astrology, diagnosis, macrocosm–microcosm, medicine.
Case/fQ/.305

"Ovromantiae, siue divinatione per vrinam, liber primvs: radicale vrinae musterium enucleans" (p. 258). BOUND WITH *Pvlsvs* and *Tractatvs*, qq.v. (Table of contents at beginning of vol. lists contents of all 3 works.)

## Fludd, Robert.

*Philosophia Moysaica. In qua sapientia & scientia creationis & creaturarum sacra veréque christiana vt pote cvjus basis sive fundamentum est unicus ille lapis angularis Iesus Christus ad amussum & enucleaté explicatur.*
Gouda: Petrus Rammazenius, 1638.
Subject: Action at a distance, iatrochemistry, magnetism, sympathy and antipathy.
Case/FL/O/.305

BOUND WITH his *Responsvm*, q.v. Basis of Mosaic philosophy, the unique square stone (lapis angularis), the root of all creatures, the account of essence and existence.

History of creation from primordial darkness to divine light; 4 elements (active: hot and cold, passive: humid and dry); the true plenitude and empty space; generation and corruption, regeneration (resurrection), condensation and rarefaction; 5th mystery: winds, clouds, rain, snow, lightning and thunder.

## Fludd, Robert.

*Pvlsvs, seu nova et arcana pvlsvvm, historia e sacro fonte radicaliter extracta, nec non medicorvm ethnicorvm dictis & authoritate comprobata. Hoc est portionis tertiae parts tertia, de pvlsvvm scientia.*
[Frankfurt? 1629?]
Bibliotheca Osleriana no. 2628^ De La Vallière Catalogue 1:520, no. 14^ Ebert 2:586, no. 10
Subject: Spagyric medicine.
Case/fQ/.305

BOUND WITH *Katholicon* and *Tractatvs*, qq.v.

## Fludd, Robert.

*Responsvm ad Hoplocrisma spongvm M. Fosteri presbiteri, ab ipso, ad vngventi armarii validitatem delendam ordinatvm. Hoc est, spongiae M. Fosteri presbyteri expressio sev elisio. In qua virtuosa spongiae ipsius potestas in detergendo vnguentum armarium, ex primitur, eliditur ac funditus aboletur: ac tandem immodestia & erga fratres suos incivilitas, aceto veritatis acerrimo corrigitur & penitus extinguitur.*
Gouda: P. Rammazenius, 1638.
Subject: Medicine, sympathy and antipathy, weapon salve.
Case/fL/O/.305

BOUND WITH *Philosophia Moysaica*, q.v.

## Fludd, Robert.

*Tomi secvndi, tractatvs primi, sectio secunda De technica microcosmi historia. Authore Robert Flud aliàs de Fluctibus.*
[Oppenheim? 1620?]
Subject: Astrology, macrocosm–microcosm, occult.
Case/fA/911/.2914

In 7 parts, including genethlialogia (of, or belonging to one's natal day), physiognomy, chiromancy and geomancy, geometry and prophecy, construction of memory palace, demons.

## Fludd, Robert.

*Tractatvs secvndi sectio prima: in qva integrvm morborvm, sev meteororvm insalvbrivm mysterivm: hoc est: hostilis mvnimenti salvtis invasio, sev tremenda & truculenta morborum cardinalium circa sanitatis arcem castrametatio, violenta meteororum insalubrium à quatuor mundi angulis, in propugnacula eiusdem eruptio, strictumque vitae & salutis humanae obsidium explicatur: vidi quatuor angelos stantes super quatuor angulos terrae, tenentes quatuor ventos terrae, quibus datum est nocere terrae & mari, &c.*
[Frankfurt? 1631?]
Subject: Disease, medicine, mysticism.
Case/fQ/.305

BOUND WITH *Katholicon* and *Pulsus*, qq.v.

## Fludd, Robert.

*Tractatus secundus, de naturae simia seu technica macrocosmi historia, in partes undecim divisa.*
Frankfurt: printed by Caspar Rötelius for heirs of Johann Theodor de Bry, 1624.
Subject: Geometry, mathematics, proportion.
Case/fA/911/.29

2d ed. Algorithms, numbers and numeration, surveying, mensuration, harmonics, memory, Pythagoras, instruments, optics (including anatomy of the eye), perspective, motion (levers, pulleys, inertia).

## Fludd, Robert.

*Utriusque cosmi maioris scilicet et minoris metaphysica, physica atqve technica historia. In duo volumina secundum cosmi differentiam diuisa.*
Oppenheim: printed by Hieronymus Galler for Johann Theodor De Bry, 1617.
Subject: Arithmetic, geometry, history of creation, macrocosm–microcosm.
Case/fA/911/.291

Tomus primus de macrocosmi historia in duos tractatus diuisa. 1: De metaphysico macrocosmi et creaturaum illius ortu. Physico macrocosmi in generatione & corruptione progressu. 2: De arte naturae simia in macrocosmo producta & in eo nutrita & multiplicata, cujus filias praecipuas hîc anatomiâ vivâ recensuimus, nempe.

## Fogliano, Lodovico, d. ca. 1540.

*Mvsica theorica Ludouici Foliani Mutinensis: docte simul ac dilucide pertractata: in qua quamplures de harmonicis interuallis: non prius tentatae: continentur speculationes.*
Venice: Io. Antonio & brothers de Sabio, 1529.
Subject: Harmonic and musical proportion.
Case/fV/5/.306

In 3 books.

## [Foley ———.]

*Computatio universalis seu logica rerum. Being an essay attempting in a geometrical method, to demonstrate an universal standard, whereby one may judge of the true value of every thing in the world, relatively to the person.*
London: J. Moxon, 1697.
STC II C5675
Subject: Geometry, mathematical logic, proportion, reason.
Wing/ZP/645/.M878

Application of mathematical logic to social, personal, life-style questions.

## Fontaine, Jacques, d. 1611.

*Des marques des sorciers et de la reelle possession que le diable prend sur le corps des hommes. Sur le subject du proces de l'abominable & detestable sorcier Louys Gaufridy, prestre beneficié en l'eglise parrochiale des accoules de Marseille, qui n'a guieres a esté executé à Aix par arrest de la cour de parlement de Prouence.*
Lyon: Claude Larjot, 1611.

Subject: Witchcraft.
Case/B/88/.306

## Fontenelle, Bernard le Bovier de, 1657–1757.

*Conversations on the plurality of worlds.*
London: printed for A. Bettesworth [etc.] 1715.
Subject: Astronomy, inhabitants of other planets, world in the moon.
L/9/.307

Translated from last Paris ed. 6th evening's conversation never before translated.

## Fontenelle, Bernard le Bovier de.

*Fontenelle's dialogues of the dead.*
London: J. Tonson, 1708.
Subject: Ancients vs. Moderns.
Case/Y/762/.F7313

In 3 pts.: 1. Dialogues of the ancients; 2. Of the ancients and the moderns; 3. Of the moderns. In pt. 1, sec. 5: Eristratus and Harvey—"Of what use are the modern discoveries in natural philosophy and physick?" Sec. 9: Apicius and Galileo—"New knowledge can be discover'd, but not new pleasures." Pt. 3, sec. 4: William of Cabestan and Frederick of Brandenburg—"On madness." Sec. 8: Paracelsus and Moliere—"Of imaginary sciences and comedy."

## Fontenelle, Bernard le Bovier de.

*A discovery of new worlds. From the French. Made English by Mrs. A. Behn. To which is prefixed a preface, by way of essay on translated prose: wherein the arguments of Father Taquet and others against the system of Copernicus are considered and answered.*
London: printed for William Canning, 1688.
STC II F1412
Subject: Astronomy, plurality of worlds.
Case/Y/762/.F7315

Copernicus mentioned re motion of the earth.

## Fontenelle, Bernard le Bovier de.

*The elogium of Sir Isaac Newton.*
London: printed for J. Tonson, J. Osborn, & T. Longman, 1728.
Subject: Motion, physics.
E/5/.N486

## Fontenelle, Bernard le Bovier de.

*Histoire des oracles.*
Paris: G. de Luyme, 1687.
Subject: Oracles, paganism, superstition.
B/863/.305

Rational approach to the supernatural. Based on A. Van Dale's *De oraculis ethnicorum dissertationes duae.* Oracles were not produced by demons. Oracles did not vanish with the coming of Christ but continued until the disappearance of paganism.

## Fontenelle, Bernard le Bovier de.

*Histoire du renouvellement de l'Academie Royale des Sciences en MDC. XCIX avec un discours preliminaire sur l'utilité des mathematiques & de la physique.*
Paris: Widow of Jean Boudot & Jean Boudot, son, 1708.
Subject: Mathematics, physics in anatomy, astronomy, geometry, navigation.
A/9/.6405

Préface sur l'utilité des mathematiques, et de la physique, et sur les travaux de l'academie des sciences.

## Fontenelle, Bernard le Bovier de.

*The history of oracles, and the cheats of the pagan priests.*
London, 1688.
STC II F1413
Subject: Oracles, paganism, superstition.
Case/B/863/.307

Trans. Aphra Behn.

## Fontenelle, Bernard le Bovier de.

*The history of oracles, and the cheats of the pagan priests. Written in Latin by Dr. Van-Dale.*
London: 1718.
Subject: Oracles, paganism, superstition.
Case/Y/1565/.B387

Trans. Aphra Behn. Vol. 1 of Behn's *Histories and novels,* pp. 137–292.

## Fontenelle, Bernard le Bovier de.

*The history of oracles. In two dissertations; wherein are proved, I. That the oracles were not given out by daemons; but were invented and supported by the craft of pagan priests. II. That the oracles did not cease at the coming of Jesus Christ; but subsisted four hundred years after it, till the entire abolition of paganism. Translated from the best edition of the original French.*
London: printed for D. Brown and J. Whiston, 1750.
Subject: Oracles, superstition.
B/863/.308

## Fontenelle, Bernard le Bovier de.

*Nouveaux dialogues des morts.*
Paris: C. Blageart, 1683.
Subject: Ancients vs. Moderns.
Y/762/.F 7319

2 vols. Vol. 1, 2d ed.; vol. 2, "nouvelle edition." Dialogue between Agnes Sorel and Roxelane, "sur le pouvoir des femmes"; between Cortez and Montezuma, "Quelle est la difference des peuples barbares, & des peuples polis."

## [Fontenelle, Bernard le Bovier de]

*A plurality of worlds.*
London: printed for R. Bentley and S. Magnes, 1688.
STC II F1416
Subject: Astronomy, plurality of worlds.
Case/L/9/.305

"Translated into English by Mr. Glanvill." In dialogue form, in 5 "evenings."

### Fontenelle, Bernard le Bovier de.

*A plurality of worlds.*
London: R. Bentley, 1695.
STC II F1417
Subject: Astronomy, plurality of worlds.
Case/L/9/.306

2d ed. Are the stars, moon, and planets inhabited?

### Fontenelle, Bernard le Bovier de.

*The theory or system of several new inhabited worlds, lately discover'd and pleasantly described in five nights conversation with Madam the Marchioness of * * *.*
London, 1718.
Subject: Astronomy, plurality of worlds.
Case/Y/1565/.B387

Translated by Aphra Behn in her *Histories and novels*, vol. 2. BOUND WITH *The history of oracles*, q.v.

### Forestus, Petrus, 1522–1597.

*Observationvm & curationum medicinalium.*
Leyden: Plantiniana, Raphelengius, 1593.
Subject: Disease, medicine.
UNCATALOGUED

In 5 vols. Vol. 1 (dated 1593): fevers, including epidemic fevers; vol. 2 (1611): diseases of the head and brain; vol. 3 (1611): chest and heart; vol. 4 (1596): intestines, kidneys; vol. 5 (1611): arthritis, venereal infection.

### Forestus, Petrus.

*Petri Foresti Alcmariani observationvm medicinalivm libri qvinqve.*
Leyden: Plantiniana, Raphelengius, 1611.
Subject: Medicine.
UNCATALOGUED

In 2 pts., each with separate title page: De morbis oculorum, dentium, oris ac linguae. De faucium item, gutturisque & gulae affectibus, & asperae arteriae (nos. 11–15). De pectoris, de cordis (nos. 16–17).

### Forte, Angelo de, 16th cent.

*Dialogo de gli incantamenti, e strigarie con le altre malefiche opre, quale tutta via tra le donne e huomini se esercitano, piaceuole e molto vtile a qualunque persona.*
Venice: Augustino de Bendoni, 1533.
Subject: Humors, psychology.
Wing/ZP/535/.B52

"Dialogue is between prosecutor and advocate appointed by Jove before a court held to judge of the guilt or innocence of women with regard to men. . . .An account is given of the mental characteristics supposed to accompany various complexions, etc." (from a tipped-in note).

### Fox Morzillo, Sebastian, b. 1528.

*De naturae philosophia, seu de Platonis, & Aristotelis consensione.*
Paris: Iacobus Puteanus, 1560.
Subject: Philosophical consensus on elements, senses, soul.
Case/B/14/.309

### Fracastoro, Girolamo, 1483–1553.

*Opera omnia, in vnum proxime post illius mortem collecta: quorum nomina sequens pagina plenius indicat.*
Venice: Giunta, 1574.
Subject: Medicine, sympathy and antipathy.
Case/Y/682/.F838

2d ed. Contents include: De causis criticorum dierum libellus. De sympathia & antipathia liber vnus. De contagionibus & contagiosis morbis, & eorum curatione libri tres. Syphilidis siue de morbo Gallico libri tres.

### Franchieres, Jean de, 15 cent.

*La favconnerie de Iean de Franchieres, grand prievr d'Aqvitaine, avec tovs les avtres avthevrs qui se sont peu trouuer traictans de ce subiect. De noueeau reueue, corrigee & augmentee, outre les precedentes impressions.*
Paris: Abel l'Angelier, 1602.
Subject: Falconry, veterinary medicine.
Case/*V/116/.31

A compendium of works on falconry, ornithology, and veterinary medicine, by, e.g., Guillaume Tardif du Puy en Vellay, Arthelouche de Alagona, and G. B.

### [Francisci, Erasmus] 1627–1694.

*Verwerffung des cometen-bespötts, oder gründliche erörterung der frage: ob der comet ein oder kein straff-zeichen sen: etwas oder nichts gutes oder böses bedeute? Worinnen de vorbedeutlichkeit mit unver werfflichen beweissthümern begründet wird. Die ungründe aber der widersprecher klärlich entdeckt; neben dem auch die vielfältige gedancken von dem ursprunge des cometens, mit eingeführet werden. Auf veranlassung des neulichst entstandenen wunder-grossen und unvergleichlichen comet-sterns. Zur abwarnung von roher sicherheit und beförderung christlicher butzfertigkeit herausgegeben.*
[n.p.] 1681.
Subject: Comets, symbolism.
Case/B/863/.32

### Franck, Bernhard Matthias, 1667–1701.

*Dissertatio medica inauguralis de catarrho, quam favente divini numinis clementia, auctoritate gratiosissimae facultatis medicae, pro licentia summos in medicina honores ac privilegia doctoralia.*
Altdorff: Henricus Meyer, 1690.
Subject: Medicine, rhinitis.
oRF/361/.F73/1690

### Freher, Paulus, 1611–1682.

*Theatrum virorum eruditione clarorum. In quo vitae & scripta theologorum, jureconsultorum, medicorum, & philosophorum,*

*tam in Germania superiore & inferiore, quam in aliis Europae regionibus, Grecia nempe, Hispania, Italia, Gallia, Anglia, Polonia, Hungaria, Bohemia, Dania & Suecia a seculis aliquot, ad haec usque tempora, Florentium, secundum annorum e mortalium seriem, tanquam variis in scenis repraesentantur.*
Nuremberg: heirs of Andrea Knorzi, for J. Hofmann, 1688.
Subject: Anatomy, botany, chemistry, mathematics, medicine, biographical dictionary.
Case/fE/3/.313

Biography of scholars. Entries arranged chronologically within each section. Engraved portraits. Presents list of writings for each person, in addition to biography. Sec. 3: Medicos, chymicos, botanicos, anatomicos. &c (earliest entry 1270). Sec. 4: Philosophos, philologos, historicos, mathematicos, poetas, &c. (earliest entry 1118).

### Freind, John, 1675–1728.

*Commentarii novem de febribus.*
Leyden: Joh. Arn. Langerak, 1734.
Subject: Fevers, gynecology, medicine.
Case/oRG/161/.F74/1734

### Freind, John.

*Historia medicinae a Galeni tempore usque ad initium seculi decimi sexti. In qua ea praecipue notantur quae ad praxin pertinet, Anglice scripta ad Ricardum Mead, M.D. Latine conversa a Joanne Wigan, M.D.*
Leyden: Joh. Arn. Langerak, 1734.
Subject: History of medicine.
Case/R/131/.F74/1734

Vol. 3 of *Opera Omnia.*

### Freind, John.

*The history of physick from the time of Galen, to the beginning of the sixteenth century. Chiefly with regard to practice. In a discourse written to Dr. Mead.*
London: printed for J. Walthoe, 1727.
Subject History of medicine.
Case/R/131/.F73/1727

In 2 parts. Pt. 1: containing all the Greek writers; pt. 2 (dated 1726): Arabian medicine.

### Freire de Montarroyo Mascarenhas, José, 1670–1760.

*Brados de ceo á insensibilidade dos homens: ou casos formidaveis, e horrorosos succedidos em differentes partes do mundo no anno de 1717.*
Lisbon: Pascoal da Sylva, 1718.
Subject: Earthquakes, prodigies.
Greenlee/4504/P855

### [Freire de Montarroyo Mascarenhas, José]

*Noticia da academia, ou curso de filosofia experimental novamente instituida nesta corte para instrucçaõ, e utilidade dos curiosos, e amantes das artes, e sciencias por Luis Baden natural da Graã Bretanha.*
Lisbon: Pedro Ferreyra, 1725.
Subject: Anatomy, experimental physics, medicine, surgery.
Greenlee/4504/P855

### [Freire de Montarroyo Mascarenhas, José]

*Noticia da destruiçaõ de Palermo, cabeça do reino de Sicilia, causado pelo horrivel terremoto . . . por J. F. M. M.*
Lisbon: Pedro Ferreira, 1726.
Subject: Earthquakes.
Greenlee/4504/P855

### [Freire de Montarroyo Mascarenhas, José]

*Noticia do fatal terremoto succedido no reyno de Napoles em 29 de Novembro do anno de 1732.*
Lisbon: Pedro Ferreyra, 1733.
Subject: Earthquakes.
Greenlee/4504/P855

### [Freire de Montarroyo Mascarenhas, José]

*Prodigiosas appariçoens & successos espantosos vistos no presente anno de 1716.*
Lisbon: Pascoal da Sylva, 1716.
Subject: Apparitions, prodigies.
Greenlee/4504/P855

### [Freire de Montarroyo Mascarenhas, José]

*Relaçam de hum formidavel e horrendo monstro silvestre, que foy visto, e morto nas visinhanças de Jerusalem, traduzido fielmente de huma, que se imprimio em Palermo no reyno de Sicilia, e se reimprimio em Genova, e em Turin; a que se accrescenta huma carta, escrita de Alepo sobre esta mesma materia.*
Lisbon: Joseph Antonio da Sylva, 1726.
Subject: Prodigies.
Greenlee/4504/P855

### Freitag, Arnold.

*Mythologia ethica, hoc est moralis philosophiae per fabulas brutis attributas, traditae, amoenissimum viridarium; in quo humanae vitae labyrintho demonstrato, virtutis semita, veluti Thesei filo docet.*
Antwerp: Philippus Gallaeus Christophorus Plantin, 1579.
Subject: Emblems.
Case/W/1025/.315

### Freitag, Arnold.

[*Mythologia ethica.*] *Viridarivm moralis philosophiae, per fabulas animalibus brutis attributas traditae, iconibus artificiosissimè in aes insculptis exornatum.*
Cologne: Georgius Mutingus, 1594.
Subject: Emblems.
Case/W/1025/.316

Title leaf, *Viridarivm,* is mounted in place of *Mythologia* title leaf. Onlythe next 3 leaves, recto and verso, and recto of 4th leaf are from the former; balance of the volume appears to be another copy of the latter (see preceding entry).

## Fries, Lorenz, 1491–1550.

*Ein kurtze schirmred der kunst astrologie wider etliche unverstandene vernichter auch etliche antwurt uff die reden und fragen. Martini Luthers Augustiners so er in seinene zehen geboten unformlich wider dise kunst gethon hat durch Laurentzen Friesen freier künsten von artzney doctorem.*
[Strassburg: Johanne Grüninger, 1520.]
Subject: Astrology, astronomy.
Case/B/8635/.282

## Froidmont, Libert, 1587–1653.

*Ant–Aristarchvs sive orbis-terrae immobilis. Liber vnicvs, in qvo decretum s. congregationis S.R.E. cardinal. an. 1616 aduersus Pythagorica-Copernicanos editum defenditur.*
Antwerp: Balthasar Moreti (officina Plantiniana), 1631.
Subject: Astronomy, religion and science.
Case/B/455/.317

Note on front flyleaf: "Cost 1 sh. at Dr. Hooker auction London 1703." BOUND WITH *Labyrinthvs sive de compositione continvi liber vnvs. Philosophis mathematicis, theologis, vtilis ac iucundus.* Subject: physics, motion, synthesis of science and theology.

## Frommann, Johannes Christianus, fl. 1675.

*Tractatus de fascinatione novus et singularis, in quo fascinatio vulgaris profligatur, naturalis confirmatur, & magica examinatur; hoc est, nec visu, nec voce fieri posse fascinationem probatur; fascinatio naturalis non per contagium, sed alio modo explicatur, magos de se nec visu, nec voce, nec contactu, nec alio modo laedere posse roboratur, theologis, jurisperitis et medicis, praesertim animarum sacerdotibus provincialibus, quibus cum variis superstitionum monstris saepè est pugnandum imo omnibus hoc seculô corruptô, quo non tantum pravus circa fascinum sensus simpliciorum igenia fascinat, sed & praeservatio & curatio morborum verbalis (ad quam per occasionem hic sit digressio) ad anime tendere videtur.*
Nuremberg: Wolfgang Andrea Endter & Johannes Andrea Endter heirs, 1675.
Subject: Taxonomy of fascination, witchcraft.
Case/B/88/.314

In 10 pts.

## Frönsperger, Leonhard, 1520–1575.

*Bauw ordnung von burger vnd nachbarlichen gebeuwen in stetten, wercken, flecken, dörffern, vnd auff dem land, sampt derselbigen anhangenden handwercker kosten, gebrauch vnd gerechtigkeit in drey theil verfasst vnd zusammen gezogen, welches innhalt nach der vorred weiter auff das aller kürzest zuvernemmen ist. Allen oberkeiten vnd vnderthanen nvtz vnd dienstlich zu gebrauchen vormals in druck nie aussgangen, rc. durch Leonhart Frönsperger.*
Frankfurt: Georg Raben & Weygand Hanen heirs, 1564.
Subject: Architecture, geometry, inventions, perspective, stonecutting.
Wing/fZP/547/.H152

BOUND WITH *Der architectur fürnembsten, notwendigsten, angehörigen mathematischen vnd mechanischen künst, eygentlicher bericht vnd verstendliche vnterrichtung, zu rechtem verstandt der lehr Vitruuij, in drey fürneme bücher abgetheilet. Als der newen perspectiua das I. buch.* Nuremberg: Gabriel Heyn, 1558; and *Perspectiua. Un schön nützlich büchlin underweisung der kunst des messens, mit dem zirkel, richtscheidt oder linial.* Frankfurt: Cyriacus Jacob, 1546. Also BOUND WITH Walter Rivius, q.v.

## Fuchs, Leonhard, 1501–1566.

[*De curandi ratione, libri VIII*]
Lyon: J. de Tournes et Guill. Gaseau, 1548.
Subject: Medicine.
SC/28

MS medical notebook, with Fuchs's printed text (to p. 140) inlaid and surrounded by MS notes [taken by Pierre Froissart] for medical course at Université de Dole. Notebook has MS title page: "De medendis morbis singvlarv[m] partivm corporis hvmani methodvs [Greek text] ex probatis medicis tum graecis cum Arabibus cum latinis praesertim ex Leo Fuchsio diligenter congesta, 1558." Pp. 3–14 of printed text inserted in a group. Pp. 15–140 are inserted individually, one being set into each notebook leaf. Text of Fuchs's printed work is incomplete, ending at p. 140, before the end of cap. 45. For ed. of Fuchs text see Cartier, *Bibliographie de Tournes*, 117. MS note on front flyleaf: "Pierre Froissart natif de Sellieres, auteur de ce commentaire, etoit professeur en médicine en l'Université de Dole avec Alexandro Cucinelli, Locosis et Catherin Mairos de Cesmes, qui succédérent à René Cerrosetpierre de Bergéres de Dole et autres encas offices Goll p. 164." Printed text, "Curatio melancholiae," p. 137, interleaved, with MS notes, at MS leaf 267.

## Fuchs, Leonhard.

*De historia stirpium commentarii insignes, maximis impensis et vigiliis elaborati, adiectis earvndem vivis plvsqvam quingentis imaginibus, nunquam antea ad naturae imitationem artificiosius efficitis & expresis, Leonharto Fvchsio medico hac nostra aetate longè clarissimo, autore. . . . Accessit iis succincta admodum uocum difficilium & obscurarum passim in hoc opere occurrentium explicatio. Vnà cum quadruplici indice, quorum primus quidem stirpium nomenclaturas graecas, alter latinas, tertius officinis seplasiariorum & herbarijs usitatas, quartus germanicas continebit.*
Basel: Isingriniana, 1542.
Subject: Botany, herbals, materia medica.
Wing/fZP/538/.I8

## Fuentes, Alonso de, b. 1515.

*Somma della natural filosofia di Alfonso di Fonte. Divisa in dialoghi sei, ne' qvali, oltra le cose fisiche, s'ha piena cognitione delle scienze, astronomia, et astrologia dell' anima, et dell' anotomia [sic] del corpo hvmano, novellamente tradotta di spagnolo in volgare da Alfonso di Vlloa.*
Venice: Plinio Pietrasanta, 1557.
Subject: Astrology, astronomy, natural philosophy.
Case/M/O/.318

## Fuller, Samuel, d. 1736?

*Practical astronomy in the description and use of both globes, orrery and telescopes. Wherein the most useful elements, and most valuable modern discoveries of the true astronomy are exhibited.*
Dublin: printed by and for Samuel Fuller, 1732.
Subject: Astronomy, astronomical instruments.
oQB/42/.F85

Description of the orrery (p. 121) by Tho. Wright. Doctrine of eclipses; concerning comets.

## Funck, Johann, 1518–1566.

*Chronologia hoc est, omnium temporum & annorvm ab initio mvndi, vsque ad annum à nato Christo M.D.LIII computatio: in qva methodicè envmerantvr omnium populorum, regnorumque memorabilium origines ac successiones: item omnes eorum reges, quando quisque coeperit, quam dia regnarit, quid dignum memoria gesserit: quis status populi dei fuerit: ac quemadmodum translata sint imperia à populo in populum, &c. Et si qui uiri illustres, quae facinora egregia, ac si quid amplius memoratu dignum extitit, ea omnia breuiter suis locis referuntur. Svnt'qve in hac computatione omnia tempora, tum ex sacris Biblijs, tum ex optimis quibusque autoribus, historicis, & astronomorum obseruationib. summa fide ac diligentia conciliata.*
Basel [Iacobus Parc for Joannes Oporinus], 1554.
Subject: Chronology.
Case/F/017/.316

## Funck, Johann.

*Chronologia hoc est, omnivm temporvm & annorvm ab initio mvndi, vsqve ad annvm á nato Christo 1552. In prima editione ab avtore dedvcta: pòst ab eodem recognita, aucta & in annum 1566: inde vero alijs in annum 1578: & tandem iterum ab alijs vsque; ad annum Christi 1601 producta compvtatio. In qva methodice envmerantvr omnivm popvlorvm, regnorvmque memorabilivm origines ac svccessiones... Suntque in hac computatione omnia tempora tum ex sacris Bibliis, tum ex optimis quibusque, autoribus historicis & astronomorum obseruationibus, summa fide ac diligentia conciliata.*
Wittenberg: Z. Lehmann for A. Hoffmann, 1601.
Subject: Chronology.

# G

## Gaffarel, Jacques, 1601–1681.

*Curiositez inovyes; hoc est, curiositates inauditae de figuris Persarum talismanicis, horscopo patriarcharum et characteribus coelestibus.*
Hamburg: Gothofredus Schultzen, 1676.
Subject: Astrology.
Case/B/85/.323

2 vols. in 1. BOUND WITH M. Gregorii Michaelis, *Notae in Jacobi Gaffarelli curiositates.* Hamburg: Gothofredus Schultzen & Amsterdam: Janssonio Waesbergios, 1676.

## Gaffarel, Jacques.

*Unheard-of curiosities: concerning the talismanic sculpture of the Persians; the horoscope of the patriarkes; and the reading of the stars.*
London: printed by G. D. for Humphrey Moseley, 1650.
STC II G105
Subject: Astrology.
Case/B/861/.322

Trans. Edmund Chilmead. In 4 pts.

## Galasso, Matthias.

*Apiaria vniversae philosophiae mathematicae. In qvibus paradoxa, et noua pleraque machinamenta ad vsus eximios traducta, & facilimis demonstrationibus confirmata, illvstriss. et excellentiss. D. Matthiae Galasso . . . Opvs non modo philosophis mathematicis, sed & physicis, anatomicis, militaribus viris, machinariae, musicae, pöeticae, agrariae, architecturae, mercaturae, professoribus, &c. vtilisimum; Tomvs primvs. Accessit ad finem secvndi tomi Evclides applicatvs, ex conditus ex apiarijs, indicatis vsibus eximijs praecipuarum propositionum in prioribus sex libris Euclideorum elementorum. Avthore Mario Bettino Bononiensi.*
Bologna: Io. Baptista Ferroni, 1642.
Subject: Euclidean geometry, mathematics, optics.
Case/6A/142

In 2 vols., 12 books. Vol. 1 by Galasso, vol. 2 by Bettino. Vol. 1 includes [apiarium 1] "Ars geometra"; [apiarium 5] "De varijs, & arcanis proiecturis, deformationibus, ac reformationibus imaginum e superficiebus conicis & cylindricis"; vol. 2 astronomy, gnomonics, music, "arithmetica paradoxa, et arcana" [apiarium 11].

## Galenus, Claudius, fl. 2nd cent. A.D.

*In aphorismos Hippocratis commentarii septem, recèns per Gvlielmvm Plantivm Cenomanuum Latinate donati, eiusdemque annotationibus illustrati.*
Lyon: Gulielmus Rouillius, 1554.
Subject: Aphorisms, Hippocrates, medicine.
Wing/ZP/539/.R7452

## Galenus, Claudius.

*In aphorismos Hippocratis commentaria: ex interpretatione Anvtii Foesii, & Gvlielmi Plantii, cum annotationibus eiusdem.*
Leyden: Ioannis Maire, 1633.
Subject: Aphorisms, Hippocrates, medicine.
UNCATALOGUED

"Editio novissima." Hippocrates' aphorisms in Greek, followed by commentaries in Latin.

## Galenus, Claudius.

*Liber de plenitudine.*
[Paris] C. VVechel, 1528.
Subject:Humors, medicine.
Case/5A/470/no. 2

BOUND WITH Celsus, Aurelius Cornelius and Scribonius Largus, qq.v. With the work by Galen is Polybus, *De salubri victus ratione priuatorum. Gvinterio Joanne Andernaco interprete,* and *Apuleius Platonicus, De herbarvm uirtutibus.* Having separate pagination is Antonius Benivenius, *Libellus de abditis nonnullis ac mirandis morborum & sanationum causis,* Wechel, 1528.

## Galenus, Claudius.

*Il libro di Clavdio Galeno dell' essercitio della palla.*
Milan: Francesco Moscheni, 1562.
Subject: Exercise, health and hygiene.
Case/Y/642/.G14

## Galenus, Claudius.

*De sanitate tvenda, libri sex. Thoma Linacro Anglo interprete.*
Paris: Claudius Chevallon, 1538.
Subject: Medicine.
Case/Q/.32

MS marginalia..

## Galilei, Galileo, 1564–1642.

*Dialogo di Galileo Galilei Linceo matematico sopraordinario dello stvdio di Pisa . . . sopra due massimi sistemi del mondo Tolemaico, e Copernicano; proponendo indeterminatamente le ragioni filosofiche, e naturali tanto per l'vna, quanto per l'altra parte.*
Florence: Gio. Batista Landini, 1632.
Subject: Astronomical systems, geocentric, heliocentric (Ptolemy, Copernicus).
Case/4A/876

Author's presentation copy, with his autograph on title page. Cinti, Biblioteca galileiana no. 89. The autograph has been overwritten (by a censor?) It has been authenticated by Osvaldo Cavallar, S.V.D., who says the MS marginalia throughout may also be in Galileo's hand.

### Galilei, Galileo.

*Discorsi e dimostrazioni matematiche, intorno à due nuoue scienze attenenti all mecanica & i movimenti locali.*
Leyden: Elsevier, 1638.
Subject: Mathematics, motion, physics.
Case/*L/35/.32

1st ed. In 4 "days," followed by "Appendice di alcune proposizioni & dimostrazioni attenti al centro di grauità de i solidi."

### Galilei, Galileo.

*Le operationi del compasso geometrico, e militare.*
Bologna: H. H. del Dozza, 1656.
Subject: Engineering, instrument making, practical geometry and mathematics.
L/10/.322

This so-called compass, known in English as a sector, is a calculator, used for solving practical mathematical problems, such as measurement of heights and distances, by triangulation.

### Galilei, Vincenzo, 1520?–1591.

*Dialogo di Vincentio Galilei, della mvsica antica, et della moderna.*
Florence: Giorgio Marescotti, 1581.
Subject: Arithmetic and music theory.
Case/6A/140

P. 7: "[V]n modo contrario di quello che vsa l'aritmetico nel sottrarre l'uno dall' altro suo numero; non di meno l'effeto è l'istesso pero che non considera il musico teorico, semplicemente il valore de mumeri come l'aritmetico." BOUND WITH his *Fronimo dialogo* (n.a.).

### Galilei, Vincenzo.

*Discorso intorno all' opere di messer Gioseffo Zarlino.*
Florence: Giorgio Marescotti, 1589.
Subject: Physics of music, relation of consonance to length of string.
Case/V/29/.991

### Gallo, Agostino, 1499–1570.

*Le dieci giornati della vera agricoltvra, e piaceri della villa . . . in dialogo.*
Venice: Domenico Farri, 1565.
Subject: Agriculture, herbal medicine, veterinary medicine.
Case/R/50/.322

Secondo giornata parla dell' herba medica.

### Gallo, Agostino.

*Le tredici giornate della vera agricoltvra & de' piaceri della uilla. Nvovamente ristampate con molti miglioramenti, & con aggiunta di tre giornate. Con le figvre de gli istrumenti pertinenti.*
Venice: Nicolò Beuilacqua, 1566.
Subject: Agriculture, herbal medicine, veterinary medicine.
Case/4A/2153

With woodcuts of tools, wagons, implements.

### Gallo, Agostino.

*Le vinti giornate dell' agricoltvra, et de' piaceri della villa. Di nuouo ristampate, & in molti luoghi ampliate. Con le figure de gl' istrumenti pertinenti, & con due tavole: vna della dichiaratione di molti vocaboli: & l'altra delle cose notabili.*
Tvrin: heirs of Beuilacqua, 1579.
Subject: Agriculture, herbal medicine, veterinary medicine.
Case/R/50/.323

Giornata 13: Horses and veterinary medicine; giornata 15: bee culture; giornata 16: silkworms.

### Gallo, Agostino.

*Le vinti giornate dell' agricoltvra, et de' piacere della villa. Con due tauole copiosissime.*
Venice: Benetto Miloco, 1674.
Subject: Agriculture, herbal medicine, veterinary medicine.
R/50/.325

In dialogue form.

### Gallo, Antonio, fl. 1540.

*De ligno sancto non permiscendo, Antonio Gallo autore.*
Paris: Simon Colinaeus, 1540.
Subject: Guaiacum, venereal disease.
Ayer/108/G17/1540

Contains errata leaf.

### Gallucci, Giovanni Paolo, 1538–1621?

*Theatro y descripcion vniversal del mvndo. En el qual no solo se descriuen sus partes, y se da regla en medirlas, mas con ingeniosa demonstracion, y figuras fijas, y mouibles, se verá lo mas importante dela astrologia, theorica de planetas, con el conocimiento de la esfera, la causa de crecer, y mengar de la mar, en que lugar hora, y tiempo acomodata para la geographia, nauegacion, y medicina, para los dias criticas, ò decretorios. Disinase el mundo, y se declaran sus quatro causas, eficiente, formal, material, y final, como es del cielo, de la tierra, del agua, y ayre, fuego, sol, y luna, y de los otros planetas, y estrellas fixas. Muestrase el estado y postura del cielo, para por el leuantar figuras, sin computo, numero, ni cuenta.*
Granada: Sebastian Muñoz, 1617.
Subject: Almanacs, astrology, astronomy.
Ayer/*7/G17/1617

Translated by Miguel Perez. "Vtilissimo non solo â los theologos, medicos, marineros, y laboradores, mas par otros estudios virtuosos." Has volvelles and calendars.

### Gallucci, Giovanni Paolo.

*Theatrvm mvndi, et temporis, in quo non solvm precipvae horvm partes describuntur, & ratio metiendi eas traditur, sed accomodatissimis figuris sub oculos legentium facilè ponuntur. Vbi astrologiae principia cernvntvr ad medicinam accomodata, geographica ad nauigationem, singulae stellae cum suis imaginibus, item ad medicinam, & Dei opera cognoscenda, & contemplanda, kalendarium gregorianum ad diuina officia, deisque sestos cælebrandos, et alia mvlta, qvae stvdiosvs lector facile in legendo cognoscet, ex quo sit, vt theologis, philosophis, medicis, astrologis, nauigantibus, agricolis, & ceteris bonarum artium, & scientiarum professoribus sit opus vtilissimum.*

Venice: Ioannes Baptista Somaschus, 1588.
Subject: Almanacs, astrology astronomy.
Ayer/*7/G17/1588

With volvelles.

## Galvão, Antonio, d. 1557.

*The discoveries of the world from their first originall vnto the yeere of ovr Lord 1555. Briefly written in the Portugall tongue by Antonie Galvano, gouernour of Ternate, the chiefe island of the Malucos. Corrected, quoted, and now published in English by Richard Hakluyt, sometimes student of Christchurch in Oxford.*
London: G. Bishop, 1601.
STC 11543
Subject: Discovery, navigation.
Graff/1493

Title of 1st page of text: "An excellent treatise of Antonie Galvano Portugall, containing the most ancient and moderne discoueries of the world especially by nauigation, according to the course of times from the flood until the yeere of grace 1555."

## García de Céspedes, Andrés.

*[Regimiento de nauegacion.*
Colophon: Madrid: en casa de Iuan de la Cuesta, 1606.]
Subject: Mathematics, navigation.
Ayer/8.9*/N2/G21/1606

In 2 pts. Title page to pt. 1 is wanting; text supplied from prelims, address of the king [by Iuan de Amezqueta]: "[E]l licenciado Andres Garcia de Cespedes . . . nos fue fecha relacion, que teniades escrito en lengua Castellana los libros de matematica siguentes. Vn regimiento de nauegacion que os auiamos mandado hazer, vna hydrografia general, donde se demonstraua, como los Portugueses auian peruertido los mapas, por poner dentro de su demarcacion las islas de la especeria." Pt. 2: "Segvnda parte la qve se pone vna hydrografia que mando hazer su magestad a Andres Garcia de Cespedes su cosmografo mayor."

## Garcilaso de la Vega [called El Inca], 1539–1616.

*Histoire des Yncas rois du Perou, depuis le premier Ynca Manco Capac, . . . jusqu'à Atahualpa, dernier Ynca.*
Amsterdam: Jean Frederic Bernard, 1737.
Subject: Anthropology, New World.
*F/998/.324

In 2 vols. "On a joint à cette edition l'histoire de la conquete de la Floride, par le même auteur &c." Vol. 1, bk. 2, chap. 21: Des sciences que les Yncas ont euës, & premièrement de l'astrologie"; vol. 2: "Nouvelle decouverte d'un pays plus grand que l'Europe, situé dans l'Amerique entre le Nouveau Mexique & la Mer Glaciale" [by Louis Hennepin].

## Garengeot, Rene-Jacques Croissant de, 1688–1759.

*Traité des operations de chirurgie, fondé sur la mécanique des organes de l'homme, & sur la théorie & sa pratique la plus autorisée. Enrichi de cures tres-singulieres & de figures en taille-douce représentant les attitudes des opérations.*
Paris: Huart, 1731.
Subject: Surgery.
oRD/32/.G37/1731

2d ed. 3 vols.

## Garimberto, Girolamo, 16th cent.

*Problemi natvrali e morali di Hieronimo Garimberto.*
Venice: Erasmo di Vicenzo Valgrisi, 1550.
Subject: Natural history.
Case/Y/682/.G18

Natural problems include universal generation in contrast to generation of man and compared to other animals; astronomy and science in general.

## Garmann, Christian Friedrich, 1640–1708.

*Academici curiosi de miraculis mortuorum libri tres, qvibus praemissa dissertatio de cadavere & miraculis in genere. Opus physico-medicum curiosis observationibus, experimentis, aliisqve rebvs, qvae ad elegantiores literas spectant, exornatum, diu desideratum & expetitum, beato autoris obitu interveniente editum â L. Immanuele Henrico Garmanno.*
Leipzig: Joh. Christoph. Zimmermann, 1709.
Graesse III, 29
Subject: Pathology.
4A/5536

Bk. 1: De cadavere; bk. 2: De miraculis, de contagio, de levitate & gravitate; bk. 3: De cadaverum putredine, de cadaverum resurrectione & resuscitatione.

## [Garth, Sir Samuel], 1661–1719.

*The dispensary: a poem in 6 cantos.*
London: J. Nutt, 1699.
STC II G274
Subject: Materia medica, medicine.
Case/Y/185/.G188

2d ed. Corrected by the author. First printed in 1699.

## [Garth, Sir Samuel]

*The Dispensary.*
London: J.Nutt, 1699.
STC II G275
Case/Y/185/.G189

3d ed. Corrected by the author.

## [Garth, Sir Samuel]

*The Dispensary: a poem in six canto's.*
London: printed by H. Hills, 1709.
Subject: Materia medica, medicine.
Y/185/.G19

Contains "The copy of an instrument subscribed by the president, Censor, most of the elects, senior fellows, candidates &c. of the college of physicians, in relation to the sick poor."

## [Garth, Sir Samuel]

*The Dispensary.*
London: printed for J. Tonson, 1718.
Subject: Materia medica, medicine.
Y/185/.G191

8th ed.

## Garth, Sir Samuel.

*The Dispensary.*
London: printed for J. & R. Tonson, 1741 (1735?).
Subject: Materia medica, medicine.
Case/E/5/P.798

10th ed. With letters of Alexander Pope.

## Garzoni, Tommaso, 1549?–1589.

*L'hospidale de' pazzi incvrabili. . . . Con tre capitoli in fine sopra la pazzia.*
Ferrara: Giulio Cesare Cagnacini & brothers, 1586.
Subject: Madness, psychology.
Case/Y/712/.G199

Discourse 17: De lunatici, o pazzi a tempo; 18: De' pazzi d'amore; 26: de' pazzi ostinati, come un mulo.

## [Garzoni, Tommaso]

*The hospital of incvrable fooles: erected in English, as neer the first Italian modell and platforme, as the vnskilfull hand of an ignorant architect could deuise.*
[London] printed by Edm. Bollifant, for Edward Blount, 1600.
STC 11634
Subject: Madness, psychology.
Case/Y/712/.G2

In 30 discourses, followed by "A discourse of the author, to the beholders concerning that part of the hospitall which appertaineth to women. Wherein he wittily setteth down all the former kindes of folly to be likewise resident in them." Among the discourses are Franticke and doting fooles (2d); Solitarie and melancholike fooles (3d); Stupide, forlorne, and extaticall fooles (7th); Of dottrels and shallow-pated fooles (9th).

## Garzoni, Tommaso.

*Il serraglio de gli stupori del mondo, di Tomaso Garzoni da Bagnacavallo. Diuiso in diece appartamenti, secondi i vari, & ammirabili oggetti. Cioè di mostri, prodigii, prestigii, sorti, oracoli, sibille, sogni, cvriosità astrologia, miracoli in genere, e maraviglie in spetie. Narrate da' più celebri scrittori, e descritte da' più famosi historici, e poeti, le quali talhora occorrono, considerandosi la loro probabilità, ouero improbilità, secondo la natura. Opera non meno dotta, che curiosa, cosi per theologi, predicatori, scritturisti, e legisti; come per filosofi, academici, astrologi, historici, poeti, & altri.*
Venice: Ambrosio & Bartolomeo Dei, brothers, 1613.
Subject: Astrology, magic, prodigies, supernatural.
Case/B/85/.3245

## Garzoni, Tommaso.

*Le theatre des divers cerveavx dv monde. Avqvel tiennent place, selon leur degré, toutes les manieres d'esprits & humeurs des hommes, tant louables que vicieuses, deduites par discours doctes & agreables.*
Paris: Felix le Mangnier, 1586.
Subject: Anatomy of the brain, humors, psychology.
Case/Y/712/.G196

55 discourses, among which are "Des cerueaux d'alquimistes," "Des cerueaux d'astrologues."

## Gassendi, Pierre, 1592–1655.

*Opera Omnia.*
Lyon: Lavrentius Anisson & Ioan. Bapt. Devenet, 1658.
Subject: Astronomy, natural philosophy, natural science.
Case/+B/239/.324

6 vols., each in several parts. Among them are, vol. 1: Logic and natural history; vol. 2: De rebus terrenis inanimis (no. 1), viuentibus, seu de animalibus (no. 2); vol. 3, no. 3: Fluddanae philosophiae examen (discussion of cardiovascular system, refuting Fludd's mysticism) Disquisitio metaphysica aduersus Cartesium (no. 4); vol. 4: Astronomica; vol. 5: Miscellanea; vol. 6: Epistolae.

## Gassendi, Pierre.

*Abregé de la philosophie de Mr. Gassendi par F. Bernier.*
Lyon: Clavde Mvgvet, 1676.
Subject: Generation and corruption, natural science.
Wing/ZP/639/.M/951

In 3 tracts: Des principes physiques; Des qualitez; De la generation & de la corruption.

## Gassendi, Pierre.

*Institvtio astronomica iuxta hypotheses tam veterum quam Copernici & Tychonis. . . . Accendunt ejusdem varij tractatus astronomici quorum catalogum pagina versa indicabit.*
The Hague: A. Vlacq, 1656.
Subject: Astronomy.
Ayer/8.9/A8/G25/1656

Editio ultima paulò ante mortem authoris recognita, aucta & emendata. In 3 books.

## Gassendi, Pierre.

*The mirrour of true nobility & gentility. Being the life of the renowned Nicolaus Claudius Fabricius lord of Peiresk, senator of the parliament at Aix.*
London: printed by J. Streater for Humphrey Moseley, 1657.
STC II G295
Subject: Astronomy, mathematics, natural science.
Case/E/5/.P/34948

In 6 books. Translated by W. Rand. P. 122: Peiresc's observations on fish. Nicolaus Claudius Fabricius de Peiresc, 1580–1637.

## Gassendi, Pierre.

*Three discourses of happiness, virtue and liberty. Collected from the works of the learn'd Gassendi by Monsieur Bernier.*
London: Awnsham and John Churchill, 1699.
Subject: Demons, enchantment and conjuration, magic.
Case/B/6/.326

Demons, pp. 429 ff. Rational (nonoccult) approach to dreams, p. 443.

## Gassendi, Pierre.

*Viri illustris Nicolai Clavdii Fabricii de Peiresc, senatoris Aqvisextiensis vita, per Petrum Gassendum.*
Paris: Sebastian Cramoisy, 1641.
Subject: Astronomy, mathematics, natural science.
E/5/.P/349473

## Gassendi, Pierre.

*Viri illustris Nicolai Claudij Fabricii de Peiresc vita.*
The Hague: Adrian Vlaq, 1655.
Subject: Astronomy, mathematics, natural science.
E/5/.P/349475

3d ed. In 6 books.

## Gaudenzio, Paganino, 1596–1649.

*L'Accademia disunita.*
Pisa: Francesco Tanagli, 1635.
Subject: Monsters, natural science, prodigies.
Y/712/.G23

Discorso 34: Degli animali monstrvosi; 24: Del viaggiare; 13: Dell' incendio Vesvviano; 44: Del fiume Vesara.

## [Gauger, Nicholas. 1680?–1730]

*La mechanique du feu oder kunst die würckung des feuers zu vermehren und die kosten davon zu verringern.*
Hanover: Nicolaus Förstern, 1715.
Subject: Fireplace and heating technology and engineering, physics.
Case/3A/1010

Engraved title page: La mechanique du feu, ou traite de nouvelles cheminées.

## Gaule, John, d. 1660.

*A collection out of the best approved authors, containing histories of visions, apparitions, prophesies, spirits, divinations, and other wonderful illusions of the devil wrought by magick or otherwise. Also of divers astrological predistions* [sic] *shewing as the wickedness of the former, so the vanity of the latter, and folly of trusting to them.*
London: printed for Joshua Kirton, 1657.
STC II G376
Subject: Astrology, magic, witchcraft.
Case/B/88/.326

Running head: Mag-astro-mancer posed and puzzled. Chap. 19: Of magicall and astrologicall artists, and their arts, wittliy derided, wisely rejected, and worthily contemned.

## Gaule, John.

[Greek Title] *The mag-astro-mancer, or the magicall-astrologicall-diviner posed and puzzled.*
London: printed for Joshua Kirton, 1652.
STC II G377.
Subject: Astrology, magic, witchcraft.
Case/B/88/.325

27 chapters. Chap. 5: From the vanity of science; chap. 7: From the law of nature; chap. 19: From the affinity to witchcraft; chap. 22: From the variety of miracles.

## Gaule, John.

*Select cases of conscience touching vvitches and witchcrafts.*
London: printed by W. Wilson for Richard Clutterbuck, 1646.
STC II G379
Subject: Witchcraft.
Case/–B/88/.328

12 cases on such subjects as Are there any witches? Figures and marks of a witch. Can a witch repent and be saved?

## Gauricus, Lucas (bp. of Civitate) 1476–1558.

*Tractatvs astrologiae judiciariae de nativitatibvs virorvm & mulierum, compositus per D. Lucem Gauricum Neapolitanum, ex Ptolemaeo & alijs autoribus dignissimis, cum multis aphorismis expertis & comprobatis ab eodem.*
Nuremberg: Iohan. Petreius, 1540.
Subject: Astrology, complexion, disease, humors, monsters.
Case/B/8635/.327

27 sections, e.g., De natis monstruosis (3); De complexione & infirmitate (7); De effectibus & natura signorum & planetarum (8); and De planete duodecim domos (25).

## Gauricus, Lucas (bp. of Civitate).

*Tractatvs astrologicvs in quo agitur de praeteritis multorum hominum accidentibus per proprias eorum genethliacus vaticinari poterit de futuris, quippe qui per uarios casus artem experientia fecit exemplo monstrante uiam.*
Venice: Curtio Troiano de Navo, 1552.
Subject: Astrology.
Case/B/8635/.328

In 6 tracts, e.g., no. 1: Ciuitatum & quorundam oppidorum figurae celestes & earum euentus; no. 5: De biothanalis hoc est violenta strage peremptis; no. 6: De azemenatis hoc est viciatis & in aliquo corporis membro mutilatis.

## Gautier, [Hubert] Henri, 1660–1737.

*Instruction pour les gens de guerre, où l'on traite de l'artillerie, des proportions, des renforts, des portées, des affûts, & generalement tout ce qui concerne les armes à feu, dont on se*

*sert en France, tant sur terre, que sur mer. De plus la maniere de jetter les bombes, où l'on donne a connoitre les proportions de ces machines, comme les mortiers qui servent à les chasser, avec plusieurs experiences sur ce sujet. Et enfin le moyen de composer toute sorte de feux d'artifice pour la guerre.*
Paris: Jacques Cognard, 1692.
Subject: Ballistics, engineering, gunpowder, military science.
U/444/.326

## Gautier, Joseph Gaspard.

*Dissertatio academica an dulcisoni musicae cantus non solum ad corporis & animi sanitatem tuendam, sed ad plurimorum morborum curationem inservire possint.*
Avignon: Carolus Giroud, 1717.
Subject: Music and medicine, psychology.
V/23/.328

Sympathetic effect of music on body and spirit.

## Gaza, Theodoro, fl. 1447–1455.

*Liber de mensibvs atticis.*
Basel, 1536.
Subject: Astronomy, calendar.
Case/Y/682/.V826

Trans. Ioannes Perrello. Includes (with continuous pagination): *Commentariolvs de ratione lvnae epactarvm, et mensis intercalarij per eundem interpretatem.* BOUND WITH Joachim Camerarius, *Opuscula*, 1536, q.v.

## Gaztañeta y de Yturribálzaga, Antonio, 1656–1728.

*Norte de la navegacion hallado por el quadrante de redvccion.*
Seville: Jvan Francisco de Blas, 1692.
Subject: Instrumentation, navigation.
Ayer/f8.9/N2/G291/1692

Use of the quadrant of reduction.

## Gell, Robert, d. 1665.

[Greek Title] *Or a sermon touching God's government of the world by angels. Preached before the learned societie of artists or astrologers, August 8, 1650.*
London: printed by John Legatt, 1650.
STC II G468
Subject: Astrology.
Case/B/8635/.88

Binder's title: *Tracts on astrology, 1642–1652.* BOUND WITH works by John Geree, John Raunce, and Edmund Reeve, qq.v.

## Gell, Robert.

*Stella nova, a nevv starre, leading wisemen unto Christ. Or, a sermon preached before the learned society of astrologers, August 1, 1649.*
London: printed for Samual Satterthwaite, 1649.
STC II G473
Subject: Astrology.
Case/B/8635/.88

Binder's title: *Tracts on astrology, 1642–1652.* BOUND WITH works by John Geree, John Raunce, and Edmund Reeve, qq.v.

## Gellius, Aulis, 2nd cent. A.D.

*Noctes Atticae.*
[Rome] Conradus Sweynheim & Arnold Pannartz, 1469.
H-C 7517^ BMC (XV)IV:6–7^ Pr. 3298^ Pell. III, 5010^ Still. G107^ Goff G-118
Subject: Encyclopedias, natural philosophy.
Inc./f3298

"The *Noctes Atticae* is a collection of interesting notes on grammar, public and private antiquities, history and biography, philosophy (including natural philosophy), points of law, textual criticism, literary criticism, and various other topics" (Introduction to John C. Rolfe's trans. 3 vols. New York: Putnam, 1927). Bk 2, chap. 21: "Super eo sydere quod greci [Greek word] nos septentrianos uocamus"; chap. 26: "Sermones M. Frontonis et Phaurini philosophi de generibus colorum"; 30 observations on waves of the sea, which take different forms depending on wind direction; bk. 3, chap. 6: "De ui atque natura palme arboris quod lignum ex ponderibus impositis renitatur"; bk. 4, chap. 2: Difference between disease, defect; chap. 13: On curing gout.

## Gellius, Aulis.

*Noctes Atticae.*
Venice: Andreas de Paltasichis, 1477.
H-C* 7520^ BMC (XV)V:251^ Pr. 4423^ Pell. 5011^ Goff G-121
Subject: Encyclopedias, natural philosophy.
Inc./f/4423

MS marginalia passim but indistinct. See Hans Baron, "Aulus Gellius in the Renaissance and a manuscript from the school of Guarino." (*Studies in Philology* 48, no. 2 [April 1951].) Gellius was the earliest of surviving encyclopedic writers of the late Roman period. He preserved abstracts and substantial quotations from works that subsequently perished.

## Gellius, Aulis.

*Noctes Atticae.*
Brescia: Boninus de Boninis de Ragusia, 1485.
H-C* 7521^ BMC (XV)VII:958^ Pr. 6958 Pell. 5012^ Goff G-122
Subject: Encyclopedias, natural philosophy.
Inc./f6958

MS marginalia passim in early hand.

## Gellius, Aulis.

*Noctes Atticae.*
Venice: Christophorus de Quaietis de Autegnago and Martinus de Lazaronibus de Rouado, socios, 1493.
H-C-R 7524^ BMC (XV)V:545^ Pr. 5509^ Pell. 5014^ Goff G-124
Subject: Encyclopedias, natural philosophy.
Inc./f5509

**Gellius, Aulis.**

*Noctes Atticae.*
Venice: Philippus Pincius Mantuanus, 1500.
H-C 7527^ BMC (XV)V:499^ Pr. 5324^ Pell. 5017^ Goff G-127
Subject: Encyclopedias, natural philosophy.
Inc./f5324

**Gellius, Aulus.**

*A. Gellij viri disertissimi noctium Atticarum libri XX summa accuratione Ioannis Connelli Carnotensem ad recognitionem Beroaldinam repositi.*
[Paris] Joannis parui [Jehan Petit], 1511.
Subject: Encyclopedias, natural philosophy.
Case/Y/672/.G304

**Gellius, Aulus.**

*A. Gellii noctes redditae nvper omni discvssa caligine micantissimae.*
Florence: Philip de Giunta, 1513.
Subject: Encyclopedias, natural philosophy.
Case/Y/672/.G3044

**Gellius, Aulus.**

*Noctivm Atticarvm.*
[Venice] Aldus [1515]
Subject: Encyclopedias, natural philosophy.
Case/Y/672/.G3049

**Gellius, Aulis.**

*Noctes Atticae.*
[Venice] Aldus & Andrea Socerus [1515]
Subject: Encyclopedias, natural philosophy.
Case/*Y/672/.G305

**Gellius, Aulus.**

*Avli Gelii noctivm Atticarvm commentaria per Bonsinem Asculanum summa nuper diligentia & studio recognita ac pristinae serenitate candorique restituta locis supra centum emendatis praeter ea.*
[Venice: I. de Tridino alia Tacuinum, 1517]
Subject: Encyclopedias, natural philosophy.
Case/6A/210

In 20 books. Commentary of Matthaeus Bonfinus Asculanus.

**Gellius, Aulus.**

*Noctes Atticae.*
Cologne: Ioannes Soterus, 1526.
Subject: Encyclopedias, natural philosophy.
Case/Y/672/.G3052

Imperfect (pp. 21–22 wanting). With annotations of Petrus Mosellanus.

**Gellius, Aulus.**

*Noctes Atticae.*
Lyon: Sebastian Gryphius, 1555.
Subject: Encyclopedias, natural philosophy.
Y/672/.G3054

**Gellius, Aulis**

*Noctes Atticae. Accesservnt ervditissimi uiri Petri Mosellani in easdem perdocte adnotationes.*
Cologne: Gualthero Fabricio, 1563.
Subject: Encyclopedias, natural philosophy.
Case/Y/672/.G3056

With glossary supplying Latin trans. of Greek terms in the foregoing text, arranged in numerical order (book by book) rather than alphabetically.

**Gellius, Aulis.**

*Noctes Atticae.*
Lyon: Antonivs Gryphivs, 1566.
Subject: Encyclopedias, natural philosophy.
Case/–Y/672/.G3057

With glossary.

**Gellius, Aulus.**

*Noctes atticae sev vigiliae atticae. Henrici Stephani noctes aliqvot Parisinae atticis A. Gellij noctibus seu vigilijs inuigilatae. Eiusdem H. Stephani annotationes in alios Gellij locos prodibunt cum notis Lud. Carrionis (qui vet. exemplaria contulit) prelo iam traditis.*
Paris, 1585.
Subject: Encyclopedias, natural philosophy.
Case/Y/672/.G3058

**Gellius, Aulus.**

*Noctium Atticarum commentarivs.*
Leyden: H. de Vogel, 1644.
Subject: Encyclopedias, natural philosophy.
Case/Y/672/.G306

**Gellius, Aulus.**

*Noctes Atticae cum selectis novisque commentariis, et accuratâ recensione Antonii Thysi, J.C. & Jacobi Oisell, J.C.*
Leyden: officina Petri Leffen, 1666.
Subject: Encyclopedias, natural philosophy.
Case/Y/672/.G3066

**Gellius, Aulis.**

*Noctes Atticae cum notis et emendatibus Joannis Frederici Gronovii.*
Leyden: Ioannes de Vivié, 1687.
Subject: Encyclopedias, natural philosophy.
Wing/Y/672/.G3

**Gellius, Aulus.**

*Noctium Atticarum. Perpetuis notis & emendationibus illustraverunt Johannes Fredericus et Jacobus Gronovii.*
Leyden: Cornelius Boutesteyn, Joh. du Vivié & Is. Severinus, 1706.
Subject: Encyclopedias, natural philosophy.
Y/672/.G307

## Gemma Reinerus, Frisius, 1508–1555.

*Gemmae Phrysii medici ac mathematici, de orbis divisione et insvlis, rebvsqve nuper inventis opvscvlvm, longae qvam antehac castigativs.*
Douai: Lodouici de Vvinde, 1576.
Subject: Descriptive geography.
Ayer/*7/C395 (C)/1576

Chap. 30: De America. BOUND WITH Cheyne, James, q.v.

## Gemma Reinerus, Frisius.

*De principiis astronomiae & cosmographiae, deqve vsu globi ab eodem editi. Item de orbis diuisione, & insulis, rebusque nuper inuentis.*
Louvain: Seruatius Zassen & Antwerp: Gregorius Bontius [1530].
John Carter Brown, p. 100^ Harrisse no. 156
Subject: Astronomy.
Ayer/7/G4/1530

## Gemma Reinerus, Frisius.

*De principiis astronomiae & cosmographiae, deqve vsu globi ab eodem editi. Item de orbis diuisione, & insulis, rebusque nuper inuentis.*
Antwerp: Ioannes Richard, 1544.
Harrisse no. 252
Subject: Astronomy.
Ayer/7/G4/1544

## Gemma Reinerus, Frisius.

*De principiis astronomiae & cosmographiae: deque vsu globi ab eodem editi.*
Paris: Gulielmus Cauellat, 1557.
Subject: Astronomy.
Ayer/7/G4/1557

## Gemma Reinerus, Frisius.

*Les principes d'astronomie & cosmographie auec l'vsage du globe. Le tout composé en Latin par Gemma Frizon, & mis en langage François par M. Claude de Boissiere Daulphinois. Plus est adiousté l'vsage de l'anneau astronomic, par ledict Gemma Frizon: et l'exposition de la mappemonde, composée par ledict de Boissiere.*
Paris: Guillaume Cauellat, 1557.
Subject: Astronomy.
Ayer/7/G4/1557a

## Gemma Reinerus, Frisius.

*De principiis astronomiae & cosmographia deque vsu globi ab eodem editi.*
Cologne: Maternus Cholinus, 1578.
Subject: Astronomy.
Ayer/7/G4/1578

MS commentary bound in. Following section "De compositione globi coelestis," there are a number of sections in contemporary hand, possibly astrological. Also "Commentaria in sphaeram mundi sive cosmographiam," with hand-drawn diagram. Some of this may be alchemical.

## [Génebrard, Gilbert, abp. of Aix] 1537–1597.

*Chronographia in duos libros.*
Paris: Martin Iuuenis, 1567.
Subject: Universal chronology.
Case/fF/017/.332

[Colophon: Marinvs Ivvenis, anno mundi 5655 et domini]

## Génebrard, Gilbert, abp. of Aix.

*Chronographiae libri quatuor.*
Paris: Martin Iuuenis 1580.
Subject: Universal chronology.
Case/fF/017/.334

Also contains *Chronologia Hebraeorvm maior.*

## Gentilini, Eugenio, b. 1529.

*Instrvttione de' bombardieri di Evgenio Gentilini da Este. Que si contiene l'esamina vsata dallo strenua Zaccharia Schiuina. . . . Dalle quali ogni bombardiero e capi maestri uengono à pieno instrutti di ciò, ch'alla lor professione appartiene; e per maggior lor eccellenza anco vi sono alcuni mezzi geometrici, che molto son giouenuoli communemento a bombardieri, & a ingegneri di fortezze.*
Venice: Franco de' Franceschi, Senese,1592.
Subject: Geometry, military technology, pyrotechnics, trajectories.
Wing/ZP/535/.F/86

Geometric instruction for bombardiers in fortresses. Chap. 8: "Modo di trouar la terra buona dal salnitro, & il moda d [sic] farlo"; chap. 9: "Modo & auuertenze per far la poluere."

## Gentilini, Eugenio.

*Instrvttione di artiglieri di Evgenio Gentilini da Este; doue si contiene la esamina vsata dallo strenuo Zaccaria Schiavina . . . vi sono alcuni mezi geometrici . . . aggionta fatta alla esamina di venetia.*
Venice: Francesco de' Franceschi, Senese, 1598
Subject: Geometry, military technology, pyrotechnics, trajectories.
Case/4A/138

* * *

*A geographical history of Nova Scotia, containing an account of the situation, extent, and limits thereof. . . . To which is added, an accurate description of the bays, harbours, lakes, and rivers, the nature of the soil, and the produce of the country.*
London: printed for Paul Vaillant, 1749.
Subject: Geography, medicine, natural history.
Ayer/150.7/N9/G4/1749

Origin of name, "Canada": Spanish treasure-seekers, disappointed at finding no gold in area near Gaspé Peninsula, said "Aca nada." Among subjects: "physick": pp. 46–47. natural history, pp. 48, 97–110.

* * *

*Geographische beschreibung der provinz Louisiana in Canada von dem fluss S. Lorenz bis an dem ausfluss des flusses Mississippi; samt einem kurzen bericht von dem ietzo florirenden actein-handel.*

[n.p. 1719?]
Subject: Geography.
Ayer/b150.3/F8/G34/1719

Single sheet broadside, rivers of America.

## Geoponica.

*Constantini Caesaris selectarum praeceptionum, de agricultura libri viginti, Iano Cornario medico physico interprete, recens in lucem emissi.*
Venice: D. Iacob à Bvrgofrancho, Papiensem, 1538.
Subject: Agriculture, medicinal uses of insect venoms and plants.
Wing/ZP/535/.B9

## [Geoponica]

*De re rvstica selectorum libri XX. . . . graeci, Constantino qvidem Caesari nuncupati, ac iam non libris, sed thesauris annumerandi. Item, Aristotelis de plantis libri dvo Graeci, nuper ab interitu liberati, ac studiosorum usui hac primum editione restituti.*
Basel [1539].
Subject: Agriculture, botany, herbal medicine, sympathy and antipathy.
Case/Y/642/.G285

De naturalibus sympathiis & antipathiis. Zoroastre. In Greek.

## Geoponica.

*Constantino Caesare de notevoli et vtilissimi ammaestramenti dell' agricoltura di greco in uolgare nouamente tradotto per Pietro Lavro.*
Venice: Gabriel Giolito di Ferrarii, 1549.
Subject: Agriculture, botany, herbal medicine, sympathy and antipathy.
Case/Y/642/.G287

In 20 books.

## Gerard, John, 1545–1612.

*The herball or generall historie of plantes.*
London: Iohn Norton, 1597.
STC 11750
Subject: Herbals.
Case/fN/405/.33

Engraved title page. In 3 books. Copiously illustrated. Description of each plant includes its "temperature and vertues," its several names.

## Gerard, John.

*The herball or generall historie of plantes . . . very much enlarged and amended by Thomas Johnson citizen and apothecarye.*
London: Adam Islip, Ioice Norton and Richard Whitakers, 1633.
STC 11751
Subject: Herbals.
Case/fN/405/.331

Contains appendix, "Certaine plants omitted in the former historie." Has "Index latinus, table of English names, catalogue of British names, table of the nature and vertues of all the herbes, trees, and plants, described in this present herbal."

## Gerard, John.

*The herball.*
London: Adam Islip, Ioice Norton, and Richard Whitakers, 1636.
STC 11752
Case/fN/405/.3312

2 vols. 2d ed. of revised Gerard. "A catalogue of the British [i.e., Welsh] names of plants, sent me by Master Robert Dauys of Guissaney in Flintshire" at end.

## Geree, John.

*Astrologo-mastix, or a discovery of the vanity and iniquity of judiciall astrology, or divining by the starres the successe, or miscarriage of humane affaires.*
London: printed by Matthew Simmons, for John Bartlet, 1646.
STC II G586
Subject: Astrology.
Case/B/8635/.88

Binders title: *Tracts on astrology, 1642–1652.* BOUND WITH works by John Raunce, Edmund Reeve, and Robert Gell, qq.v.

## Gerlach, Benjamin, d. 1683.

*Unvorgreifliches urtheil von der cometen. Würckung und bedeutung.*
Schweidnitz: Christian Okeln, 1681.
Subject: Comets.
Case/B/863/.334

## Gerson Joannes, 1363–1429.

*Opera.*
[Strassburg: Grüninger? 1488]
Goff G-186
Subject: Astrology, magic, medicine, occult.
Inc./534–36

Includes *Trigilogium astrologie theologisate* [tract no. 40]; *Contra superstitiosam observationem dierum* [tract no. 41]; *Contra superstitionem cuiusdam medici* [tract no. 42]; *De erroribus circa artem magicam & articulis reprobatis* [tract no. 43].

## Gerson, Joannes.

*Opera.*
Basel: Kessler, 1489.
Goff G-187
Subject: Astrology, medicine, occult.
Inc./7622

## Gerson, Joannes.

*Opus tripartitum de praeceptis decalogi, de confessione, et de arte moriendi.*
[Cologne: Ulrich Zell, ca. 1467]

Pell. 5190^ Pr. 875^ BMC (XV)I: 190^ Goff G-239
Subject: Astrology, passions.
Inc./875

Binders title, *Works of Jean Gerson.* Among the tracts included in this vol. are *De simplificatione, Trigilogium astrologiae, De passionibus animae, De meditationibus*, qq.v., and others, out of scope of this checklist.

## Gerson, Joannes.

*De cognitione castitatis et pollutionibus diurnis.*
Cologne: Ulrich Zell, 1470.
H-7692*^ C 2689^ BMC(XV)I:184^ Goff G-195
Subject: Medicine.
Inc./*831

## Gerson, Joannes.

*[M]alleorvm qvorvndam maleficarvm, tam vetervm, qvam recentivm avtorum.*
Frankfurt: Nicolaus Bassaeus, 1582.
Subject: Exorcism, witchcraft.
Case/3A/2297

Probably contitutes vol. 2 of *Malleus maleficarum.* Cf. Caillet 7056. 7 tracts, each by different author: M. Bernhardus Basin, *Opusculum de artibus magicis, ac magorum maleficiis*; Ulrich Molitor, *Dialogum de lamijs & pythonicis mulieribus*; Ieron. Menghi, *Flagellum daemonum: seu exorcismi efficacissimi, & remedia probatissima ad malignos spiritus expellendos*; D. Iohannis Gerson, *Libellum de probatione spiritum*; Thomas Murner, *Libellum de pythonico contractu*; Felix Hemmerlin, *Malleoli . . . Tractatus duos exorcismorum*; Bartholomaeus de Spina, *Quaestionum de strigibus seu maleficis. Item eiusdem apologiam quadruplicem de lamijs: contra Io. Franciscum Ponzinibium.*

## Gerson, Joannes.

*Le mirouer de l'homme et de la femme.*
Paris: Symon Vostre [15—?].
Subject: Body and spirit, soul.
Case/3A/531

1st ed. In 7 chapters: Du monde, du corps humaine, de la mort, d'enfer, du jugement, de la gloire des saincts, de la gloire des corps.

## Gerson, Johannes.

*De passionibus animae.*
[Cologne:. Ulrich Zell, ca 1472]
H-C 7677^ Pell. 5208^ BMC (XV)I:190^ Goff G-248
Subject: Passions.
Inc./871

Shelved under Inc. 875. Binders title: *Works.*

## Gerson, Joannes.

*De pollutionibus nocturnis.*
[Esslingen: Conrad Fyner, 1473?]
H 7699^ Pell. 5216^ BMC (XV)II:512^ Pr. 2470^ Goff G-259
Subject: Medicine (nocturnal emissions).
Inc./2470

## Gerson, Joannes.

*[Tractatus de simplificatione, stabilicione, sive mundificacione cordis.]*
[Cologne: Ulrich Zell, ca. 1472.]
H-C 7681*^ Pell. 5230^ Pr. 873^ BMC (XV)I:190^ Goff G-270
Subject: Astrology, passions of the mind.
Inc./873

Title supplied from Goff. Shelved under Inc. 875. Binders title: *Works.*

## Gerson, Joannes.

*De meditatio cordis opusculum astrologie theologisate editum.*
[Cologne: Ulrich Zell, ca. 1470]
Goff G-231
Subject: Astrology, passions.
Inc./836

Shelved under Inc. 875. Binders title: *Works.*

## Gesner, Konrad, 1516–1565.

*Bibliotheca institvta et collecta primvm a Conrado Gesnero deinde in epitomen redacta & nouorum librorum accessione locupletata, iam vero postremo recognita & in duplum post priores editiones aucta, per Iosiam Simlerum Tigurinum.*
Zurich: Christophorus Froschouer, 1574.
Subject: Encyclopedias, scientific authors.
Case/Z/9/.34

"Habes hic, optime lector, catalogum locupletissimum omnium ferè scriptorum, à mundi initio ad hunc vsque diem, extantium & non extantium, publicatorum & passim in bibliothecis latitantium."

## Gesner, Konrad.

*Bibliotheca . . . tertio recognita, & in duplum post priores editiones aucta, per Iosiam Simlerum: iam verò postremo aliquot mille, cum priorum tum nouorum authorum opusculis, ex instructissima Viennensi.*
Zurich: Christophorus Froschouer, 1583.
Subject: Encyclopedias, scientific authors.
Case/fZ/9/.342

## Gesner, Konrad.

*Bibliotheca vniuersalis siue catalogus omnium scriptorum locupletissimus, in tribus linguis, Latina, Graeca, & Hebraica: extantium & non extantium, ueterum & recentiorum in hunc noque diem, doctorum & indoctorum, publicatorum & in bibliothecis latentium. Opus nouum, studiosis omnibus cuiuscunque artis aut scientiae ad studia melius formanda utilissimum: authore Conrado Gesnero Tigurino doctore medico.*
Zurich: Christophorus Froschouer, 1545.
Subject: Science bibliography.
Case/fZ/9/.338

1st ed. 4 vols. of which Newberry has vols. 1 and 2. Vol. 1 is alphabetical by author; vol. 2, *Pandectarvm*, alphabetical by subject, has separate entry below.

## Gesner, Konrad.

*Epitome bibliothecae Conradi Gesneri, conscripta primum à Conrado Lycosthene Rubeaquensi.*
Zurich: Christophrus Froschouer, 1555.
Subject: Encyclopedias, bibliography, science.
Case/fZ/9/.339

With index; paralipomena (additions).

## Gesner, Konrad.

*The newe jewell of health, wherein is contayned the most excellent secretes of phisicke and philosophie, deuided into fower bookes. In the which are the best approued remedies for the diseases as well inwarde as outwarde, of all the partes of mans bodie: treating very amplye of all dystillations of waters, of oyles, balmes, quintessences, with the extraction of artificiall saltes, the vse and preparation of antimonie, and potable gold. Gathered out of the best and most approued authors, by that excellent doctor Gesnerus. Also the pictures and manner to make the vessels, furnaces, and other instrumentes therevnto belonging. Faithfully corrected and published in Englishe, by George Baker, chirurgian.*
London: printed by Henry Denham, 1576.
STC 11798
Subject: Alchemy, spagyric medicine.
Case/N/405/.34

## Gesner, Konrad.

*Pandectarvm sive partitionum uniuersalium Conradi Gesneri Tigurini, medici & philosophiae professoris, libri XXI.*
Zurich: Christophorus Froschouer, 1548.
Subject: Encyclopedia of science by subject.
Case/fZ/9/.338

Vol. 2 of *Bibliotheca uniuersalis.* Among subjects are arithmetic, geometry, music, astronomy, astrology, divination, geography, physics, metaphysics.

## Gesualdo, Filippo, d. 1619.

*Plutosofia ... nella quale si spiega l'arte della memoria, con altre cose notabili, pertinenti tanto alla memoria naturale quanto all' artificiale.*
Vicenza: heirs of Perrin for P. Bertelli, 1600.
Subject: Mnemonics.
Case/B/528/.335

20 "lettioni."

## Ghaligai, Francesco, fl. ca. 1515.

*Pratica d' arithmetica.*
Florence: B. Giunta, 1548.
Subject: Arithmetic.
Case/L/130/.335

In 13 books.

## Gherardo, Paolo, fl. 1543–1560.

*Il portolano del mare, nel qval si dichiara minvtamente del sito di tvtti porti, qvali sono da Venetia in Leuante, & in Ponente: & d'altre cose vtilissime, & necessarie à i nauganti.*
Venice: Daniel Zanetti, 1576.
Subject: Navigation.
Ayer/*130/G41/1576

## Giambelli, Cipriano, fl. 16th cent.

*Il trattato dell' anima.*
Treviso: Domenico Amici, 1594.
Subject: Psychology, senses, soul.
Case/B/51/.335

4 books, in dialogue form: dell' anima vegetatiue, dell' odorato & del gusto, del tatto, si dichiara principalmente natura, e l'operationi del senso commune, e della fantasia.

## Giambulari, Pietro Francesco 1495–1555.

*Del sito forma & misure dello inferno di Dante.*
Florence: Neri Dortelata [pseud.] 1544.
Subject: Cosmography, mensuration.
Case/Y/D/72/.332

Printer is Cosimo Bartoli, 16th cent. P. 142: "Sono adunque le distanzie queste che seguono—Da Ierusalem a'l Limbo—miglia 406, Bràc. 750."

## Gibel, Otto, 1612–1682.

*Propositiones mathematico, das ist: etliche fürnehme und gar nützliche musicalische aussgaben auss der mathesi demonstriret, und nach beschaffenheit in beygefügten kupfferstüchen künstlich repraesentiret und für augen gestellet.*
Minden an der Weser: Johann Ernst Heydorn, Heddewigs Wittiben, 1666.
Subject: Mathematics and music.
Case/4A/1025

## Gilbert, William, 1540–1603.

*De mundo nostro sublunari philosophia nova.*
Amsterdam: Elzevir, 1651.
Subject: Astronomy, elements, magnetism of the earth, motion, physics.
Case/*G/117/.34

In 5 books. "Physiologiae nova contra Aristotelem."

## Gilles de Corbeil [Aegidius Corboliensis], fl. 1200.

*Opus excellentissimi magistri Egidii: de vrinis et pulsu cum expositione clarissimi magistri Gentilis de Fulgineo.*
[Lyon: Martin Havard, 1500?]
C 34^ R 1431^ GW I, 272^ Still. A87^ Goff A-96
Subject: Diagnosis, medicine.
Inc./8658/.7

Diagnosis and treatment of urinary and coronary disorders.

## Gilles, Pierre, 1490–1555.

*Petri Gyllii de topographia Constantinopoleos, et de illivs antiqvitatibvs libri qvatvor.*
Lyon: Gvlielmvm Rovillivm, 1561.
Subject: Geography, topography.
Case/F/0259/.34

## Giovio, Paolo, bp. of Nocera, 1483–1552.

*Descriptio Britanniae, Scotiae, Hyberniae, et Orchadvm.*
Venice: M. Tramezinus, 1548.
Subject: Descriptive geography.
Ayer/7/G6/1548

Beginning p. 45: "Virorum alquot in Britannia, qui nostro seculo eruditione, & doctrina clari, memorabilisque fuerunt, elogia. Per Georgium Lilium, Britannum." (Includes Thomas Linacre, John Colet.)

## Giovio, Paolo, bp.of Nocera.

*Descriptio Britanniae.* [Another copy]
Case/G/4494/.34

MS marginalia, esp. in table of new and ancient place names, pp. 42 ff.

## Giovio, Paolo, bp. of Nocera.

*Descriptiones, quotquot extant, regionum atque locorum.*
Basel, 1561.
Subject: Anthropology, descriptive geography.
Case/G/4494/.35

Ed. Johann Heroldt. 2 vols. in 1. Contains "De piscibus romanis." Vol. 2: Pavli Iovii . . . Nucerini Moschouia, in qua situs regionis antiquis incognitus, religio gentis, mores, &c. fidelissimè referuntur.

## Giovio, Paolo, bp. of Nocera.

*Pavli Iovii Novocomensis, De piscibvs marinis, lacvstribvs, flvvia tilibvs, item de testaceis ac salsamentis liber.*
Rome: F. Mintius Calvi, 1527.
Subject: Fish, marine biology.
Ayer/8.9*/Z9/G4/1527

In 42 chaps., each describing a particular species of marine life. Includes, e.g., ostreis & alijs testaceis.

## Giovio, Paolo, bp. of Nocera.

*The worthy tract of Paulus Iouius.*
London: printed for S. Waterson, 1585.
STC 11900
Subject: Emblems.
Case/W/1025/.339

"[A]n inuention or *impresa* (if it be to be accounted current) ought to have these fiue properties, first iust proportion of body and soule. Secondly that it be not obscure, that it need a sibilla to enterprete it, nor so apparant that euery rusticke may vnderstand it. Thirdly, that it haue especially a beautifull shewe which makes it become more gallant to the view, interserting it with starres, sunes, moones, fire, water, greene trees, mechanicall instruments, fantasticall birds. Fourthly that it haue no human forme. fifthly it must haue a posie which is the soule of the body. . . . Six[th], it is requisite also to be briefe" (sig. Biii verso).

## Giuntini, Francesco, 1523?–1590.

*Speculum astrologiae, qvod attinet ad ivdiciariam rationem nativitatvm atque annuarum reuolutionum: cum nonnullis approbatis astrologorum sententiis.*
Lyon: Philippi Tinghi, 1573.
Subject: Astronomy, judicial astrology.
Case/4A/1509

For collation see Baudrier VI:459–60. MS notes on 3 leaves at back of book. In 9 parts, including 1) Defensionem contra astrologiae aduersarios; 2) De natiuitatum iudiciis; 5) Compendium stellarum fixarum; 9) Tabulas resolutas supputandi motus omnium planetarum secundum obseruationem Nicolai Copernici.

## Glanvill, Joseph, 1636–1680.

*A blow at modern sadducism in some philosophical considerations about witchcraft. To which is added, the relation of the fam'd disturbance by the drummer in the house of Mr. John Mompesson: with some reflections on drollery, and atheisme. By a man of the Royal Society.*
London: printed by E. C[otes] for J. Collins, 1668.
STC II G799
Subject: Witchcraft.
Case/B/88/.339

Defends the existence of witches. Asserts that the Devil wants to persuade men that witches do not exist. Critical of Scot's *Discovery of Witchcraft*, q.v. Contains 3 tracts: "A philosophical endeavour in the defence of the being of witches and apparitions"; "Palpable evidence of spirits and witchcraft"; "A whip for the droll, fidler to the atheist."

## Glanvill, Joseph.

*A blow at modern sadducism in some philosophical considerations about witchcraft.*
London: printed by E. Cotes for J. Collins, 1668.
STC II G800
Subject: Witchcraft.
Case/B/88/.34

4th ed. corrected and enlarged.

## [Glanvill, Joseph]

*A blow at modern sadducism.* [Another copy]
Case/B/88/.3401

## Glanvill, Joseph.

*Essays on several important subjects in philosophy and religion.*
London: printed by J. D. for J. Baker, 1676.
STC II G809
Subject: Witchcraft.
Case/B/0/.34a

Contains 7 essays, including "Against sadducism in the matter of witchcraft."

## Glanvill, Joseph.

*Philosophia pia; or, a discourse of the religious temper, and tendencies of the experimental philosophy which is profest by*

*the Royal Society. To which is annext a recommendation, and defence of reason in the affairs of religion.*
London: printed by J. Macock for James Collins, 1671.
STC II G817
Subject: Experimental philosophy, religion and science.
Case/B/7/.34

1. God is to be praised for his works; 2. His works are to be studied by those that would praise him for them; 3. The study of nature and God's works is very serviceable to religion; 4. Ministers of religion should promote knowledge of nature.

## Glanvill, Joseph.

*Plus ultra: or the progress and advancement of knowledge since the days of Aristotle. In an account of some of the most remarkable late improvements of practical, useful learning.*
London: printed for James Collins, 1668.
STC II G820
Subject: Biology, medicine, physical sciences, scientific positivism.
Case/L/0114/.344

Advances in physical, medical, and biological sciences, astronomy, mathematics, chemistry, anatomy, geometry, natural history, technology (optick glasses), instruments. "But of all the modern discoveries . . . the noblest is that of the circulation of the blood" (p. 15).

## Glanvill, Joseph.

*Sadducismus triumphatus: or, full and plain evidence concerning witches and apparitions. In two parts. The first treating of their possibility, the second of their real existence. And an authentick but wonderful story of certain Swedish witches; done into English by Anth. Horneck.*
London, 1681.
STC II G822
Subject: Witchcraft.
Case/B/88/.3403

Includes testimony of witnesses who claim to have seen witches, etc.

## Glanvill, Joseph.

*Sadducismus triumphatus: or, a full and plain evidence, concerning witches and apparitions. In two parts. The first treating of their possibility. The second of their real existence. Also two authentick, but wonderful stories of certain Swedish witches. Done into English by Dr. Horneck.*
London: printed for A. Bettesworth and J. Batley [etc.] 1726.
Subject: Witchcraft.
B/88/.3405

4th ed., with additions. With some account of Mr. Glanvill's life and writings.

## Glanvill, Joseph.

*Scepsis scientifica: or, confest ignorance, the way to science; in an essay of the vanity of dogmatizing, and confident opinion. With a reply to the learned Thomas Albius.*
London: printed by E. Cotes for Henry Eversden, 1665.
STC II G827
Subject: Occult, sympathetic medicine.
Case/B/245/.339

Chap. 24: Three instances of reputed impossibilities (which likely are not so). (1) Power of imagination. (2) Secret conveyance. (3) Sympathetick cures. BOUND WITH Reply to Thomas White, "Scire tuum nihil est" [same imprint].

## Glanvill, Joseph.

*Some philosophical considerations touching the being of witches and witchcraft. Written in a letter to the much honour'd Robert Hunt esq.; by J. G. a member of the Royal Society.*
London: printed by E. C. for J. Collins, 1667.
STC II G832
Subject: Witchcraft.
Case/B/88/.3408

Penciled note on front flyleaf:" The first editions of this work quarto 1666 were almost all destroyed in the fire of London and a good copy of the second edition 1667 is very scarce." Glanvill mentions Scot's *Discovery of witchcraft.*

## Glareanus, Henricus, 1488–1563.

*Ausz Glareani musick ein vssug, mit verwillung und hilff Glareani allem Christlichen kirchen alt vnd götlich gsang zulernem auch zü verstan ganz nutzlich vnd denen zu hilff so der mathematick un vyls licht der latinischen sprach nitt ganz vnderricht.*
[Basel: Heinrich Petr., 1559]
Subject: Mathematics and music.
Case/3A/733

Chap. 7: Von schlag oder mensur in der musick.

## Glareanus, Henricus.

*Mvsicae epitome ex Glareani dodecachordo vnà cvm qvinqve vocvm melodiis svper eivsdem Glareani panegyrico de Helveticarum XIII urbium laudibus, per Manfridum Barbarinum Coregiensem.*
Basel [1599].
Subject: Mathematics and music.
Case/3A/732

MS marginalia passim.

## Glareanus, Henricus.

*De geographia liber vnvs.*
Basel [Joannes Faber Emmevs, 1528].
Subject: Astronomy, earth sciences, geography, mensuration.
Ayer/7/G7/1528

Chap. 1: De geometriae principiis ad sphaerae astronomicae; chap. 7: De compositione materialis sphaerae, atqve inibi de circulorum theoria; chap. 15: De mensvra et eivs parti.

## Glareanus, Henricus.

*De geographia.*

Freiburg im Briesgau: Ioannes Faber, 1530.
Subject: Astronomy, geography, geology, mensuration.
Ayer/*7/G7/1530

Scattered MS notes, esp. on final leaf.

## Glareanus, Henricus.

*De geographia liber unus, ab ipso avthore iam tertio recognitvs.*
Freiburg im Briesgau: Joannes Faber Emmevs, 1536.
Subject: Astronomy, geography, geology, mensuration.
Ayer/*7/G7/1536

## Glareanus, Henricus.

*De geographia liber unus ab ipso authore iam tertio recognitus.*
Venice: Ioan. Ant. de Nicolinis de Sabio for Melchior Sessa, 1538.
Subject: Astronomy, geography, geology, mensuration.
Ayer/*7/G7/1538

## Glareanus, Henricus.

*De geographia liber vnvs, ab ipso avthore iam novissime recognitvs.*
Freiburg im Briesgau: Ioannis Faber Emmevs, 1539.
Subject: Astronomy, geography, geology, mensuration.
Ayer/*7/G7/1539

## Glareanus, Henricus.

*De geographia liber unus, ab ipso authore iam nouissime recgonitus.*
Freiburg im Briesgau: Stephanus Melechus Grauius, 1543.
Subject: Astronomy, geography, geology, mensuration.
Ayer/*7/G7/1543

## Glareanus, Henricus.

*Geographia liber unus ab ipso authore iam tertio recognitus.*
Venice: Petrus & Io. Maria, brothers, & Cornelius nephew of Nicolino de Sabio, for Melchior Sessa, 1549.
Subject: Astronomy, geography, geology, mensuration.
Ayer/*7/G7/1549

## Glareanus, Henricus.

*Liber de asse & partibus eius.*
Basel: Mich. Isingrinius, 1551.
Subject: Mensuration, solid geometry.
Case/fW/5636/.344

MS notes. Chap. 7: De vasis geometricis apvd avtores celebratis, in quadrum certum ac indubitatum per pedis Romani mensuram cogendis.

## Glauber, Johann Rudolf, 1604–1668.

*Consolatio navigantium: in quâ docetur, & deducitur, quomodo per maria peregrinantes à fame ac siti immò etiam morbis, qui longinquo ab itinere ipsis contingere possunt, sibi providere ac suppetiari liceat.*
Amsterdam: Joannes Janssonius, 1657.
Subject: Iatrochemistry, spagyric medicine.
U/83/.343

## Glissenti, Fabio, ca. 1550–1620.

*Discorsi morali . . . contra il displacer del morire, detto athanatophilia. Diuisi in cinque dialoghi, occorsi in cinque giornate. Ne' quale si discorre quanto ragioneuolmente si dourebbe desiderar la morte; e come naturalmente la si uada fuggendo. Con trenta vaghi, & vtili ragionamenti, come tante piaceuoli nouelle interposti; cauati da gli abusi del presente viuer mondano; et vn molto curioso trattato della pietra de' filosofi.*
Venice: Domenico Farri, 1596.
Subject: Death, philosophers stone.
Case/B/79/.346

## Goddard, Jonathan, 1617?–1675.

*A discourse setting forth the unhappy condition of the practice of physick in London, and offering some means to put it into a better; for the interest of patients, no less, or rather much more than of physicians.*
London: printed by John Martin & James Allestry, 1670.
STC II G914
Subject: Medical practice.
Case/I/045/.88/no. 2

A group of 4 tracts, binders title *Tracts on physick, 1665–70.* See Merrett, Christopher; also "Lex talionis."

## Godelmann, Johann Georg, 1559–1611.

*Tractatus de magis, veneficis et lamiis, deqve his recte cognoscendis et puniendis.*
Frankfurt: Ioannis Sauer for Nicolaus Bassaeus, 1601.
Subject: Incantations, natural magic, necromancy, witchcraft.
Case/B/88/.348

In 3 books. 1: De malitia & odio diabolo aduersus deum & homines; 2: Lamiis; 3: De magis et veneficis.

## Godfridus.

*The knowledge of things unknown, shewing the effects of the planets, and astronomical constellations. With the strange events that befall men, [wo]men, and children born under them. Together with the husbandmans practice: o[r], prognostications for ever; as teacheth Albert, Alkind, and Ptolomy. With the shepherds prognostication for the weather, and Pythagoras his wheel of fortune.*
London: Printed for W. Thackeray, 1679.
Subject: Almanacs, astronomy, judicial astrololgy.
Case/B/8635/.348

## Godolphin, John, 1617–1676.

[Greek : sunegoros thallassios] *A view of the admiral jurisdiction.*
London, 1661.
STC II G952
Subject: History of navigation, laws of the sea.
Case/U/745

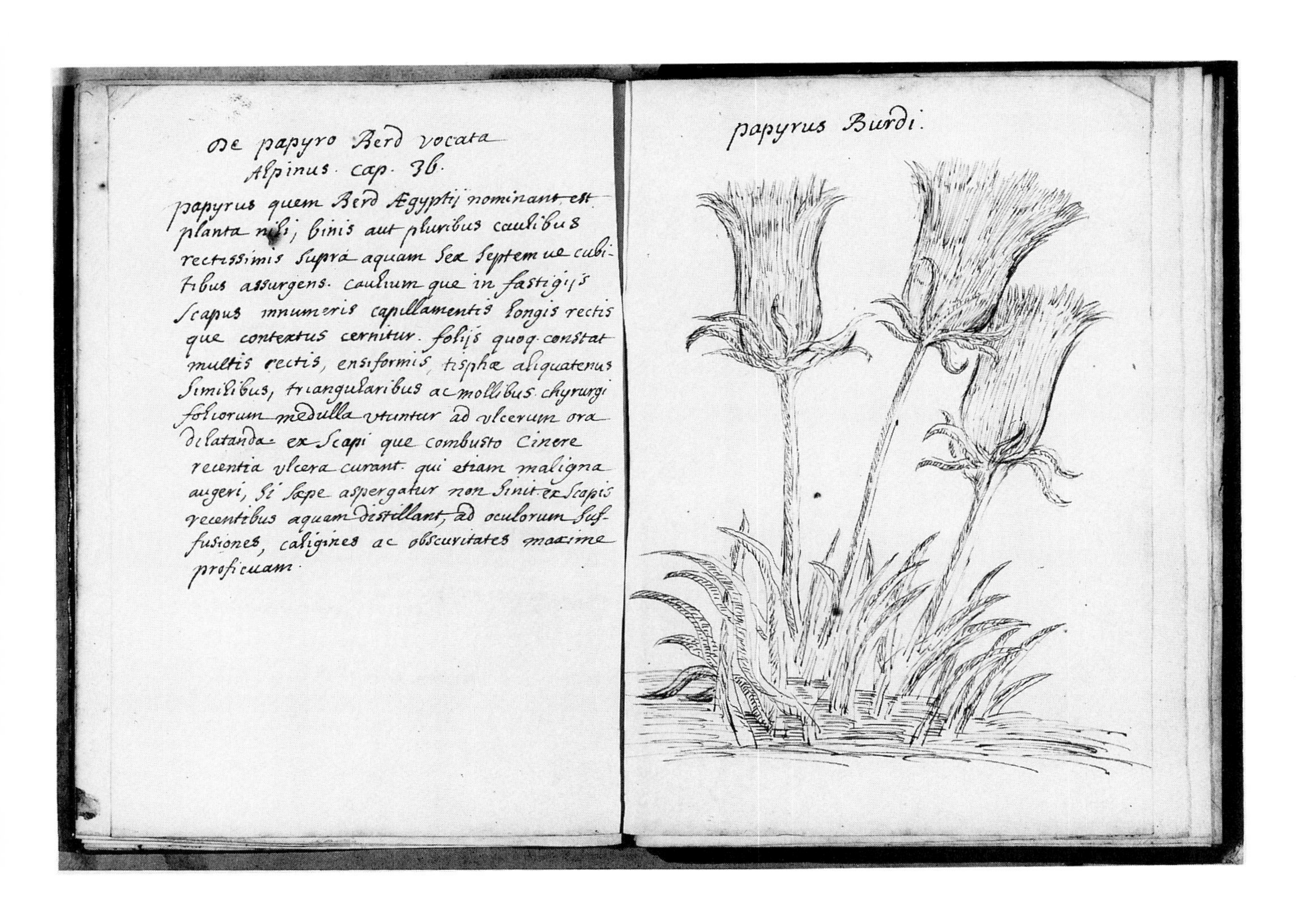

De papyro Berd vocata
Alpinus cap. 36.

papyrus quem Berd Ægyptij nominant est
planta nili; binis aut pluribus caulibus
rectissimis supra aquam sex septem ve cubi-
tibus assurgens. caulium que in fastigijs
scapus innumeris capillamentis longis rectis
que contextus cernitur. folijs quoq constat
multis rectis, ensiformis, tisphæ aliquatenus
similibus, triangularibus ac mollibus. chyrurgi
foliorum medulla vtuntur ad vlcerum ora
dilatanda. ex scapi que combusto cinere
recentia vlcera curant. qui etiam maligna
augeri, si sepe aspergatur non sinit. ex scapis
recentibus aquam destillant, ad oculorum suf-
fusiones, caligines ac obscuritates maxime
proficuam.



1. Melchior Guilandini (i.e., Wieland), *De Stirpibvs.* Padua: Gratiosus Perchacinus, 1558. This manuscript note and possibly the drawing accompanying it are in Joseph Justus Scaliger's hand. Once in his library, the volume has notes in his neat and legible script scattered throughout.

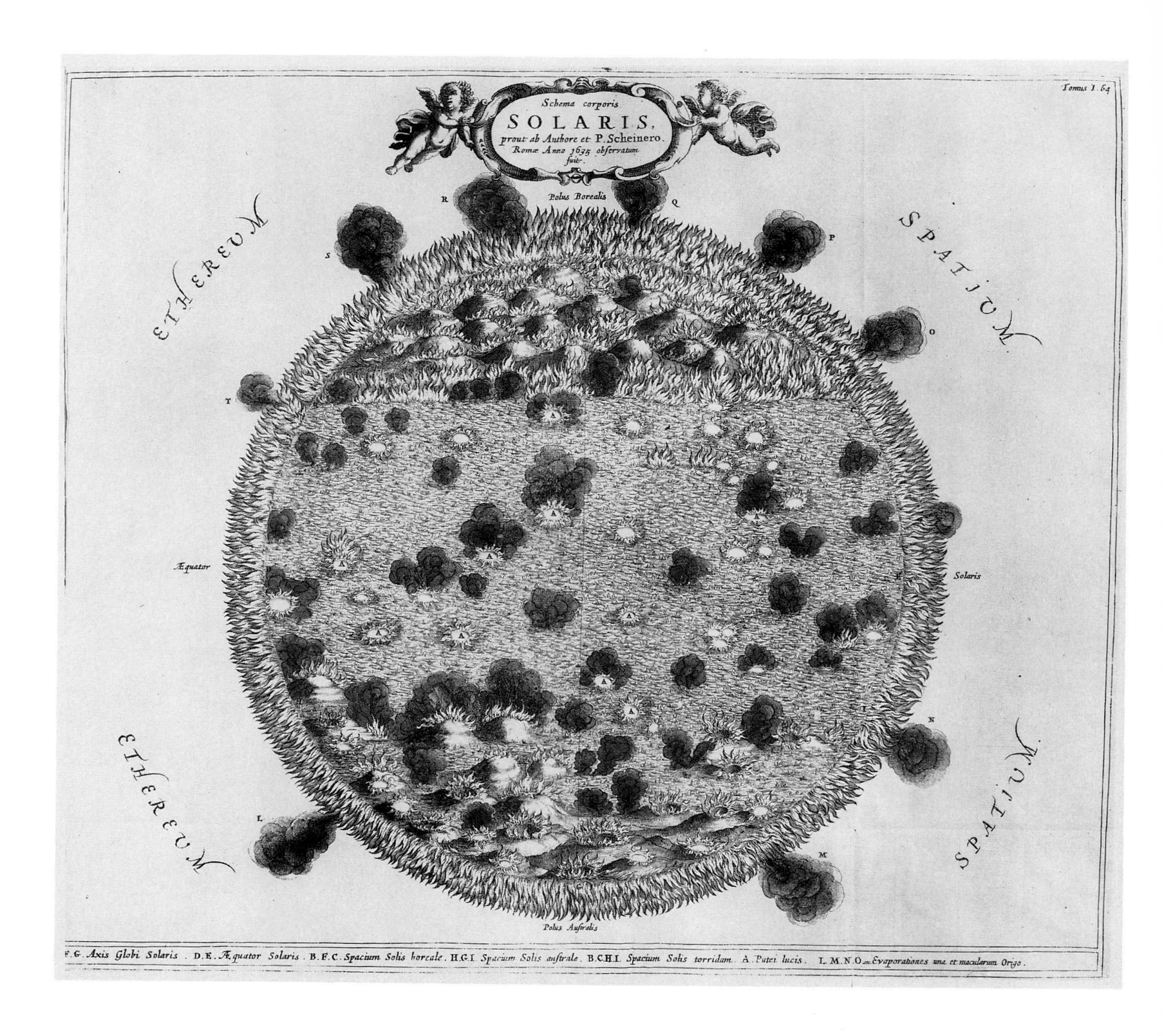


2. Athanasius Kircher, *Mundus Subterraneus.* Amsterdam: Joannes Janssonius à Waesberge and widow E. Weyerstraet, 1668. This portrait of the living sun shows its wells of light, flares of steam, and its spots. The balance of day and night and the progression of the seasons, all under the sun's aegis, are evidence of God's providence and solicitude.

EMBLEMA L. *De ſecretis Naturæ.*

Draco mulierem, & hæc illum interimit, ſimulque ſanguine perfunduntur.

EPIGRAMMA L.

*A Lta veneno ſo fodiatur tumba Draconi.*

3. Michael Maier, *Atalanta Fugiens*. Oppenheim: Hieronymus Galler for Johann Theodor de Bry, 1618. De Bry's engraving combines classical and mythological elements with visual metaphors for the processes of alchemy. In this example, with its Coliseum-like ruins in the background, the love/death alchemical union is represented by the poisonous serpent embracing the woman. Many of the Latin epigrams have a German translation and a musical setting on the facing page as this one does.

4. Gaspar Tagliacozzi, *De Curtorum Chirurgia per Insitionem.* Venice: Gaspar Bindonus, junior, 1597. Gift of Dr. Stanton A. Friedberg. Reconstructive plastic surgery for victims of swordplay required an ingenious device to keep the graft nourished. This handbook for surgeons supplies detailed information on the instruments used for the surgery as well as on the construction of this vest and headpiece.

Icon Octaua.

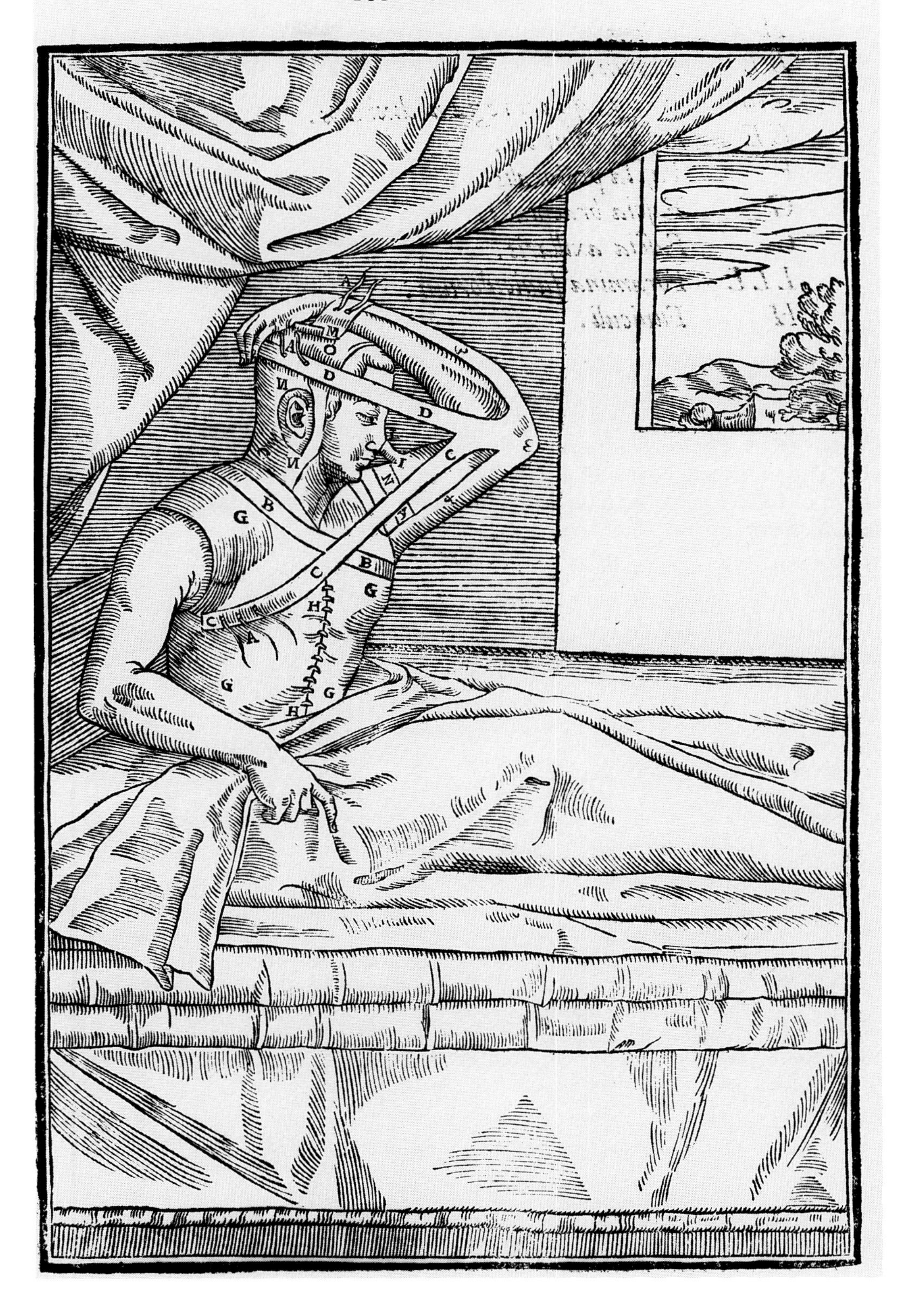

5. Mapheus Vegius, *Philalethes.* [Basel:] Michael Furter, not after 1492. Third edition. Frontispiece. The markings covering the body of Veritas may be eyes (cf. the depiction of "Fama" in *Emblemata,* p. 1536. I have been unable to discover what plant Philalethes is offering (?) and why it is upside down.

Philalethes Veritas

Alli spiriti gentili & virtuosi
Giouābatista Verini Fiorentino.

Chi bene, & presto cercha dimparare
Scriuer, di varie sorte di ualore
Chi vole il tempo sparso raquistare
E lignoranza schacciar dal suo core
Che cercha la virtu a se tirare
Chi da richezza & fama con honore
Chi vuol con vil suo, prender diletto
Studi del fiorentin questo libretto

Alle fanciulle virtuose & da bene
Giouābatista Verini Fiorentino.

Donne galante & voi gentil pulzelle
Che virtu seguitate a tutte le hore
Acioche meglio vi fondate in quelle
Cōposto ho questo sol per vostro amore
El qual insegna far littere belle
Che fia per voi vtile, & honore
Scriuere imparerete bene, & presto
Senza douere stare sotto al Maestro



Chi ha virtu e richo ben assai
El mondo puo cerchare inogni parte
nō gli auanza: e nō gli mancha mai

A ii

6. Giovanni Battista Verini, *Luminario libro 7.* [Milan: P(ietro) Paulo Verini, 1536.] Handwriting books and the instructors like Verini who produced them promoted literacy (and possibly economic opportunity as well) among boys of modest background by teaching basic scribal and arithmetical skills. Verini, a schoolmaster, bookseller, and hack writer, may have produced this vernacular textbook for sale in his own shop.

### Goeurot, Jean.

*Trattato e reggimento, vtilissimo per seruirsene nel tempo della peste. Con vna breue raccolta di auuertimenti per guardarsi da essa, & altri rimedii approuati.*
Siena: Luca Bonetti, 1576.
Subject: Epidemics, medicine, pest tracts.
Case/4A/1530

Translated from the French, by Camillo Spannocchi.

### Gohory, Jacques [Parisien. Leo Suavius, pseud.], d. 1576.

*Instrvction svr l'herbe petvm ditte en France l'herbe de la royne ou medicée: et sur la racine Mechiocan principalement (auec quelques autres simples rares et exquis) exemplaire à manier philosophiquement tous autres vegetaux.*
Paris: Galiot du Pré, 1572.
Subject: Herbal medicine, materia medica.
Ayer/*8.9/B7/G6/1572

In 2 pts. Pt. 2: Contenant un brief traitte de la racine Mechoacan, venue de l'Espagne nouuelle: medicine tres-excellent du corps humain, (blasonée en mainte region la reubarbe des Indes).

### Gohory, Jacques [Parisien. Leo Suavius, pseud.]

*De vsu & mysteriis notarum liber. In quo vetusta literarum & numerorum ac diuinorum ex Sibylla nominum ratio explicatur.*
Paris, 1550.
Subject: Secret writing, occult.
Case/Z/216/.36

### Goldschmidt, Peter, d. 1713.

*Verworffener hexen- und zauberer-advocat. Das ist: wolgegründete vernichtung des theörichten vorhabens hn ... Christiani Thomasii ... und aller derer welche durch ihre superkluge phantasie-grissen dem teufflischen heren-geschmötz das wort reden wollen in dem gegen die selbe aus dem unsviedersprechlichem göttl. worte.*
Hamburg: Gottfried Liebernickel, 1705.
Subject: Demonology, divination, witchcraft.
B/88/.353

Chap. 4: Die crystall-seheren wie auch specula magica, oder die zauber-spiegeln behaupten das teufflische zauber-wesen.

### Goodcole, Henry, 1586–1641.

*The wonderfull discouerie of Elizabeth Sawyer a witch, late of Edmonton, her conuiction, condamnation, and death. Together with the relation of the diuels accesse to her, and their conference together.*
London: printed for W. Butler, 1621.
STC 12014
Subject: Witchcraft, possession stories.
Case/B/8845/.354

### Goropius, Joannes Becanus (1518–1572).

*Opera ... hactenus in lucem non edita: nempe, Hermathenu, Hieroglyphica, Vertumnus, Gallica, Francica, Hispanica.*
Antwerp: Plantin, 1580.
Subject: Geography, sciences and medicine.
Case/F/02/.358

Edited by Laevinus Torrentius, bp. of Antwerp.

### [Goulart, Simon] 1543–1628.

*A learned summary upon the famous poems of William of Saluste, Lord of Bartas. Wherein are discovered all the excellent secrets in metaphysicall, physicall, morall, and historicall knowledge. Fitt for the learned to refresh their memories, and for younger students to abreviat and further theire studies: wherein nature is discovered, art disclosed, and history layd open.*
London: printed for Iohn Grismond, 1621.
STC 21666
Subject: Encyclopedias, epistemology, history of man.
Case/fY/762/.D8355

Translated by T. L. D. M. P. See summary of the first day of the second week for list of illnesses that beset Adam and his children. This is a condensation of Du Bartas's *Divine weekes and workes*, q.v.

### Grassetti, Giovanni Battista, 1609–1684.

*Metodo facile per conoscere la vera dalla falsa astrologia con l'aggiunta della vera e della falsa chiromanzia.*
Rome: Giuseppe San–Germano Coruo, 1698.
Subject: Astrology, chiromancy.
B/8635/.36

### Grassetti, Giovanni Battista.

*La vera, e falsa astrologia con l'aggiunta della vera, e della falsa chiromanzia. Opera di Gio: Battista Tasgresti* [anagram of Grassetti].
Subject: Astrology, chiromancy.
Rome: Giuseppe Corvo, 1683.
B/8635/.362

In 2 books. 1: "Trattato della vera astrologia" (material sphere, zodiac, the fixed stars, origins of stars); 2: "Trattato della falsa astrologia" (authority of Scripture, ecclesiastical and imperial law against judicial astrology, against astrological predictions, genethliacus); followed by "Breve trattato della vera e falsa chiromanzia."

### Grataroli, Guglielmo, 1516–1568.

*Opuscula.*
Basel: Nicolaus Episcopius, jun., 1554.
Subject: Mnemonics, physiognomy, weather.
Case/B/56/.364

Includes *De memoria reparanda, augenda, conservandaque, ac de reminiscentia: tutoria omnimoda remedia, praceptionesque optimae* [local and artificial memory]. *De praedictione morum naturarumque hominum, cum ex inspectione partium corporis, tum alijs modis* [physiognomy]. *De temporum omnimoda mutatione, perpetua & certissima signa & prognostica* [weather prediction]. MS marginalia. "Omnia ab autore correcta, aucte satis, & ultimo edita."

## Grataroli, Guglielmo.

*De litteratorvm et eorvm qvi magistratvm gervnt, conservanda valetudine liber omnibus, quibus secunda valetudo cura est, apprime utilis & necessarius.*
Frankfurt: Nicolaus Hofmann for Iona. Rhodius, 1604.
Subject: Health, hygiene, medicine.
Case/3A/357

In 4 parts. Includes Grataroli, *De memoria reparanda.* BOUND WITH Heinrich Rantzau, *De conservanda valetvdine liber*, q.v.

## Gratia Dei Esculanus [Graziadei d'Ascoli], Joannes Baptista de, d. 1341.

*Questiones* [in libros physices Aristotelis] *fratris Gratia Dei de Esculo ordinis predicatorum per ipsum in Florentissimo studio Patauino disputate excellentissima sacre pagine doctoris.*
Venice: Joannes Mocenigo for Antonio de Regio, 1484.
H 7877^ BMC (XV)V:356^ Pr. 7413^ Goff G-358
Subject: Physics.
Inc./f7168/Inc./4785.5/Inc./5017.5/Inc./5110.5

BOUND WITH tracts by Walter Burley; Thomas Aquinas; Gaetano de Thienis; and Joannes de Janduno, qq.v.

## Grévin, Jacques, 1538–1570.

*Deux livres des venins, ausquels il est amplement discouru des bestes venimeuses, theriaques, poisons & contrepoisons: par Iaqves Grevin de Clermont en Beauuais; medecin à Paris. Ensemble, les oeuures de Nicandre, medecin & poëte Grec, traduictes en vers François.*
Antwerp: Christofle Plantin, 1568.
Subject: Medicine, natural history, poisons, antidotes.
Wing/ZP/5465/.P701

1st ed. In 2 parts: (1) nature of poisonous animals and antivenom remedies; (2) nature of poisons and "contrepoisons." BOUND WITH Nicander, q.v.

## [Grick, Friedrich, (Agnostus Irenaeus, pseud.)] fl. 1621.

*Fons gratiae; das ist, kurtze anzeyg und bericht, wenn, zu welcher zeit und tag der jenigen, so von der heiligen gebenedeyten fraternitet des Rosencreutzes, zu mit brüdern auffgenommen, völlige erlösung und perfection anfangen und hergegen, wessen sie sich in principio dess Heyls vnnd der Gnaden zu verhalten haben.*
[N.p.] 1619.
Subject: Rosicrucians.
K/975/.372

Also attributed to Johann Valentin Andrea. Closely trimmed; top of title page mutilated. BOUND WITH (also by Grick) *Thesaurus fidei*, 1619; *Frater non frater*, 1619; and *Speculum constantiae*, 1618, qq.v.

## [Grick, Friedrich, (Agnostus Irenaeus, pseud.)]

*Frater non frater das is eine hochnot dürfftige verwarnung an die gottselige fromme discipul der h. gebenedeyten societet dess Rosencreutzes.*
[N.p.] 1619.
Subject: Rosicrucians.
K/975/.372

## Grick, Friedrich [Agnostus Irenaeus, pseud.]

*Prodromus Fr. R.C. Das ist, ein vorgeschmack und beylauffige anzeig der grossen aussfürrlichen apologi* [Greek words] *welche baldtfolgen sol, gegen vnd wider den zanbrecher vnd fabelprediger Hisaiam sub cruce zu steiffer unwider treiblicher defension, schützung und rettung hochgedachter heyliger gotteseliger gesellschäffe in eil neben andern wichtigen überhaufften geschäfften auss sonderbarem gnädigen geheiss un befelch verfertiget.*
[N.p.] 1620.
Subject: Rosicrucians.
K/975/.373

## Grick, Friedrich [Agnostus Irenaeus, pseud.]

*Speculum constantiae: das ist, eine nohtwendige vermahnung an die jenige so ihre namen bereits bey der heiligen gebendeiten Fraternitet dess Rosencreutzes angegeben.*
[n.p.] 1618.
Subject: Rosicrucians.
K/975/.372

## [Grick, Friedrich (Agnostus Irenaeus, pseud.)]

*Thesaurus fidei, das ist, ein notwendiger bericht vnnd verwarnung en die novitios, oder junge angehende discipel.*
[n.p.] 1619.
Subject: Rosicrucians.
K/975/.372

## Grillandus, Paulus.

*Tractatus de hereticis et sortilegiis omnifariam coitu: eorumque penis item de questionibus & torturae ac de relaxationem carceratorum.*
Lyon: Jacobus Giunta, 1545.
Subject: Witchcraft.
Case/B/88/.372

See Dorbon, *Bibliotheca esoterica*, no. 1989. In 4 books: de hereticis, de sortilegiis, de penis omnifarium coitus, de questionibus & torturae.

## Grisley, Gabriel, fl. 1655–1661.

*Desengano para a medicina ou botica para todo o pay de familias. Consiste na declaração das qualidades, & virtudes de 260 hervas, con o uso dellas.*
Coimbra: Joseph Autunes da Silva, 1714.
Subject: Herbals.
Greenlee/4532.1/M48/G86/1714

Supplies Latin name of plants, and, for many, a reference to Dioscorides, e.g., "No. 124: Mastruco = nasturtium, Diosc. 1.2.C146." Not illustrated.

## [Grosse, Henning] 1553–1621, ed.

*Magica de spectris et apparitionibus spiritum, de vaticiniis, divinationibus &c.*
Leyden: Franciscus Hackius, 1656.
Subject: Demons, divination, magic, supernatural, witchcraft.
B/863/.374

In 2 books.

## Grosseteste, Robert, bp. of Lincoln, d. 1253.

*Commentaria Roberti Linconiensis in libros posteriorum Aristotelis. Cum textu seriatim inserto. Scriptus Gualterii Burlei super eodem libros posteriorum.*
Venice: Bonetus Locatellus for Octavianus Scotus, 1494.
H-C *10105^ BMC (XV)V:444^ Pr. 5057^ Still. R198^ Goff R-204
Subject: Epistemology, syllogistic reasoning.
Inc./f5057

Leaves 32–35: "Gualterii Burlei scriptum super Posteriora." MS marginalia.

## Grotius, Hugo, 1583–1645.

*Hvgonis Grotii De origine gentivm Americanarum dissertatio.*
[n.p.] 1642.
Subject: Anthropology, New World.
Ayer/104/G8/1642

"Haec sunt quae de Americanarum gentium origine, partim ex traditione, partim ex coniectura colligere potui: quibus si quis habeat certiora, fruar lucro permutationis, & pro eo lucro reddam gratias" (p. 15).

## Grotius, Hugo.

*Hugonis Grotii De origine gentium Americanarum dissertatio altera, adversus obtrectatorem, opaca quem bonum facit barba.*
Paris: Sebastian Cramoisy, 1643.
Subject: Anthropology, New World.
Ayer/104/G8/1643

Speculation on origins of indigenous population of the West Indies & America. Both beginning & ending of text very different from 1642 ed.

## Grube, Hermann, 1637–1698.

*De ictu tarantulae & vi musices in ejus curatione, conjecturae physico- medicae.*
Frankfurt: D. Paulli, 1679.
Subject: Music and medicine.
Case/V/23/.375

Cure of illness by music, esp. tarantula bite (tarantella). Symptoms of tarantula bite, and music cure. BOUND WITH Bartholin, Erasmus, q.v.

## Grynaeus, Johann Jakob, 1540–1617.

*Epistolarum selectarvm (qvae svnt ad pietatem veram incentiuum) libri dvo. Quorvm primo historicae, paraleticae, nutheticae, &c. altero physiologiae theologicae continentur.*
Offenbach: M. Georgius Beatus for Iona Rhosa, 1612.
Subject: Natural theology, physiology.
Case/Y/682/.G931

Collected, edited, and illustrated by Abraham Scultetus. Bl. II is physiology.

## Gualtieri, Niccolo, 1680–1744.

*Index testarvm conchyliorvm qvae ad servantvr in mvseo Nicolai Gvaltieri.*
Florence: Caietani Albizzini, 1742.
Subject: Conchology, marine biology.
+O/8/.376

110 engravings of shells divided by classes: terrestrial, fresh, and salt water.

## Guazzo, Francesco Maria.

*Compendivm maleficarvm in tres libros distinctum ex plvribvs avthoribvs.*
Milan: heirs of August. Tradati, 1608.
Subject: Witchcraft.
Case/B/88/.375

Bk. 1: Artificial magic; bk. 2: sleep-inducing magic, death caused by demons; bk. 3: Signs by which to recognize devils and witches; supernatural and divine remedies against witchcraft (holy water, crucifix, eucharist).

## Guibelet, Jourdain.

*Examen de l'examen des esprits par Iovrdain Gvibelet, docteur en medicine, & medecin du roy à Evreux.* [par Juan Huarte de San Juan]
Paris: Michel Soly, 1631.
Subject: Epistemology, medicine, physiology, psychology.
B/593/.424

Chap. 12: "Brieue description des parties du cerueau" [Are parts of the spirit lodged in the brain?]. Chap. 31: Theory and practice of medicine.

## Guiffart, Pierre, 1597–1658.

*Discovrs dv vvide, svr les experiences de monsievr Paschal et le traicté de Mr Pierius. Auquel sont rendües les raisons des mouuemens des eaux, de la generation du feu, & des tonnerres, de la violence & des effects de la poudre à canon, de la vitesse & du poids augmenté par la cheute des corps graues.*
Roven: Chez Iacqves Besongne, 1648.
Subject: Astronomy, natural science, physics.
Case/L/3/.376

## Guilandini, Melchior [Melchior Wieland], 1520–1589.

*Apologiae adversvs Petr. Andream Mattheolvm liber primvs qui inscribitvr Theon.*
Padua: Gratiosus Perchacinus, 1558.
Subject: Botany.

Case/Wing/Z/30523/.376

BOUND WITH 2 other works by Guilandini, *De stirpibus* and *Papyrus*, qq.v.

### Guilandini, Melchior.

*Melchioris Gvilandini Papyrvs, hoc est commentarivs in tria C. Plinij Maioris de papyro capita. Accessit Hieronymi Mercvrialis repugnantia, qua pro Galeno strenuè pugnatur. Item Melchioris Gvilandini assertio sententiae in Galenum â se pronunciatae.*
Venice: M. Antonius Vlmus, 1572.
Subject: Botany, papyrus.
Case/Wing/Z/30523/.376

1st ed. MS marginalia said to be in the hand of Joseph Justus Scaliger as this was his copy. "It was read by Josephus Justus Scaliger (1540–1609), 'the greatest scholar of modern times.' He covered most margins with a wealth of information and interesting remarks which were evidently written down after 1592 (there are references to Prospero Alpini's *De plantis Aegypti* of 1592). Scaliger's notes are in a very clear hand" (from bookseller's catalogue copy laid into vol.). A pen and ink drawing of papyrus (artist not known) is bound in at end of vol.

### Guilandini, Melchior.

*In C. Plinii Maioris capita aliqvot, vt difficilima ita pulcherrima et vtilissima commentarius, variâ & non vulgari eruditione refertus; vbi Matheoli errores non pauci deteguntur.*
Lavsanne: Franciscus le Preux, 1576.
Pritzel 3967
Subject: Natural history of papyrus.
Wing/ZP/538/.L54

### Guilandini, Melchior.

*De stirpibvs aliqvot, epistolae V. Melchioris Gvilandini Borvssi R. IIII. Conradi Gesneri Tigvrini I. Adiecta est Andreae Patricii ad Gabrielem Falloppivm praefatio.*
Padua: Gratiosus Perchacinus, 1558.
Subject: Herbals.
Case/Wing/Z/30523/.376

Pen drawing (by J. J. Scaliger?) interleaved following p. 27.

### Gvillemeav, Iacqves.

*Anatomie generale dv corps hvmain, composée en tables methodiques avec les portraits et figures de chacvne des parties: Et declarations d'icelles. Diuisee en sept liures.*
Paris: Iean Charron, 1586.
Subject: Anatomy, medicine.
UNCATALOGUED

Title page wanting. In 7 books, followed by "Methodique division et denombrement des maladies qvi adviennent a tovtes les parties dv corps hvmain."

# H

## Hakewill, George, 1578–1649.

*An apologie or declaration of the power and providence of God in the government of the world. Consisting in an examination and censvre of the common errovr tovching natvres perpetuall and vniversall decay, divided into fovr bookes.*
Oxford: printed by William Turner, 1630.
STC 12612
Subject: Natural science, natural religion.
Case/fC/558/.379

2d ed. revised. "So though there be many changes and variations in the world, yet all things come about one time or another to the same point againe" (from "The argvment of the front and of the vvorke").

## Hakewill, George.

*An apologie or declaration of the power and providence of God in the government of the world.*
Oxford: printed by W. Turner, 1635.
STC 12613
Subject: Natural science, natural religion.
Case/fC/558/.38

3d ed. revised and "whole sections augmented by the author; besides the addition of two entire bookes not formerly published." The world does not decay but "will have an end by fire, and by it bee intirely consumed."

## Hakluyt, Richard, 1552?–1616.

*Divers voyages touching the discouerie of America, and the ilands adiacent vnto the same, made first of all by our Englishmen, and afterward by the Frenchmen and Britons: and certiane notes of aduertisements for obseruations, necessarie for such as shall hereafter make the like attempt. With two mappes annexed hereunto for the plainer vnderstanding of the whole matter.*
London: Thomas VVoodcocke, 1582.
Church 128^ John Carter Brown 346
Subject: Discovery, voyages.
Ayer/*110/H2/1582

Maps supplied in facsimile. Note laid in: "Richard Hakluyt compiled this first collection of voyages to the new world for an English-reading public. There are numerous accounts of native inhabitants; they vary exceedingly in accuracy and charm."

## Hakluyt, Richard.

*The principall navigations and discoveries of the English nation, made by sea or ouer land to the most remote and farthest distant quarters of the earth at any time within the compasse of these 1500 yeeres: deuided into three seuerall parts, according to the positions of the regions whereunto they were directed.*
London: George Bishop and Ralph Newberie, deputies to Christopher Barker, 1589.
STC 12625
Subject: Discovery, voyages.
Ayer/*110/H2/1589

In 3 pts. Pt. 3: "Including the English valiant attempts in searching almost all the corners of the vaste new world of America. Whereunto is added the last most renowned English nauigation, round about the whole globe of earth."

## Hakluyt, Richard.

*The principal navigations, voiages, traffiqves and discoueries, of the English nation, made by sea or ouer-land, to the remote and farthest distant quarters of the earth.*
London: George Bishop, Ralph Newberie, and Robert Barker, 1598 [1600].
Subject: Discovery, voyages.
STC 12626
Ayer/*110/H2/1598

3 vols. in 2. Vol. 3 includes discovery and exploration of "all the yles both small and great lying before the cape of Florida, the bay of Mexico. . . . Together with the two . . . voyages of Sir Francis Drake and M. Thomas Cavendish."

## Hakluyt, Richard.

*The principal navigations, voiages, traffiqves and discoueries, of the English nation, made by sea or ouerland, to the remote and farthest distant quarters of the earth, at any time within the compasse of these 1600 yeres.*
London: George Bishop, Ralph Newberie, and Robert Barker, 1599.
Subject: Discovery, voyages.
Case/G/12/.38

3 vols. in 2.

## Hale, Sir Matthew, 1609–1676.

*A collection of modern relations of matter of fact, concerning witches and witchcraft upon the persons of people.*
London: printed for John Harris, 1693.
STC II H224
Subject: Witchcraft.
Case/F/456/.88

Part 1. Binder's title: *Tracts 1670–1724.*

## Hale, Sir Matthew.

*The primitive origination of mankind considered and examined according to nature.*
London: printed by William Godkid for William Shrowsbery, 1677.
STC II H258
Subject: Natural religion.

Case/fC/257/.38

Reasonableness of divine hypothesis; man not eternal; heavenly bodies in motion.

## Hale, Sir Matthew.

*A tryal of witches, at the assizes held at Bury St. Edmonds for the county of Suffolk on the tenth day of March, 1664.*
London: printed for Will. Shrowsbery, 1682.
STC II H260^ T224
Subject: Witchcraft.
Case/B/8845/.38

BOUND WIITH a group of treatises on witchcraft, including also *Pleas of the crown: or a methodical summary of the principal matters relating to that subject.* London: Richard & Edward Atkyns for D. Brown, 1707. Page 3 of this latter tract deals with heresy, p. 7 with witchcraft. It is heavily annotated.

## Hall, Francis.

*An explication of the diall sett up in the Kings Garden at London, an. 1669. In which very many sorts of dyalls are conteined; by which besides the houres of all kinds diversly expressed, many things also belonging to geography, astrology, and astronomy, are by the sunnes shadow made visible to the eye. Amongst which, very many dialls, especially the most curious, are new inventions, hitherto divulged by none.*
Liege: printed by Guillaume Henry Streel, 1673.
STC II H332
Subject: Chronology, instrumentation, technology.
Case/R/25/.382

## Halley, Edmund, 1656–1742.

*Atlas maritimus & commercialis: or, a general view of the world, so far as relates to trade and navigation: describing all the coasts, ports, harbours, and noted rivers, according to the latest discoveries and most exact observations . . . To which are added sailing directions for all the known coasts and islands on the globe; with a sett of sea-charts, some laid down by Mercator, but the greater part according to a new globular projection, adapted for measuring distances (as near as possible) by scale and compass . . . The use of the projection justified by Dr. Halley.*
London: printed for James and John Knapton [etc.], 1728.
Phillips 3298
Subject: Navigation.
Ayer/135/A83/1728

[Typed note laid in] Often known as Halley's atlas because of his commentary on the projection employed, but probably the work of John Harris, John Senex, and Henry Wilson. In 2 pts. Pt. 2: *A general coasting pilot; containing directions for sailing into, and out of, the principal ports and harbours thro'out the known world . . . To all which are prefix'd, directions to mariners, whereby the use of the globular charts is made plain and easy to the meanest capacity; sundry examples laid down on the said charts, and the arithmetical computations added, whereby the mariners will be able to judge of the truth of the said charts; and the errors of the plain chart shewn, that so the mariner may avoid them. By Nathaniel Cutler.*

## Happel, Eberhard Werner, 1647–1690.

*Mundus mirabilis tripartitus, oder wünderbare welt in einer kurtzen cosmographia fürgestellet: Also dass der erste theil handelt von dem himmel beweg-und unbeweglichen sternen samt ihrem lauff und gestalt cometen jahre eintheilung lufft meteoris, meer und dessen beschaffenheit seen insuln ebb und fluth.*
Ulm: Matthaeus Wagner, 1687.
Subject: Astronomy, comets, cosmography, zoology.
Ayer/*7/H2/1687

*Der andere theil von den menschen und thieren der erden allerhand dignitäten potentaten religionen estaats, maximen macht intraden kriegs-arten regiments geschichten, wahl ceremonien . . . Der dritte theil von den universitaten.*

## Harderwijk, Gerard van, d. 1503.

*In epitomata totius philosophiae naturalis Aristotelis.*
Cologne: Heinrich Quentel, 1496.
H-C *8362^ BMC (XV)I:285^ Pr. 1333^ Vouillième 441^ Still. G153^ Goff G-168
Subject: Natural philosophy, natural science.
Inc./1333

BOUND WITH *De coelo et mundo e natura scientia fere plurima videtur circa corpora & magnitudines; Epitomataque et reparationes appellantur in libros de generatione qui a Boetio greca lingua perigeneseos.*

## Hariot, Thomas, 1560–1621.

*Admiranda narratio fida tamen, de commodis et incolarvm ritibvs Virginiae nvper admodvm ab Anglis qvi à Dn. Richardo Greinville eqvestris ordinis viro eò in coloniam anno 1585 dedvcti svnt inventae, svmtvs faciente Dn. VValtero Raleigh.*
Frankfurt: printed by Ioannis Wechel for Theodor de Bry, 1590.
Subject: Exploration, New World.
Ruggles Coll./No. 153

Pt. 1, "De mercatvrae commoditatibvs," contains a description of plants native to Virginia. 2d section: "Vivae imagines et ritvs incolarvm eivs provinciae in America, qvae Virginia appellata est ab Anglis. . . . omnia diligenter observata, et ad viuum expressa a Joanne With [John White], eius rei gratia in illam prouinciam annis 1585 & 1588."

## Hariot, Thomas.

*Admiranda narratio fida tamen, de commodis et incolarvm ritibvs Virginiae.*
Frankfurt am Main: printed by Joannes Wechel for T. de Bry, 1590.
Church 143
Subject: Exploration, New World.
Ayer/*110/B9/1590a/vols. 1-3

2d ed., 2d impression (p. 29 is 1st ed., 2d issue). Included in this vol. (not by Hariot): "Brevis narratio eorvm qvae in Florida."

### Hariot, Thomas.

*Admiranda narratio fida tamen, de commodis et incolarvm ritibvs Virginiae.*
Frankfurt am Main: printed by Joannes Wechel for T. de Bry, 1590.
Cf. Church 140.
Subject: Exploration, New World.
Ayer/110/B9/1590/v. 1

1st ed., 1st issue (except for title page and pts. 11 and 12, which are 1st ed., 2d issue).

### Hariot, Thomas.

*Admiranda narratio fida tamen, de commodis et incolarvm ritibvs Virginiae.*
Frankfurt am Main: printed by Joannes Wechel for Theodor de Bry, 1590.
Subject: Exploration, New World.
Ayer/110/B9/1590b/v. 1

### Hariot, Thomas.

*A briefe and true report of the new found land of Virginia. Of the commodities and of the nature and manners of the naturall inhabitants. Discouered by the English colony there seated by Sir Richard Greinuile knight in the yeere 1585. This fore booke is made in English by Thomas Hariot.*
Frankfurt: Ioannis Wechel for Theodor De Bry, 1590.
STC 12786
Subject: Exploration, New World.
Ayer/*f150.5/V7/H2/1590

Engravings by De Bry from John White's drawings. Pt. 1: "Of marchantable commodities"; pt. 2: "The trve pictvres and fashions of the people in that parte of America now called Virginia. Translated out of Latin into English by Richard Hakluyt. Diligently collected and draowne by Iohn White who was sent thiter [sic] speciallye and for the same purpose by the said Sir Walter Ralegh the year abouesaid 1585, and also the year 1588. Now cutt in copper and first published by Theodore de Bry att his wone charges." [According to sale catalgue description laid in, this is the 2d original English work on America. Only vol. in English of De Bry's voyages.]

### Hariot, Thomas.

*Merveillevx et estrange rapport, tovtesfois fidele, des commoditez qvi se trovvent en Virginia, des facons des natvrels habitans d'icelle.*
Frankfurt: Joannes Wechel for Theodor de Bry, 1590.
Subject: Exploration, New World.
Ayer/*110/B9/1590e

1st pt. consists of brief history of Virginia (in French; see Ruggles above for gen. descr.). 2d part (in Latin) consists of map and engravings of natives.

### Hariot, Thomas.

*Von dem auch new erfundenen lande, Virginia.*
In Dresser, Matthaeus *Historien und bericht,* q.v.
Ayer/*150.5/V7/D7/1598

### Harris, John, 1667?–1719.

*The description and uses of the celestial and terrestrial globes; and of Collins's pocket quadrant.*
London: printed by E. Midwinter for D. Midwinter & T. Leigh, 1703.
Subject: Astronomy, navigation.
Ayer/8.9/N2/H314/1703

### Harris, John.

*Lexicon technicum: or, an universal English dictionary of arts and sciences explaining not only the terms of art, but the arts themselves.*
London: printed for Dan. Brown, etc., 1704, 1710.
Subject: Encyclopedias, sciences, technology.
Case/Wing/folio/oAE/5/.H37/v. 1, 2

2 vols. Each vol. has complete alphabetical sequence. Each vol. is 1st ed. Has Newton's only published chemical work (on acids). Technical articles were commissioned by Harris from such experts as Newton and Halley. Among articles based on Halley are those on gunnery, variation of the compass, rainbows, springs and fountains, wind, heat, comets. Articles based on Newton include attraction, color, comets, fluxions, light, motion, sun, moon, planets, tides. Articles on quadrature provide 1st English translation of "Two treatises" in the *Opticks*. Flamsteed's table of natural days, fluxions, hyperbola, hydrostatics. Vol. 2, 1710, Introduction: "The reader will find here many parts of natural philosophy and anatomy largely treated of, which were either but just nam'd, as it were, or entirely omitted in the former." Contains astronomical tables. Has added a catalogue of books, table of logarithms, table of artificial sines, tangents, and secants to every degree and minute of the quadrant. Table of altitude, article on arteries and veins. From Pickering & Chatto catalogue no. 665: This is the first modern scientific encyclopedia.

### Harris, John.

*Lexicon technicum.*
London: printed for Dan. Brown, etc., 1716.
Subject: Encyclopedias, sciences, technology.
Case/Wing/folio/oAE/5/.H37/1716

3d ed. of vol. 1; 1st ed. of vol. 2.

### Harsnett, Samuel, abp. of York, 1561–1631.

*A discovery of the fraudulent practises of Iohn Darrel . . . in his procedings concerning the pretended possession and dispossession of William Somers at Nottingham: of Thomas Darling . . . and of Katherine Wright . . . and of his dealings with one Mary Couper at Nottingham, detecting in some sort the deceitfull trade in these latter dayes of casting out deuils.*
London: Iohn Wolfe, 1599.
STC 12883
Subject: Possession, sorcery, witchcraft.
Case/B/88/.444

BOUND WITH James, King of England, q.v.

## Hartlib, Samuel, d. ca. 1670.

*Chymical, medicinal and chyrurgical addresses: Made to Samuel Hartlib.*
London: printed by G. Dawson for Giles Calvert, 1655.
STC II H978
Subject: Alchemy, iatrochemistry, metallurgy, spagyric medicine.
Case/B/8633/.16

9 tracts: (1) Whether the vrim & thummin were given in the Mount, or perfected by art; (2) Sir George Ripley's epistle; (3) Gabriel Plats caveat for alchymists; (4) A conference concerning the philosophers stone; (5) Philaretvs to Empyricus: An invitation to a free and generous communication of secrets and receits in physick; (6) [One of Monsieur Renaudot's French conferences] A conference concerning this qvestion: Whether or no, each several disease hath a particular and specifical remedy [followed by "another conference of Monsieur Renaudots"]; (7) A new and easie method of chirurgery, for the curing of all fresh wounds or other hurts; (8) The appendix, containing Mr. Gerard Malynes philosophy; (9) A translate of the eleventh chapter from a theosophicall German treatise Postilion or a new almanack. "Necessary considerations for all learned and experienced men who deal in chyrurgery either as practitioners or teachers thereof" (p. 159).

## Hartlib, Samuel.

*A rare and new discovery of a speedy way, and easie means, found out by a young lady in England, she having made full proofe thereof in May, anno 1652. For the feeding of silkworms in the woods, on the mulberry-tree leaves in Virginia: who after fourty dayes time present the most rich golden-coloured silken fleece, to the instant wonderfull enriching of all the planters there, requiring from them neither cost, labour, or hindrance in any of their other employments whatsoever. And also to the good hopes that the Indians, seeing and finding that there is neither art, skill or pains in the thing: they will readily set upon it being by the benefit thereof inabled to buy of the English (in way of truck for their silk-bottomes) all those things that they most desire. So that not only their civilizing will follow, thereupon, but by the infinite mercie of God, their conversion to the Christian faith.*
[London] printed for Richard Wodenothe, 1652.
STC II H996
Subject: Natural history, silk manufacture, silkworms.
Ayer/*150.5/V7H3/1652

## Hartlib, Samuel.

*Samuel Hartlib his legacie: or an enlargement of husbandry used in Brabant and Flaunders; wherein are bequeathed to the common-wealth of England more outlandish and domestick experiments and secrets in reference to universall husbandry.*
London: printed by H. Hills for R. Wodenothe, 1651.
STC II H989
Subject: Agriculture.
Ayer/*152.4/E7H33/1651

Beekeeping, silkworms, "diseases of cattel and their cures." MS marginalia passim.

## Hartmann, Johann, 1568–1631.

*Praxis chymiatrica edita à Iohanne Michaelis.*
Geneva: Johannes de Tournes & Jacobus de la Pierre, 1635.
Subject: Iatrochemistry, materia medica, medicine, therapeutics.
oRS/78/.H37/1635

Imperfect: pp. 295–98 wanting. "Hoc postremi editioni adiecti sunt, propter affinitatem materiae, tres tractatus noui. 1. De oleis variis chymice distillatis; 2. Basilica antimonij Hameri poppij Thallini; 3. Marci Cornachini Methodus, que omnes humani corporis affectiones ab humoribus copia, vel qualitate peccantibus, chymicè & Galenicè curantur."

## Hartmann, Johann Ludwig, 1640–1684.

*Hochzeit-Predigten . . . sampt angehängter decade concionum singularium und vierfachen vollständigen register. Welche so woll ins gemein auff alle gelegenheit und fälle; als auch auff unterschiedene personen, jungen und alten, ledigen und verwittibten. Sonderlich aber auff all stände, facultäten, handwercker und zauff-nahmen gerichtet sind, bey copulation geistlicher juristen medicorum, cantorum, kauffleuten, büchändler, apothecker, krämer, balbierer, bader, becker, färber, gerber, häffner, maurer, müller, schmidt, schneider, wagner, weber, zimmerlen, & ziegler & wie auch tasterhaffter personen, welche um anzucht willen kirchen.*
Giessen: Albrecht Otto Fabers, by Johann Eberhard Petri & Christian Liebenstein, 1679.
Subject: Sermons for, e.g., doctors and patients.
C/826/.387

Sermon 13 (for doctors and patients): Ein weisser mann schaffet mit seinen rath nutz und triffts &c. [p. 144]; sermon 14: Die kunst des artzes erhöhet ihn &c. [p. 157]; p. 176: Die Christ-löbliche apothekerkunst nach der irdischen geistlichen und himmlischen apothecken; sermon 17: Bey copulation eines balbieres und wundartzes.

## Harvey, Gideon, 1640?–1700?

*The third edition of the rarities of philosophy & physick . . . convenient to be perused by divines and students in any faculty; but more necessarily by physicians. . . . There is also now added, a third medicine without which the design of this treatise would be imperfect. Offering moreover at [sic] different hypotheses in metaphysicks, natural and moral philosophy; also in the art of physick, almost throughout the whole body, and particular relating to indigestion, and other diseases of the stomach.*
London: printed for A. Roper and R. Basset, 1702.
Subject: Circulation of the blood, kidneys, medicine, metaphysics.
oR/128.7/.H3/1702

## [Harvey, John] 1563?–1592.

*A discovrsive probleme concerning prophesies, how far they are to be valued, or credited, according to the surest rules, and directions in diuinitie, philosophie, astrologie, and other learning: by I. H. physition.*

London: printed by Iohn Iackson for Richard Watkins, 1588.
STC 12908
Subject: Divination, prognostication, rational approach to astrology.
Case/B/863/.391

## Harvey, William, 1578–1657.

*Exercitationes de generatione animalium quibus accedunt quaedam de partu: de membranis ac humoribus vteri: & de conceptione.*
Amsterdam: Ludovic Elzevir, 1651.
Subject: Generation of animals, physiology, zoology.
Case/*Y/682/.H26

Conceptu cervarum & damarum. Additamenta: De partu, de vteri membranis & humoribus, de placenta, de vmbilico, de conceptione.

## Harward, Simon, 1572–1614.

*A new orchard and garden: or the best way for planting, grafting, and to make any ground good for a rich orchard. . . . With the country housewifes garden for herbes of common vse, their vertues, seasons, profits, ornaments . . . As also the husbandry of bees with their seuerall vses and annoyances.*
London: printed by I. H. for Roger Iackson, 1623.
Subject: Agriculture, gardening, home remedies.
Case/R/50/.552

Not in STC. A group of tracts on aspects of agriculture. Binders title: *A way to get wealth.* BOUND WITH Gervase Markham, William Lawson, qq.v. Also another treatise by Harward, "The art of propagating plants" (STC 15330).

## Hauber, Eberhard David, 1695–1765.

*Bibliotheca acta et scripta magica. Gründliche sachrichten und urtheile von solchen büchern und hand jungen welche die macht des teufels in leiblichen dingen betressen.*
Lemgo: Joh. Heinrich Meyer, 1738–45.
Subject: Magic, occult.
Case/B/85/.386

36 pt. in 3 vols. See Graesse, 118–130. Vols. 1 and 2 dated 1739, though vol. 2 has parts dated 1740. Vol. 3, 1741.

## Haunoldt, Abraham, fl. 1588–1594.

*De gravi avditv et svrditate, de qvibvs, divina aspirante gratia, in celeberrima VVitebergensivm academia, praeside.*
[Wittenberg?] 1592.
Subject: Deafness, medicine.
UNCATALOGUED

Thesis. "Clarissimo viro M. Hieronymo Nymanno Torgense, pro ingenii modulo, respondebit Abrahamvs Havnoldt Uratislaviensis . . . loco & tempore consuetis. 1592."

## Hegenitius, Gottfried, 17th cent.

*Itinerarivm Frisio-Hollandicvm et Abr. Ortelii itinerarivm Gallo-Brabanticum.*
Leyden: Elzeviriana, 1630.
Subject: Descriptive geography, travel.
Case/-G/467/.395

## Hegius, Alexander, ca. 1433–1498.

*Opuscula et tractatus.*
[Strassburg: Johann Prüss, ca. 1493–1500.]
H *8425^ BMC (XV)I:129^ Pr. 581^ Still. H13^ Goff H-16
Subject: Exorcism, witchcraft.
Inc./f581

## Hegius, Alexander.

*Carmina et grauia et elegantia: cum ceteris eius opusculis que subijciunt. De scientia et eo quod scitur contra academicos de triplici anima uetegabili, [sic] sensili, et rationali de uera pasche inueniendi ratione quam ex Isaac Argyro greco excepisse apparet. De rhetorica, de arte et inertia, de sensu et sensili, de moribus, de philosophia, de incarnationis misterio, erotemata.*
Deventer: R. Pafraet, 1503.
Subject: Natural philosophy.
Case/Y/682/.H35

Lists 11 tracts including, besides the above, De physicus. Only extant ed. at Deventer. See Wellcome catalogue. MS marginalia passim in Greek and Latin.

## Heide, Antonius de, d. 1646.

*Anatome mytuli, Belgicè mossel, structuram elegantem ejusque motum mirandum exponens. Subjecta est centuria observationum medicarum.*
Amsterdam: Janssonio-Waesberg, 1684.
Subject: Marine fauna, medicine.
oQL/430.7/.U6/H45/1684

## Heister, Lorenz, 1683–1758.

*Institvtiones chirvrgicae in qvibvs qvicqvid ad rem chirvrgicam pertinet, optima et novissima ratione, pertractatvr atque in tabulis multis aeneis praestantissima ac maxime necessaria instrumenta itemque artificia, sive encheirises praecipuae & vincturae chirurgicae repraesentantur. Opus triginta annorvm, nunc demum, post aliquot editiones germanica lingua evulgatas, in exterorum gratiam Latine publicatum.*
Amsterdam: Janssonio-Waesbergios, 1739.
Subject: Surgery.
UNCATALOGUED

3 pts. in 2 vols.

## [Heller, Johann]

*Neuw jag vnnd weydwerck buch das is ein grundtliche beschreibung vom anfang de jagten, auch vom jäger, seinem horn vnd stimm, hunden, wie die zu allerley wildpret abzurichten, zu pfneischen, vnd vor der wüt vnd andern zufällen z bewahren. . . . Item von adelichen weydwerck der falcknerey, beyssen vnd federspiel, auch wie die falcken zu tragen, zu haben, zulocken, ässen, vnd auff den raub anzubringen, vnd wie man allerley feld vnd wassergeflügel, als krasnich, rephüner, wachteln, reyger, wilde gäns, vnd antvögel, r.c. veyssen vnd fangen sol. Dessgleichen vom fisch, krebs, otter und biber fang wie mans mit netzen, reusen,*

*angeln, kasten, otter vnd biberhunden vnd allerley darzu gehöriger gelegenheit fahen sol.*
Frankfurt: printed by Johann Feyerabendt for Sigmund Feyerabendt, 1582.
Subject: Falconry, fishing.
Wing/fZP/547/.F428

2 vols. in 1.

### Helmont, Franciscus Mercurius van, 1614–1699.

*Alphabeti verè naturalis Hebraici brevissima delineatio. Quae simul methodum suppeditat, juxta quam qui surdi nati sunt sic informari possunt, ut non alios saltem loquentes intelligant, sid & ipsi ad sermonis usum perveniant.*
Sulzbach: Abraham Lichtenthaler, 1657 [i.e., 1667].
Subject: Deafness, medicine, sign language.
Case/oQP/306/.H46/1667

### Helvetius, Johann Friedrich, d. 1709

*Diribitorium medicum de omnium: morborum accidentiumque & in externorum definitionibus ac curationibus, ex saporibus, odoribus foetoribs[q]ve [sic] provenientibus a fermentorum, effervescentiarum aut putrefactionum salibus, sulphuribus vel mercuriis, quae male inveniuntur in succis alibilibus bene constitutis omnium ventriculorum, glandularum, vasorumque lymphaticorum totius corporis.*
Amsterdam: Janssonius à Waesberge, 1670.
Osler 2939
Subject: Spagyric medicine.
Wing/ZP/646/.W123

BOUND WITH his *Microscopium* and *Vitulus Aureus*, qq.v.

### Helvetius, Johann Friedrich.

*Microscopium physiognomiae medicum, id est, tractatus de physiognomia, cujus opere non solum animi motus simul ac corporis defectus interni, sed & congrua iis remedia noscuntur per externorum lineamentorum, formarum, colorum, odorum, saporum, domiciliorum ac signaturarum intuitum, qui harmonicarum hominis constitutionem & medicandi notitiam ex simplicibus indicat.*
Amsterdam: Janssonius-Waesberge, 1676.
Osler 2940
Subject: Astrological medicine, doctrine of signatures, physiognomy.
Wing/ZP/646/.W123

BOUND WITH his *Vitulus Aureus* and *Diribitorum*, qq.v. Preface in German, text in Latin.

### Helvetius, Johann Friedrich.

*Vitulus Aureus, quem mundus adorat & orat, in quo tractatus de rarissimo naturae miraculo transmutandi metalla, nempe quomodo tota plumbi substantia vel intra momentum ex qua vis minima lapidis veri philosophici particula in aurum obryzum commutata fueri Hagae Comitis.*
Amsterdam: Johannes Janssonius à Waesberge, & widow Elizei Weyerstraet, 1667.
Subject: Alchemy, Paracelsian and spagyric medicine.
Wing/ZP/646/.W123

Bound with his *Diribitorum* and *Microscopium*, qq.v.

### Helwig, Christoph, 1587–1617.

*Chronologia Universalis ab origine mundi per qvatvor svmma imperia (qvas monarchias appellant,) ad praesens tempus compendiosè deducta, cum praecipuis synchronismis virorum celebrium, eventorum, & politicarum, seu regnorum caeterorum.*
Gessen: Caspar Chemlinus, 1618.
Subject: Universal chronology.
F/006/.393

### [Helwig, Christoph] Helvicus, Christopher.

*The historical and chronological theatre of Chr. Helvicus, distributed into equal intervals of tens, fifties and hundreds: with an assignation of empires, kingdoms, governments, kings, electours, princes, Roman popes, Turkish emperors, and other famous and illustrious men, prophets, divines, lawyers, physicians, philosophers, oratours, poets, historians, hereticks, Rabbins, councils, synods, academies, &c. and also of the usual epochaes. Faithfully done into English according to the two best editions, viz., that of Francofurt and that of Oxford.*
Oxford: printed by M. Flesher, for George West and John Crosley, 1687.
STC II H1411
Subject: Universal chronology.
F/006/.395

### Helwig, Christoph.

*Theatrvm historicvm: sive chronologiae systema novum, aeqvalibvs denariorvm, qvinqvagenariorum & centenariorum intervallis; cum assingatione imperiorum intervallis; cum assignatione imperiorum regnorum . . . medicorum . . . ut vniversa temporum et historiarvm series, à mvndi origine ad praesentem annvm 1629.*
Marburg: Nicolavs Hampelius, 1629.
Subject: Universal chronology.
fF/006/.394

### Hemmerlin, Felix, 1389–1461.

*Opuscula et tractatus.*
[Strassburg: Johann Prüss, ca. 1493–1500.]
H *8425^ BMC (XV)I:129^ Pr. 581^ Still H13^ Goff H-16
Subject: Demons, exorcism.
Inc./f581

"Tractatus de exorcismis," fol. 74. "Alius tractatus exorcismorum seu adiurationum," fol. 76. "De crudelitate [*sic*] demonibus exhibenda," fol. 79.

### Hemmerlin, Felix.

*Opuscula et tractatus.*
[Strassburg, 1497.]
H-C *8424^ BMC (XV)I:172^ Pr. 482^ Still. H14^ Goff H-17
Subject: Demons, exorcism.
Inc./779/.5

"Tractatus de exorcsismus sub littera r"; "Alius tractatus exorcismorum seu adiurationum sub littera s"; "De credulitate demonibus exhibenda sub littera t." MS marginalia throughout, with additional MS material on verso of title page and on back flyleaf.

**Hemminga, Sixtus ab, 1533–1586.**

*Astrologiae, ratione et experientia refvtatae liber: continens breuem quandam apodixin de incertitudine & vanitate astrologica, & particularium praedictionum exempla triginta: nunc primùm in lucem editus contra astrologos; Cyprianum Leouitium; Hieronymum Cardanum; & Lucam Gauricum.*
Antwerp: Christophorus Plantin, 1583.
Subject: Astrology, horoscope.
Case/B/8635/.386

MS marginalia.

**Hennepin, Louis, 1640?–1701.**

*Aenmerckelycke historische reys-beschryvinge door verscheyed landen veel grooter als die van geheel Europa onlanghs ontdeckt.*
Utrecht: Anthony Schouten, 1698.
Subject: Discovery, exploration, New World.
Ayer/*123/H5/1698a

**Hennepin, Louis.**

*Beschreibung der landschafft Lovisiana welche, auf befehl des königs in Frankreich neulich gegen sudwesten new-Frankreichs in America entdecket worden. Nebenst einer land-carten, und bericht von den sitten und lebens-art der wilden in selbiger landschafft.*
Nuremberg: Andrea Otto, 1689.
Subject: Discovery, exploration, New World.
Ayer/*123/H5/1689

**Hennepin, Louis.**

*Beschryving van Louisiana, nieuwelijks ontdekt ten zuid-westen van nieuw-vrankryk . . . met de kaart des landts, en een nauwkeurige verhandeling van de zeden en manieren van leeven der wilden. . . . Mits gaders de geographische en historische beschrijving der kusten van noord-America, met de naturlijke historie des landts.*
Amsterdam: Jan ten Hoorn, 1688.
Subject: Discovery, exploration, New World.
Ayer/*123/H5/1688

In 2 pts. Trans. by Denys, whose "Geographische en historische beschrijving" follows. Note on front pastedown states that Denys translated from the French of 1672.

**Hennepin, Louis.**

*Description de la Louisiane, nouvellement decouverte au sud'oüest de la Nouvelle France . . . avec la carte du pays: des moeurs & la maniere de vivre des sauvages.*
Paris: Widow of Sebastien Huré, 1683.
Subject: Discovery, exploration, New World.
Graff/* 1858

Pt. 2 contains description of fertility of the land, mythological origins, physical description of the natives, medicinal remedies, clothing, marriage customs, feasts, games, hunting and fishing, tools, burial customs, superstitions, "croyances ridicules."

**Hennepin, Louis.**

*Description de la Louisiane.* [Another ed.]
Paris Amable Auroy, 1684.
Ayer/123/H5/1684

**Hennepin, Louis.**

*Description de la Louisiane.* [Another ed.]
Paris: Amable Auroy, 1688.
Ayer/*123/H5d/1688

**Hennepin, Louis.**

*Description de la Louisiane.* [Another copy]
Case/G/876/.39

**Hennepin, Louis.**

*Descrizione della Lvigiana, paese nuouamente scoperto nell' America settentrionale, . . . con la carta geografica del medesimo, costumi, e maniere di viuere di que' selnaggi.*
Bologna: Giacomo Monti, 1686.
Subject: Discovery, exploration, New World.
Ayer/123/H5/1686

**Hennepin, Louis.**

*A discovery of a large, rich, and plentiful country, in the north America; extending above 4000 leagues. Wherein, by a very short passage, lately found out, thro' the Mer-Barmejo into the South-Sea; by which a considerable trade might be carry'd on, as well in the northern as the southern parts of America.*
London: printed for W. Boreham [n.d.]
Subject: Discovery, exploration, New World.
Ayer/123/H5/1720

**Hennepin, Louis.**

*De gedenkwaardige West-Indise voyagien, gedaan door Christoffel Columbus, Americus Vesputius, en Lodewijck Hennepin.*
Leyden: Pieter Vander Aa, 1704.
Subject: Discovery, exploration, New World.
Ayer/110/B4/1704

Separate pagination for Hennepin voyage: *Aenmerkelyke voyagie gedaan na 't gedeel 'te van noorder America, behelzende een nieuwe ontdekkinge van een seer groot land, gelegen tusschen Nieuw Mexico en de Ys-Zee. . . .* Illustrations include bison, Niagara.

**Hennepin, Louis.**

*Neue entdeckung vieler sehr grossen landschafften in America zwischen Neu-Mexico und dem ensmeer gelegen.*
Bremen: Philip Gottfr. Saurmann, 1699.
Subject: Discovery, exploration, New World.
Ayer/*123/H5n/1699

With illustration of Niagara Falls.

**Hennepin, Louis.**

*A new discovery of a vast country in America, extending above four thousand miles, between New France and New Mexico; with a description of the great lakes, cataracts, rivers, plants, and animals. Also, the manners, customs, and languages of the several native Indians.*
London: printed for M. Bentley, J. Tonson, H. Bonwick, T. Goodwin, and S. Manship, 1698.
STC II H1451
Subject: Discovery, exploration, New World.
Ayer/*123/H5/1698e

**Hennepin, Louis.**

*A new discovery of a vast country in America.* [Another ed.]
London: printed by for [sic] Henry Bonwicke, 1699.
STC II H1452
Ayer/*123/H5/1699

BOUND WITH *A continuation of the new discoverie of a vast country in America* [same imprint]. With illustrations, e.g., taking of Quebec by the English.

**Hennepin, Louis.**

*Nieuwe ontdekkinge van een groot land, gelegen in America, tusschen nieuw Mexico en de Ys-Zee. Behelzonde de gelegenheid der zelve nieuwe ontdekte landen, de rivieren en groote meeren, en voor al de groote rivier Meschasipi genaamd: . . . met een korte aanmerkinge over de zo genaamde Straat Aniam; en't middel om door een korte weg, zonder de Linie Aequinoctiaal te passeeren, na China en Japan te komen.*
Amsterdam: Andries van Damme, 1702.
Subject: Discovery, exploration, New World.
Ayer/123/H5/1702

**Hennepin, Louis.**

*Nouveau voyage d'un pais plus grand que l'Europe avec les reflections des entreprises du Sieur de la Salle, sur les mines de St. Barbe, &c.*
Utrecht: Antoine Schouten, 1698.
Subject: Discovery, exploration, New World.
Graff/1863

**Hennepin, Louis.**

*Nouveau voyage.* [Another copy]
Ayer/* 123/H5/1698b

**Hennepin, Louis.**

*Nouvelle decouverte d'un tres grand pays situé dans l'Amerique, entre le nouveau Mexique et la mer glaciale, avec les cartes, & les figures necessaires, & de plus l'histoire naturelle & morale, & les avantages, qu'on peut tirer par l'etablissement des colonies.*
Utrecht: Guillaume Broedelet, 1697.
Subject: Discovery, exploration, New World.
Ayer/*123/H5/1697

Illustrations of bison, pelican, opossum following p. 186; chap. 30: "Description de la chasse."

**Hennepin, Louis.**

*Nouvelle decouverte.* [Another copy]
Graff/*1864

Leaves in this copy are untrimmed. Chap. 34: "Construction d'un fort, que nous fîmes bastir sur la riviere des Illinois, noumé Chécagou par les barbares, & par nous le Fort de Crevecour."

**Hennepin, Louis.**

*Nouvelle decouverte.* [Another ed.]
Amsterdam: Abraham van Someren, 1698.
Ayer/*123/H5/1698d

**Hennepin, Louis.**

*Relacion de un pais que nuevamente se ha descubierto en la America septentrional de mas estendido que es la Europa.*
Brussels: Lamberto Marchant, 1699.
Subject: Discovery, exploration, New World.
Ayer/*123/H5r/1699

Contains also "Breve tratado de geographia divido en tres partes. Descripcion del rio y imperio de las Amazones Americanas, con su carta geographica. Lo que poseen Franceses y Ingleses &c. en el nuevo mundo. Il estado presente del imperio del gran mogor y reyno de Siam" (Brussels: Lamberto Marchant, 1700).

**Hennepin, Louis.**

*Voyage en un pays plus grand que l'Europe, entre la mer glaciale & le Nouveau Mexique.*
Amsterdam: Jean Frederic Bernard, 1720.
Subject: Discovery, exploration, New World.
Ayer/150.5/M6 B5/1720

General title page: *Relations de la Louisiane, et du fleuve Mississipi.* Contains 6 tracts, the first 4 of which are continuously paginated. They include "Relation de la Louisianne ou Mississipi; Relation de la Louisianne et du Mississipi par le Chevalier de Tonti; Voyage en un pays plus grand que l'Europe (by Hennepin); Relations des voyages de Gosnol, Pringe et Gilbert à la Virginie en 1602 & 1603; Relation du Detriot et la Baie de Hudson par Monsieur Jeremie; Les trois navigations de Martin Frobisher, pour chercher un passage a la Chine et au Japon par la mer glaciale, en 1577 et 1578 (these last two separately paginated).

**Hennepin, Louis.**

*Voyage en un pays plus grand que l'Europe.*
Case/G/81/.091

Contains the first 4 (continuously paginated) tracts, as entry above; the last 2 (separately paginated tracts) are wanting.

**Hennepin, Louis.**

*Voyage curieux du R. P. Louis Hennepin, . . . qui contient une nouvelle decouverte d'un tres-grand pays situé dans l'Amerique. . . . Outre cela on a aussi ajoûté ici un voyage qui contient une relation exacte de l'origine, moeurs, coûtumes,*

*religion, querres, voyages des Caraibes . . . faite par le Sieur de la Borde.*
Leyden: Pierre vander Aa, 1704.
Subject: Discovery, exploration, New World.
Ayer/*123/H5/1704

## Hennepin, Louis.

*Voyage ou nouvelle decouverte d'un tres-grand pays. . . . Avec un voyage des Caraibes faite par le Sieur de la Borde.*
Amsterdam Adrian Braakman, 1704.
Subject: Discovery, exploration, New World.
Ayer/*123/H5v/1704

## Hennepin, Louis.

*Voyage ou nouvelle decouverte d'un très-grand païs, dans l'Amerique, entre le Nouveau Mexique et la Mer Glaciale. . . . Avec . . . une relation . . . des Caraibes.*
Amsterdam: Jacques Desbordes, 1712.
Subject: Discovery, exploration, New World.
Ayer/*123/H5/1712

## Henrion, Denis, d. ca. 1640.

*Cosmographie, ov traicté general des choses tant celestes qv'elementaires, auec les accidens & proprietez plus remarquables d'icelles.*
Paris: Author, 1626.
Subject: Astronomy, cosmography.
Ayer/*7/H38/1626

2d ed. In 5 books.

## Henriquez, Francisco da Fonseca, 1665–1731.

*Medicina lusitana e soccorro delphico a os clamores da natureza humana, para total profligaçaõ se seus males.*
Amsterdam: Miguel Diaz, 1710.
Subject: Medicine.
Greenlee/f4532.1/M48/H51/1710

In 3 parts. 1: Vida de homen antes de nacer. 2: Arte de criar e curar meninos, desde que nacem até serem adultos; e o verdadeyro methodo racional, de curar a mayor parte dos males que padecem os homens en qualquer idade. 3. Tratado de febres.

## Hera y de la Varra, Bartolome Valentin, 16th cent.

*Reportorio del mvndo particvlar, de las spheras del cielo y orbes elementales, y delas significaciones, y tiempos correspondientes a su luz, y mouiento:* [sic] *con los eclipses, y lunario, desde este año* [1583] *hasta el de 1604 añadido el prognostica temporal, de las mudanças y passiones de ayre.*
Madrid: Gvillermo Druy, 1584
Subject: Almanacs, astrology, astronomy, climate, prognostications.
Ayer/8.9/H 53/1584

In 3 pts.: 1: "Del mvndo particvlar"; 2: "De los orbes elementales de la tierra y de la agua"; 3: "De los tiempos y fiestas correspondientes a los mouimientos del sol y de la luna." Note laid in: "Palau, vol. IV, the tables of polar elevations include towns in Spanish America."

* * *

*Herbarius* (Hortus Sanitatis minor).
Augsburg [Johann Schönsperger], 1485.
H *8949^ BMC (XV)II:365^ Pr. 1763^ Still. G88^ Goff G-98 [s.v. Gart der gesundheit]
Subject: Herbals.
Inc./f*1763

"Dis büch wirt geteylt in fünff teÿl" [sig. aij]. Hand-colored woodcuts throughout.

* * *

*Herbarius* (Hortus Sanitatis minor).
Venice: Simon Bevilaqua, 1499.
H-C *1807^ BMC (XV)V:524^ Pr. 5415^ Still. H63^ Goff H-69
Subject: Herbals.
Inc./5415

Plants arranged alphabetically in 7 parts: (1) De uirtutibus herbarum ad appoteca spectantium modum antidotorum dispensatarum; (2) De simplicibus laxatiuis & linitiuis; (3) De simplicibus confortatiuis seu speciebus aromaticis; (4) De fractibus & seminibus & radicibus; (5) De gummis & eis similibus (6) De generibus salis & mineris lapidibus; (7) De animalibus et prouenientibus ab eis.

## Herbert of Cherbury, Edward Herbert, 1st baron, 1583–1648.

*De veritate, provt distingvitvr a revelatione. a verisimili, a possibili, et a falso.*
London: Augustinus Matthevs, 1633.
STC 13180
Subject: Epistemology, metaphysics.
Case/B/4/.395

Said to be the first purely metaphysical book in the English language.

## Hermann, Paul, 1646–1695.

*Horti academici Lugduno-Batavi catalogus exhibens plantarum omnium nomina, quibus ab anno 1681 ad annum 1686 hortus fuit instructus ut & plurimarum in eodem cultarum & à nemine hucusque editarum descriptiones & icones.*
Leyden: C. Boutesteyn, 1687.
Subject: Botany.
Ayer/*8.9/B7/H55/1687

Hermann was a professor of botany and medicine. Some plants described and illustrated, some merely catalogued, with references to botanical vols. in which more detailed descriptions would presumably be found.

## Hermes Trismegistus.

*Contenta in hoc volvmine, Pimander: Mercurij Trismegisti liber de sapientia & potestate dei. Asclepius: Eiusdem Mercurij liber de voluntate diuina. Item Crater Hermetis, a Lazarelo Septempedano.*
[Paris: Simon Colinaeus, 1522.]
Subject: Alchemy, Hermeticism, neo-Platonism.
Case/Y/642/.H23965

Edited by Jacques Le Fèvre d'Etaples.

**Hermes Trismegistus.**

*Divinvs Pymander Hermetis Mercvrii Trismegisti cvm commentariis R. P. F. Hannibalis Rosseli.*
Cologne: Officina Choliniana for Petricholini, 1630.
Subject: Alchemy, Hermeticism, neo-Platonism.
Case/Y/642/.H2397

6 vols. in 1. Each section is in dialogue form or has statements of Hermes followed by commentary.

**Hermes Trismegistus.**

*Mercvrii Trismegisti liber de potestate et sapientia Dei Graeco in Latinvm tradvctvs a Marsilio Ficino Florentino.*
Venice: Lucas Dominici, 1481.
H-C 8458^ BMC (XV)V:280^ Pr. 4490^ Goff H-79
Subject: Alchemy, Hermeticism, neo-Platonism.
Inc./4490

From unidentified booksellers' catalogue copy pasted in front flyleaf: "Written in Egypt in 3rd century; a cult of Egyptian chemists, alchemists of the neo-Platonic school."

**Hermes Trismegistus.**

*Mercvrii Trismegisti liber de potestate et sapientia Dei per Marsilivm Ficinvm tradvctvs.*
Venice: Damianus de Mediolano, 1493.
H-C *8461^ BMC (XV)V:543-44^ Pr. 5514^ Goff H-81
Subject: Alchemy, Hermeticism, neo-Platonism.
Inc./5514

**Hermes Trismegistus.**

*Mercurii Trismegisti Poemander seu de potestate ac sapientia diuina. Aescvlapii definitiones ad ammonem regem.*
Paris: A. Turnebus, 1554.
Subject: Alchemy, Hermeticism, neo-Platonism.
Wing/ZP/539/.T852

In Greek. Followed by *Mercvrii Trismegisti liber de potestate et sapientia Dei, cui titulis Poemander, Marsilio Ficino Florentino interprete.* [Colophon: Excudebat Parisiis Guil. Morelius, 1554.]

**Hermes Trismegistus.**

*Mercvrii Trismegisti Pymander, de potestate et sapientia Dei. Eivsdem Asclepivs, de uoluntate Dei. Iamblichvs de mysterijs Aegyptiorum, Chaldaeorum, & Assyriorum. Proclvs in Platonicvm Alcibiadem, de anima & daemone. Idem de sacrificio & magia.*
Basel [M. Isingrinivs], 1532.
Subject: Alchemy, Hermeticism, neo-Platonism.
Case/Y/642/.H23966

Edited by Jacques Le Fèvre d'Etaples. Translated by Marsilio Ficino.

**Herrera y Tordesillas, Antonio de, 1559–1625.**

*Novvs Orbis sive descriptio Indiae occidentalis, auctore Antonis de Herrera . . . Accesserunt & aliorum Indiae Occidentalis descriptiones, & navigationis nuperae Australis Jacobi le Maire historia uti & navigationum omnium per Fretum Magellanicum succincta narratio.*
Amsterdam: Michael Colinivs, 1622.
Subject: Descriptive geography.
Case/G/80/.39

Tracts separately paginated: *Descriptio Indiae Occidentalis, quis nempe regionum ac provinciarum singularum sit situs; quas quantasque vel auri vel argenti opes quaeque habeant, & quomodo per eas iter institui debent, avthore Petro Ordonnez de Cevallos;* and *Brevis ac succincta Americae, sive novi orbis omniumque, quae in eo sunt, regionum hactenus exploratarum descriptio, excerpta è tabulis geographicis, P. Bertii.*

**Heurne, Johan van, 1543–1601.**

*Opera omnia: tam ad theoriam, qvam ad praxin medicam spectantia, quorum elenchus ante indicem capitum habetur. Iuxta Otthonis Hevrnii, auctoris filij, . . . recensionem ac oeconomiam fideliter expressa, ac duos in tomos tributa.*
Lyon: Antonii Hvgvetan, & Marci Antonii Ravavd, 1658.
Subject: Practice of medicine.
UNCATALOGUED

**Heurne, Otto van, 1577–1652.**

*Babylonica, Indica, Aegyptia, &c. Philosophiae primordia.*
Leyden: Ioannes Maire, 1619.
Subject: Ancient natural philosophy, astrology.
Case/–B/1/.398

Includes material on Zoroaster. Pt. 2, p. 205: Quo modo Aegypti duodecim zodiaci signa deprehenderint.

**Heydon, Sir Christopher, d. 1623.**

*An astrologicall discourse, manifestly proving the powerful influence of planets and fixed stars upon elementary bodies, in justification of the verity of astrology. Together with an astrological judgment upon the great conjunction of Saturn and Jupiter 1603.*
London: printed by Iohn Macock for Nathaniel Brooks, 1650.
STC II H1663A
Subject: Astrology, astronomy, climate.
Ayer/*8.9/A8/H615/1650

Foreword by William Lilly: "This treatise having been near 40 years detained in private hands, is now . . . made publike; it being the one, and only copy of this subject extant in the world: . . . [T]hat very thing which all antagonists cry out for, viz. *where's the demonstration of the art?* is hear in this book by undeniable mathematical demonstrations so judiciously proved, that the most scrupulous may receive full satisfaction." An astronomical explanation for climate and seasons (position and motion of the planets).

**Heydon, Sir Christopher.**

*A defence of ivdiciall astrologie, in answer to a treatise lately published by M. Iohn Chamber. Wherein all those places of scripture councells, fathers, schoolmen, later divines, philosophers, histories, lawes, constitutions, and reasons drawne out of Sixtus Empiricus, Picus, Pererius, Sixtus ab Hemminga and others, against this arte, are particularly

*examined: and the lawfulnes thereof by equivalent proofs warranted.*
[Cambridge] Printed by Iohn Legat, 1603.
STC 13266
Subject: Astrological medicine, critical days, judicial astrology.
Case/B/8635/.39

For astrological medicine, see pp. 443 ff.

### Heylyn, Peter, 1599–1602.

*Cosmographie in four bookes. Containing the chorographie and historie of the whole world, and all the principall kingdomes, provinces, seas, and isles thereof.*
London: printed for Henry Seile, 1652.
STC II H1689
Subject: Chorography, cosmography.
Case/fF/09/.4

Begins with Old Testament account of creation. Each book covers a group of nations (e.g., Bk. 2: Belgivm, Germanie, Denmark, Swethland, Rvssia, Poland, Hvngarie, Slavonica, Dacia, and Greece). Appendix on "vnknown parts of the world. Especially of Terra Australis incognita, or the southern continent."

### Heylyn, Peter.

*Cosmographie in foure bookes.*
London: printed for Henry Seile, 1657.
STC II H1690
Case/folio/oG/114/.H61/1657

2d ed. General introduction "Containing the creation of the world by almighty God, and the plantation of the same by the sons of men; the necessary use of history and geography."

### Heylyn, Peter.

*Cosmographie the fourth book. Part II containing the chorographie and historie of America, and all the principal kingdoms, provinces, seas, and islands of it.*
London: printed for A. S[eile], 1668.
Subject: Chorography, cosmography.
Ayer/7/H4/1668

### Heylyn, Peter.

*Cosmographie in four bookes.*
London: printed for Anne Seile, 1669.
STC H1692A
Ayer/7/H4/1669

5th ed. corrected and enlarged by the author.

### Heylyn, Peter.

*Cosmography, in four books.*
London: Printed for P.C.T. Passenger, 1682.
STC H1696
Greenlee/4890/H61/1682

6th ed. Corrected and enlarged by the author.

### Heylyn, Peter.

*Cosmography in four books. . . . Improv'd with an historical continuation to the present times, by Edmund Bohun.*
London: printed for E. Brewster, R. Chiswell, B. Tooke, T. Hodgkin, & T. Bennett, 1703.
Subject: Chorography, cosmography.
Ayer/7/H4/1703

7th ed. Corrected and enlarged by Edmund Bohun. Begins with creation of the world by sons of Noah. Has "Table of the climates belonging to the three sorts of inhabitants."

### Heylyn, Peter.

*Mikrokosmos. A little description of the great world.*
Oxford: printed by I. L. & W. T. for William Turner & Thomas Huggins, 1627.
STC 13278
Subject: Descriptive geography.
Case/G/117/.388

3d ed. revised.

### Heylyn, Peter.

[*Mikrokosmos*]
Oxford: printed by W. Turner for William Turner & Thomas Huggins, 1629.
STC 13279
Subject: Descriptive geography.
Case/G/117/.401

4th ed. revised.

### Heylyn, Peter.

*Mikrokosmos.*
Oxford: printed for William Turner & Robert Allott, 1631.
STC 13280
Subject: Descriptive geography.
Case/G/117/.4

5th ed.

### Heylyn, Peter.

[*Mikrokosmos*]
Oxford: printed for William Turner & Robert Allott, 1633.
STC 13281
Subject: Descriptive geography.
Case/G/117/.403

6th ed. Note on front pastedown: "Forty pages devoted to America."

### Heytesbury, William, fl. 1340.

*Tractatus Gulielmi Hentisberi de sensu compositio & diuiso.*
Venice: Bonetus Locatellus for Octavianus Scotus, 1494.
H *8437^ BMC (XV)V:443^ Pr. 5054^ Still. H51^ Goff H-57
Subject: Local motion, natural philosophy.
Inc./f5054

Includes "Questio messini de motu localicum expletione Gaetani."

## Heywood, Thomas, d. 1641.

*The hierarchie of the blessed angells. Their names, orders and offices. The fall of Lucifer with his angells.*
London: printed by Adam Islip, 1635.
STC 13327
Subject: Witchcraft.
Case/f*C/559/.4

1st ed. Includes some curious stories concerning witches. In 9 books, in verse. Bk. 6: Fall of Lucifer; bk. 7: classes of fallen angels; bk. 8: Satan; bk. 9: about witches. Each of the other books deals with a specific angel, e.g., bk. 5. Haniel: "The consonance and sympathie betwixt the angels hierarchie, the planets and coelestiall spheares, and what similitude appeares 'twixt one and other."

## Hierocles of Alexandria, 5th cent.

*Hieroclis philosophi commentarius in aurea Pythagoreorvm carmina. Ioan Curterio interprete.*
Paris: Nicolaus Niuellius, 1583.
Subject: Medicine, mysticism.
UNCATALOGUED

## Hierocles of Alexandria.

*Hieroclis philosophi commentarius in aurea Pythagoreorum carmina. Joan. Curterio interprete.*
London: printed by J. R. for J. Williams, 1673.
STC II H1935
Subject: Mysticism.
Y/642/.H/605

2 vols. in 1. Greek and Latin on facing pages. BOUND WITH *Hierocles de providentia & fato: una cum fragmentis ejusdem; et Lilii Gyraldi interpretatione symbolorum Pythagorae; notisque Merici Casauboni ad commentarium Hieroclis in aurea carmina.* P. 120: Sal apponendus.

* * *

*The high-German doctor, with many additions and alterations, To which is added a large explanatory index.*
London: printed for J. Baker, 1715.
Subject: Medicine and medical terminology as a synonym for politics.
UNCATALOGUED

Tract no. 21 (July 13, 1714), p. 126: "Tho physick, and divination by wand are the two illustrious branches of my art, by which I recommend my-self to the world, and gain ground daily vpon regular physicians." (Political satire in which medical quackery is equated with political chicanery?)

## Hill, Thomas, fl. 1590.

*The arte of gardening, whereunto is added much necessarie matter, with a number of secrets: and the phisicke helps belonging to each hearb, which are easily prepared. Heer-vnto is annexed two proper treatises, the first intituled the meruailous gouernment, propertie, & benefite of bees, with the rare secrets of the honie and waxe: the other, the yearly coniectures, very necessary for husband-men. To these is likewise ioyned a treatise of the arte of graffing and planting of trees.*
London: Edward Allde, 1608.
STC 13497
Subject: Apiculture, botany, gardening, herbals.
Case/*R/57/.405

[Separately paginated] *A profitable instrvction of the perfect ordering of bees, with the marveilovs natvre, property, and gouernment of them....* London: H. B., 1608. [Another tract paginated consecutively with the one on bees] "Husbandly coniectures." BOUND WITH this is *The gardners labyrinth. Also, the physick benefit of each herb, plant, and flowre, with the vertues of the distilled waters of euery* [one] *of them, as by the sequel may further appeare. Gathered out of the best approued writers of gardening, husbandrie, and physicke: by Dydimus Mountain.* London: printed by Henry Ballard, 1608. STC 13489

## Hill, Thomas.

*A pleasant history: declaring the whole art of physiognomy, orderly vttering all the speciall parts of man, from the head to the foot.*
[London] Printed by W. Iaggard, 1613.
STC 13483
Subject: Physiognomy, chiromancy.
Case/B/56/.41

"The iudgement of the anckles, the 53 chapter. . . .Rasis vttereth, that when the anckles shall be bigge, as thorow a fleshines, do argue such a person to be a dullard, and unshamefast."

## [Hill, Thomas]

[*The proffitable arte of gardening now the third time set fourth to which is added much necessary matter, and a number of secrettes with the physick helpes belonging to eche herbe, and that easie prepared. To this annexed, two proper treatises, . . . beese . . . and the other, the yerely coniectures meete for husbandmen to knowe.*
[London: Thomas Marshe, 1586.]
STC 13491
Subject: Apiculture, botany, gardening, herbals.
Case/R/570/.38

2 pts. in 1 vol. Imperfect. Title page, preliminaries up to leaf Bii, and leaf 191 are photocopies. Fol. 167: "Of the falling sicke, in any of the weeke daies out of that auncient physician Hypocrates."

## Hippocrates, 460?–377 B.C.

[Title and text in Greek]
Paris: Martinus Iuuenis, 1556.
Subject: Medicine.
Case/Y/642/.H7188

MS marginalia, leaves 14v, 15r and v, 16r.

## [Hippocrates]

*Manuale medicorum: seu* [Greek word] *aphorismorum Hypocratis praenotionum, coacarum, & praedictionum,*

*secundum propriam morborum omnium nomenclaturam, alphabetico digesta ordine.*
London: printed by Tho. Roycroft for Jo. Martin, Ja. Allestry, & Tho. Dicas [etc.], 1659.
STC II H2075
Subject: Medicine.
UNCATALOGUED

"Labore & industria D. Honorati Bicaissii"

## Hippocrates.

*Les trois premiers livres de la chirurgie d'Hippocrates mis de grec en françois par le Feure, & illustrez des commentaires de Vidus Vidius medecin Florentin, faits de Latins [sic] François par ledict François le Feure natif de Bourges en Berry.*
Paris: Iacques Keruer, 1555.
Subject: Medicine.
Case/–Y/.H71872

1: Des vlceres; 2: Des fistvles; 3: Des playes de teste.

* * *

*A history of nature, in two parts. Emblematically express'd in near a hundred folio copper plates. Wherein are also represented all the operations, facultys, and passions of the mind, &c. according to the manner of the most celebrated poets and philosophers.*
London: printed for D. Browne, etc., 1720.
Subject: Emblems, symbolism.
Case/W/1025/.408

Emblems with descriptions and explanations of their symbolism (animals, insects, occult qualities, medicine).

## Hobbes, Thomas, 1588–1679.

*Decameron physiologicum; or ten dialogues of natural philosophy. To which is added the proportion of a straight line to half the arc of a quadrant. By the same author.*
London: printed by J. C. for W. Crook, 1678.
STC II H2226
Subject: Geometry, natural philosophy.
Case/L/3/.412

## Hobbes, Thomas.

*Elementorum philosophiae sectio prima de corpore.*
London: Andrew Crook, 1655.
STC II H2230
Subject: Natural philosophy, physics.
Case/B/245/.4078

In 4 parts: Logica; philosophica prima; de rationibus motuum & magnitudum; physica sive de naturae phaenomenis.

## Hobbes, Thomas.

*Elements of philosophy, the first section, concerning body . . . now translated into English. To which are added six lessons to the professors of mathematics of the institution of Sr. Henry Savile in the University of Oxford.*
London: printed by R. & W. Leybourn for Andrew Crooke, 1656.
STC II H2232.
Subject: Mathematics, natural philosophy.
Case/B/245/.408

Half title: "Six lessons to the professors of mathematiques, one of geometry, the other of astronomy: in the chaires set up by . . . Sir Henry Savile."

## Hobbes, Thomas.

*Humane nature: or, the fundamental elements of policie. Being a discoverie of the faculties, acts, and passions of the soul of man, from their original causes, according to such philosophical principles as are not commonly known or asserted.*
London: T. Newcomb for Fra: Bowman, 1650.
STC II H2242
Subject: Dreams, imagination, passions of the mind, psychology.
Case/–B/245/.4084

Chap. 3: causes of sleep and dreams, phantoms. Chap. 9: Emotions. "Sense proceedeth from the action of external objects upon the brain" (p. 120).

## Hobbes, Thomas.

*De mirabilibus pecci: being the wonders of the peak in Derbyshire, commonly called the Devil's arse of peak. In English & Latine. The Latine by Thomas Hobbes. The English by a person of quality.*
London: printed for William Crook, 1678.
STC H2224.
Subject: Geology, natural history.
Case/Y/185/.H645

4th ed. In verse, Latin and English on facing pages. Euxtoris (hot sulphur baths), pp. 68–69.

## Hobbes, Thomas.

*Opera philosophica, quae Latinè scripsit, omnia.*
Amsterdam: Ioannes Blaeu, 1668.
Subject: Geometry, mathematics, physics.
Case/B/245/.4018

8 pts. in 1 vol. Includes Examinatio et emendatio mathematicae (6 dialogues); Dialogus physicvs de natvra aeris conjectura sumpta ab experimentis nuper Londini habitis Collegio Greshamensi. Item de duplicatione cubi; Problemata physica; De principiis et ratiocinatione geometrarvm.

## Hobbes, Thomas.

*A supplement to Mr. H. his works. . . . Being a 3rd vol. containing 1. De mirabilibus pecci. 2. Three papers to the Royal Society against Dr. Wallis. 3. Lux mathematica. 4. Prima partis doctrinae Wallisianae. 5. Rosetum geometricum. 6. Principia & problemata aliquot geometrica. 7. Catalogue of authors' works.*
London: printed by J. C. for W. Crooks, 1675.
STC II H2262
Subject: Geometry, mathematics.
Case/B/245/.4101

## Hobbes, Thomas.

*Hobbs's tripos. In three discourses: the first humane nature, or the fundamental elements of policy. Being a discovery of the faculties, acts and passions of the soul of man, from their original causes, according to such philosophical principles as are not commonly known, or asserted.*
London: printed for Matt. Gilliflower, Henry Rogers, etc., 1684.
STC II H2266
Subject: Passions of the mind, psychology.
Case/B/245/.409

3d ed.

## Hoffmann, Friedrich, 1660–1742.

*Medicina consvltatoria, worinnen unterschiedliche über einige schwere casvs ausgearbeitete consilia und responsa facultatis medicae enthalten, und in fünff decvrien eingetheilet, dem publico zum besten heraus gegeben.*
Halle: Renger, 1732.
Subject: Consilia, medicine.
UNCATALOGUED

Vol. 1 wanting. Vols. 2 and 3 (1738) bound together.

## Hoffmann, Friderich.

*Medicinae rationalis systematicae.*
Halle, 1729.
Subject: Anatomy, medicine, pathology.
UNCATALOGUED

In 4 vols, each divided into several books, variously dated 1729–1740. Vol. 1: *Qvo philosophia corporis hvmani vivi et sani ex solidis physico-mechanicis et anatomicis principiis methodo plane demonstrativa*; vol. 2: *Qvo philosophia corporis hvmani morbosi*; vol. 3: *Qvo vera therapiae fvndamenta medendi methodvs et leges tam natvrae qvam artis; vol. 4: Qvo specialis morborvm pathologia.*

## Holder, William, 1616–1698.

*A treatise of the natural grounds and principles of harmony.*
London: printed by J. Hepinstall, 1694.
STC II H2389
Subject: Acoustics and physics of music.
Case/V/54/.4148

Relevant chaps.: 1. Sound in general; 3. Of consonancy and dissonancy; 4. Of concords; 5. Of proportion.

## Holinshed, Raphael [Stow continuator], d. ca. 1580.

*The first and second volumes of chronicles, comprising the description and historie of England, . . . Ireland, . . . Scotland: First collected and published by Raphaell Holinshed, William Harrison, and others: Now newlie augmented and continued (with manifold matters of singular note and woorthie memorie) to the yeare 1586 by John Hooker aliàs Vowell Gent. and others.*
London: [printed for Iohn Harison, etc., 1587]
STC 13569(?)
Subject: Topography.
Case/fF/45/.415

## Holland, Henry, d. 1604.

*A treatise against vvitchcraft: or, a dialogue, wherein the greatest doubts concerning that sinne, are briefly answered: a sathanicall operation in the witchcraft of all times is truly prooued: very needful to be knowen of all men, but chiefly of the masters and fathers of families, that they may learn the best meanes to purge their houses of all vnclean spirits, and wisely to auoide the dreadfull impieties and greate daungers which come by such abhominations. Wherevnto is also added a short discourse, containing the most certen meanes ordained of God, to discouer, expell, and to confound all the sathanicall inuentions of witchcraft and sorcerie.*
Cambridge: printed by John Legatt, 1590.
STC 13590
Subject: Witchcraft.
Case/B/88/.402

## Holwell, John, 1649–1686?

*Catastrophe mundi: or, Evrope's many mutations until the year 1701. Being an astrological treatise of the effects of the tripple conjunction of Saturn and Jupiter 1682, and 1683, and of the comets 1680 and 1682, and other configurations concomitant. Wherein the fate of Europe for the next 20 years is (from the most rational grounds of art) more than probably conjectured. Likewise, for the benefit of the sons of art, there is a table of all the years, in which the conjunctions of Saturn and Jupiter happen in any of the four triplicities from 3958 before Christ, to the year of our redemption 1702, with a table of the year in which the six primary planets changed their aphelions from the beginning of the world for many years to come, all deduced from the Caroline tables. Also, an ephemeris of all the comets that have appeared from the year of our Lord 1603, to the year 1682, with a new way how to find in what year they will most operate. Whereunto is annexed, the hieroglyphicks of Nostradamus (published by Mr. Lilly in the year 1651).*
London: printed for the author, 1682.
STC II H2516
Subject: Almanacs, astrology, prognostication.
Case/B/863/.886

Binders title: *Tracts on astrology*. BOUND WITH Ferrier, Auger; Melton, John, qq.v.

## Honold, Jacob.

*Dissertatio astronomica, de cometis & speciatim eo, qui mense Novembri anni 1680 apparuit.*
Strassburg: Johannes Pastorius, 1682.
Subject: Astronomy, comets.
Case/B/863/.415

## Honter, Johannes, 1498–1549.

*Rvdimenta cosmographica.*
[Kronstadt (Brasov), 1542.] Antwerp: Ioann. Richard [1560].
Subject: Cosmography, geography.
Ayer/*7/H7/1542

Cosmography, geography, in 4 books, in verse. BOUND WITH Boemus, Joannes, q.v.

## Honter, Johannes.

*Rvdimentorvm cosmographicorum Ioan. Honteri Coronensis libri III cum tabellis geographicis elegantissimis.*
Zurich: Froschouer, 1549.
Subject: Cosmography, geography.
Ayer/*7/H7/1549

In verse. With "De uariarum rerum nomenclaturis per classes."

## Honter, Johannes.

*Rvdimentorvm cosmographicorum.* [Another ed.]
Zurich: Froschouer, 1552.
Subject: Cosmography, geography.
Ayer/*7/H7/1552

With maps, including "Circvli sphaerae cvm v. zonis."

## Honter, Johannes.

*Rvdimentorvm cosmographicorum.* [Another ed.]
Antwerp: Ioannes Richardus, [1560].
Subject: Cosmography. geography.
Ayer/7/H7/1560

## Honter, Johannes.

*Rvdimentorvm cosmographicorvm.* [Another ed.]
Zurich: Froschouer, 1578.
Subject: Cosmography, geography.
Ayer/*7/H7/1578

Described (in typed note attached to last flyleaf) as earliest schoolbook on geography with atlas.

## Horapollo, 4th cent.

*Orvs Apollo de Aegypte de la signification des notes hieroglyphiques des Aegyptiens, c'est a dire des figures par les quelles ilz escripuoient leurs mysteres secretz, & les choses sainctes & diuines.*
Paris: Iacques Keruer, 1543.
Subject: Hieroglyphics.
Case/*W/1025/.415

In 2 books. "En ce second liure ie te donneray uraye raison du de mourant & adiouxteray premierement ce qui na aucunement este explicque ny declare es autres liures."

## Horapollo.

*Ioannes Frobenivs stvdiosis S. D. Damus nunc uobis Orum Apollinem Niliacum de hieroglyphicis notis à Bernardino Trebatio Veicetino latinate donatum. In quo ueteris Aegyptiorum sapientiae thesaurum reperietis, uulgo haud dum cognitum, & miras rerum animatium naturas ac proprietates.*
[Basel: Frobenius, 1518.]
Subject: Hieroglyphics.
Case/W/1025/.4142

No plates or illustrations; explanatory text only, describing the signification of symbols and emblems.

## Horapollo.

*De sacris Aegyptiorvm notis, Aegyptiacè expressis. libri dvo, iconibus illustrati & aucti.*
Paris: Galleotus à Prato, & Ioannes Ruellius, 1574.
Subject: Hieroglyphics.
Case/*W/1025/.4145

Double colums, Latin and French. Argument of book 2: "En ce mien second volume ie vous diray la vraye & idoine raison de ce qui reste, & que i'estime necessaire, consideré qu'il ne fut iamais exposé par aucuns autheurs auant moy."

## Horapollo.

*De sacris notis & sculpturis libri duo, vbi ad fidem vetusti codicis manu scripti restituta sunt loca permulta, corrupta ante ac deplorata.*
Paris: Iacobus Keruer, 1551.
Subject: Hieroglyphics.
Case/W/1025/.414

## Horn, Georg, 1620–1670.

*Arca Mosis sive historia mundi.*
Leyden and Rotterdam: Hackiana, 1669.
Subject: Alchemy, iatrochemistry.
L/0114/.417

"Tabula secunda rerum naturalium sub septem planetis ex libris Arabum & Hebraeorum."

## Horn, Georg.

*Historiae naturalis & civilis, ad nostra usqve tempora libri septem.*
Leipzig: Johann Frideric Lüdervvald. 1671.
Subject: Encyclopedia, natural history.
F/09/.418

Subjects include alchemy, demons, eclipses, elements, magic, magnets, man, mathematics, monsters, occult qualities, the senses.

## Horozco y Covarrubias, Juan de, fl. 1589–1608.

*Emblemas morales de Don Ivan de Horozco y Couaruuias Arcediano de Cuellar en la santa yglesia de Segouia.*
Segovia: Iuan de la Cuesta, 1591.
Subject: Emblems, history of symbols.
Greenlee/5100/H81/1591

In 2 books. 1: "En que se declara que cosa son emblemas, empresas, insignias, diuisas, symbolos, pegmas, y hieroglyphicos"; 2: "De la primer insignia o empresa que huuo en el mundo"; cap. 31: De los symbolos de Pythagoras, y la declaracion dellos. Bk. 2 has woodcut emblems.

## Horozco y Covarrubias, Juan de.

*Emblemas morales de Don Ivan de Horozco y Covarvvias.*
Saragossa: Alonso Rodriguez, [for Iuan de Bonilla] 1603–4.
Subject: Emblems, history of symbols.
Case/W/1025/.416

In 2 books. Chap. 29 [i.e., 19]: "Delos hieroglyphicos de los Egypcios, y de que manera significauan a dios, Osiris, Isis, Apis, Ocho, Anubis, Horo"; chap. 20: "Del sol y sus mouimientos, de la luna, del cielo, y de la tierra habitabile"; chap. 26: "Del amigo de edificar, del carpintero, y otros muchos"; chap. 35: "De las colores, yo de lo que por ellas se significaua."

### Horst, Jakob, 1537–1600.

*Epistolae philosophicae et medicinales, cum vita Georgii Horsti.*
[Leipzig?] Valentin Vögelin, 1596.
Subject: Practice of medicine.
Case/Q/.42

* * *

Hortus Sanitatis (maior).
Mainz: Jakob Meydenbach, 1491.
H-C *8944^ BMC (XV)I:44-45^ Pr. 160^ Still. H416^ Goff H-486
Subject: Herbals.
Inc./f160

Copiously illustrated with woodcuts. Manuscript notes especially in section "De urinis." Sections include Tractatus de animalibus vitam in terris ducentium; tractatus de auibus; de piscibus; de lapidibus; tabula super tractan de herbis [and other indexes].

### Host von Remberch, Johann, fl. 1485–1532.

*Congestorium artificiose memorie. . . . Regularis obseruantie predicatorie. Omnium de memoria preceptiones aggragatim complectens. Opus omnibus theologis; predicatoribus & confessoribus; aduocatis & notariis; medicis; philosophiis. Artius liberalium professoribus. In super mercatoribus mantijs & tabellarijs pernecessarium.*
[Venice: Melchior Sessa, 1533.]
Subject: Mnemonics.
Case/B/528/.4

Woodcut, p. 12, profile head, showing locations of mental faculties, e.g., "sensus, fantasia, imagina, logitativa, estimativa, memorativa."
ALL/ Mnemonics

### Hoste, Paul, 1652–1700.

*Théorie de la construction des vaisseaux, qui contient plusieurs traitez de mathématiques sur des matières nouvelles & curieuses. Par le P. Paul Hoste de la compagnie de Jesus, professeur des mathématiques dans le séminaire Royal de Toulon.*
Lyon: Anisson, & Posuel, 1697.
Subject: Military technology, plane and solid geometry.
Case/F/U/7/.417

In 2 books: "De la figure du vaisseau en général"; "De la solidité des vaisseaux." BOUND WITH *L'art des armées navales, ov traité des evlutions navales.*

### [Hotman, Antoine] 1525?–1596.

*Traicté de la dissolvtion dv mariage par l'impuissance & froideur de' l'homme ou de la femme.*
Paris: Mamert Pattison, chez R. Estienne, 1581.
Subject: Medicine, psychology.
Case/K/75/.416

P. 8: "Le mariage nous est concedé à fin de procreer des enfans, mais principalement pour esteindre la chaleur & bruslement de nature."

### Huarte de San Juan, Juan, 16th cent.

*Essame de gl'ingegni de gl'hvomini, per apprender le scienze: Nel quale, scoprendosi la varietà delle nature, si mostra, a che professione sia atto ciascuno, & quanto profitto habbia fatto in essa: de Gio. Hvarte: Tradotto dalla lingua Spagnuola da M. Camillo Camilli.*
Venice: Aldus, 1586.
Subject: Education, humors, medicine, psychology.
Case/Y/722/.H867

### Huarte de San Juan, Juan.

*Examen de ingenios. The examination of mens wits. In which, by discouering the varietie of natures, is shewed for what profession each one is apt, and how far he shall profit therein.*
London: printed by Adam Islip, 1596.
STC 13893
Subject: Education, humors, medicine, psychology.
Case/Y/722/.H859

Chap. 15: In what manner parents may beget wise children, and of a wit for learning. Sec. 1: By what signs we may know, in what degree of hot and drie, euery man resteth. Sec. 2: What women ought to marry with what men, that they may haue children. Sec. 3: What diligence is to be vsed, that children male, and not female may be borne.

### Huarte de San Juan, Juan.

*Examen de ingenios. The examination of mens wits. In which, by discovering the varietie of natures, is shewed for what profession each one is apt, and how far he shall profit therein. Translated out of the Spanish tongue by M. Camillo Camilli. Englished out of the Italian by R. C.*
London: printed by Adam Islip for Thomas Adams, 1616.
STC 13895
Subject: Education, humors, medicine, psychology.
Case/Y/722/.H86

Chap. 5: "It is prooued that of the three qualities, hot moyst, and drie, proceed all the differences of mens wits."

### Hughes, Griffith, fl. 1750.

*The natural history of Barbados.*
London: printed for the author, 1750.
Subject: Biology, disease, natural history.
Ayer/1000.5/B22/H89/1750

In 10 books. Bk. 1: Air, soil, climate (esp. hurricane of 1665); 2: Diseases peculiar to the West Indies; 3: Land animals, including insects; 4–8: Vegetables, trees, plants; 9: Shores of the island; 10: Fish.

### Hughes, William, fl. 1665–1683.

*The American physitian; or, a treatise of the roots, plants, trees, shrubs, fruit, herbs, &c. growing in the English plantations in*

*America. Describing the place, times, names, kindes, temperature, vertues and uses of them, either for diet, physick, &c. Whereunto is added a discourse of the cacao-nut-tree, and the use of its fruit; with all the ways of making chocolate. The like never extant before.*
London: printed by J. C. for William Crook, 1672.
STC II H3332
Subject: Herbals.
Ayer/109.9/B6/H8/1672

"Explanation," pp. 156–59, contains Indian names of a few items.

## Hulsius, Levinus, ed., d. 1608.

*Sammlung von sechs und zwanzig schiffahrten in verschiedene fremde länder durch Levinus Hulsius und enige andere, aus dem Holländischen ins deutsche übersetzt und mit allerhand anmerkungen versehen.*
Frankfurt [and other locations]: various publishers, 1608–1650.
Subject: Discovery, exploration, New World, shipwrecks, voyages.
Ayer/*110/H9/1608

26 parts in 3 vols. No general title page, but usually catalogued under this factitious title. Cf. Church catalogue, v. 2, p. 601. These accounts are various eds. and issues. Includes accounts of voyages of Magellan, Drake, Cavendish, Hudson, and others.

## Hume, David, 1711–1776.

*Philosphical essays concerning human understanding.*
London: printed for A. Millar, 1748.
Subject: Experimental reasoning, natural philosophy, psychology.
B/243/.417

Chap. 10: Of miracles; chap. 11: Of the practical consequences of natural religion; chap. 20: Of the reason of animals.

## Hutcheson, Francis, 1694–1746.

*An essay on the nature and conduct of the passions and affections. With illustrations on the moral sense.*
London: printed by J. Darby & T. Browne, 1728.
Subject: Passions, psychology.
B/529/.424

In 2 pts. Pt. 1: An essay on the nature and conduct of the passions; pt. 2: Illustrations upon the moral sense [truth and virtue].

## Hutcheson, Francis.

*Metaphysical synopsis: ontologiam, et, pneumatologium complectens.*
N.p., 1742.
Subject: Metaphysics, soul.
B/4/.428

Pt. 1: De ente; Pt. 2: De mente humana; Pt. 3: De deo. Has sections on De amimae viribus, de voluntate, de animi imperio in corpus suum.

## Hutcheson, Francis.

*Reflections upon laughter and remarks upon the Fable of the Bees.*
Glasgow: printed by R. Urie for Daniel Baxter, 1750.
Subject: Passions of the mind, psychology.
Case/B/5293/.428

## Hutchinson, Francis, bp. of Down and Connor, 1660–1739.

*An historical essay concerning witchcraft. With observations upon matters of fact; tending to clear the texts of the sacred scriptures, and confute the vulgar errors about that point.*
London: printed for R. Knaplock [etc.], 1718.
Subject: Reason vs. superstition, witchcraft.
Case/B/88/.429

Asserts that witches do not exist. Work also contains 2 sermons, one in proof of Christian religion, one concerning good and evil angels.

## Hutchinson, Francis, bp. of Down and Connor.

*An historical essay concerning witchcraft.* [Another ed.]
London: printed for R. Knaplock [etc.], 1720.
Subject: Witchcraft.
B/88/.43

## Hutchinson, Francis, bp. of Down and Connor.

*Historischer versuch von der hexerey, in einem gespräch zwischen einem geistlichen, einem schottländischen advocaten und englischen geschwornen.*
Leipzig: Johann Christian Martini, 1726.
Subject: Witchcraft.
B/88/.432

Translated by Theodor Arnold. "Verzeichnisz derer in diesem werck nur teutsch angeführten Englischen autorum." A list of English works on witchcraft, this appears to be a translation of Hutchinson's 1718 *Historical essay* (see above). List includes Dr. Harsenet's Declaration of popish imposture; Meric Casaubon, Of credulity and incredulity; his Preface before Dr. Dee's transactions of spirits; Mr. Baxter's Certainty of the world of spirits; More's and Glanvil's collections; Synclare's Satan's invisible world; A blow to modern saducism, by a member of the Royal Society; several tracts by Increase and Cotton Mather; and The compleat history of magick, sorcery and witchcraft.

## Hutten, Ulrich von, 1488–1523. [attributed author]

*Dvo volvmina epistolarvm obscvrorvm virorvm ad D.M. Ortuinum Gratium, Attico lepôre referta denuò excusa à mendis repurgata, quibvs ob stili & argumenti similtudinem adiecimus in calce dialogum mirè sestiuum, eruditis salibus refertum.*
[N.p.], 1557.
Subject: Defense of Johann Reuchlin.
Case/Y/682/.E585

Attributed to Crotus Rubeanus and Ulrich von Hutten. Defense of Reuchlin and criticism of monastic pedantry. BOUND WITH De generibvs ebriosorvm, et ebrietate vitanda. Cvi adiecimvs de meretricvm in svos amatores, & concubinarum in sacerdotes fide. [N.p.] 1557.

## Hutten, Ulrich von.

*Dvo volvmina epistolarvm obscvrorvm virorvm.* [Another ed.] [N.p.] 1570.
Case/Y/682/.E586

Does not include the tract on drunkenness (see above).

## Hutten, Ulrich von.

*Epistolae obscurorum virorum.*
[Venice (Cologne?), 1516?]
Subject: Defense of Johann Reuchlin.
Case/Y/682/.E582

[In impressoria A. Minutii, 1516?] 2d ed. of 1st pt. (See leaf e ii verso for references to medicine and leaf e[6] verso on Reuchlin.)

## Hutten, Ulrich von. [attr. auth.]

*Epistolae obscurorum virorum.*
London: H. Clements, 1710.
Subject: Defense of Johann Reuchlin.
Y/682/.E588

Numerous references to Reuchlin, whom Hutten defended in this work. See, e.g., pp. 195–204.

## Huygen, Pieter, 17th cent.

*De beginselen van gods koninkryk in den mensch.*
Amsterdam: Jacob ter Beek, 1740.
Subject: Emblems.
Case/W/1025/.427

Met konst-plaaten door Jan Luyken.

## Huygens, Christian, 1629–1695.

*The celestial worlds discover'd: or, Conjectures concerning the inhabitants, plants and productions of the worlds in the planets.*
London: printed for Timothy Childe, 1698.
STC II H3859
Subject: Astronomy, cosmology, extraterrestrials, plurality of worlds.
Case/L/9/.427

## Huygens, Christian.

*The celestial worlds discover'd: or conjectures concerning the inhabitants, plants and productions of the worlds in the planets. Written in Latin by Christianus Huygens, and inscribed to his brother Constantine Huygens.*
London: printed for James Knapton, 1722.
Subject: Astronomy, cosmology, extraterrestrials, plurality of worlds.
L/9/.429

2d ed. corrected and enlarged.

## Huygens, Constantin, 1596–1687.

Ghebruik, en onghebruik van't orghel, in de kerken der vereenighde Nederlanden.
Amsterdam: Arent Gerritsz vanden Heuvel, 1660.
Subject: Music and the passions.
Case/3A/746

Sidenote, pp. 83–84: "The reason hereof is an admirable facilite which musicke hath to expresse and represent to the minde, . . . the turnes and varities of all passions whereunto the minde is subject." Sidenote, pp. 86–87: "So that we lay altogether aside the consideration of dittie or matter, the very harmonie of sounds being framed in due sort, and carryed from the eare to the spirituall faculties of our soules, is by a natiue puissance and efficacie greatly auaileable to bring to a perfect temper wharsour is there troubled, apt as well to quicken the spirits, as to allay that which is too eager, soueraigne against melancholy and despaire, forcible to draw forth teares of devotion, if the minde be such as can yeeld them, able both to moue and to moderate all affections."

## Hyginus, C. Julius.

*Fabvlarvm liber. Eiusdem poeticon astronomicon libri quatuor.*
Paris: Joannes Parant, 1578.
Wellcome 3381
Subject: Astronomy.
Case/Y/672/.H9057

Includes F. Fulgentius, Albricius, Aratus (*Phaenomena*), Apollodorus, Lilius, Gyraldus, and Proclus (*De sphaera*, in Greek and Latin).

## Hyginus, C. Julius.

*De mundi et sphaeri ac vtriusque partium declaratione cum planetis et variis signis historiaris.*
Venice: Melchior Sessa & Peter de Rauanis, 1517.
Subject: Astronomy.
Case/*Y/672/.H9051

Verso of last leaf: "In nomine Domini: A terra vsque ad lunam est distantia 126,600 stadiorum. 1,115825. miliariorum. . . . circuitus terrae est 180,000 stadiorum. 1.225500 miliariorum.

## Hyginus, C. Julius.

*Poeticon astronomicon.*
Venice: Erhard Ratdolt, 1482.
H *9062^ BMC (XV)V:286-87^ Pr. 4387^ Still. H487^ Goff H-560
Subject: Astronomy.
Inc./*4387

In 2 books. 1: De mundi & spherae; 2: De signorum celestium historijs. Descriptionum formarum coelestium astronomicon. De sole & luna: ac ceteris planetis.

# I

## Iamblichus, of Chalcis, 250–330.

*De mysteriis Aegyptiorum, Chaldeorum, Assyriorum.*
Venice: Aldus Manutius, 1497.
H-C *9358^ BMC (XV)V:557^ Pr. 5559^ Goff J-216
Subject: Hermeticism, magic, mysticism, neo-Platonism, supernatural.
Inc./f5559

Translated by Marsilio Ficino. In addition to tract by Iamblichus, the work includes Proclus, "In Platonicum Alcibiadem de anima atque daemone"; Proclus, "De sacrificio & magia"; Porphyrius, "De diuinis atque daemonibus"; Synesius Platonicus, "De somnii"; Psellus, "De daemonibus"; "Expositio Prisciani & Marsilii in Theophrastum de sensu phantasia & intellectu"; Alcinoi Platonici philosophi, "Liber de doctrinam Platonis"; Speusippi Platonis disciputi, "Liber de Platonis diffinitionibus"; Pythagoras, "Philosophi aurea uerba"; "Simbola Pythagorae philosophi"; Xenocrates philosophi Platonici, "Liber de morte"; Marsilio Ficino "Liber de uoluptate."

## Iamblichus, of Chalcis.

*De mysteriis Aegyptiorvm. Nunc primùm ad uerbum de Graeco expressus. Adiecta de uita & secta Pythagorae Flosculi, ab eodem Scutellio ex ipso Iamblicho collecti.*
Rome: Antonio Blado, d'Assola, for D. Vincentius Luchrini, 1556.
Subject: Hermeticism, magic, mysticism, neo-Platonism, supernatural.
Wing/ZP/535/.B57

BOUND WITH *Pythagorae vita ex Iamblicho collecta*, by Nicolao Scvtellio (same imprint).

## Iamblichus, of Chalcis.

*Iamblicvs De mysteriis AEgyptiorum, Chaldaeorum, Assyriorum.*
Lyon: Ioan. Tornaesius, 1570.
Subject: Hermeticism, magic, mysticism, neo-Platonism, supernatural.
Case/Y/6409/.29

*Proclvs in Platonicum Alcibiadem de anima, atque daemone. Idem desacrificio & Magia. Porphyrivs De diuinis atque daemonibus. Psellvs dedaemonibus. Mercurii Trismegisti Pimander. Eiusdem Asclepius.*

## Iciar, Juan de, 1523–ca.1575.

*Arte subtilissima por la qual se enseña a escreuir perfectamente, hecho y experimentado, y agora: de nueuo añadido por Jan de' Yciar Vizcayno.*
Saragossa: Pedro Bernuz, 1550.
Subject: Ink and paper chemistry, geometry and proportion.
Wing/ZW/540/.I164

Recipe for making ink, for coloring paper, proportions of letters.

## Indagine, Joannes ab, 16th cent.

*Chyromantia Ioanis Indagine.*
[Utrecht: Jan Berntsz, 1536.]
Subject: Astrology, chiromancy, humors, physiognomy.
Wing/*fZP/546/.B45

"Enn dit boer leert van drie naturlike consten als physiognomia, astrologia naturalis, chiromania." Contains also "regulen van cranckheyden."

## Indagine, Joannes ab.

*Introductiones apotelismaticae in physiognomiam, astrologiam naturalem, complexiones hominum, naturas planetarum, cum periaxiomatibus de faciebus signorvm et canonibvs de aegritudinibus hominum . . . quibus ob similem materiam accessit Gvlielmi Grataroli Bergomatis Opvscvla de memoria reparanda augenda, conservanda; de praedictione morum naturarumque hominum; de mutatione temporum, ejusque signis perpetuis. Et Pomponi Gavrici Neapolitani tractatus De symmetrijs, lineamentis & physiognomia ejusque speciebus, &c.*
Strassburg: Lazarus Zetzner, 1630
Zinner 5181
Subject: Astrology, chiromancy, humors, physiognomy.
Ayer/3A/377

## Indagine, Joannes ab.

*Introductiones apotelesmaticae in physiognomiam complexiones hominum; astrologiam naturalem; naturas planetarum. Cum periaxiomatibus de faciebvs signorvm et canonibvs de aegritudinibvs hominum. Quibus ob similem materiam accessit Guilielmi Grataroli Bergomatis opuscula de memoria reparanda . . . de praedictione morum . . .de mutatione temporum . . . et Pomponii Gaurici Neapolitani tractatvs de symmetriis, lineamentis & physiognomia.*
Strassburg: Simon Pauli, 1663.
Subject: Astrology, chiromancy, humors, physiognomy.
Case/B/56/.437

With woodcut diagrams and illustrations.

## Indagine, Joannes ab.

*Die kunst der chiromantzey, urz besehung der bend. Physiognomey, urz anblick des menschens. Natürlichen astrologey noch dem lauff der sonnen. Complexion eins yegklichen menschens. Natürlichen ynflüssz de planeten. Der zwölff zeichen angesychten. Ettliche canones/ zü erkant nüsz der menschen kranckheiten, solich weiss normals nye beschriben oder gedruckt.*

[Strassburg: Joannes Schott, 1523.]
Subject: Astrology, chiromancy, humors, physiognomy.
Wing/fZP/547/.S382

## Ingegneri, Giovanni.

*Fisionomia natvrale . . . nella quale con ragioni tolte dalla filosofia, dalla medicina, & dall' anatomia, si dimostra come dalle parti del corpo humano, per la sua naturale complessione, si possa agevolmente conietturare quali siano l'inclinationi de gl'huomini.*
Padua: Pietro Paolo Tozzi, 1623.
Subject: Physiognomy.
B/56/.714

BOUND WITH Porta, Giovanni della; and Polemo, Antonius, qq.v.

## Institorus, Henricus, d. ca. 1500.

*Mallevs maleficarvm, maleficas et earvm haeresim frame â conterens* [i.e., continens], *ex variis avctoribvs compilatvs, & in quatuor tomos iustè distributus qvorvm dvo priores vanas daemonvm versutias, praestigiosas eorum delusiones, superstitoisas, strigimagarum ceremonias, horrendos etiam cum illis congressus.*
Lyon: Clavdivs Bovrgeat, 1669.
Subject: Witchcraft
Case/oBF/1569/.A2/158/1669

In 2 vols., the 1st containing works by Ioannis Nider, Bernardus Basin, Thomas Murner, Bartholomaeus de Spina, Paulus Grillandus, Ioannes Gerson, Alphonsus à Castro, Bernardus Comensis, Ambrosius de Vignate, Io. Laurentius Anania, Io. Franciscus Leonis. Vol. 2: *Daemonastix, sev adversvs daemones et maleficos,* and *Fvstis daemonum ad malignos spiritvs effvgandos de oppressis corporibvs, Fvga satanae, authore Petro Antonis,* and *Artis exorcistica.*

## Irenicus, Franciscus, 1495–1559(ca.)

*Germania exegeseos volvmina dvodecima . . . Vrbis Norinbergae descriptio, Conrado Celte enarratore.*
[Hagenau: T. Anshelm for I. Kobergius, 1518.]
Subject: Descriptive geography, Nuremberg.
Case/F/0247/.446

Bk. 8: Descriptoribus fluminum Germaniae; de Germaniae humiditate; de piscibus Germaniae. Bk. 10: De mathematicali descriptione totius Germaniae.

## Isidorus, Saint, bp. of Seville, 560?–636.

*Etymologiae.*
[Augsburg] Günther Zainer, 1472.
H *9273^ BMC (XV)II:317^ Pr. 1532^ Goff I-181
Subject: Encyclopedias.
Inc./f1532

Liberal arts, man and his parts, animals, medicine, physical geography, human geography. Bk. 16, chaps. 1–24 a lapidary; chap. 16 "de metallis."

## Isidorus, Saint, bp. of Seville.

*Etymologiae.*
[Strassburg: Johann Mentelin, ca. 1473.]
H-C *9270^ BMC (XV)I:57^ Pr. 227^ Goff I-182
Subject: Encyclopedias.
Inc./+227

## Isidorus, Saint, bp. of Seville.

*Etymologiae. De summo bono.*
Venice: Peter Löslein, 1483.
H-C *9279^ BMC (XV)V:379–80^ Pr. 4904^ Goff I-184
Subject: Encyclopedias.
Inc./4904

De summo bono has MS marginalia passim. Bk. 3: arithmetic; bk. 4: medicine; bk. 11: of man; bk. 14: of the earth. BOUND WITH *Juris vocabularium* (out of scope).

## Isidorus, Saint, bp. of Seville.

*Etymologiae.*
Basel: [Michael Furter], 1489.
H *9274^ BMC (XV)III:787^ Pr. 7580^ Goff I-185
Subject: Encyclopedias.
Inc./7718.5

## Isidorus, Saint, bp. of Seville.

*Etymologiae. De summo bono.*
Venice: Bonetus Locatellus for Octavianus Scotus, 1493.
H *9280^ BMC (XV)V:442^ Goff I-186
Subject: Encyclopedias.
Inc./f*5049

MS marginalia, bk. 8.

## Isidorus, Saint, bp. of Seville.

*Etymologiae. De summo bono.*
[Venice: Bonetus Locatellus, 1500?]
H-C *9277^ Goff I-188
Subject: Encyclopedias.
Inc./f*5104.5

Goff dates this after 1500. Some MS marginalia bk. 16.

# J

## Jacques de Chevanes, d'Autun, Capuchin friar.

*L'incredulité scavante, et la credvlité ignorante: au sujet des magiciens et des sorciers. Auecque la response à vn liure [par Gabriel Naudé] intitulé Apologie pour tous les grands personnages, qui on esté faussement soupçonnés de magie.*
Lyon: Iean Molin, 1671.
Subject: Magic, witchcraft.
Case/B/88/.443

Divided into 3 pts. 1: Qu'il y a des magiciens & des sorciers (poets and demons to cure illnesses—discours 46). 2: Diuers moyens pour connoître les magiciens et les sorciers. 3: De l'obligation de punir les magiciens & les sorciers. The "Apologie" to Naudé's apology includes magic and its kinds, and tracts on Orpheus, Pythagoras, Democritus, and Empedocles (justly suspected of magic), Peter of Abano, Paracelsus, Cardanus, H. C. Agrippa, Lull, and others.

## Ja' far ibn Muhammad (Abū Ma'shar), al-Balkhī, 805–886.

*Flores Astrologiae.*
Augsburg: Erhard Ratdolt, 1488.
H-C *609^ BMC (XV)II:382^ GW I:837^ Pr. 1877^ Still. A323^ Goff A-356
Subject: Astrology.
Inc./1877

## Ja' far ibn Muhammad (Abū Ma'shar), al-Balkhī.

[*Albumasar de magnis conjunctionibus annorum revolutionibus.*]
Venice: Jacob Pentius de Leucho, 1515.
Subject: Astrology.
Case/B/8635/.453

Title page lacking. Information supplied from catalogue of Bibliothèque Nationale. "Ex libris Gilbert Redgrave." Penned note on front flyleaf, dated 1909: "The famous firm of Sessa did not scruple to produce pirated issues of the works of Ratdolt. This would seem, from the account given by the Duc de Rivoli, to be founded on Ratdolt's issue of 1489, which he published at Augsburg, after his departure from Vienna. The same printer, Jacobus Pentius de Leucho, was working for M. Sessa in 1506, in which year he printed Albumasar's Introductorium Astronomicam."

## Jacob von Jüterbogk, 1381–1465.

*De animabus e corporis exutis.*
Leipzig: Wolfgang Stöckel, 1496.
H *9352^ BMC (XV)III:653^ Pr. 3052^ Goff J-24a
Subject: Body and soul.
Inc./3052

8 tracts among which are "De animabus exutis à corporibus editus"; "De egressu animarum humanarum a corporibus per sententiam mortis"; "De receptaculis et locis ad que perducuntur post egressum et de habitudinibus eorum"; "De apparitionibus que fiunt ab animabus exutis ad homines viuentes." MS marginalia. Only copy listed in Goff.

## James I, King of England, 1566–1625.

*Daemonologie, in forme of a dialogve, diuided into three bookes.*
London: printed for William Cotton and Will. Asple, according to the copie printed at Edenburgh, 1603.
Subject: Occult, witchcraft.
Case/B/88/.444

Bk. 1: Description of magic in general; bk. 2: Description of sorcery and witchcraft; bk. 3: Description of all these kindes of spirites that troubles men or women (4 main kinds: spectra, incubi and succcubi, demoniacks, and phairies). Also has material on the trial and punishment of witches. BOUND WITH [Harsnett,] q.v.

## Jarava, Juan de.

*I qvattro libri della filosofia natvrale di Gioan Sarava. Doue Platonicamente, & Aristotelicamente, si discorreno tutte le principali materie fisiche, le prime cagioni e gli effetti loro, & i fini. Et in particolare si ragiona del mondo, delle meteorologie, de metalli, & uirtù, & proprietà delle pietre.*
Venice: A. Rauenoldo, 1565.
Subject: Alchemy, astronomy, earth sciences, weather.
Case/B/42/.442

Translated from Spanish into Italian by Alfonso Vlloa.

## [Jean de Meun.] 13th cent.

*Le plaisant iev dv dodechedron de fortune, non moins recreatif, que subtil & ingenieux.*
Paris: Nicolas Bonfons, 1577.
Subject: Astrology, geometry, prognostication.
Case/B/863/.452

"Avertisement ac lecteur: Entre tous les ieux & passetemps de fortune, . . . cestuy . . . es le plus subtil & artificiel: car il procede selon les reigles. & demonstrations de l'astrologie iudiciare."

## Jessen, Johann von, 1566–1621.

*Zoroaster.*
Wittenberg: ex officina Cratoniana, 1593.
Subject: Elements, mysticism, Hermeticism.
Case/B/942/.45

## Johannes, Canonicus.

*Quaestiones super physica Aristotelis.*
Venice: Octavianus Scotus, 1481.
H-C *4345^ BMC (XV)V:276^ Pr. 4569^ Still. J231^ Goff J-263
Subject: Physics.
Inc./f4569

Edited by Franciscus de Benzonibus.

## Joannes XXI, Pope, ca 1210–1277.

*The treasury of healthe conteyning many profitable medycines gathered out of Hypocrates, Galen and Auycen, by one Petrus Hyspanus and translated into Englysh by Humfree Lloyde who hath added thereunto the causes and sygnes of euerye dysease, wyth the aphorismes of Hypocrates . . . and a compendiouse table conteynyng the purging and confortatyue medycynes.*
[London: Wyllyam Coplande, 1550? (n.d.)]
STC 14652a
Subject: Diagnosis, materia medica, medicine.
Case/B/8615/.45

## Joannes de Janduno, d. 1328.

*Acutissimae qvaestiones in dvodecim libros metaphysicae ad Aristotelis . . . cum Marci Antonii Zimarae in eiasdem. De individvatione naturae & de triplici causalitate intelligentiarum ad Aristotelis & Auerrois mentem subtiliter examinatae.*
Venice: Hieronymus Scottus, 1560.
Subject: Metaphysics.
Case/fY/642/.A8564

## Joannes de Janduno

*Ioannis de Iandvno in libros Aristotelis De coelo et mvndo qua extant quaestiones. . . . qvibvs nvper consvlto adiecimus Averrois sermonem de svbstantia orbis.*
Venice: Heirs of Luca Antonius Iunta, 1552.
Subject: Astronomy, cosmography, natural history.
Case/fY/642/.A83932

## Joannes de Janduno.

[Incipit] *Super libro de substantia orbis Joannis de Gandauo philosophi preclarissimi.*
[Vicenza: Henricus de Sancto Ursio, Zenus, 1486]
H-15504^ Pr. 7168^ BMC (XV)VII:1046^ Goff G-27
Subject: Astronomy, earth science.
Inc./f4785.5/and/f7168

Has leaves 92–108 only. Leaves 109–123, including colophon, wanting. BOUND WITH tracts by Walter Burley; Thomas Aquinas, Gratia Dei Esculanus, and Gaietanus de Thienis, qq.v.

## Joannes de Janduno.

*Super octo libris Aristotelis de physico avditv svbtilissimae qvaestiones: in quarum singularum capite tituli earum . . . Eliae etiam Hebraei cretensis quaestiones: uidelicet de primo motore, de mundi efficientia, de esse & essentia, & vno cum eiusdem in dictis Auerrois super eosdem libros qua'm castigtissimae leguntur.*
Venice: Heirs of Luca Antonius Iunta, 1551.
Subject: Physics.
Case/fY/642/.A8673

## Joannes de Sancto Geminiano, fl. 1296–1332.

*Liber de exemplis et similitudinibus rerum.*
Deventer: Richard Pafraet, ca. 1477.
Goff J-427
Subject: Encyclopedias.
Inc./f8955

MS marginalia and notes on front flyleaf. Bks. 5-10 only. Bk.5: De animalibus terrestribus; bk. 6: De hominis et membris eius; bk. 9: De artificibus et rebus artificialibus (emotions, also mechanic arts, inventions); bk. 10: De actibus et moribus humanis.

## Joannes de Sancto Geminiano.

*Liber de exemplis et similitudinibus rerum.* [Another ed.]
[Cologne: Johann Koelhoff the Elder, ca. 1485.]
Goff J-428
Subject: Encyclopedias.
Inc./f1061

In 10 books, among which are bk. 1: De celo & elementis; bk. 2: De metallis et lapidibus; bk. 3: De vegetabilibus et plantis; bk. 4: De natatilibus & volatilibus; bks. 5–10 as in 1477 ed. above.

## John of Salisbury, d. 1180.

*Polycratius sive de nugis curialium.*
[Brussels: Brothers of the Common Life, ca. 1479–81]
H-C *9430^ Pr. 9337^ Still. J383^ Goff J-425
Subject: Astrology, magic.
Inc./f9337

In 8 books.

## Jonstonus, Joannes, 1603–1675.

*An history of the wonderful things of nature: set forth in ten severall classes. Wherein are contained I. the wonders of the heavens. II. Of the elements. III. Of meteors. IV. Of minerals. V. Of plants. VI. Of birds. VII. Of four-footed beasts. VIII. Of insects. IX. Of fishes. X. Of man.*
London: printed by John Streater, 1657.
STC II J1017
Subject: Natural history, taxonomy.
Case/L/0114/.45

## Josephus, Flavius, 37–100(?).

*Opera.*
Verona: Petrus Maufer, 1480.
H-C *9452^ BMC(XV)VII:951^ Still. J436^ Goff J-484
Subject: Medicine.
Inc./f6918

## Josephus, Flavius

*Opera.*

Venice: Joannes Rubeus, 1486.
H-C *9454^ BMC(XV)V:415-16^ Goff J-486
Subject: Medicine.
Inc./f5118

Trans. Rufinus Tyrannius; ed. Hieronymus Squarzaficus. MS notes. BOUND WITH Lactantius, *Opera* (Inc. f5262), q.v.

## Josephus, Flavius.

*Josephi judei historici preclara opera.*
Paris: [Ioannis Barbier: for Francisci Reynault & Ioannis Petit, 1513–14.]
Subject: Medicine.
Case/5A/432

Ed. R. Goullet.

## Josephus, Flavius.

*Works* [Greek]. *Flavii Josephi Opera.*
Basel: [H. Frobenius], 1544.
Subject: Medicine.
Wing/fZP/538/.F925

## Josephus, Flavius.

*Flavii Josephi Opera.* [Greek]
Basel: [H. Frobenius], 1544.
Subject: Medicine.
Case/6A/255

Ed. Arnoldus Peraxylus Arlenius.

## Josephus, Flavius.

*Opera quae reperiri potuerunt, omnia. Ad Codices fere omnes . . . recensuit, nove versione donavit, & notis illustravit Joannes Hudsonus.*
Oxford: Sheldonian Theatre, 1720.
Subject: Medicine.
fC/18/.456

2 vols.

## Josselyn, John fl. 1630–1675.

*New England's rarities discovered: in birds, beasts, fishes, serpents, and plants of that country. Together with the physical and chyrurgical remedies wherewith the natives constantly use to cure their distempers, wounds, and sores.*
London: printed for G. Widdowes, 1672.
Subject: Materia medica, natural history, New World.
Ayer/*150.5/N4/J8/1672

Descriptions of birds, beasts [raccoon], fish; parts of certain fish have properties to heal cuts, cure the stone, stop heavy menstrual bleeding.

* * *

*Le Journal des scavans de l'an M.DC.LXV–[M.DC.LXVI].*
Cologne: Pierre Michel, 1665–66.
Subject: Astronomy, biology, chemistry, physics.
Case/A/54/.538

2 vols. in 1. Issues for 1665: Par le sieur de Hedouville [pseud. of Dennis de Sallo]; for 1666: Par le sieur G. P. Reissue of a journal founded in Paris in 1665 by de Sallo. Pt. 2, p. 53: A review of Novum lumen chimicum auctore Ioh. Rud. Glaubero. Amsterdam.

* * *

*Journal des scavans.*
Amsterdam: Pierre le Grand, [les Janssons à Waesberge until 1749].
Subject: Astronomy, biology, chemistry, physics.
X/007/.459

156 vols. in 150 (tomes 1–169). Wanting: vols. 80, 138–143, 145–146; 148, 155–156. Vol. 1 (transferred from the John Crerar Library, issues for the years 1665–66 [Amsterdam: Pierre le Grand, 1684]), p. 431: Four suns that appeared at Chartres, 1666; vol. 2 (issues for the years 1667–71 [Amsterdam: Pierre le Grand, 1685]), p. 553: "Description anatomique d'un caméleon, d'un castor, d'un dromadaire, d'un ours, & d'un gazelle"; pp. 624–29 and 631–38: astronomy and physics; vols. 3 and 4 (Amsterdam: Pierre le Grand, 1683). Vol. 3 contains issues for years 1672–74: "Conferences sur les sciences, 1673"; vol. 4 contains issues for 1675–76: Monday, 6 January 1676, p. 3: Anatomy of plants.

# K

## Kaempfer, Engelbert, 1651–1716.

*Amoenitatum exoticarum politico-physico-medicarum fasciculi V, quibus continentur variae relationes, observationes, & descriptiones rerum persicarum & ulterioris Asiae.*
Lemgo: Heinrich Wilhelm Meyer, 1712.
Subject: Botany, Japan, Persia.
Case/G/635/.46

Fasciculus 3: natural history; fasc. 4: Relationes botanico-historicas; fasc. 5: Plantarum japonicarum.

* * *

[Kalendrier des bergers] *Le grant kalendrier & compost des bergiers auecq leur astrologie. Et plusieurs aultres choses.*
Lyon:[Claude Nourry, 1524]
Subject: Almanacs, astrological medicine, astrology, astronomy, calendar.
Case/*Y/76095/.14

* * *

[Kalendrier des bergers] *The shepheards kalendar: newly augmented and corrected.*
London: printed by Robert Ibbitson, 1656.
Subject: Almanacs, astrological medicine, astrology, astronomy, calendar.
Case/Y/1094/.4065

Bookplate: "From the library of William Morris."

## [Ketham, Johannes de] 15th cent.

*Fasciculus medicinè houdende in hem dese nauolghende tractaten de allen cyrurginen enn andere menschente wetene seere profitelijc enn nootsakelijczijn.*
Antwerp: N. de Grave, 1529.
Subject: Anatomy, astrological medicine, pest tracts.
Wing/*fZP/5465/.G78

In 13 tracts. Tract 1: Die viere naturen ende complexien der menschen; 2: Die aderen'te latene oft te flobothomerene; 3: Die juditien de anderenenndes bloets metvele schoone cautelen; 4: Een tractaet om te Latene die astronomie die ghebreken der vrouwen. Een speciael tractaet teghen die pestilentie van heere Ramitius. . . . Een preseruatijf regiment ende curatijf regiment teghen die pestilentie van meester Petrus Tansignano gemaect dat all regimenten te bonen gaet. Een alder excellentste ende expertste anothomie van meester Mondinus.

## Kircher, Athanasius, 1602–1680.

*Ad Alexandrum VII . . . . obelisci aegyptiaci, nuper inter Isaei romani rudera effossi, interpretatio hieroglyphica Athanasii Kircheri.*
Rome: Tyopgraphica Varesii, 1666.
Subject: Hieroglyphics, translation and interpretation.
Case/fF/0271/.465

## Kircher, Athanasius.

*Ars magna lucis et umbrae in X libros digesta. Quibus admirandae lucis & umbrae in mundo atque adeò universa natura, vires effectusque uti nova, ita varia novorum reconditiorumque speciminum exhibitione, ad varios mortalium usus panduntur.*
Amsterdam: Joannes Janssonius à Waesberge & heirs of Elizei Weyerstraet, 1671.
Subject: Alchemy, astronomy, colors, experimental science, optics.
Case/G/66/.466

BOUND WITH *China monumentis*, q.v.

## Kircher, Athanasius.

*Ars magna sciendi in XII libros digesta, qua nova & universali methodo per artificiosum combinationum contextum de omni reproposita plurimis & prope infinitis rationibus disputari, omniumque summaria quaedam cognitio compari potest.*
Amsterdam: J. Jansonius à Waesberge, & widow Elizei Weyerstraet, 1669.
Subject: Alchemy, epistemology, mysticism.
Case/fB/49/.466

Bk. 3, pt. 3: "Clavis arcanorum"; tome 2, bk. 7: "Paradigmata artesnostre ex metaphysicâ, logicâ, seu dialecticâ, physicâ, medicinâ, exhibet."

## Kircher, Athanasius.

*Artis magnae de consono & dissono ars minor; das ist philosophischer extract und auszug ans dess welt-berühmten teutschen Jesuitens Athanaei Kircheri von Fulda musurgia universali, in sechs bücher verfasset. . . . So wol in einer ieden kunst-facultät der ganzen encyclopediae philosophicae, als absonderlich in der philosophi, rhetoric, poetic, physic, metaphysic, mathematic, astronomi, ethic, politic, chymic, medicin, mechanic, &c. So dann auch der theologi natürlichen magi und echotectonic &c. eröfnet gewisen und vor augen gestellet wird.*
Schw. Hall: Hans Reinh. Laidigen, 1662.
Subject: Acoustics, alchemy, encyclopedias, magic, sympathy and antipathy.
Case/V/5/.4663

## Kircher, Athanasius.

*China monumentis quà sacris quà profanis nec non variis naturae & artis spectaculis, aliarumque rerum memorabilium argumentis illustrata.*
Amsterdam: Joannes Janssonius à Waesberge & Elizeum Weyerstraet, 1667.
Subject: Engineering, biology, geography.
Case/G/66/.466

Includes chaps. on plants, animals, fish. Interpretation of monuments & translation of symbols. Pt. 4, "China curiosis naturae & artes miracules illustrata." Pt. 5, engineering, bridges. BOUND WITH *Ars magna lucis*, q.v.

### Kircher, Athanasius.

*Magnes sive de arte magnetica opus tripartitum quo vniversa magnetis natura, eiusque in omnibus scientijs & artibus vsus, noua methodo explicatur: ac praeterea è viribus & prodigiosis effectibus magneticarum, aliarumque abditarum naturae motionum in elementis, lapidibus, plantis, animalibus, elucescentium, multa hucusque incognita naturae arcana, per physica, medica, chymica, & mathematica omnis generis experimenta recluduntur.*
Rome: V. Mascardi, 1654.
Subject: Magnetism, physics, occult.
Case/fL/7/.469

Contains "magnetismus elementorum," "magnetismus metallicorvm," also "magnetismus animalivm, medicinalivm, musicae," and "geographia magnetica." Pp. 377–82: "De occultis naturae operationibus, earumque causis."

### Kircher, Athanasius.

*Mundus subterraneus in XII libros digestus; quo divinum subterrestris mundi opificium, mira ergasteriorum naturae in eo distributio. . . . Universae denique naturae majestas & divitiae summa rerum varietate exponuntur. Abditorum effectuum causae acri indagine inquisitae demonstrantur; cognitae per artis & naturae conjugium ad humanae vitae necessarium usum vario experimentorum apparatu, necnon novo modo, & ratione applicantur.*
Amsterdam: Joannes Janssonius à Waesberge & widow Eliza Weyerstraet, 1668.
Subject: Alchemy, earth sciences, experimental science, generation of plants and insects.
Ayer/7/K58/1668

2 vols. in 1.

### Kircher, Athanasius.

*Mvsvrgia vniversalis sive ars magna consoni et dissoni in X libros digesta. Quà vniuersa sonorum doctrina, & philosophia, musicaeque tam theoricae, quam practicae scientia, summa varietate traditur; . . . tum in omnipoenè facultate, tum potissimùm in philologià, mathematicà, physicà, mechanicà, theologià, aperiuntur & demonstrantur.*
Rome: heirs of Franciscus Corbellettus, 1650.
Subject: Acoustics, magic, music, mysticism.
Case/fV/5/.466

2 vols. in 1. Bks. 1–4: physiologicus arithmeticus, geometricus; vol. 2 (bks. 8–10), has separate title page with imprint: Rome: Ludouici Grignani, 1650). Includes magicus, harmoniam mundi.

### Kircher, Athanasius.

*Neue hall-und thonkunst oder mechanische gehaim verbindung der kunst und natur durch stimme und hall-wissenschaft gestifftet, vorinn ingemein der stimm, thons, hall-und schalles natur eigenschaft, krafft und wunder-würkung . . . vorgestellt werden.*
Nördlingen: printed by Friderich Schultes for Arnold Heylen, 1684.
Subject: Acoustics, music, physics.
Case/fV/5/.46645

Translation of *Phonurgia nova.*

### Kircher, Athanasius.

*Oedipus Aegyptiacvs. Hoc est vinuersalis hieroglyphicae veterum doctrinae temporum iniuria abolitae instavratio. Opus ex omni orientalium doctrina & sapientia conditum, nec non viginti diuersarum linguarum authoritate stabilitum.*
Rome: Vitalis Mascardi, 1652–54.
Subject: Hieroglyphics, mysticism.
Case/fX/0141/.465

3 vols. in 4 parts (vol. 2 has 2 pts.). Vol. I (1652): Egyptian chorography, politics, religion compared and contrasted with Hebrew and others; vol. 2, pt. 1 (1653): "Gymnasium sive phrontisterion hieroglyphicum in duodecim classes distrubutum"; pt. 2: "Mathematica hieroglyphica, mechanica sive architectonica"; vol. 3 (1654): "Theatrum hieroglyphicvm . . . interpretatio iuxta sensum physicum, tropologicum, mysticum, historicum, politicum, magicum, medicum, mathematicum, cabalisticum, hermeticum, sophicum, theosophicum; ex omni orientalium doctrina & sapientia demonstrata."

### Kircher, Athanasius.

*Phonurgia nova, sive conjugium mechanico-physicum artis & natvrae paranympha phonosophia concinnatum.*
Kempten: Rudolph Dreherr, 1673.
Subject: Acoustics, music, physics.
Case/fV/5/.4665

Nature and property of sound, experiments in hearing. Sec. II, chap. 4, pp. 204–16: tarantism.

### Kircher, Athanasius.

*Physiologia Kircheriana experimentalis, qua summa argumentorum multitudine & varietate naturalium rerum scientia per experimenta physica, mathematica, medica, chymica, musica, magnetica, mechanica comprobatur atque stabilitur.*
Amsterdam: Janssonio-Waesbergiana, 1680.
Subject: Experimental science.
Case/6A/144

Drawn from Kircher's works by Joannes Stephan Kestler. Includes accounts of Kircher's experimental work in natural motion, light, magnetism, sound, mechanical engineering, and pyrotechnics.

### Kircher, Athanasius.

*Prodromvs Coptvs sive Aegyptiacvs. . . . in quo cum linguae coptae, sive Aegyptiacae, quondam pharaonicae, origo, aetas, vicissitudo, inclinatio, tùm hieroglyphicae literaturae instauratio, vti per varia variarum eruditionum, interpretationumque difficilimarum specimina, ita noua quoque & insolita methodo exhibentur.*
Rome: S. Cong. de Propag. Fide, 1636.

Subject: Hieroglyphics, mysticism.
Case/X/42/.46

Chap. 5, p. 147: names of 7 planets in Coptic, Latin, Hebrew, and Arabic; chap. 8: "De utilitate linguae Coptae seu Aegyptiacae."

## Kircher, Athanasius.

*Prodromvs Coptvs sive Aegyptiacvs.* [Another copy]
Wing/ZP/635/.C/2842

Title page vignette differs from the one in copy above.

## Kircher, Athanasius.

*Sphinx mystagoga, sive diatribe hieroglyphica, qua mumiae, ex Memphiticis pyramidum adytis erutae, & non ita pridem in Galliam transmissae, juxta veterum hieromystarum mentem, intentionemque, plena fide & exacta exhibetur.*
Amsterdam: Janssonio-Waesbergiana, 1676.
Subject: Hieroglyphics.
Case/fX/041/.465

In 3 parts: De scopo; Oedipus Sphingi; De hieroglyphicorum. Chap. 1: "De metmepsychosi sive revolutione animarum, quod & placitum fuit proprium veterum Aegyptiorum." Included in pt. 2, chap. 5, paragraphus 2: "De herbis & lapidibus ab Hieromantis in Adytis usurpatis."

## Kircher, Athanasius.

*Turris Babel, sive archontologia qua primo priscorum post diluvium hominum vita, mores rerumque gestarum magnitudo, secundo turris fabrica civitatumque exstructio, confusio linguarum, & inde gentium transmigrationis, cum principalium inde enatorum idiomatum historia, multiplici eruditione describuntur & explicantur.*
Amsterdam: Janssonio-Waesbergiana, 1679.
Subject: Archeology, comparative linguistics, Hermeticism.
Case/fF/092/.464

P. 177: "Primaeva literarum Aegyptiarum fabrica, & institutio facta à tauto sive Mercurio Trismegisto."

## Koebel, Jacob, ca. 1470–1533.

*Kalender new geordent. Mit vielen underweisungen der himmelischen leüff der zeit, der Christlichen gesage. Auch kurzwilig (gereympt) unnd lüftig mitt exempelnn und figuren getruckt.*
Oppenheim [1512?].
Subject: Almanacs, astrology, astrological medicine, zodiac.
Wing/ZP/547/.K807

In verse. Pt. 1, calendar; pt. 2, "Eyn inleytung unnd anrede Jacob Cöbels."

## Krönsperger, Leonhart.

*Bauw ordnung. Von burger und nachbarlichen gebeuwen in stetten, werckten, flecken, dörffern, und auff dem land sampt derselbigen anhangenden handwerker kosten gebrauch und gerechtigkeit in drey theil verfast vnd zusammen gezogen welches innhalt nach der vorred weiter auff das aller kürtzest zuvernemmen ist.*
Frankfurt: Georg Raben and heirs of Weygand Hanen, 1564.
Subject: Architecture.
Wing/fZP/547/.H152

## Kulmus, Johann Adam, 1689–1745.

*Tabulae anatomicae, in quibus corporis humani omniumque ejus partium structura & usus brevissimè explicantur. Accesserunt majoris perspicuitatis causâ: Annotationes et tabulae aeneae.*
Amsterdam: Janssonio-Waesbergios, 1732.
Subject: Human anatomy.
oQM/21/.K26/1732

# L

## La Charrière, Joseph de, d. 1690.

*Traité des operations de la chirurgie, dans lequel on explique les causes des maladies qui les précedent, fondées sur la structure de la partie; leurs signes & leurs symptômes: & dans lequel on a introduit plusieurs nouvelles remarques après chaque opération, & un traité des playes avec la méthode de les bien panser. Augmenté des bandages & appareils à la fin de chaque opération.*
Paris: Horthemels brothers, 1727.
Subject: Diagnosis, medicine, surgery, wounds.
RD/32/.L23/1727

"Nouvelle edition revüe & corrigée par l'auteur."

## Lacinius, Janus.

*Pretiosa margarita novella de thesavro, ac pretiosissimo philosophorvm lapidae. Artis huius diuinae typus, & methodus: collectanea ex Arnoldo, Rhaymundo, Rhasi, Alberto, & Michaele Scoto; per Ianum Lacinium.*
[Venice: Aldus sons, 1546.]
Subject: Alchemy, philosophers stone, transmutation.
Case/B/8633/.472

1st ed. Includes sections, pro arte, contra artem; on alchemicalmethods and processes.

## Lactantius, Lucius Caecilius Firmianus, 250–317?

*Opera.*
Subiaco [Conrad Sweynheim & Arnold Pannartz], 1465.
H-C *9806^ BMC (XV)IV:2^ Pr. 3288^ Goff L-1
Subject: Hermeticism, natural philosophy, origin of man, true and false knowledge.
Inc./f3288

Errata leaves in MS at back of vol. Pt. 1 has MS subheads and MS marginalia.

## Lactantius, Lucius Caecilius Firmianus.

*Opera.*
Rome: Conrad Sweynheim & Arnold Pannartz, 1468.
H *9807^ BMC (XV)IV:4^ Pr. 3291^ Goff L-2
Subject: Hermeticism, natural philosophy, origin of man, true and false knowledge.
Inc./f*3291

Section "De falsa sapientia" includes "De antipodibus quos ideo esse finxerunt: que opinati sunt mundum esse rotundum"; chap. 24: "Quid illi qui esse contrarios"; chap. 6: "Contra Epicurum qui ex athomis fortuito concurrentibus compactum fabricatumque hominem disputant." Also "De opificio di vel formatione hominis. BOUND WITH table of contents and errata list from 1478 ed. (Inc. f*4332), q.v.

## Lactantius, Lucius Caecilius Firmianus.

*Opera.*
[Venice] Adam von Ambergau, 1471.
H *9809^ BMC (XV)V:188^ Pr. 4144^ Still. L4^ Goff L-4
Subject: Hermeticism, natural philosophy, origin of man, true and false knowledge.
Inc./f4144

## Lactantius, Lucius Caecilius Firmianus.

*Opera.*
Rome: Ulrich Han & Simon Nicolai Chardella, 1474.
H 9811^ BMC (XV)IV:24-25^ Pr. 3360^ Goff L-6
Subject:Hermeticism, natural philosophy, origin of man, true and false knowledge.
Inc./f3360

MS notes. Chap. 6: "Quo sapientiam medio errantum fuerit, & academici contra physicos. & physica contra academicos dimicauerint."List of contents annotated in early MS hand, and MS marginalia passim.

## Lactantius, Lucius Caecilius Firmianus.

*Opera.*
[Venice: Johannes de Colonia & Johann Manthen, 1478.]
H-C *9814^ BMC (XV)V:233^ Pr. 4332^ Goff L-9
Subject:Hermeticism, natural philosophy, origin of man, true and false knowledge.
Inc./f*4332

[Consists of table of contents and errata list ONLY.] BOUND WITH Lactatntius *Opera*, 1468 (Inc. f*3291), q.v.

## Lactantius, Lucius Caecilius Firmianus.

*Opera.*
Venice: Andreas de Paltasichis Catarensis & Boninus de Boninis Sociis "1478" [1479].
H-C *9813^ BMC (XV)V:251^ Pr. 4425^ Goff L-8
Subject: Hermeticism, natural philosophy, origin of man, true and false knowledge.
Inc./f4425

## Lactantius, Lucius Caecilius Firmianus.

*Opera.*
Venice: Theodorus de Ragazonibus de Asula, 1490.
H-C 9815^ BMC (XV)V:477^ Pr. 5262^ Goff L-10
Subject: Hermeticism, natural philosophy, origin of man, true and false knowledge.
Inc./f5262

MS marginalia. BOUND WITH Josephus, Flavius (Inc. 5118), q.v.

## Lactantius, Lucius Caecilius Firmianus.

*Opera.*

Venice: Vincentius Benalius, 1493.
H-C*9816^ BMC (XV)V:525^ Pr. 5376^ Goff L-11
Subject: Hermeticism, natural philosophy, origin of man, true and false knowledge.
Inc./f5376

Half title: Lactantii Firmiani de diuinis institutionibus libri septem: de ira dei: et opificio hominis cum epithomon eiusdem foeleciter incipiunt.

### Lactantius, Lucius Caecilius Firmianus.

*Opera.*
Venice: Simon Bevilaqua, 1497.
H-C*9818^ BMC (XV)V:522^ Pr. 5401^ Goff L-13
Subject: Hermeticism, natural philosophy, origin of man, true and false knowledge.
Inc./5401

### Laet, Joannes de, 1593–1649.

*Joannis de Laet Antuerpiani notae ad dissertationem Hvgonis Grotii de origine gentium Americanarum; et observationes aliqvot ad meliorem indaginem difficillimae illus quaestionis.*
Paris: Widow of Gvilielmus Pelé, 1643.
Subject: Anthropology, philology.
Ayer/104/L2/1643

In addition to Grotius's text, vol. includes comparative linguistics, e.g., comparison of Huron and Mexican words for common objects (pp. 173–85). Theory of origins of native peoples of the Americas based on comparison of their languages.

### Laet, Joannes de.

*Ioannis de Laet Antwerpiani responsio ad dissertationem secundam Hvgonis Grotii, de origine gentium Americanarum.*
Amsterdam: Ludovic Elsevir, 1644.
Subject: Anthropology, philology.
Ayer/*104/L2/1644

### Lafitau, Joseph François, 1691–1746.

*Mémoire presenté a son altesse Royale Monseigneur le duc d'Orléans, regent du royaume de France: concernant la précieuse plante du gin seng de Tartarie, découverte en Canada par le P. Joseph François Lafitau, de la compagnie de Jesus, missionaire des Iroquois du Sault Saint Louis.*
Paris: Joseph Mongé, 1718.
Subject: Descriptive botany, herbal medicine.
Ayer/*150.6/L2/1718

Discovery of ginseng in Canada and its medicinal uses.

### Lafitau, Joseph François.

*Moeurs des sauvages ameriquains, comparées aux moeurs des premiers temps.*
Paris: Saugrain the elder; Charles Estienne Hochereau, 1724.
Subject: Anthropology; medical practices, New World.
Case/F/801/.47

In 2 vols. Among subjects treated (vol. 1): "Idée & caractere des sauvages en general"; "Des mariages & de l'education"; (vol. 2): "Malade jonglé"; diseases and medicine, pp. 359–86; death, funeral practices.

### Lafitau, Joseph François.

*Moeurs des sauvages ameriquains.* [Another copy]
Ayer/301/L2/1724

### Lafitau, Joseph François.

*Moeurs des sauvages ameriquains.* [Another copy]
Graff/5127

4 vols. Vol. 2: marriage and education; vol. 3: occupations of men and women; vol. 4: hunting and fishing, diseases and medicine, death, funeral practices.

### Lahontan, Louis Armand de Lom d'Arce, baron de, 1666–1715?

*Dialogues de M. le baron de Lahontan et d'un sauvage, dans l'Amerique. Avec les voyages du même au Portugal & en Danemarc.*
Amsterdam: Widow of Boeteman, 1704.
Subject: Anthropology, New World, voyages.
Ayer/123/L16s/1704a

See Wright Howes, *USiana*, 26. Binders title: *Voyages dans l'Amerique.* Pp. 82–84: Comparison of illnesses and physical condition of Huron Indians with French. BOUND WITH *Conformité des coutumes des Indiens orientaux, avec celles des Juifs & des autres peuples de l'antiquité, par Mr. de la C****.* Brussels: George de Backer, 1704. Contains chapters on circumcision, metempsychosis, magic spells, unctions, division of days into units of time. From Howes: "The dialogues [were] a diatribe against Christianity.... French editions of all three volumes were, after 1704, so changed and polished as to verify the suspicion that editorial work on them was done by the apostate monk Nicolas Guedeville. Some authorities even believe him to be the author of the 'Dialogues.'"

### Lahontan, Louis Armand de Lom d'Arce, baron de.

*New voyages to North America. Containing an account of the several nations of that vast continent; their customs, commerce, and way of navigation upon the lakes and rivers; . . . A geographical description of Canada, and a natural history of the country. . . . Also a dialogue between the author and a general of the savages.*
London: printed for H. Bonwicke, etc., 1703.
Subject: Anthropology, New World, voyages.
Ayer/*123/L16n/1703

2 vols. in 1. Vol. 2: "A view of the diseases and remedies of the savages," pp. 45–55. See Howes 25: "Lahontan's narrative [was] of considerable value when confined to his actual sojourneyings [*sic*] in the Lake region."

**Lahontan, Louis Armand de Lom d'Arce, baron de.**

*New voyages to North-America.* [Another copy]
Graff/2364

2 vols.

**Lahontan, Louis Armand de Lom d'Arce, baron de.**

*Nouveaux voyages de Mr. le baron de la Hontan, dans l'Amerique septentrionale.*
The Hague: l'Honoré brothers, 1703.
Subject: Anthropology, New World, voyages.
Graff/2365

2 vols. Vol. 1: 25 letters; vol. 2: *Memoires de l'Amerique septentrionale* includes material on natural history, hieroglyphics, and a "Dictionnaire de la langue du païs." Howes 25.

**Lahontan, Louis Armand de Lom d'Arce, baron de.**

*Nouveaux voyages de Mr. le baron de Lahontan.* [Another copy]
Graff/2366

**Lahontan, Louis Armand de Lom d'Arce, baron de.**

*Nouveaux voyages de Mr. le baron de Lahontan.* [Another copy]
Case/G/82/.47

2 vols. in 1. Vol. 2, p. 144: "Maladies & remédes des sauvages."

**Lahontan, Louis Armand de Lom d'Arce, baron de.**

*Nouveaux voyages de Mr. le baron de Lahontan, dans l'Amerique septentrionale.*
The Hague: L'Honoré bros., 1704.
Subject: Anthropology, New World, voyages.
Ayer/123/L16/1704/v.1

Library has vol. 1 only.

**Lahontan, Louis Armand de Lom d'Arce, baron de.**

*Nouveaux voyages de Monsieur le baron de Lahontan dans l'Amerique septentrionale, qui contiennent une relation des differens peuples qui y habitent.*
The Hague: Isaac Delorme, 1707–8.
Subject: Anthropology, New World, voyages.
Ayer/123/L16/1707

Vols. 1 and 2 bound together; vol. 3 wanting.

**Lahontan, Louis Armand de Lom d'Arce, baron de.**

*Nouveaux voyages de Monsieur le baron de Lahontan, dans l'Amerique septentrionale.*
The Hague: L'Honoré brothers, 1709.
Subject: Anthropology, New World, voyages.
Ayer/123/L16/1709

2 vols. Frontispiece to vol. 2: "La cérémonie du mariage."

**Lahontan, Louis Armand de Lom d'Arce, baron de.**

*Suite du voyage de l'Amerique, ou dialogues de monsieur le baron de Lahontan et d'un sauvage, dans l'Amerique. Contenant une description exacte des moeurs & des coutumes de ces peuples sauvages.*
Amsterdam: Widow Boeteman, 1704.
Subject: Anthropology, New World, voyages.
Ayer/123/L16s/1704

See Howes 26. Binders title: *Voyages dans l'Amerique,* tome 3.

**Lahontan, Louis Armand de Lom d'Arce, baron de.**

*Voyages du baron de La hontan dans l'Amerique septentrionale, qui contiennent une rélation de différens peuples qui y habitent.*
Amsterdam: François l'Honoré, 1705.
Subject: Anthropology, New World, voyages.
Ayer/123/L16/1705

2d ed. 2 vols. See Howes 25. Revised, corrected, and enlarged ed. Contains description of fishing and hunting, natural history (animals and trees native to area); "Conversations de l'auteur de ces voyages avec Adario, sauvage distingué."

**Lahontan, Louis Armand de Lom d'Arce, baron de.**

*Voyages du baron de la Hontan dans l'Amerique septentrionale.*
The Hague: Charles Delo, 1706.
Subject: Anthropology, New World, voyages.
Ayer/123/L16v/1706

2d ed. 2 vols. Howes 25.

**Lahontan, Louis Armand de Lom d'Arce, baron de.**

*Nouveaux voyages de Monsieur le baron de Lahontan, dans l'Amerique septentrionale.*
The Hague: Isaac Delorme, 1712.
Subject: Anthropology, New World, voyages.
Ayer/123/L16/1712

2 vols.

**Lahontan, Louis Armand de Lom d'Arce, baron de.**

*Nouveaux voyages de Monsieur le baron de Lahontan, dans l'Amerique septentrionale.*
The Hague: L'honoré brothers, 1715.
Subject: Anthropology, New World, voyages.
Ayer/123/L16/1715

2 vols.

## Lahontan, Louis Armand de Lom d'Arce, baron de.

*Voyages du baron de Lahontan dans l'Amerique septentrionale.*
Amsterdam: Widow Boeteman, 1728.
Subject: Anthropology, New World, voyages.
Ayer/123/L16/1728

3 vols. in 2.

## Lahontan, Louis Armand de Lom d'Arce, baron de.

*Voyages du baron de Lahontan dans l'Amerique septentrionale.*
Amsterdam: François l'Honoré, 1741.
Subject: Anthropology, New World, voyages.
Ayer/123/L16/1741

3 vols. Vol. 1: 15 letters; vol. 2: *Suite des voyages du baron de Lahontan dans l'Amerique septentrionale* (letters 16–25 and "explication de quelques termes"); Lahontan asks the natives whether a beaver can live entirely out of water; they are surprised that he doubts this and assure him that it can live on land like a dog (illustration of a beaver opposite p. 8). Vol. 3: *Memoires de l'Amerique septentrionale, ou la suite des voyages.*Contains sections on natural history, hieroglyphics, and the "Dictionnaire de la langue des sauvages."

## [Lamy, Bernard] 1640–1715.

*Entretiens sur les sciences, dans lequels on apprend comme l'on doit étudier les sciences, & s'en servir pour se faire l'esprit juste, & le coeur droit.*
Lyon: Jean Certe, 1706.
Subject: Education, religion and science.
L/O/.475

Study of sciences for the greater glory of God. Intellectual discipline of scientific study.

## Lamzweerde, Jan Baptist van, fl. 17th cent.

*Historia naturalis molarum uteri: in qva de natura feminis, ejusque circulari in sanguinem regressu, accuratius disquiritur.*
Leyden: Petrus vander Aa, 1686.
Subject: Gynecology, medicine.
Case/oRG/591/.L25/1686

## Lancre, Pierre de, d. 1630.

*L'incredulité et mescreance dv sortilge plainement convaincve. Ou il est amplement et cvrievsement traicté, de la vertié ou illusion du sortilege, de la fascination, de l'attouchement, du scopelisme, de la diuination, de la ligature ou liaison magique, des apparitions: et d'vne infinité d'autres rare et noueueaux subjects.*
Paris: Nicolas Buon, 1622.
Subject: Divination, fascination, witchcraft.
Case/B/88/.478

Scopelisme = use of charmed stones. Sec. 1: "S'il y a quelque verité et certitude au sortilege"; sec. 8: "Des Jvifs, apostats et athées. MS note on front flyleaf asserts that Lancre believes in witches and mentions his other work, *Tableau*, q.v.

## Lancre, Pierre de.

*Tableauv de l'inconstance des mavvais anges et demons, ov il est amplement traicté des sorciers & de la sorcelerie. Livre tres-vtile et necessaire, necessaire non seulement aux iuges, mais à tous ceux qui viuent soubs les loix Chrestiennes.*
Paris: Iean Berjon, 1612.
Subject: Demonology, witchcraft.
Bonaparte/Coll./No./3051

In 6 books. Discusses why there are more female than male sorcerers.

## Landi, Ortensio, ca. 1512–ca. 1553.

*Commentario delle piu notabili & mostruose cose d'Italia, & altri luoghi: di lingui Aramea in Italiana tradotto con un breue catalogo de gli inuentori delle cose si mangiano & beueno, nouamente ritrouato.*
Venice: Bartholomeo Cesano, 1553.
Subject: Houses, cities in Italy and their commodities (e.g., food).
Case/Y/712/.L234

## Lando, Gio Giacomo.

*Aritmetica mercantile di Gio. Giacomo Lando Genovese. Nella quale si vede, come si hanno da fare li conti, per li cambi, che si fanno nelle città principali della Christianità; Il modo di raguagliare le piázze, di aggiustare ogni sorte di comissioni de cambi, & mercantie, & formare arbitrij.*
Naples: Tarquinio Longo, 1604.
Subject: Tradesman's arithmetic; conversion of coinage.
Case/L/130/.478

## Lando, Gio Giacomo.

*Aritmetica mercantile. Nella qvale si vede, come si hanno da fare li conti, per li cambri, che si fanno nelle città principali della Christianità; Il modo di raggvagliare le piazze, di aggiustare ogni sorte di comissioni de cambi, & mercantie, & formare arbitrij.*
Venice: Ghirardo Imberti, 1640.
Subject: Tradesman's arithmetic; conversion of coinage.
L/130/.479

## Langlet Dufresnoy, Nicolas 1674–1755.

*Histoire de la philosophie hermetique accompagnée d'un catalogue raisonné des ecrivains de cette science. Avec le véritable Philalethe, revû sur les originaux.*
The Hague: Pierre Gosse, 1742.
Subject: Hermeticism.
B/8633/.499

3 vols. Vol. 1: History of hermetic philosophy; vol. 2: transmutation of metals; vol. 3: history of alchemy and chemistry and catalogue.

## La Perriere, Guillaume de, 1499–1565.

*Les considerations des qvatre mondes, à sauoir est: divin, angeliqve, celeste, & sensible: comprinses en quatre centuries*

*de quatrains, contenans la cresme de divine & humaine philosophie.*
Lyon: Macé Bonhomme, 1552.
Subject: Atoms, elements, Epicureanism, natural philosophy.
Case/W/1025/.484

Includes material on "le monde sensible," the humors, the 4 qualities (hot, cold, dry, moist), the sun, the soul. BOUND WITH *Picta poesis*, Lyon: Mathias Bonhomme, 1552 (emblems with aphoristic verses), and *La morosophie, contenant cent emblemes moraux*. Lyon: M. Bonhomme, 1553.

### La Perriere, Guillaume de.

*Le theatre des bons engins, auquel sont contenus cent emblemes.*
Paris: Denis Ianot [1539].
Subject: Emblems.
Wing/ZP/539/.J2625

### La Perriere, Guillaume de.

*Le theatre des bons engins, auquel sont contenuz cent emblemes moraulx. Composé par Guillaume de la Perriere Tolosain: et nouuellement par iceluy limé, reueu, & corrigé.*
[Paris:] Denis Ianot, [1539]
Subject: Emblems.
Case/W/1025/.486

### La Primaudaye, Pierre de, b. ca. 1545.

*Academie francoise, en laquelle il est traité de l'institution des moeurs, & de ce qui concerne le bien & heureusement viure en tous estats & conditions: par les preceptes de la doctrine, & les exemples de la vie des anciens sages, & hommes illustres.*
Lyon: Iean Veirat, 1591, 1593.
Subject: Natural history, psychology.
Case/Y/762/.L298

4th ed. 2 vols. Vol. 2: *Suite de l'academie francoise: en laquelle est traitté de l'homme: &, comme par vne histoire naturelle du corps & de l'ame, est discouru de la creation, matiere, composition, forme, nature, des causes naturelles de toutes affections, des vertus & des vices: item, de la nature, puissances, oeuures, & immortalité de l'ame.* [Geneva:] Iaques Chouët, 1593.

### La Primaudaye, Pierre de.

*The French academie.... Fvlly discoursed and finished in foure bookes.*
London: printed for Thomas Adams, 1618.
STC 15241
Subject: Natural history, psychology.
Case/fY/762/.L3

Bk. 1: "The institution of manners and callings of all estates"; bk. 2: "Concerning the soule and body of man" (includes anatomical descriptions, e.g., "kernels in the body & their sundry vses; especially the breasts of women, of their beauty and profit in the nourishing of children, & of the generation of milk"); bk. 3: "A notable description of the whole world, &c."; bk. 4: "Christian philosophie instructing the true and onely meanes to eternal life."

### La Quintinie, Jean de, 1626–1688.

*The compleat gard'ner; or, directions for cultivating and right ordering of fruit-gardens and kitchen-gardens; with divers reflections on several parts of husbandry. In six books.*
London: printed for Matthew Gillyflower, 1693.
STC II L431
Subject: Gardening, implement making.
Case/R/57/.479

Translated by John Evelyn. Includes "The dictionary. An explication of the terms of gard'ning"; pt. 2: "Proofs of a good soil"; pt. 4, chap. 10: "Of the tools that are necessary for pruning"; and in final section, "Reflections upon the influence of the moon in its wain and full &c."

### La Ramée, Pierre de, 1515–1572.

*Scholae in liberales artes.*
Basel: Eusebius & Nicolaus F., Episcopius heirs, 1578.
Subject: Metaphysics, physics.
Case/fB/239/.184374

Includes 8 bks. on physics and 14 on metaphysics. (Mathematics a separate vol.)

### La Ramée, Pierre de.

*Scholarvm mathematicarvm, libri vnvs et triginta.*
Basel: Evsebivs Episcopivs & heirs of brother Nicolavs, 1569.
Subject: Mathematics.
Case/L/10/.484

"Argumentum scholarum mathematicarum: Tres primi libri continent proemium mathematicum, id est, exhortationem ad mathematicas artes ad Catharinam Mediceam.... Duo proximi disputant praecipua quaedam capita arithmeticae reliqui ex ordine disserunt de quindecim libris euclideae."

### Lavater, Ludwig, 1527–1586.

*De spectris, lemvribvs et magnis atqve insolitis fragoribus variisque praesagitionibus, quae plerumque obitum hominum, magnas clades, mutationésque imperiorum praecedunt, liber vnvs. Lvdovico Lavatero Tigvrino avtore.*
Geneva: Evstathivs Vignon, 1580.
Subject: Ghosts, occult, premonitions, supernatural, superstition, witchcraft.
Case/B/893/.494

### Lavater, Ludwig.

*Trois livres des apparitions des esprits, fantosmes, prodiges & accidens merueilleux qui precedent souuantesfois la mort de quelque personnage renommmé, ou vn grand changement és choses de ce monde.*
[Geneva?] François Perrin for Iean Durant, 1571.
Subject: Occult, premonitions, witchcraft.
Case/B/893/.492

"Plus trois questions proposees & resolues par M. Pierre Martyr . . . lesquelles conuiennent à ceste matiere." (Summary, p. 234.) 3 questions: 1. Who appeared to Saul summoned by the sorceress? 2. To learn whether the devil can materialize; will he reply to questions? 3. Is it legal to ask advice from the devil and to serve him?

## Law, William, 1686–1761.

*The case of reason, or natural religion, fairly and fully stated.*
London: printed for W. Innys, 1731.
Subject: Natural religion, reason.
B/72/.878

## Lawson, John, d. 1712.

*A new voyage to Carolina; containing the exact description and natural history of that country: together with the present state thereof. And a journal of a thousand miles travel'd thro' several nations of Indians. Giving a particular account of their customs, manners, &c.*
London, 1709.
Subject: Anthropology, descriptive geography, natural history.
Ayer/150.5/N74/L2/1709

Lawson was surveyor-general of North Carolina.

## Lazzarelli, Luigi, 1450–1500.

*Crater Hermetis. Contenta in hoc volumine Pimander: Mercurii Trismegisti liber de sapientia & potestate dei. Asclepius: Eiusdem Mercurij liber de voluntate diuina. Item Crater Hermetis a Lazarelo Septempedano.*
Paris: Simon Colinaeus, 1522.
Subject: Hermeticism.
Case/Y/642/.H23965

Edited by Jacques Le Fèvre d'Etaples.

## [Le Blon, Jakob Christoffel] 1667–1741.

*Preparation anatomique des parties de l'homme servant le generation.*
London: 1712.
Subject: Anatomy, medicine.
Wing/fZP/746/.L12

Anatomical illustrations (in color) with explanatory text in Latin and Dutch.

## Le Boë, Franciscus de, called Sylvius, 1614–1672.

*Opera medica, tam hactenus inedita, quàm variis locis & formis edita; nunc verò certo ordine disposita, & in unum volumen redacta.*
Amsterdam: Daniel Elsevir and Abraham Wolfgang, 1680.
Subject: Medicine.
oR/128.7/.S9/1680

"Editio altera correctior & emendatior."

## Lebrija, Elio Antonio de, 1441?–1522.

*Lexicon latino catalanvm sev Dictionarium Aelij Antonij Nebrissensis. . . . Accessit etiam eiusdem auctoris medicum dictionarium in sexcentis penè locis nunc denuò emendatum, & in fine adiectum: vt hac dispositione rei herbariae studiosos aliqua ex parte leuaremus.*
Barcelona: Clavdivs Bornativs, 1560–63.
Subject: Medicine, medical dictionaries.
Bonaparte/Coll./No./4524

Special title page (p. 118): "Dictionarivm medicvm olim salmanticae ab Antonio Nebrissensi commentvm, privs Antverpiae excussvm, nunc maiori cura, ac vigilantia, quam vnqvam fuerit, emendatum: & in sexcentis locis restitutum," 1561.

## Le Cat, Claude-Nicolas, 1700–1768.

*Traité des sens.*
Amsterdam: J. Wetstein, 1744.
Subject: Physiology, senses.
oQM/501/.L43/1744

"Nouvelle edition, corrigée, augmentée, & enrichie de figures en taille douce."

## Le Clerc, Daniel, 1652–1728, ed.

*Bibliotheca anatomica sive recens in anatomia inventorvm thesaurus locupletissimus, in qvo integre atque absolutissima totius corporis humani descriptio, ejusdémque oeconomia è praestantissimorum quorúmque anatomicorum tractibus singularibus, tum hactenus in lucem editis, tum etiam ineditis, concinnata exhibetur. Adiecta est partium omnium administratio anatomica, cum variis earundem praeparationibus curiosissimis. Digesserunt, tractatvs svpplevervnt, argumenta, notulas, & observtiones anatomico-practicas addiderunt Daniel Le Clerc & I. Iacobvs Mangetvs.*
Geneva: Joannis Anthonii Chovët, 1685.
Subject: Anatomy, medicine.
UNCATALOGUED

In 2 vols. Title page of vol. 2 mutilated.

## Le Clerc, Daniel.

*Histoire de la medecine, où l'on voit l'origine & les progrès de cet art, de siecle en siecle; les sectes, qui s'y son formées; les noms des médecins, leurs découvertes, leurs opinions, & les circonstances les plus remarquables de leur vie.*
Amsterdam: George Gallet, 1702.
Subject: History of medicine.
oR/131/.L43/1702

## Le Clerc, Daniel.

*The history of physick, or, an account of the rise and progress of the art, and the several discoveries therein from age to age. With remarks on the lives of the most eminent physicians. Written originally in French by Daniel Le Clerc, M.D. and made English by Dr. Drake and Dr. Baden.*
London: printed for D. Brown [etc.], 1699.
STC II L811
Subject: History of medicine.
Case/oR/135/.L613/1699

## Le Clerc, Sebastien, 1637–1714.

*Pratique de la geometrie sur le papier et sur le terrain. Ou par une methode nouvelle & singuliere l'on peut avec facilité & en peu de tems se perfectionner en cette science.*
Amsterdam: widow of P. de Coup & G. Kuyper, 1735.

Subject: Plane and solid geometry.
Wing/ZP/746/.C856

2 vols. Vol. 1: "Principes, axiomes"; vol. 2: "Traité de geometrie sur le terrain."

## Le Dran, Henri François, 1685–1770.

*The operations in surgery of Mons. Le Dran with remarks, plates of the operations, and a sett of instruments, by William Cheselden.*
London: printed for C. Hitch . . . & R. Dodsley, 1749.
Subject: Medicine, surgery.
UNCATALOGUED

Trans. Thomas Gataker.

## L'Écluse, Charles de, 1526–1609.

*Aliqvot notae in Garciae aromatum historiam. Eivsdem descriptiones nonnullarum stirpium, & aliarum exoticarum rerum, que à generoso viro Francisco Drake . . . & his obseruatae sunt, qui eum in longa illa nauigatione, qua proximis annis vniuersum orbem circumiuit, comitati sunt: & quorundam peregrinorum fructuum quos Londini amicis accepit.*
Antwerp: Christophorus Plantin, 1582.
Subject: Botany.
Ayer/*8.9/B7/L46/1582

Illustrations of trees (coconut, jasmine), fruits and nuts (cacao, fig), breadfruit. Text is more descriptive than medicinal or pharmaceutical.

## L'Écluse, Charles de.

*Caroli Clvsii Atrebatis rariorum aliquot stirpium, per Pannoniam, Austriam, & vicinas quasdam prouincias obseruatarum historia, qvatvor libris expressa.*
Antwerp: Christophorus Plantin, 1583.
Subject: Herbals.
Bonaparte/Coll./No./2587

Bk. 1: "Arbores & frutices"; bk. 2: "Bulbosis"; bk. 3: "Bulbosis & cornarias"; bk. 4: "Prvnella et bvgvla"; p. 758: "De medica lvteo flore"; Appendix: "Qvae ab avctore svnt observata." Index has separate title page: "Stirpivm nomenclator pannonicus. Antverpiae: Christophori Plantini, 1584."

## L'Écluse, Charles de.

*Svmmi botanici Caroli Clvsi Galliae Belgicae corographica descriptio posthvma.*
Leyden: Iacobus Marcus, 1619.
Subject: Chorography, geography.
G/465/.496

Edited by Ioachim Mors.

## Le Fèvre, Jacques, d'Étaples, 1450?–?1537.

*In hoc opere continentur totius philosophiae naturalis paraphrases, a Francisco Vatablo, insigni philosopho, ac linguae hebraicae apud Parisios professore regio, recognitae, adiectis ad literam scholijs declaratae, & hoc ordinae digestae.*
Paris: P. Vidoue, 1533.
Subject: Natural philosophy.
Case/fY/642/.A8249

Includes physicorum Aristotelis; de caelo & mundo; de generatione & corruptione; de meteorum; de anima; de sensa & sensato; de memoria & reminiscentia; de somno & vigilia; de diuinatione; de longitudine & breuitate vitae; dialogi quatuor ad metaphysicorum intelligentiam introductorij.

## Le Fèvre, Jacques, d'Étaples.

*Introductorium astronomicum theorias corporum coelestium duobus libris complectens: adiecto commentario declaratum.*
Paris: Henricus Stephanus, I., 1517.
Subject: Astronomy.
Case/fL/900/.498

MS marginalia up to leaf 17, bk. 1. "Iudoci Clichtoue Neoportuensis adiecto commentario declaratus."

## Le Fèvre, Nicolas, 1544–1612.

*Nicolai Fabri . . . Opvscvla.*
Paris: Petrus Chevalier, 1618.
Subject: Medicine.
Y/682/.M822

P. 311 (sig. Ddij): "Responce a vne qvestion sur l'epistre de Sainct Bernard. S'il est licite aux personnes religieuses d'vser de medecines."

## Legati, Lorenzo.

*Mvseo Cospiano annesso a qvello del famoso Vlisse Aldrovandi e donato alla sua patria dall' i illustrissimo signor Ferdinando Cospi.*
Bologna: Giacomo Monti, 1677.
Subject: Biology, natural history.
fM/O/.498

## Le Grand, Antoine, d. 1699.

*An entire body of philosophy, according to the principles of Renate des Cartes, in three books; I. The institution. II. The history of nature in 9 parts. III. A dissertation of the want of sense & knowledge in brute animals—in 2 parts.*
London: printed by Samuel Roycroft, 1694.
STC II L950
Subject: Elements, natural history, natural philosophy.
Case/Wing/B/239/.22

Bk. 1 includes natural theology, physics, or natural philosophy, metals & meteors; bk. 2, things dug out of the earth, plants, animals, man. Trans. Richard Blome.

## Le Grand, Antoine.

*Men without passion: or, the wise stoick, according to the sentiments of Seneca.*
London: printed for C. Harper and J. Amery, 1675.
STC II L958
Subject: Passions, psychology.
Case/B/529/.498

"Englished by G. R." Treatise 2: nature of the passions. P. 13: "It's true that physick came to succour the academia" (p. 13).

## Le Grand, Jacques, d. ca. 1425.

*Sophologium.*
[Strassburg: Adolf Rusch, ca. 1476]
H-C* 10471^ BMC (XV)I:62^ Pr. 240^ Goff M-43
Subject: Astronomy, geometry, history of science, medicine.
Inc./240

Contents include "De arismetrica" [*sic*] (6); "De geometria & eius inuentoribus" (7); "De medicina & eius inuentoribus" (10).

## Le Grand, Jacques.

*Sophologium.*
Paris: Martin Crantz, Ulrich Gering, & Michael Friburger, 1477.
H-R 10478^ Still. M35^ BMC (XV)VII:9^ Goff M-44
Subject: Astronomy, geometry, history of science, medicine.
Inc./7848.5

In 2 books. Bk. 1, chap. 16 (fol. 18): "Quomodo magice artes sunt inutiles ac improbande"; bk. 2 chaps. 6-11: history of sciences. MS marginalia passim.

## Le Grand, Jacques.

*Sophologium Magister Jacobi Magni.*
Lyon: Jehan de Vingle [1495].
H 10479^ BMC (XV)VII:311^ Pr. 8643A^ Still. M38^ Goff M-48
Subject: Astronomy, geometry, history of science, medicine.
Inc./8643A

## Le Grand, Jacques.

*Sophologium sapientie magistri Jacobi Magni.*
[Paris: Felix Baligault for Jean Richard, 1498?]
H-C 10480^ BMC (XV)VIII:177^ Still. M40^ Goff M-49
Subject: Astronomy, geometry, history of science, medicine.
Inc./8290.3

Table of contents bound in at back of vol.

* * *

*Le grand calendrier & compost des bergers composé par le berger de lagrand montaigne. Auquel sont adioustez plusieurs nouuelles tables & figures.*
Paris: Nicolas Bonfons [1589]
Subject: Almanacs, astrology, astronomy, horoscope, zodiac.
Wing/ZP/539/.B66

Has signs of the zodiac, anatomical and astronomical figures. Contents include: "La table pour cognoistre chacun iour en quel signe est la lune"; "L'anathomie de tout le corps humain"; "Des quatre complexions"; "Le reiglement du berger, comme il doit guerir ses moutons, des maladies qui peuuent leur suruenir: comme aussi se garder de tous sorts des enchanteurs & mechans sorciers."

## Leigh, Charles, 1662–1701?

*The natural history of Lancashire, Cheshire, and the Peak, in Derbyshire: with an account of the British, Phoenician, Armenian, Greek, and Roman antiquities in those parts.*
Oxford: printed for the author, etc., 1700.
STC II L975
Subject: Archaeology, local geography.
Case/fF/O24/548/.52

## Leigh, Charles.

*The natural history of Lancashire.*
[Another copy]
Case/fF/O245/.487

## Leigh, Valentine, fl. 1652.

*The moste profitable and commendable science, of surueying of lands, tenementes, & hereditamentes: drawen & collected by the industrie of Valentine Ligh, whereunto is also annexed by the same aucthor a right necessary treatise of the measurig [sic] of all kindes of landes, . . . and that as well by certaine easie, and compendious rules.*
London: printed by John Windet for Robert Dexter, 1592.
STC 15419
Subject: Geometry, surveying.
Case/H/94/.5

## Le Loyer, Pierre, Sieur de la Brosse, 1550–1634.

*Discovrs, et histoires des spectres, visions et apparitions des esprits, anges, demons, et ames, se monstrans visibles aux hommes. Divisez en huict livres. . . . Avssi est traicté des extases et rauissemens. . . . plus des magiciens & sorciers.*
Paris: Nicolas Bvon, 1605.
Subject: Demons, magic, occult, witchcraft.
B/893/.496

## Le Loyer, Pierre, Sieur de la Brosse.

*IIII livres des spectres ov apparitions et visions d'esprits, anges et demons se monstrans sensiblement aux hommes.*
Angers: Georges Nepueu, 1586.
Subject: Demons, magic, occult, witchcraft.
Case/B/893/.498

## [Le Loyer, Pierre, Sieur de la Brosse]

*A treatise of specters or straunge sights, visions and apparitions appearing sensibly vnto men. Wherein is delivered, the nature of spirites, angels, and divels: their power and properties: as also of witches, sorcerers, enchanters, and such like. Newly done out of French into English.*
London: printed by Val. S. for Matthew Lownes, 1605.
STC 15448
Subject: False perceptions of apparitions, supernatural.
Case/B/893/.5

Epistle of the French author signed Peter de Loire.

## Lémery, Louis, 1677–1743.

*A treatise of all sorts of foods, both animal and vegetable: also of drinkables: giving an account how to chuse the best sort of all*

*kinds; of the good and bad effects they produce; the principles they abound with; the time, age and constitution they are adapted to. Wherein their nature and use is explain'd according to the sentiments of the most eminent physicians and naturalists antient and modern.*
London: printed for T. Osborne, 1745.
Subject: Diet.
3a/600

Translated by D. Hay, M.D. 1. Foods made of vegetables or plants; 2. Foods prepared of animals; 3. Drinkables. P. 154, "Of cloves: You ought to chuse those that are large, plump, fresh, easy to be broken, and of a pleasant aromatic taste and smell. They fortify the parts, stop vomiting, resist the malignity of the humours, ease the tooth-ace [sic], help digestion, and sweeten the breath."

## Lemnius, Levinus, 1505-1568.

*Levini Lemnii medici Zirizaei de habitv et constitvtione corporis, quam Graeci* [Greek word] *triuiales complexionem vocant, libri II omnibus secunda valetudo curae est, apprimè necessarij, ex quibus cuique procliue erit corporis sui conditionem, animíque motus, ac totius conseruandae sanitatis rationem adamussim cognoscere.*
Erffurdt: Esaias Mechlerus, 1582.
NLM 2766
Subject: Complexion, humors, medicine.
UNCATALOGUED

## Lemnius, Levinus.

*De miracvlis occvltis naturae libri IIII. Item de vita cvm animi et corporis incolvmitate recte instituenda, liber vnus.*
Frankfurt: printed by Nicolaus Hofmann for Ioannes Rhodius, 1604.
Subject: Humors, medicine (preventive), monsters, occult, supernatural.
Case/–B/85/.498

Bk. 1: obstetrics, gynecology; bk. 2: humors (melancholic, maniacal, phrenetic) epilepsy, newborn syphillis, physiognomy, astrology, alchemy.

## Lemnius, Levinus.

*Occulta naturae miracula. Wunderbarliche geheimnisse der natur in des menschen leibe und seel auch in vielen andern natürlichen dingen als steinen ersst gewechs und thieren. Allen frommen hausswirthen verstendigen hausfrawen, fleissigen naturkündigern guten hausärtzten liebhabern der gesundheit und gemeinem vaterland zum besten-nicht allein aus dem Latein in Deutsche sprach gebracht, sondern auch zum dritten mal vermehret und eines grossen theils von newes selbs geschrieben durch Iacobvm Horstivm.*
Leipzig [Michael Lantzenberger for Valentin Vögelin], 1592.
Subject: Humors, medicine, supernatural.
Case/Q/.498

## Lemnius, Levinus.

*Les occvltes merveilles et secretz de natvre, avec plvsievrs enseignements des choses diuerses, tant par raison probable, que par coniecture artificielle: exposées en deux liures, de non moindre plaisir que profit au lecteur studieux.*
Paris: Galiot du Pré, 1574.
Subject: Humors, medicine, supernatural.
Case/Q/.496

Trans. Jacques Gohory. Bk 1, chap. 16: "Les humeurs & les viandes manifestement changent la disposition du corps & l'estat de l'ame. & que de là procede la source des passions, & les remors de conscience; incidemment, quel est l'effet de la melancholie, & comme on y peut remedier."

## Lemnius, Levinus.

*De gli occulti miracoli, & uarii ammaestramenti delle cose della natura, con probabili ragioni & artificiosa congiettura confermati.*
Venice: Lodouico Auanzo, 1563.
Subject: Humors, medicine, supernatural.
Case/B/85/.497

Cap. 5, car. 12: "Dello strano appetito delle donne grauide, & del desiderio che elle hanno di molte cose, lequali essendo loro negate, stanne in pericolo di conciarsi, & diperdere"; bk. 2, cap. 46, car. 146: "Come si possano guarire i uaiuoli, & le rosole de' bambini, & quai siano le cose appropriate a questi mali"; cap. 48, car. 148: "Che le tempeste di mare si possono antiuedere col toccare con mano li acquamirina."

## Lemnius, Levinus.

*The touchstone of complexions. Generally applicable, expedient and profitable for all such, as be desirous and carefull of theyr bodyly health. Contayning most easy rules & ready tokens, whereby euery one may perfectly try, and throughly knowe, as well the exacte state, habite, disposition, and constitution of his body outwardly: as also the inclinations, affections, motions, and desires of his mynde inwardly.*
London: Thomas Marsh, 1581.
STC 15457
Subject: Humors, medicine, supernatural.
Case/Q/.499

Englished by Thomas Newton.

## [Lemos, Antonio Correia de] 1680–1747?

*A fenix das tempestades . . . um discurso sobre a origem dos ventos, composta, e ordenada por hum anonymo.*
Lisbon: Joseph Antonio da Sylva, 1732.
Subject: Astrology.
Greenlee/4504/P855

A list of astrological occurrences through history or of occurrences with astrological components.

## Le Moyne de Margues, Jacques, d. 1588.

*Brevis narratio eorum qvae in Florida Americae provincia Gallis acciderunt secunda in alliam nauigatione, duce Renato de Laudonniere classis praefecto: Anno MDLXIIII qvae est secvnda pars Americae. . . . Auctore Iacobe le Moÿne, cui cognomen de Morgues, Laudonnierum in ea navigatione sequnto.*

Frankfurt: Joannis Wechel for Theodor de Bry, 1591.
Subject: Anthropology, discovery, exploration.
Ruggles/No. 220

Second title page: Indorum Floridam provinciam inhabitantium eicones.

## Lencker, Hans, d. 1585.

*Perspectiva hierinnen auffs kürtzte bescrieben mit exempeln eröffnet vnnd an tag gegeben wird ein newer besonder kurtzer doch gerechter vnnd sehr liechter weg wie allerley ding es sehen corpora gebew gebem oder was möglich zuerdencken vnd in grund zulegen ist verruckt oder unuerruckt ferner in die perspectyf gebracht werden mag.*
Nuremberg: Dietrich Gerlatz, 1571.
Subject: Perspective.
Case/Wing/fZA/3/.5

BOUND WITH Euclides, *Die sechs erste bücher* [1562], q.v.

## Lencker, Hans.

*Perspectiva, in welcher ein leichter weg allerley ding es sehen corpora gebew und was müglich in grund zulegen ist veruckt oder unveruckt durch gar geringe instrument in die perspectiv zubringen gezeiget wirdt.*
Vlm: Johann Meder for Stephan Michelspachers, 1616.
Subject: Perspective.
Case/Wing/fZA/3/.501

## Lenglet Dufresnoy, Nicolas, 1674–1755.

*Histoire de la philosophie hermetique. Accompagnée d'un catalogue raisonné des ecrivains de cette science. Avec le véritable Philalethe, revû sur les originaux.*
The Hague: Pierre Gosse, 1742.
Subject: Alchemy, Hermeticism.
B/8633/.499

In 3 vols. Vol. 1: chronology of most celebrated Hermetic philosophers; vol. 2: alchemy ("Transmutations métalliques"); vol. 3: catalogue.

## Lenglet Dufresnoy, Nicolas.

*Methode pour etudier la geographie dans laquelle on donne une description exacte de l'univers, tirée des meilleurs auteurs, & formée sur les observations des messieurs de l'Académie Royale des Sciences.*
Amsterdam: at the expense of the Company, 1718.
Subject: Geography.
Ayer/290/L485/1718

In 4 vols. Vol. 1, bk. 1, chap. 9: "De l'usage du globe"; vols. 2 and 3: Europe, Asia, Africa, America; vol. 3, bk. 6: "Des continens inconnus ou connus seulement le long des côtes"; vol. 4: ancient geography.

## Lenglet Dufresnoy, Nicolas.

*Methode pour etudier la geographie. Où l'on donne une description exacte de l'univers, formée sur les observations de l'Académie Royale des Sciences, & sur les auteurs originaux.*
Paris: Rollin son & De Bure, senior, 1741–42.
Subject: Geography.
G/117/.499

3d ed. 7 vols. in 8. Vol. 2: geography for children. Only vols. 1 and 7 carry imprint date 1742.

## Le Normant, Iean, Sieur de Chiremont.

*Histoire veritable et memorable de ce qvi c'est passé sovs l'exorcisme de trois filles possedées és païs de Flandre, en la descouuerte & confession de Marie de Sains, soy disant princesse de la magie; & Simone Dourlet complice, & autres. Ov il est avssi traicté de la police du sabbat, & secrets de la synagogue dos magiciens & magiciennes.*
Paris: Nicolas Buon, 1623.
Subject: Exorcism, possession, witchcraft.
B/88/.503

With *Les confessions de Didyme sorciere, de Maberthe sorciere, de Loyse seduite par le diable; conuersion admirable d'vn ieune homme magicien* (has separate pagination but no title page).

## Le Normant, Jean, sieur de Chiremont.

*Vera ac memorabilis historia de tribvs energvmenis in partibvs Belgii, et de quibvsdam aliis magiae complicibvs. De fine mundi. . . . De vocatione magorum & magarum in genere, & in particulari.*
Paris: Nicolas Buon, 1623.
Subject: Exorcism, possession, witchcraft.
Case/B/88465/.492

## Leo, Hebraeus, d. 1535.

*De amore, dialogi tres.*
Venice: Franciscus Senensis, 1564.
Subject: Astrology, natural philosophy.
Case/Y/712/.L56

"Diuersae astrologorum sententiae in planetis disponendis & collocandus" (car. 123). Also sententiae of Averroës, Avicenna, Algazelus, Raby, Moses.

## Leo, Hebraeus.

*Dialoghi di amore, composti per Leone medico, di natione Hebreo et dipoi fatto Christiano.*
[Venice: Aldus sons, 1545.]
Subject: Astrology, natural philosophy.
Case/*Y/712/.L556

## Leo, Hebraeus.

*Dialoghi di amore.* [Another ed.]
Venice: Aldus sons, 1552.
Subject: Astrology, natural philosophy.
Case/Y/712/.L557

## Leo, Hebraeus.

*Dialoghi di amore di Leone Hebreo medico. Di nvovo corretti et ristampati.*
Venice: Domenico Giglio, 1558.
Subject: Astrology, natural philosophy.
Case/Y/712/.L5572

## Leo, Hebraeus.

*Dialogi d'amore di maestro Leone medico Hebreo.*
[Rome: Antonio Blado d'Assola, 1535]
Subject: Astrology, natural philosophy.
Case/Y/712/.L554

## Leonicus Thomaeus, Nicolaus, 1456–1531?

*N. Leonici Thomaei de animorvm essentia dialogvs.*
[Venice: Io. Ant. Sabio and brothers, 1530]
Subject: Immortality of the soul.
Case/B/511/.502

## Leonicus Thomaeus, Nicolaus.

*Nicolai Leonici Thomaei opvscvla nvper in lvcem aedita qvorvm nomina proxima habentur pagella.*
[Venice: Bernardinus Vitalis, Venetus, 1525]
Subject: Mechanics, motion.
Case/Y/682/.L554

Note on front flyleaf: "Most important his Latin version with commentary and woodcut illustrations of Aristotle's Mechanics." Contents include "Paraphrasis in commentariolvm Aristotelis de animalivm motione"; "Conversio mechanicarvm qvaestionum Aristotelis cvm figvris et annotationibvs qvibusdam"; "Procli lytii explicatio Platonis ex Timaeo vbi de animorvm generatione agitvr, in Latinvm conversa, cvm nvmerorvm harmoniarvmque mvltiplici figvra." Leaf 41: dentist's forceps with scientific explanation of its action.

## Leopold of Austria, 1176–1230.

*Compilatio Leupoldi ducatus Austrie filij de astrorum scientia decem continens tractatus.*
Augsburg: Erhard Ratdolt, 1489.
H-C *10042^ BMC (XV)II:382^ Pr. 1879^ Still. L161^ Goff L-185
Subject: Astrology, astronomy, meteorology, zodiac.
Inc./*1879

Said to have early (or first?) example of 2-color printing. See Houzeau and Lancaster, vol. 1, pt. 1, sec. 2, p. 764, no. 4702.

## Leopold of Austria.

*Compilatio Leupoldi ducatus Austrie filij de astrorum scientia decem continentis tractatus.*
[Venice: Melchior Sessa & Petrus de Rauanis, partners, 1550]
Subject: Astrology, astronomy, meteorology, zodiac.
Wing/ZP/535/.55

* * *

*Le plaisant discovrs d'vn medecin savoyart, emprisonné pour auoir donné aduis au Duc de Savoye de ne croire son deuin.*
[N.p.] 1600.
Subject: Astrology.
Bonaparte/Coll./No. 3973

In verse.

* * *

*Le premier livre de la description philosophale de la natvre et condition des animaux, tant raisonnables que brutz, auec le sens moral comprins sus le naturel & condition d'iceux: ensemble plusieurs augmentations de diuerses & estranges bestes, outre la precedente impression.*
Paris: Magdaleine Boursette, 1554.
Subject: Animal symbolism, emblems, natural history.
Case/*W/1025/.71

## Levret, André, 1703–1780.

*Observations sur la cure radicale de plusieurs polypes de la matrice, de la gorge et du nez opérée par de nouveaux moyens inventés par M. Levret.*
Paris: Delaguette, 1749.
Subject: Pathology, surgery, tumors.
oRC/254/.L48/1749

In 2 parts: gynecological surgery; "polyps" in the head.

* * *

*Lex talionis sive vindiciae pharmacoporvm: or a short reply to Dr. Merrett's book; and others, written against the apothecaries: wherein may be discovered the fravds and abvses committed by doctors professing and practising pharmacy.*
London: printed by Moses Pitt, 1670.
STC II S6055
Subject: Apothecary vs. physician, medical specialties.
Case/I/045/.88/no. 3

Binders title: *Tracts on physick, 1665–70.* BOUND WITH tracts by Goddard and Merrett, qq.v. Written by Henry Stubbs (or Stubbe)?

## Leybourn, William, 1626–1700?

*The compleat surveyor: containing the whole art of surveying of land, by the plain table, the odolite, circumferentor, and peractor: . . . together with the taking of all manner of heights and distances. . . . Hereunto is added, the manner how to know whether water may be conveyed from a spring head to any appointed place or not, and how to effect the same: with whatsoever else is necessary to the art of surveying.*
London: printed by R. & W. Leybourn for E. Brewster & G. Sawbridge, 1653.
STC II L1907
Subject: Geometry, instrumentation, surveying.
Case/5A/386

## Libavius, Andreas, ca. 1560–1616.

*D. O. M. A. Wolmeinendes bedencken von der fama vnd confession der brüderschafft dess rosen creutzes, eine uniuersal reformation und umbkehrung der gantzen welt.*
Erffurdt: Johann Rhöbock, 1616.
Subject: Rosicrucians.
K/975/.036

BOUND WITH *Fama fraternitatis*, q.v., and other Rosicrucian tracts, e.g., *Confessio fraternitatis; Send brieff oder bericht—Ivlianvs de Campis; Bericht von einem fratre dess ordens RC.*

## Liceti, Fortunio, 1577–1657.

*Hieroglyphica, siue antiqva schemata gemmarvm anvlarivm, qvaesita moralia, politica, historica, medica, philosophica & sublimiora.*
Padua: Sebastiani Sardi, 1653.
Subject: Emblems, hieroglyphics on signet rings, symbols.
Case/W/1025/.504

## Lichtenberger, Johann, 15th cent.

*Prognosticatio Latina anno lxxxviii. ad magnum coniunctionem Saturni & Jouis que fuit anno lxxxiiii. ad eclipsim solis anni sequentis.*
[Mainz: Jakob Meydenbach, 1492.]
H 10082*^ Pr. 161^ Schreiber 4500^ Schramm XV, p. 7^ Goff L-205
Subject: Astrology.
Inc./f161

Cap. 11: "Quomodo celestis influxus habet alterare & imutare corpora & spiritus virtutum & inducere egritudines & mortalitates"; cap. 12: "Quo effectus super celestium & applicationes astrorum durant multis annis depost."

## Lichtenberger, Johann.

*Prognosticatio Ioannis Liechtenbergers, qvam olim scripsit super magna illa Saturni ac Iouis coniunctione, quae fuit anno MCCCCLXXXIIII. praeterea ad eclipsim solis anni sequentis videlicet LXXXV. in annum adhuc usque durans MDLXVII. iam iterum, mendis quibusdam haud modicis sublatis.*
[Cologne, 1526.]
Subject: Astrology.
Case/*B/8635/.496

## Lichtenberger, Johann.

*Prognosticatio Iohannis Liechtenbergers iam denuo sublatis mendis, quibus scatebat pluribus, quam diligentissima excussa.*
[Cologne: Petrus Quentel, 1528.]
Subject: Astrology.
Wing/ZP/547/.W842

See Merlos, Kölnische künstler, col. 987–89 (Holzschnitte no. 1–38).

## Lilio, Zaccaria, bp., d. ca. 1522.

*Orbis brevarium.*
Naples: Ayolfus de Cantono, 1496.
H-C *10102^ BMC (XV)VI:874^ Pr. 6744^ Still. L191^ Goff L-219
Subject: Geographical dictionary.
Inc./6744

## Lilio, Zaccaria.

*De origine et laudibus scientiarum.*
Florence: Francesco Buonaccorsi for Piero Pacini, 1496.
H-C 10103^ BMC (XV)VI:675^ Pr. 6316^ Still. L193^ Goff L-221
Subject: Epistemology, theory and history of knowledge.
Inc./6316

"In hoc volvmine continentvr hi libri. 1. De origine; 2. Contra antipodes; 3. De miseria hominis & contemptu mundi; 4. De generibus uentorum; 5. Vita Caroli Magni." Leaves [b7v] and [b8]: arithmetic, geometry; sig. ci: astrology; ciii[verso] medicine; ciiii[verso] herbaria.

## Lilly, William, 1602–1681.

*Anglicus, peace or no peace, 1645. An exact ephemeris of the daily motions of the planets; with an easie introduction to the use thereof. Monethly-observations. A table of houses, and explanations thereof.*
London: printed by J. R. for John Partridge and Humphrey Blunden, 1645.
STC II L2207
Subject: Almanacs, astrology, astronomy.
Case/B/8635/.4976

Contains two other tracts (continuously paginated): "An ephemeris for 1645," London: printed by J. Raworth, 1645; "The errours of Master Whartons prognostication. 1645. Refelled, refuted, and retorted"; p. 40: "William Lilly his reply to Master Whartons objections against supernaturall sights &c. Published at Oxford, 1644."

## Lilly, William.

*Anima astrologiae: or, A guide for astrologers. Being the considerations of the famous Guido Bonatus faithfully rendered into English. As also the choicest aphorisms of Cardans seaven segments . . . with a new table of the fixed stars. . . . A work most useful and necessary for all students, and recommended as such to the sons of art.*
London: printed for B. Harris, 1676.
STC II L2208
Subject: Astrology.
Case/B/8635/.4978

## Lilly, William.

*Annus tenebrosus, or the dark year. Or astrologicall judgements upon two lunar eclipses and one admirable eclips of the sun, all visible in England, 1652. Together with a short method how to judge the effects of eclipses.*
London: printed for the Company of Stationers, and H. Blunden, 1652.
STC II L2209
Subject: Astrology, astronomy.
Case/B/8635/.498

Title page and pp. 57 and 58 are photocopies. Has a section at back, "Astrologicall aphorismes."

## Lilly, William.

*An astrologicall prediction of the occurrances in England, part of the yeers 1648, 1649, 1650. Concerning these particulars viz. 1. The effects depending on the late conjunction of the two malevolent planets, Saturn and Mars. . . . 6. What may succeed the apparition of three suns in Lancashire, seen of many, the 28 Feb. last.*
London: printed by T. B. for John Partridge and Humfrey Blunden, 1648.
STC II L2211

Subject: Astrology, prognostication.
Case/B/8635/.51

### [Lilly, William]

*Catastrophe mundi: or Merlin reviv'd, in a discourse of prophecies & predictions. . . . with Mr. Lilly's hieroglyphicks exactly cut; By a learned pen.*
London: J. How [etc.], 1683.
STC II L2214
Subject: Astrology.
Case/B/863/.154

Has a section of woodcuts, emblems (pp. 23–59) followed by "Notes or observations on the foregoing figures, and the nature of hieroglyphics in general."

### Lilly, William.

*Christian astrology modestly treated of in three books. First the use of an ephemeris, the erecting of a scheam of heaven; nature of the twelve signs of the zodiack, of the planets; with a most easie introduction to the whole art of astrology. 2nd how to judge or resolve all manner of questions contingent unto man, viz, of health, sicknesse, riches, marriage, preferment, journies, &c. 3rd whereby to judge upon nativities.*
London: printed by Tho. Brudenell for John Partridge and Humph. Blunden, 1647.
STC II L2215
Subject: Astrology.
Case/B/8635/.511

Has catalogue of astrology books at end.

### Lilly, William.

*A collection of ancient and moderne prophesies concerning the present times, with modest observations thereon. The nativities of Thomas, earle of Strafford, and William Lavd.*
London: printed for John Partridge and Humphrey Blunden, 1645.
STC II L2217
Subject: Astrology.
Case/B/8635/.513

BOUND WITH his *Englands propheticall Merline* and *Merlinus Anglicus junior*, qq.v.

### Lilly, William.

*An easie and familiar method whereby to iudge the effects depending on eclipses, either of the sun or moon.*
London: printed for the Company of Stationers, and H. Blunden, 1652.
STC II L2219
Subject: Astrology.
Case/L/9651/.505

Includes "Part of the second book of Ptolomey, concerning the judging of eclipses."

### Lilly, William.

*Englands propheticall Merline, foretelling to all nations of Europe untill 1663 the actions depending upon the influence of the conjunction of Saturn and Jupiter 1642/3. The progress and motion of the comet 1618. under whose effects we in England, and most regions of Europe now suffer. What kingdomes must yet partake of the remainder of the influence, viz. of war, plague, famine, &c. The beginning, and end of the watry trygon: an entrance of the fiery triplicity, 1603. The nativities of some English kings.*
London: printed by John Raworth for John Partridge, 1644.
STC II L2221
Subject: Astrology.
Case/B/8635/.513

BOUND WITH his *Merlinus Anglicus, junior*, q.v.

### Lilly, William.

*Lillies prophetic occurences or, an extract of some passages in Mr. Lilies astrological judgment for the year 1677.*
[London: printed for L. Curtis, 1682.]
STC II L2241
Subject: Astrology, prognostication.
Case/fF/455/.88/no./93

Astrological explanation of historic events, e.g., fire of London (1666), plague (1665), and astrological predictions. Folio sheet in bound vol., *Tracts, 1648–1716*."

### Lilly, William.

*Merlinus Anglicus junior: The English Merlin revived; or, his prediction upon the affaires of the English common-wealth, and of all or most kingdomes of Christendome this present yeare, 1644. [astrologically handled]*
London: printed by R. W. for T. V., 1644.
STC II A1919
Subject: Almanacs, astrology, prognostication.
Case/B/8635/.513

BOUND WITH his *Englands propheticall Merline*, q.v.

### Lilly, William.

*Merlinus Anglicus junior: The English Merlin revived; or, a mathematicall prediction upon the affairs of the English common-wealth.*
London: printed by R. W. for T. V., 1644.
STC II A1919A
Subject: Almanacs, astrology, prognostication.
Case/B/8635/.5131

2d ed.

### Lilly, William.

*Monarchy or no monarchy in England. . . . Aenigmaticall types of the future state and condition of England.*
London: printed for H. Blunden, 1651.
STC II L2228
Subject: Astrology, prognostication.
Case/F/4552/.51

"Severall observations upon the life and death of Charles late king of England."

### Lilly, William.

*Mr. Lillie's new prophecy. or, sober predictions of peace between the French and the Dutch, and their allies, speedily to be*

*concluded. Drawn from some astrological considerations of the general assembly of planets happening in Sagittary, in the month of December, this present year 1675, &c.*
[London] printed for John Clarke [1675].
STC II L2232
Subject: Astrology, prognostication.
Case/B/8635/.5135

## Lilly, William.

*The starry messenger; or, an interpretation of that strange apparition of three suns seen in London, 19 Novemb. 1644. being the birth day of King Charles. The effects of the eclipse of the sun, which will be visible in England, 11 August 1645, whose influence continues in force from January 1646. to Decemb. 1647. almost two whole yeares; and cannot but be the fore-runner of some extraordinary mutation in most common wealths of Europe, but principally in England. With an answer to an astrologicall judgement.*
London: printed for John Partridge & Humphry Blunden, 1645.
STC II L2245
Subject: Astrology, astronomy.
Case/B/8635/.514

## Lilly, William.

*Svpernatvrall sights & apparitions seen in London June 30 1644. interpreted. With a mathematicall discovrse of the now imminent conjunction of Iupiter and Mars, 26 Iuly, 1644. the effects which either here or in some neere counties from thence may be expected.*
London: printed for T. V., 1644.
STC II L2249
Subject: Astrology, astronomy.
Case/B/8635/.5142

## Lilly, William.

*The vvorlds catastrophe, or, Europes many mutations until, 1666. . . . Government of the vvorld under God by the seven planetary angels; their names, times of government. An exact type of the 3 suns seen in Cheshire & Shropshire 3 April 1647. Their significance and portent, astrologically handled.*
London: printed for John Partridge and Humphrey Blunden, 1647.
STC II L2252
Subject: Astrology, astronomy.
Case/B/8635/.515

## Lipen, Martin, 1630–1692.

*Bibliotheca realis universalis omnium materiarum, rerum et titulorum. . . . seu speciales bibliothecas theologicam, juridicam, medicam, et philosophicam divisa, ordine alphabetica disposita.*
Frankfurt am Main: Johannes Friderich, 1679–85.
Subject: Medicine, bibliography.
fZ/9/.508

6 vols. Vol. 4 (1679): *Bibliotheca realis medica, omnium materiarum, rerum, et titulorum, in universa medicina occurrentivm. Ordine alphabetico sic disposita, vt primo statim intvitv tituli, et sub titulis autores medici.*

## L'Obel, Matthias de, 1538–1616.

*Plantarvm, sev stirpivm historia.*
Antwerp: Christophorus Plantin, 1576.
Nissen, p. 72
Subject: Botany, herbals, plant classification.
Wing/fZP/5465/.P7018

Illustrations of flowering plants, grasses, marine flora, and trees.

## Locke, John, 1632–1704.

*An abridgement of Mr. Locke's essay concerning humane understanding.*
London: printed for A. & J. Churchill, & Edward Castle, 1696.
STC II L2735
Subject: Natural philosophy.
Case/B/245/.51041

Starts with bk. 2, ideas; chap. 9: of perception, the senses; chap. 16: of numbers; bk. 3, words and language; bk. 4, knowledge in general.

## Locke, John.

*A collection of several pieces.*
London: printed by J. Bettenham for R. Francklin, 1720.
Subject: Natural history, natural philosophy.
Case/B/245/.5087

P. 179: elements of natural philosophy (matter and motion, air and atmosphere, vegetables, plants, animals, 5 senses). P. 249: "A letter to Mr. [Henry] Oldenburg secretary of the Royal Society" on poisonous fish in the Bahamas.

## Locke, John.

*An essay concerning humane understanding.*
London: printed by Eliz. Holt for Thomas Basset, 1690.
Subject: Natural philosophy, experimental and rational science.
STC II L2738
Case/B/245/.5104

In 4 books. Observations on sleeping, dreaming, of memory, space, duration, cause and effect.

## Locke, John.

*An essay concerning humane understanding.*
London: printed for Awnsham and John Churchil, & Samuel Manship, 1694.
STC II L2740
Subject: Natural philosophy, experimental and rational science.
Case/fB/245/.510405

2d ed. In 4 books. With large additions.

## Locke, John.

*An essay concerning humane understanding.*
London: printed for Awnsham & John Churchil, 1700.
STC II L2742
Subject: Natural philosophy, experimental and rational science.

Case/fB/245/.510415

4th ed. In 4 books. With large additions.

## Locke, John.

*A letter to Edward, Ld Bishop of Worcester, concerning some passages relating to Mr. Locke's essay of humane understanding: in a late discourse of his lordships in vindication of the Trinity.*
London: printed for A. & J. Churchill, 1697.
STC II L2748A
Subject: Epistemology, operations of the mind.
Case/B/245/.51058

2 vols. Binders title: *Locke's letters, v. 1.*

## Lomazzo, Giovanni Paolo, 1538–1600.

*Idea del tempio della pittvra . . . nella quale egli discorre dell' origine, & fondamento delle cose contenute nel suo trattato dell' arte della pittura.*
Milan: Paolo Gottardo Pontio [1590]
Subject: Proportion, skiagraphy (projection of shadows).
Case/W/00/.51

Perspective, proportion, color, light. Lomazzo studied with Raphael andMichaelangelo.

## Lomazzo, Giovanni Paolo.

*A tracte containing the artes of curious paintinge carvinge & buildinge written first in Italian by Jo Paul Lomatius painter of Milan and Englished by R. H. student in physik.*
[Oxford: Ioseph Barnes for R. H(aydocke), 1598]
STC 16698
Subject: Proportion, skiagraphy.
Case/fW/00/.52

In 5 books. Contains a section on actions and gestures, passions of the mind, humors, and a "Discovrse of the artificiall beauty of women."

## [Lonitzer, Adam] 1528–1586.

[*Kreutterbuch.*]
[Frankfurt: Christian Egenolph, 1546]
Subject: Herbals, materia medica, natural history.
Case/fM/0/.522

Hand-colored frontispiece portrait (of Lonitzer?), no title page. One index in Latin and one in German, plus "Index zu allen kranckheiten und gebresten artznei und rath inn eil zufinden; an welchem blat und under welchem nebenge setzten büchstaben." Subjects include distilling, animals, birds (including phoenix), and fish, each group arranged alphabetically. Bk. 1: "Von naturen eygenschafften wirckungen lebendiger creaturen und thier"; bk. 2: "Folgt hernach von natur eygenschafft unnd wirckung der edelgesteyn ertz metalls, erden und gummi. Von polierung allerhand edelgestein rechter kunst"; bk. 3: "Folget das kreuterbuch von allem erdgewechs, pflantzen unnd kreutern." MS notes on back flyleaf, recto and verso.

## Lonitzer, Adam.

[*Kreuterbuch*] *Kunstliche conferteytunge der baume, standen hecken kreuter, getryde, gewürtze. mit eigentlicher beschreibung derselben namen in sechserley sprachen, nemlich Grieschisch, Lateinisch, Italiänisch, Frankösisch, Teutsch, und Hispanisch und derselben natürlicher krafft und wirckung. Sampt künstlichem und artlichem bericht dess distillierens. Item von fürnembsten gethieren der erden, vögelin, unnd fischen. Dessgleichen von metallen, ertze, edelgesteinen, gummt und gestandenen safften. Jetzo auffs fleissigst zum letztmal von newen ersehen und durchauss an vielen orthen gebessert.*
Frankfurt: Johann Saur for Christian Egenolff [1598].
Subject: Herbals, materia medica, natural history.
Case/6A/176

Some illustrations hand colored. MS notes throughout. In 3 books. Bk. 1: Distilling, gardens, trees; bk. 2: Fruit; bk. 3: Natural history.

## Lopez, Gregorio, 1542–1596.

*Tesoro de medicinas, para diversas enfermedades. Dispvesto por el venerable Gregorio Lopez. Añadido, corregido, y enmendado, con notas de los doctores Mathias de Salazar Mariaca, y Joseph Diaz Brizuela, con tres indices muy copiosos de diversos achaques, de yervas, y simples, y de sus virtudes, y calidades.*
Madrid, 1708.
Subject: Materia medica, medicine.
Ayer/*657.4/M4/L86/1708

3d impression. Includes "Libro de medicina por orden alphabetico."

## Lucretius Carus, Titus, 96?–55 B.C.

*An essay on the first book of T. Lvcretivs Carvs de rerum natura. Interpreted and made English verse by J. Evelyn esq.*
London: Gabriel Bedle & Thomas Collins, 1656.
STC II L3446
Subject: Natural history, natural science.
Case/Y/672/.L906

The argument: "The poet invocates Venus [Goddess of nature or rather, Nature itself]." In the persons of Venus and Mars the poet speaks of generation and corruption. In Latin and English verse on facing pages.

## Lucretius Carus, Titus.

*De rerum natura.*
Frankfurt: heirs of Andrea Wechel, 1583.
Subject: Natural history, natural science.
Case/Y/672/.L905

Body and soul, sun and moon, elements, earth, acoustics. Bk. 5 is natural history.

## Ludovico degli Arrighi, Vicentino, fl. 1522.

*Essemplario de scrittori il qvale insegna a scrivere diuerse sorti di lettere. Col modo di temprare le penne secondo le lettere, & conoscer la bontà di quelle, e carte, e far inchiostro, verzino, cenaprio, & vernice, con molti altri secreti pertinenti alli scrittori . . . con una ragione d'abbaco breue, & utilissima.*

Rome: Valerio & Luigi Dorici brothers, 1557.
Subject: Arithmetic, ink and varnish making.
Wing/ZW/535/.L96

Colophon: "Stampata in Roma per inuentione di Ludouico Vicentino. Resvrrexit Vgo da Carpi."

### Ludovico degli Arrighi, Vicentino.

*La operina di Ludouico Vicentino, da imparare di scriuere littera cancellarescha.*
Rome, 1522.
Subject: Ink making, writing instruments.
Wing/ZW/535/.L9611

With "Il modo de' temperare le penne con le uarie sorti de littere ordinato per Ludouico Vicentino" (Rome, 1523).

### Ludovico degli Arrighi, Vicentino.

*La operina di Ludouico Vicentino da imparare di scriuere littera cancellarescha . . et una bellissima ragione di abbacho molto necessario à chi imparara à scriuere & fare conto Vgo scr.* [i.e., Ugo da Carpi scrisse]
[Rome: Carpi, 1525]
Subject: Arithemetic, writing instruments.
Wing/ZW/535/.L9613

Appended: Angelus Mutinen, *Ragione di abacho.*

### Luigini, Luigi, b. 1526, ed.

*Aphrodisiacus, sive de lue venerea; in duos tomos bipartitus, continens omnia quaecumque hactenus de hac re sunt ab omnibus medicis conscripta ubi de ligno Indico, salsa perilla, radice Chynae, argento vivo; caeterisque rebus omnibus ad hujus luis profligationem inventis, diffusissima tractato habetur. Opus hac nostra aetate, qua morbi Gallici vis passim vagatur, apprimé necessarium: ab excellentissimo Aloysio Luisino.*
Leyden: Johan. Arnold. Langerak and Johan. & Herm. Verbeek, 1728.
Subject: Aphrodisiacs, medicine, venereal disease.
UNCATALOGUED

### Lull, Ramòn, ca. 1235–1315.

*Raymvndi Lvllii Opera ea qvae ad inventam ab ipso artem vniversalem scientiarvm artivmqve omnium breui compendio, firmaque memoria apprehendendarum, locupletissimaque vel oratio ex tempore pertractandarum, pertinent.*
Strassburg: for Lazarus Zetzner, 1598.
Subject: Alchemy, mnemonics, mysticism, universal knowledge.
Case/Y/722/.L95

Quaritch catalogue 1070, no. 81: 1st ed. of 1st collected works. [Typed note on front flyleaf: "Mystic alchemist. Inventor of Ars magna Lulli or the Lullian art of universal knowledge. With interpretations by Bruno and Cornelius Agrippa."]

### Lull, Ramòn.

*Opera ea quae ad inventam ab ipso artem universalem, scientiarum artiumque omnium brevi compendio firmaque; memoria apprehendendarum, locupletissimaque vel oratione ex tempore pertractandarum, pertinent.*
Strassburg: for the heirs of Lazarus Zetzner, 1651.
Subject: Alchemy, mnemonics, mysticism, universal knowledge.
Case/Y/722/.L952

In 6 books with 6 books of interpretation. 1: Ars brevis; 2: De auditu kabbalistico seu kabbala; 3: Duodecim principia philosophiae Lullianae; 4: Dialectica seu logica; 5: Rhetorica; 6: Ars magna. Interpret Jordanus Bruno de specierum scrutinio; Commentaria Agrippae in artem brevem Lullian. De lampade combinatoria Lulliana (by Bruno).

### Lull, Ramòn. [Spurious and doubtful works]

*Opusculum Raymundinum de auditu kabbalistico siue ad omnes scientias introductorium. Incipit libellus de kabbalistico auditu in via Ramundi Lullii.*
Paris: Aegidius Gorbinus, 1578.
Subject: Cabala, mathematics.
Case/-C/13/.528

P. 50: "Mathematicheitas siue mathematica est actus mathematici: ratione cuius mathematicus non agit nisi mathematicum."

### Narcissus Luttrell collection of broadsides.

*A satyr against coffee* (no. 49).
[1679] Dated by Luttrell.
*On man. A satyr. By a person of honour* [John Wilmot, Earl of Rochester, 1647–1680] (no. 64). [1679] Dated by Luttrell.
Case/6A/158

A collection of broadsides, boxed. Those listed here have some (marginal) relevance as commentaries on medicine, hygiene, popular science.

### Narcissus Luttrell collection of broadsides.

*A satyr upon musty-snuff* (no. 64).
[1707] Dated by Luttrell.
*The pretended Prince of Wales's lamentation for the small pox* (no. 66).
[1708] Dated by Luttrell.
Case/6A/159

### Narcissus Luttrell collection of broadsides.

*A particular relation of the sickness and death of his late majesty K. William the third, (of ever blessed memory) who departed this life on Sunday the 8th of March 1701/2, in the 51st year of his age, and the thirteenth of his reign.*
London: printed for A. Roper, March the 9th, 1702.
Subject: Description of illness, death.
Case/6A/160/no. 74

### Narcissus Luttrell collection of broadsides.

*A description of a new kinde of artificiall bathes lately inuented* (no.16).
n.d.
Case/6A/162

## Luyts, Jan [praeses], 1655–1721.

*Exercitatio physica, de mundi duratione.*
Utrecht: Franciscus Halma, 1690.
Subject: Natural theology.
B/42/.528

## Lycosthenes, Conrad, ca. 1518–1561.

*Prodigiorvm ac ostentorvm chronicon, quae praeter naturae, ordinem, motum, et operationem, et in svperioribus & his inferioribus mundi regionibus, ab exordio mundi usque ad haec nostra tempora, acciderunt. Quod portentorum genus non temerè euenire solet, sed humano generi exhibitum, seueritatem iramque Dei aduersus scelera, atque magnas in mundo uicissitudines portendit. Partim ex probatis fidéque dignis authoribus Grecis, atque Latinis: partim etiam ex multorum annorum propria obseruatione, summa, fide, studio, ac sedulitate, adiectis etiam rerum omnium ueris imaginibus, conscriptum per Conradvm Lycosthenem Rvbeaqvensem.*
Basel: Henricvs Petri, 1557.
Subject: Monsters as prognosticators or emblems of the extraordinary.
Case/B/862/.52

Relation between appearance of monsters and the occurrence of cataclysmic events. Signs, prodigies, portents related to divination and medical curiosa. Exotic fauna (elephant, rhino, camel, baboon).

## Lydiat, Thomas, 1572–1646.

*Emendatio temporum compendio facta ab initio mundi ad anno MDCVIII. Qua, praeter alia plurima restituta sunt nativitas & baptisma & cruciatis Domini Christi ad annos fere quaternos post vulgi calculo: Confutatis opinionibus Scaligerana Baronianaque, pariter ac Iesuita Joannis Derkerii & Laurentii Suslygae, atque Ioannis Kepleri Caesari mathematici.*
The Hague: Samuel Broun, 1654.
Subject: Chronology.
C/588/.542

# M

## Maclaurin, Colin, 1698–1746.

*An account of Sir Isaac Newton's philosophical discoveries in four books.*
London: printed for the author's children, 1748.
Subject: Astronomy, gravity, natural philosophy, physics.
E/5/.N4864

Bk. 1: Newton's account of the system of the world, controversies in natural philosophy; bk. 2: Theory of motion or rational mechanics; bk. 3: Gravity, astronomy; bk. 4: Effects of the general power of gravity deduced synthetically.

## Macrobius, Ambrosius Theodosius, 4th–early 5th cent. A.D.

*In somnium Scipionis.*
[Venice: Nicolas Jenson, 1472.]
H-C-R 10426^ BMC (XV)V:172^ Pr. 4085^ Still. M4^ Goff M-8
Subject: Creation of world, harmony of the spheres, mysticism, natural, moral, and rational philosophy, zodiac.
Inc./f4085

From penciled note on front flyleaf: "Editio princeps. It contains the first printed text of parts of Homer and Lucretius." With *Conviviorum primi diei Satvrnaliorvm.*

## Macrobius, Ambrosius Theodosius.

*Somnium Scipionis ex Ciceronis libro de republica excerptum.*
Brescia: Boninus de Boninis, 1485.
H-C *10428^ BMC (XV)VII:969^ Pr. 6962^ Goff M-10
Subject: Creation of world, harmony of the spheres, mysticism, natural, moral, and rational philosophy, zodiac.
Inc./f6962

With *Conviviorum primi diei Satvrnaliorvm.* [see Wellcome]

## Macrobius, Ambrosius Theodosius.

*Somnium Scipionis ex Ciceronis libro de repvblica excerptvm.*
Venice: Joannes Rubeus, 1492.
H-C *10429^ BMC (XV)V:417^ Pr. 5131^ Goff M-12
Subject: Creation of world, harmony of the spheres, mysticism, natural, moral, and rational philosophy, zodiac.
Inc./f5131

## Madeira Arrais, Duarte [Edward].

*Arbor vitae; or, a physical account of the tree of life in the Garden of Eden. By Edward Madeira Arrais M.D. physician to John the IV. king of Portugal. Translated out of the Latine. A piece usefull for divines as well as physicians.*
London: printed for Tho. Flesher, 1683.
Subject: Natural history, natural theology.
Case/B/245/.0707

Trans. Richard Browne. Not in STC. BOUND WITH Bacon, Roger, *The cure of old age*, q.v.

## Maffei, Giovanni Camillo, n.d.

*Scala natvrale, overo fantasia dolcissima di Gio. Camillo Maffei da Solofra, intorno alle cose occulte, e desiderate nella filosofia.*
Venice: Gio. Varisco & associates, 1564.
Subject: Astrology, cosmology, earth sciences, flora, natural history, zodiac.
Case/M/O/.544

Fourteen "gradi" (or "progressions") including 4 elements, celestial objects, planets, tides.

## Maffei, Raffaele, of Volterra, 1451–1522.

*Commentariorvm vrbanorvm liber duodequadraginta.*
[Rome: Ioannes Besicken Alemanus, 1506.]
Subject: Ancient geography, biology, science encyclopedia.
Ayer/106/M2/1506

Includes ancient geography; anthropology; philology (a digest of Aristotle's natural history, de animalibus, descriptive anatomy, medicine, etc.); hygiene; balneology; digestion; "de ratione & causis plantarum ex Aristotele," also Theophrastus, "de metallicis, de pigmentis, de lapidibus, de scientiis mathematicis ac primum arythmetica, de harmoniacis, . . . de optice & catoptrice," geometry, proportion, astrology. BOUND WITH *Oeconomicum Xenophantis* (12 leaves).

## Maggi, Lucilio, 1510–ca.1578.

*Lvcilli Philalthaei, philosophiae medicinaeque professoris pvblici, in IIII libros Aristotelis de caelo & mundo, commentarij. Quibus accedunt quàm plurimi icones, ad astronomiam, & totam mathematicen spectantes: qui quanto futuri sint iuuamento omnibus liberalium artium professoribus, ab ijsdem inter legendum facilè percipi poterit.*
Venice: Vincentivs Valgrisivs, 1565.
Subject: Astronomy.
Case/6A/414

In 4 books.

## Magini, Giovanni Antonio, 1555–1617.

*Io. Antonii Patavini . . . de astrologica ratione ac vsu dierum criticorum, seu decretoriorum; ac praeterea de cognoscendis morbis ex corporum coelestium cognitione. Opus duobus libris . . . quorum primus complectitur commentarium in Claudij Galeni librum tertium de diebus decretorijs. Alter agit de legitimo astrologiae in medicina vsu. His additus de*

*annui temporis mensura in directionibus: & de directionibus ipsis ex Valentini Naibodae scriptis.*
Venice: heirs of Damianus Zenarius, 1607.
Subject: Astrological medicine, astrology.
Case/B/8635/.545

1st ed.

## Maier, Michael, 1568?–1622.

*Arcana arcanissima, hoc est hieroglyphica aegyptiograeca, vulgo necdum cognita, ad demonstrandam falsorum apud antiquos deorum, dearum, heroum, animantium & institutorum pro sacris receptorum, originem, ex vno Aegyptiorum artificio, quod aureum animi et corporis medicamentum peregit, deductam, vnde tot pöetarum allegoriae, scriptorum narrationes fabulosae & per totam encyclopaediam errores sparsi clarissima veritatis luce manifestantur suaque tribui singula restituuntur, sex libris exposita authore.*
[Oppenheim? 1614?]
Subject: Emblems, hieroglyphics, occult.
Case/B/8633/.548

Bk. 1: Hieroglyphics; bk. 2: "De hieroglyphicis Graecorvm, ac primó de allegorijs auro magis conspicvis."

## Maier, Michael.

*Atalanta fugiens, hoc est emblemata nova de secretis naturae chymica.*
Oppenheim: H. Galler for J. Theodore de Bry, 1618.
Subject: Alchemy, emblems.
Case/*B/8633/.55

BOUND WITH the following alchemical tracts: Maier's *Viatorium, hoc est de montibvs planetarvm septem seu metallorum.* Oppenheim: Hieronymus Galler for Joh. Theodor de Bry, 1618; *Tripvs avrevs, hoc est tres tractatvs chymici selectissimi, nempe (1) Basilii Valentini . . . Practica via cum 12.clauibus & appendice Germanico; (2) Thomae Nortoni, crede mihi seu ordinale; (3) Cremeri cvivsdam abbatis Westmonasteriensis Angli testamentum; Michael Maieri.* Frankfurt: Pauli Iacobi for Lvgae Iennis, 1618; and *Basilica antimonii in qva antimonii natvra exponitvr et nobilissimae remediorum formulae, quae pyrotechnica arte ex eo elaborantur, quam accurate traduntur: manvali experientia comprobata & conscripta ab Mamero Poppio Thallino, phylochymico.* Frankfurt: Antonivs Hvmmivs, 1618.

## Maillet Benoît de, 1656–1738.

*Telliamed: or, discourses between an Indian philosopher and a French missionary, on the diminution of the sea, the formation of the earth, the origin of men and animals, and other curious subjects, relating to natural history and philosophy. Being a translation from the French original of Mr. Maillet.*
London: printed for T. Osborne, 1750.
Subject: Cosmogony, natural history.
A/911/.55

Dedicated to Cyrano de Bergerac, "author of the imaginary travels thro' the sun and moon." Divided into 6 "days."

## Malpighi, Marcello, 1628–1694.

*Epistolae anatomicae, virorum clarissimorum Marcelli Malpighii et Caroli Fracassati.*
Amsterdam: Caspar Commelinus, 1669.
Subject: Anatomy.
Case/oR/128.7/.M27/1669

## Mandelslo, Johann Albrecht von, 1616–1644.

*Des hoch edelgebornen Johan Albrechts von Mandelslo morgenländische reyse-beschreibung worinnen zugleich die gelegenheit vnd heutiger zustand etlicher fürnehmen Indianischen länder, provincien, städte vnd insulen, sampt derer einwohner leben, sitten, glauben vnd handthierung: wie auch die beschaffenheit der seefahrt über das oceanische meer. Heraus gegeben durch Adam Olearium mit desselben unterscheidlichen notis oder anmerckungen, wie auch mit vielen kupffer platen gezieret.*
Schleswig: Johan Holwein, 1658.
Subject: Voyages.
Ayer/2062/M27/1658

## Mandelslo, Johann Albrecht von.

*Morgenländische reise beschreibung. Worinnen zugleich der zustand der fürnembsten ost-Indianischen länder, städte und der einwohner leben, sitten, handthierung und glauben: wie auch die gefärliche schifffahrt über das oceanische meer berichtet wird. Zum andern mahl heraus gegeben und mit etlichen denckwürdigen, vermehrten notis oder anmerckungen. . . . durch Adam Olearium.*
Schleswig: Johan Holwein, 1668.
Subject: Voyages.
Case/6A/19

## Mandelslo, Johann Albrecht von.

*Voyages celebres & remarquables, faits de Perse aux Indes Orientales, par le Sr. Jean-Albert de Mandelslo, . . . contenant une description nouvelle & très-curieuse de l'Indostan, de l'empire du Grand-Mogol . . . mis en ordre & publiez, après la mort de l'illustre voyageur, par le Sr. Adam Olearius.*
Amsterdam: Michel Charles Le Céne, 1727.
Subject: Voyages.
Greenlee/4725/I55/M27/1727

2vols. in 1. Trans. A. de Wicquefort.

## Mandeville, Sir John, d. 1372.

[*Itinerarius*] *Johannes de Monte Villa Itinerarium in partes Iberosolimitanas. Et in ulteriores transmarinas.*
[Cologne: Cornelius de Zierikzee? ca. 1500.]
C II, 3832^ Variant of BMC (XV)I:309^ Pr. 1498^ Voullième 682^ Goff M-162
Subject: Descriptive geography, travel.
Inc./1498

Goff: "Copy dated 1495 in contemporary hand."

### Mandosio, Prospero, 1650–1709.

*In quo maximorum Christiani orbis pontificum archiatros Prosper Mandosivs . . . spectandos exhibet.*
Rome: Franciscus de Lazaris, 1696.
Subject: Medicine.
Case/4A/716

A bibliography or list of medical books.

### Manetti, Giannozzo, 1396–1459.

*De dignitate & excellentia hominis libri IIII.*
Basel: [And. Cratander] 1532.
Subject: Physiology, psychology.
Case/C/57/.548

In 4 bks.: Body; spirit, soul (anima); rational intelligence; misery, mortality.

### Manfredi, Girolamo, d. 1492.

*Liber de homine: cvivs sunt libri dvo, primus liber de conservatione sanitatis capitvlvm primvm de cavsis & natvris omnivm eorvm quae svmvntvr in cibo. Quesita LXX.*
Bologna: Ugo Rugerius and Dionysius Bertochus, 1474.
H-R 10689^ C 2623^ BMC (XV)VI: 805–6^ Pr. 6529^ Still. M166^ Goff M-191
Subject: Disease, materia medica, medicine.
Inc./f*6529

### Manfredi, Girolamo.

*Libro intitolato Il perché, tradotto di Latino in Italiano, dell' eccellente medico, & astrologo, M. Geronimo de' Manfredi, di nuouo ristampata & ripurgata da quelle cose, che hanessero potuto offendere il simplice animo del lettore.*
Venice: Guerra, 1607.
Subject: Medicine.
Q/.548

In 8 parts.

### Manilius, Marcus, late 1st cent. B.C. or early 1st cent. A.D.

*M. Manili Astronomicon libri qvinqve. Iosephvs Scaliger., Ivl. Caes. F. recensvit, ac pristino ordini suo restituit.*
Paris: printed by Robertus Stephanus for Mamertus Patisson, 1579.
Subject: Astrology, astronomy.
Case/Y/672/.M293

### Manilius, Marcus.

*Astronomicon libri qvinqve. Iosephvs Scaliger Ivl. Caes. F. recensvit ac pristino ordini suo restituit. Eiusdem Ios. Scaligeri commentarius in eosdem libros, & castigationum explicationes.*
[Heidelberg] Sanctandreana, 1590.
Subject: Astrology, astronomy.
Case/Y/672/.M295

Also bound with these tracts (half title only, but separate pagination)*In Manilii astronomicon variae lectiones.*

### Manilius, Marcus.

*Astronomicon.*
Strassburg: Ioannis Ioachim Bockenhoffer, 1655.
Subject: Astrology, astronomy.
Y/672/.M296

Edited by J. J. Scaliger. In 5 books, in verse. Has additions by Thomas Renesi (Animadversions on Manilius) and by Ismael Bvllialdi, 1605–94.

### Manni, Dominico Maria, 1690–1788.

*De Florentinis inventis commentarium.*
Ferrara: Bernardini Pomatelli, 1731.
Subject: Discoveries, inventions.
R/2/.548

Chap. 16: "De Florentinis in medica arte inventis"; chap. 26: "De inventore thermometri."

### Marchetti, Domenico de, 1626–1688.

*Anatomia cui responsiones ad Riolanum, anatomicum Parisiensem.*
Harderwijk: Societatis typographica, 1656.
Subject: Anatomy.
QM/21/.M35/1656

"Editio altera."

### Marcolini, Francesco [F. Marcolino da Forli], ca. 1500–1559.

*Le ingeniose sorti composte per Francesco Marcolini da Forli. Intitvlate Giardino di pensieri.*
Venice, 1550.
Subject: Fortune telling using playing cards.
Wing/fZP/535/.M3315

"Qvesiti pertinenti a hvomini"; "Qvesiti pertinenti a donne"; "Qvesiti pertinenti a hvomini et a donne."

### Marcuccius, Gaspar, fl. 1624–1644.

*Quadripartitvm melancholicvm, Gasparis Marcucij . . . qvo variae qvaestiones de melancholia morbo, essentia, differentijs, causis, prognosi, curatione habentur. Et plvra de morbo hypocondriaco veris melancholicorum somnijs, & amantium melancholia innotescunt.*
Rome: Andrea Phaei, 1644.
Subject: Diagnosis and cure of melancholy, health, psychology.
Case/B/529/.552

Pt. 1: definition; pt. 2: origin, hypochondria; pt. 3: hygiene and health; pt. 4: melancholic elements and humors.

### Marinati, Aurelio, d. 1650.

*La prima parte della somma di tvtte le scienze nella qvale si tratta delle Sette arti liberali . . . grammatica, rettorica, logica [called dialettica], musica, aritmetica, geometria & astrologia.*
Rome: Bartholmeo Bonfadino, 1587.
Subject: Arithmetic, astrology, geometry, quadrivium.
Case/Y/712/.M352

### Marius, Simon, 1570–1624.

*Neuer und alter schreib calender auff das schalt jahr, mit dem stand, lauff und aspecten sonnen, monds, vnnd der andern planeten auch den gemeinen astrologischen erwehlungen calculteret und beschrieben durch Simonem Marium, Guntzenhausanum Fr. astronomiae & medicinae studiosum.*
Nuremberg: Christoff Lochner, Johann Lauers, 1612.
Subject: Almanacs, astrological calendar.
Wing/ZP/647/.L812

BOUND WITH Marius's *Prognosticon astrologicum, das ist aussfürliche beschreibung dess gewitters/ sampt andern natürlichen zufällen auff das jar nach unsers herrn und seligmachers geburt* [only copy listed in *NUC*].

### [Markham, Gervase C.] 1568?–1637.

*Cavelarice or the tracconer, contayning the arte and secrets which belong to ambling horses.*
London: printed for Ed. White, 1607.
STC 17334
Subject: Veterinary medicine.
Case/R/50/.552

BOUND WITH the following (by Markham): *Markham's farewell to hvsbandry.* London: printed by M. F. for Roger Iackson, 1625 (STC 17373); *Cheape and good hvsbandry.* 3d ed. London: printed by T. S. for Roger Iackson, 1623 (STC 17338); *Covntrey contentments, or the English hussvvife.* London: printed by I. B. for R. Iackson, 1623 (STC 17343); *A nevv orchard and garden.* London: printed by I. H. for Roger Iackson, 1623 (STC 15330); Simon Harward, (title page wanting) [*A most profitable new treatise*] (STC 12921; part of STC 15330); William Lawson (fl. 1618), *A new orchard and garden.* London: printed by I. H. for Roger Iackson, 1623. All of these are BOUND WITH Markham's *A way to get wealth,* q.v.

### [Markham, Gervase C.]

*Cheape and good hvsbandry for the well-ordering of all beasts, and fowles, and for the generall cure of their diseases. Contayning the natures, breeding, choise, vse, feeding, and caring of the diseases of all manner of cattell, . . . And diuers good and well-approued medicines, for the cure of all the diseases in hawkes, of what kinde soeuer. Together with the vse and profit of bees. . . . Gathered together for the generall good and profit of this whole realme, by exact and assured experience from English practises, both certaine, easie, and cheape: differing from all former and forraine experiments, which eyther agreed not with our clime, or were too hard to come by, or ouer-costly, and to little purpose.*
London: printed by T. S[nodham] for Roger Iackson, 1623.
STC 17338.
Subject: Veterinary medicine.
Case/R/50/.547

3d ed.

### Markham, Gervase.

*The inrichment of the weald of Kent.*
London: printed by Eliz. Purslow, for John Harrison, 1649.
STC II M637
Subject: Agriculture.
Case/R/50/.5514

### Markham, Gervase.

*Markham's farewell to hvsbandry.*
London: printed by M. F. for Roger Iackson, 1625.
STC 17373
Subject: Agriculture.
Case/R/50/.55

"Newly reuiewed, corrected, and amended."

### Markham, Gervase.

*Markham's farewell to hvsbandry.*
London: printed by W. Wilson for E. Brewster and George Sawbridge, 1656.
STC II M639
Subject: Agriculture.
Case/R/50/.551

"Now newly the sixth time revised, corrected, and amended."

### Markham, Gervase.

*Markham's maister-peece: containing all knowledge belonging to a smith, farrier, or horse-leach, touching the curing of all diseases in horses, drawne with great paine, and most approued experience from the publicke practice of all the forraigne horse-marshalls in Christendome, and from the priuate practice of all the best farriers of this kingdome. Being diuided into two bookes. The first containing all cures physicall: the second all belonging to surgery, with an addition of 130 principall chapters and 340 most excellent medicines neuer written of nor mentioned in any other author whatsoeuer. Together with the true nature, vse and quality of euery simple spoken of through the whole worke. Now newly imprinted, corrected and augmented, with diuers most assured and approued medicines, which without all faile (by Gods grace) will cure those diseases which all our farriers hold impossible to be cured.*
London: Nicholas Okes, 1615.
STC 17377
Subject: Veterinary medicine.
Case/R/674/.55

2 parts in 1 vol.

### Markham Gervase.

*A vvay to get vvealth, by approued rules of practice in good husbandry and huswiferie. Containing the foure principall offices which svpport and maintaine a familie. As I. The husbanding and inriching of all sorts of grounds. . . . Also the preseruation of graine. II. The ordering and curing of all sorts of cattell and fowle. III The office of the English housewife in physicke, surgerie. IIII. The office of planting and grafting: the office of gardening.*
London: printed for Roger Iackson, 1625.
STC 17395.7
Subject: Gardening, household medicine, veterinary medicine.
Case/R/50/.552

4 vols. in 1. "The first three bookes gathered by G. M. The last by Mr. William Lawson." A group of tracts on agriculture, animal husbandry, and gardening, primarily by Markham. Most have individual title pages and are separately paginated. BOUND WITH Markham's *Cavelarice*, etc. qq.v.

## Markham, Gervase.

*A way to get wealth. Containing six principal vocations, or callings, in which every good husband or house-wife may lawfully employ themselves.*
London: printed by E. H. for George Sawbridge, 1676.
STC II M681
Subject: Medicine, veterinary medicine.
Case/R/50/.553

13th ed. Among the six tracts mentioned on general title page are *The office of a house-wife, in physick, chyrurgery, extraction of oyles, . . . conceited secrets, distillations, malting, brewing, baking* (1675); *The inrichment of the weald of Kent* (1675); [with a separate title page] *Markham's farewel to husbandry* 10th time revised (1676); and [with separate title pages] *Cheap and good husbandry,* 13th ed. (1676), STC II M617; *The English house-wife (1675); A new orchard & garden,* 6th ed. (1676), STC II L736.

## Markham, Gervase.

*The way to get wealth: or, a new and easie way to make twenty three sorts of wine equal to that of France; . . . to which is added, A help to discourse, giving an account of trade of all countries, and inventers of arts and sciences. . . . and many other curiosities.*
London: printed for G. Conyers, 1702.
Subject: Health, hygiene, inventions.
Case/R/999/.545

2 vols. in 1. Includes "To strengthen the memory, so that you may remember all that you read or do all your life after."

## Marmol Carvajal, Luis del, fl. 1575.

*L'Afriqve de Marmol, de la traduction de Nicolas Perrot sieur d'Ablancovrt.*
Paris: Louis Billaine, 1667.
Subject: Descriptive geography, natural history.
Greenlee/4725/A45/M35/1667

3 vols.

## Marstaller, Gervasius, d. 1518.

*Artis diuinatricis, qvam astrologiam sev iudiciariam vocant, encomia & patrocinia, quorum catalogum sequens pagella continet.*
Paris: Christian Wechel, 1549.
Subject: Astrology.
Case/4A/465

Series of tracts, including one by Melanchthon—his preface to Johann Schoner's book on judicial astrology (nativities). Also *Trapezontii libellvs cur astrologorum iudicia vt plurimum sint falsa & dialogvs Ioviani Pontani.*

## Martianus Capella, 5th cent., A.D.

*De nuptiis philologiae et Mercurii libidus.*
Vicenza: Henricus de Sancto Ursio Zenus, 1499.
H*4370^ BMC (XV)VII:1048^ Pr.7174^ Still. 105^ Goff C-117
Subject: Arithmetic, astronomy, geometry.
Inc./f7174

On the 7 liberal arts.

## Martianus Capella.

*De nuptiis philologiae et Mercurii.*
Modena: Dionysius Bertochus, 1500.
H* 4371^ BMC (XV)VII:1068^ Pr. 7215^ Still. C 106^ Goff C-118
Inc./f7215
Subject: Arithmetic, astronomy, geometry.

## Marzio, Galeotto, ca. 1427–1497.

*Galeoti Martii Narniensis de homine libri dvo. Georgii Mervlae Alexandrini in Galeotvm annotationes.*
[Basel: Joannes Frobenius, 1517.]
Subject: Anatomy and physiology of man.
Case/Q/.554

Bk. 1.: "De homine exterior"; bk. 2.: "De homines interiori." Merula's commentary criticizes lack of systematic division of body into parts.

## Marzio, Galeotto.

*Della varia dottrina. Tradotto in volgare fiorentino per M. Francesco Serdonati con la giunta d'alcune breui annotatzioni.*
Florence: Filippo Giunti, 1615.
Subject: Astrology, medicine, natural science.
L/0114/.551

Errors of Avicenna, description of the medicine of Galen and of metals.

## Mascall, Leonard, d. 1589.

*A booke of the arte and manner how to plant and graffe all sorts of trees, how to sette stones and sow pepins, to make wild trees to graffe on, as also remedies & medicines.*
London: T. East for Thomas VVight, 1592.
STC 17578
Subject: Botany, gardening, materia medica.
Case/R/570/.55

## Mason, James.

*The anatomie of sorcerie. Wherein the wicked impietie of charmers, inchanters, and such like, is discouered and confuted.*
London: printed by I. Legatte, 1612.
STC 17615
Subject: Witchcraft, sorcery.
Case/B/88/.554

Religious implications of witchcraft. True and false miracles (p. 17).

## Massé, Pierre, fl. 1579.

*De l'impostvre et tromperie des diables, devins, enchantevrs, sorciers, novevrs d'esguillettes, cheuilleurs, necromanciens, chiromanciens, & autres qui par telle inuocation diabolique, ars magiques & superstions abusent le peuple.*
Paris: Iean Poupy, 1579.
Subject: Amulets, divination, magic, witchcraft.
Case/B/88/.556

## Mather, Cotton 1663–1728.

*The Christian philosopher: a collection of the best discoveries in nature, with religious improvements.*
London: Eman. Matthews, 1721
Subject: Astronomy, natural philosophy.
Case/B/72/.56

Said to be the first general work on science published in North America[*World who's who in science*]. Introduction: "Philosophy is no enemy, but a mighty and wondrous incentive to religion."

## Mather, Cotton.

*Memorable providences, relating to witchcrafts and possessions. A faithful account of many wonderful and surprising things that have befallen several bewitched and possessed persons in New-England.*
Boston: R. P., 1689.
Subject: Witchcraft.
Case/B/8884/.197

## Mather, Cotton.

*The wonders of the invisible world: Being an account of the tryals of several witches, lately executed in New-England: and of several remarkable curiosities therein occurring.*
Boston: printed and London reprinted for John Dunton, 1693.
STC II M1174
Subject: Witchcraft.
Ayer/*150.5/N4/M4/1693

"Together with, I. Observations upon the nature, the number, and the operations of the devils. . . . IV. A brief discourse upon those temptations which are the more ordinary devices of Satan." 1st ed. [Imperfect: 1/2 title wanting.]

## Mather, Cotton.

*The wonders of the invisible world: Being an account of the tryals of several witches lately executed in New-England: and of several remarkable curiosities therein occurring.*
Boston: printed and London reprinted for John Dunton, 1693.
Subject: Witchcraft.
STC II M1175
Case/B/8884/.55

2d ed. Reprinted and abridged.

## Mather, Increase, 1639–1723.

*Cometographia [in Greek] or, a discourse concerning comets; wherein the nature of blazing stars is enquired into: With an historical account of all the comets which have appeared from the beginning of the world unto this present year, M. DC. LXXXIII. Expressing the place in the heavens, where they were seen, their motion, forms, duration; and the remarkable events which have followed in the world, so far as they have been by learned men observed.*
Boston: printed by S. G. for S. S., 1683.
Subject: Astronomy, history of comets.
Case/B/863/.558

## [Mathews, Richard]

*The unlearned alchymist his antidote: or a more full and ample [e]xplanation of the use, w[o]rt[h]e and benefit of my pill. Entitluled, an effectual diaphoretick, diuretick purgeth by sweating, urine. Whereunto is added sundry cures and experiences, with p[ar]ticular direction unto particular diseases and distempers. Also, sund[r]y plain and easie receits, which the ingenious may prepare for their own health. Together with a precious pearl in the midst of a dunghil, being a true and faithful receit of Mr. [Ri]chard Mathews's pill, according to his own p[ra]ctice recorded in writing under his own h and, . . . Presented to the world by M[r]is Anne Mathews, amongst many sad complaints of wrongs done to her, and the community, and her deceased husband.*
London: printed for Joseph Leigh, 1663.
STC II M1291
Subject: Materia medica.
UNCATALOGUED

Title page mutilated.

## Matthiolis, Matthiolus de, Perusinus, [Matthew of Perugia] d. ca. 1470.

*Tractatus clarissimi philosophi et medici Matheoli perusini de memoria.*
[Rome? 1480?]
BMC (XV)IV:100^ H-R 10907^ Pr. 3779^ Graesse iv, 441^ Goff M-360
Subject: Mnemonics, psychology.
Ayer/*107.56/1493/1493a

Chap. 1: "De artificio memorie"; chap. 2: "Conseruare vel augere memoriam cum medicinalibus opus est laborissimum." Goff gives imprint [Rome Stephan Planck, ca. 1490]. BOUND WITH Christopher Columbus, *Epistola* [Rome, 1493], and several other works; only colophon (for another work in vol.) dated 1483. 1st section of the volume is in MS.

## Mauro of Florence, Servite, 1495–1556.

*Annotationi sopra la lettione della Spera del Sacro Bosco doue si dichiarano tutti e principii mathematici & naturali che in quella si possan' desiderare. Con alcune quistioni notabili a ditta spera necessarie, & altri notandi & rari segreti, che in quella son acasti. Con le infrascritte cose, cio, e, vna nuoua & fedele (ad verbum) traduttione di detta spera. Vna spera theologica diuina & Christiana. Vna spera platonica, con alcune eccitationi mathematiche, theologiche & diuine. Vna nuoua inuentione & astronomico instrumento, per subitamente fabricare le dodici case celesti della figura astronomica, senza altri canoni, o calculo.*

[Florence: Lorenzo Torrentino, 1550.]
Subject: Astronomical computation and instrumentation.
Ayer/6/S2/M45/1550

### Maurolico, Francesco, 1494–1575.

*Cosmographia in tres dialogos distincta: in quibus de forma, situ, numerorumque tam coelorum quae elementorum, alijsque rebus ad astronomica rudimenta spectantibus satis disseritur.*
Venice: heirs of Lucantonius Iunta, Florentine, 1543.
Harrisse 142
Subject: Arithmetic, astronomy, plane and solid geometry, instrument making.
Ayer/*7/M2/1543

In 4 sections, with material on the sphere, the astrolabe, the quadrant.

### Maurolico, Francesco.

*Cosmographia Francisci Mavrolyci Messanensis Sicvli, in tres diàlogos distincta: in quibus de forma, situ, numerorumque tam coelorum quàm elementorum, aliisque rebus ad astronomica rudimenta spectantibus satis disseritur.*
Paris: Gulielmus Cauellat, 1558.
Subject: Arithmetic, astronomy, plane and solid geometry, instrument making.
Ayer/7/M2/1558

Euclid elements (conic sections, sphere, Archimedes); 2d dialogue in 2 parts, 2d part on astrology, zodiac.

### [Maynwaring, Everard], 1628–1699?

*Inquiries into the general catalogue of diseases shewing the errors and contradictions of that establishment. With a new scheme representing more truly, and essentially; the various diseased state of human nature.*
[London, 1691.]
STC II M1496
Subject: Disease, medicine.
Q/16/.562

"When I consider the fatality of sickness [in the army and navy] to be greater than that of the sword, I cannot but reflect upon the methods and provision of medicines" (p. 8).

### Mazzolini, Silvestro, da Prierio, 1460–1523.

*De strigimagarvm demonvmqve mirandis libri tres.*
[Rome: Antonius Bladus de Asula, 1521]
Caillet 8974^ Graesse V-443
Subject: Witchcraft.
Case/oBF/1565/.M38

[Note from sale catalogue copy on front pastedown] "Mazzolini was one of the first writers against Luther, became one of his judges, and later was appointed an inquisitor."

### Mazzolini, Silvestro, da Prierio.

*De strigimagarvm, daemonumque mirandis, libri tres.*
Rome: Po. Ro. [Populi Romani], 1575.
Subject: Witchcraft.
Case/B/88/.562

### Medina, Pedro de, 1493?–1567?

*Arte de nauegar en que se contienen todas las reglas, declaraciones, secretos, y auisos, que a la buena nauegacion son necessarios, y se deuen saber, hecha por el maestro Pedro de Medina.*
[Valladolid: Francisco Fernandez de Cordoua, 1545.]
Harrisse 266^ John Carter Brown I:142
Subject: Navigation.
Ayer/*7/M4/1545

### Medina, Pedro de.

*L'arte del navegar in laqval si contengono le regole, de chiarationi, secreti & auisi, alla bona nauegation necessarij. Composta per l'eccel. dottor M. Pietro da Medina, & tradotta de lingua Spagnola in volgar Italiano, à beneficio, & vtilità de ciascadun nauigante.*
Venice: printed by Aurelio Pincio for Gioanbattista Pedrezano, 1554.
Subject: Navigation.
Ayer/*7/M4/1554

In 8 books, including (4) use of the sun's declination for navigation.

### Megerlin, Peter, 1623–1686.

*Astrologische muthmassungen von der bedeuttung des jüngst entstandenen cometen.*
[Basel:] Johana Rudolph Genath, 1665.
Subject: Comets.
Case/B/863/.565

"Consideratio mathematica & astrologica cometae anno 1664."

### Megerlin, Peter.

[*Discursus mathematica de cometa.*]
[Basel, 1661.]
Subject: Comets.
Case/B/863/.566

"Duorum celebratissimorum theologorum judicia de cometarum significatione."

### Meibom, Marcus, d. 1711.

*Antiqvae mvsicae avctores septem. Graece et Latine. Marcvs Meibomivs restituit ac notis explicavit.*
Amsterdam: Ludovic Elzevir, 1652.
Subject: Harmonics, music.
Case/Y/6492/.56

A group of tracts on music in double columns, Greek and Latin. Includes Euclid, *Introdvctio harmonica interprete Marco Meibomio; Nicomachi Gorasseni Pythagorici Harmonices manvales;* Martianus Capella and others.

### Meier, Samuel.

*De panvrgia lamiarvm, sagarvm, e strigum ac veneficarum totiusque chohortis magicae cacodaemonia libri tres. Dat ys: Nödige vnderrichtinge I van der töuerschen geschwinden list*

*vnd geschicklichiet quadt tho donde. Il vnde, dat töuerye eine düuelsch sünde sy.*
Hamburg [Hans Binder], 1587.
Subject: Witchcraft.
Case/B/88/.564

## Meinders, Hermann Adolph.

*Unvorgreifliche gedancken und monita, wie ohne blinden eyfer und ubereilung mit denen hexen-processe[n] und der inquisition wegen der Zauberey.*
Lemgo: Heinrich Wilhelm Meyer, 1716.
Subject: Possession, witchcraft.
4A/2481

## Mela, Pomponius, 1st cent. A.D.

[*Cosmographia*]
*Pomponii Mellae cosmographi geographia: prisciani quoque ex Dionysio Thessalonicensi de situ orbis interpretatio.*
Venice: Erhard Ratdolt, 1482.
H-C *11019^ BMC (XV)V:286^ Pr. 4385^ Still. M389^ Goff M-452
Subject: Cosmography, geography.
Inc./4385

In 3 books. Followed by *Prisciani Cesariensis interpretatio ex Dionysio de orbis situ* (in verse).

## Mela, Pomponius.

*Cosmographia.*
[Venice: Christophorus de Pensis, after 1493?]
H-C 11013*^ BMC (XV)V:476^ Pr. 5259^ Goff M-453
Subject: Cosmography, geography.
Inc./5259

Edited by Ermolao Barbaro. Has insert of 4 leaves consisting of hand-colored MS map on vellum leaf (with countries identified by number) and 3 paper leaves in early hand. MS marginalia passim.

## Mela, Pomponius.

[*Cosmographia*]
*Cosmographia Pomponij Mele: authoris nitidissimi tribus libris digesta: pernuo quedam compendio Joannis Coclei Norici adaucta quo geographie principia generaliter comprehenduntur.*
Nuremberg: J. Weyssenburger, 1512.
Subject: Cosmography, geography.
Wing/ZP/547/.W549

## Mela, Pomponius.

*Pomponius Mela. Ivlivs Solinvs. Itinerarivm Antonini Avg. vibivs seqvester. P. Victor de regionibus urbis Rome. Dionysius Afer de situ orbis prisciano interprete.*
Venice: Aldus & Andrea Socerus, 1518.
Subject: Cosmography, geography.
Case/*Y/672/.M605

## Mela, Pomponius.

*Pomponivs Mela. Ivlivs Solinvs. Itinerarivm provinciarvm Antonini Avg. vibivs seqvester. [Virgiliano filio]. P. Victor de regionibus urbis Romae. Dionysius Afer de situ orbis prisciano interprete.*
Venice: Aldus & Andrea Socerus, 1518.
Subject: Cosmography, geography.
Case/Y/672/.M6051

## Mela, Pomponius.

*Pomponii Melae Hispani, libri de situ orbis tres, adiectis Ioachimi Vadiani Heluetii in eosdem scholiis: Addita quoque in geographiam catechesi: & epistola Vadiani ad Agricolam digna lectu.*
Vienna: printed by Ioannes Singrenivs for Lucas Alantse, 1518.
Subject: Cosmography, geography.
Ayer/*6/P9/M5/1518

Edited by Franciscus Asulanus. MS notes in Mela's text, which is printed in center of page and surrounded by Vadianus's commentary.

## Mela, Pomponius.

*Libri de situ orbis tres.* [Another copy]
Case/Y/672/.M60511

## Mela, Pomponius.

*Pomponius Mela de situ orbis ab Hermolao Barbaro & Jo. Camerte castigatus.*
Venice, 1520.
Subject: Cosmography, geography.
Ayer/*6/M5/1520

In 3 books.

## Mela, Pomponius.

*Pomponii Melae de orbis sitv libri tres, accvratissime emendati, unà cum commentariis Ioachimi Vadiani Heluetii castigatioribus, & multis in locis auctioribus factis: id quod candidus lector obiter, & in transcursu facile deprehendet.*
Basel: Andreas Cratander, 1522.
Harrisse 112
Subject: Cosmography, geography.
Ayer/f*6/M5/1522

## Mela, Pomponius.

*Pomponii Melae de orbis sitv libri tres, accvratissime emendati, una cum commentariis Ioachimi Vadiani Heluetii castigatioribus, & multis in locis auctoribus factis: id quod candidus lector obiter, & in transcursu facile deprehendet. Adiecta sunt praeterea loca aliquot ex Vadiani commentarijs summatim repetita, & obiter explicata: in quibus aestimandis censendisque doctissimo uiro Ioanni Camerti. . . . Rvrsum, epistola Vadiani, ab eo penè adulescente ad Rudolphum Agricolam iuniorem scripta, non indigna lectu, nec inutilis ad ea capienda, quae aliubi in commentarijs suis libare magis, qùam longius explicare uoluit.*
Basel: Andreas Cratander, 1522.
Subject: Cosmography, geography.
Case/fY/672/.M6052

### Mela, Pomponius.

*[De orbis sitv]* [Another copy]
Greenlee/5100/M51/1522

Title page wanting in this copy.

### Mela, Pomponius.

*Pomponii Melae de sitv orbis.*
Florence: printed by heirs of Philippus Iunta, 1526.
Subject: Cosmography, geography.
Ayer/*6/M5/1526

Half title: "Pomponivs Mela. Ivlivs Solinvs [Polyhistor]. Itinerarivm Antonini Avg. [Itinerarivm provinciarvm Antonini Avgvsti]. Vibivs seqvester. P. Victor de regionibus urbis Romae. Dionysius Afer de situ orbis prisciano interprete iterum castigatus & cum Graeco exemplari collatus [Dionysius Afer poëma de orbis sitv prisciano interprete]."

### Mela, Pomponius.

*Pomponii Melae de orbis situ libri tres, accuratissime emendati, vna cvm commentariis Ioachimi Vadiani Helvetii castigatioribus, & multis locis auctoribus factis: id quod candidus lector obiter, & in transcursu facile deprehendet.*
Paris: Ioannes Roigny, 1540.
Subject: Cosmography, geography.
Ayer/*6/M5/1540

### Mela, Pomponius.

*Pomponii Melae de orbis situ libri tres, accuratissime emendati, vna cvm commentariis Ioachimi Vadiani Heluetii castigatioribus, & multis in locis auctoribus factis: id quod candidus lector obiter, & in transcursu facile deprehendet. Adiecta sunt praeterea loca aliquot ex Vadiani commentariis summatim repetita, & obiter explicata: in quibus aestimandis censendisqùe, doctissimo uiro Ioanni Camerti. . . . Rvrsvm epistola Vadiani, ab eo penè adolescente ad Rudolphum Agricolam iuniorem scripta, non indigna lectu, nec inuitlis ad ea capienda, quae aliubi in commentarijs suis libare magis, quàm longius explicare uoluit.*
Paris: Christian Wechel, 1540.
Subject: Cosmography, geography.
Case/fY/672/.M6054

### Mela, Pomponius.

*Pomponii Melae de orbis situ.* [Another copy]
Ayer/*7/S7/1520

BOUND WITH Solinus, *Enarrationes* (1520), q.v.

### Mela, Pomponius.

*De orbis situ libri III & C. Ivlii Solini, polyhistor, quorum ille descriptionem singularum orbis terreni partium atque regionum: Hic verò praeter eadem, quae vbique memorabilia sint loca, animantia, plantae, gemmae, & similia, compendiosè enarrat.*
Basel: Sebastianvs Henricpetri [1595].
Subject: Cosmography, geography.
Case/Y/672/.M6059

### Mela, Pomponius.

*Pomponii Melae libri tres de situ orbis, nummis antiquis & notis illustrate ab Jacobo Gronovio. Julii Honorii oratoris excerpta cosmographiae, ab eodem primum ex MS. edita. Cosmographia, falso aethicum auctorem praefereus, cum variis lectionibus ex MS. Ravennas Geographus, ex MS. Lugdunensi suppletus.*
Leyden: Jordanus Luchtman, 1696.
Subject: Cosmography, geography.
Case/Y/6782/.37

### Mela, Pomponius.

*De situ orbis libri tres ad omnium Angliae & Hiberniae codicum MSS. fidem, summa cura & diligentia recogniti & collati; . . . Accedunt 1. Notae partim criticae, partim geographicae: 2. Dissertatio de Ariané. 3. Synopsis Pomponiana, qua illius vetera faciliùs a Tyrunculis mente accipi possint, & memoria teneri: atque è novis committuntur clariora & magis explorata.*
Exeter: Farleanis for Philip Bishop, 1711.
Subject: Cosmography, geography.
Case/5A/540

### Melton, John d. 1640.

*Astrologaster or, the figvre-caster. Rather the arraignment of artlesse astrologers, and fortune-tellers that cheat many ignorant people vnder the pretence of foretelling things to come, of telling things that are past, finding out things that are lost, expounding dreams, calculating deaths and natiuities, once again brought to the barre.*
STC 17804
London: Barnard Alsop for Edward Blackmore [1620].
Subject: Astrology, prognostication.
Case/B/863/.886

Binders title: *Tracts on astrology.*

### Memmo, Giovanni Maria, d. 1553.

*Tre libri della sostanza et forma del mondo. Ne quali per modo di dialogo si disputano molte acutissime questioni, & sono poi risolute con le ragioni de i piu faui philosophi, & de i piu dotti astrologi antichi.*
[Venice: Giouanni de Farri & brothers, 1545.]
Subject: Astrology, astronomy, cosmography.
Case/B/42/.566

### Menavino, Giovanni Antonio.

*I cinque libri della legge religione et vita de'Turchi. . . . Oltre cio vna prophetia de' Mahometani.*
Venice: V. Valgrisi, 1548.
Subject: Anthropology.
Case/F/59/.566

Prophecy of Mohammed trans. by Lodovico Domenichi. "Cio che si fa de Christiani che non hanno arti mechaniche" (p. 209).

### Menavino, Giovanni Antonio.

*Trattato di costvmi et vita de Turchi.*
Florence, 1548.

Subject: Anthropology.
Case/F/59/.568

In 5 bks.: (1) "Quel che si fa delle fanciulle & altre donne"; (2) "Cio che si fa de Christiani, che non hanno arti mechaniche"; (3) "Del viuere degli animali; (4) "Della agricoltura"; (5) "Della diuersità de gli animali." Includes "delli medici & barbieri." Penciled note on front flyleaf: "Designed to put all the Christian princes on their guard against the Turk." BOUND WITH Bartolomeo Giorgieuits [Gjorgjevc, Bartholomaeus], *Prophetia di Maometani.*

## Menestrier, Claude François, 1631–1705.

*La philosophie des images enigmatiques, ou il est traité enigmes, hieroglyphiques, oracles, sorts, divinations, loteries, talismans, songes, centuries de Nostradamus, de la baguette.*
Lyon: Hilaire Baritel, 1694.
Subject: Divination, emblems, occult.
B/85/.565

## Menestrier, Claude François.

*La philosophie des images enigmatiques, ou il est traité des enigmes, hieroglyphiques, oracles, propheties, sorts, divinations, loteries, talismans, songes, centuries de Nostradamus, de la baguette.*
Lyon: Jaques Guerrier, 1694.
Subject: Divination, emblems, occult.
B/85/.566

"Des chifres et enigmes numerales" (p.51); "Des enigmes suspectes, decriées et dangereuses" (p.249); "Monuments enigmatiques" (p. 187).

## Menghi, Girolamo, d. 1610.

*Compendio dell' arte essorcistica, & possibilità delle mirabili & stupende operationi delli demoni & de i malefici. Con gli rimedij opportuni all' infirmitadi maleficiali.*
Bologna: G. Rossi, 1584.
Subject: Demonology, exorcism, witchcraft.
Case/B/88/.565

## Menghi, Girolamo.

*Flagellvm daemonvm. Exorcismos terribiles, potentissimos, et efficaces. Remediaque probatissima, ac doctrinam singularem in malignos spiritus expellendos, facturasque, & maleficia fuganda de obsessis corporibus complectens; cum suis benedictionibus, & omnibus requisitis ad eorum expulsionem.*
Bologna: Ioannes Rossius, 1589.
Subject: Demonology, exorcism, witchcraft.
Case/B/88/.566

BOUND WITH *Remedia efficacissima in malignos spiritvs expellendo* and *Fvstis daemonvm, adiurationes formidabiles, potentissimas, & efficaces in malignos spiritus fugandos de oppressis corporibus humanis.* This tract is also included in Gerson, *[M]alleorvm maleficarvm*, and in *Thesaurus exorcismorum*, qq.v.

## Mentzel, Christian, 1622–1701.

*Index nominum plantarum universalis, diversis terrarum, gentiumque linguis quotquot ex auctoribus ad singula plantarum nomina excerpti & juxta seriem A.B.C. collocati potuerunt, ad unum redactus videlicet.*
Berlin: Rungiana, 1682.
Subject: Botany, herbals, medicinal plants.
Bonaparte/Coll./No. 1938

In Latin, Greek, and German. Indicates medicinal properties of plants, cites Pliny, Dioscorides, Theophrastus. Plants from Euope, Asia, Africa, Egypt, America. Pugillus, *Plantarum rariorum cum figuris aliquot aeneis, quibus intertextus indiculus plantarum nonnullarum Brasiliae, nondum editarum. cum quibusdam clar. Jacobi Breynii rarioribus, in prodromo suo fasciculi rariorum plantar*, bound in at end.

## Mentzel, Christian.

*Lexicon plantarum polyglotton universale, ex diversis, Europaeorum, Asiaticorum, Africanorum & Americanorum, antiquis & modernis linguis, earumque dialectis variis . . . in quo plantarum, genera, species, colorum & quarumvis partium differentiae, ab eruditis hactenus adnotae, legitimo ordine collocantur . . . accessit Pugillus plantarum rariorum cum figuris aliquot aeneis and brevibus descriptionibus.*
Berlin: Christoph Gottlieb Nicolai, 1715.
Subject: Botany, herbals, medicinal plants.
Bonaparte/Coll./No. 1939

Pugillus bound in at front.

## Mercator, Gerardus, 1512–1594.

*Atlas; or a geographicke description of the regions countries and kingdomes of the world.*
Amsterdam: Henry Hondius, 1630.
Phillips 449^ STC 17827
Subject: Cosmography.
Wing/+ZP/646/.H755

10th ed. Translated by Henry Hexham. Hand-colored engraved title page and maps.

## Mercator, Gerardus.

*Atlas sive cosmographicae meditationes de fabrica mvndi et fabricati figvra.*
[Dusseldorf: A. Busius, 1595.]
Subject: Cosmography.
Ayer/*135/M5/1595

Has a section on creation of the world. Second part: *Atlantis pars altera. Geographia nova totius mundi.* Penciled note on front pastedown: "Rare 1st edition of third part."

## Mercator, Gerardus.

*Chronologia, hoc est, temporvm demonstratio exactissima, ab initio mvndi, vsqve ad annvm domini M.D.LXVIII. Ex eclipsibvs et observationibvs astronomicis omnium temporum, sacris quoque Biblijs, & optimis quibusque.*
Cologne: heirs of Arnold Birckmann, 1569.
Subject: Astronomical observations, chronology, eclipse.
Case/fF/817/.568

## Mercuriale, Girolamo, 1530–1606.

*Artis gymnasticae apvd antiqvos celeberrimae, nostris temporibvs ignoratae, libri sex in quibus exercitationum*

*omnium vetustarum, genera, loca, modi, facultates & quicquid denique ad corporis humani exercitationes pertinet, diligenter explicatur.*
Venice: Iunta, 1569.
Durling Cat. NLM 3087^ Choulant, p. 160^ Wellcome I:4223^ Bibl. Osleriana 3387
Subject: Exercise, gymnastics, hygiene, medicine.
Case/V/17/.568

Bk. 1, chap. 1, "De principijs medicinae." Some MS marginalia.

## Mercuriale, Girolamo.

[*Artis gymnasticae*]
*De arte gymnastica libri sex in quibus exercitationum omnium vetustarum genera, loca, modi, facultates, & quidquid denique ad corporis humani exercitationes pertinet diligenter explicatur.*
Amsterdam: Andrea Frisius, 1672.
Subject: Exercise, gymnastics, hygiene, medicine.
UNCATALOGUED

## Mercuriale, Girolamo.

*Hieronymi Mercvrialis medici hac tempestate clarissimi tractatvs, de compositione medicamentorum. De morbis oculorum & aurium.*
Venice: Ivntas, 1601.
Subject: Diseases of the eye and ear, preparation of medicines.
UNCATALOGUED

Ed. Michaele Columbo. BOUND WITH his *De venenis, et morbis venenosis* (Venice, 1601); and tracts by Marsilio Cagnato, *De morta cavssa partvs*, another on guiacum, and one on unguents, aromatics, and other medicines (Rome: Aloysius Zannetti, 1602).

## Mercuriale, Girolamo.

*In omnes Hippocratis aphorismorum libros praelectiones Patauinae. In quibus innumeri penè ipsius Hippocr. obscuriores loci, ac sententiae elucidantur, problemataque permulta obstrusiora facili methodo enodantur.*
Bologna: Hieronymus Tamburinus, 1619.
Subject: Medical aphorisms and commentary.
fY/642/.H7186

In 9 books; bk. 8: "Pro spvrio Hippocratis habetvr."

## Mercurio, Scipione, d. 1615.

*De gli errori popolari d'Italia libri sette divisi in dve parti* . . . [1] *modo nel gouerno de gl'infermi* . . . [2 ] *nelle cause delle malatie.*
Padua: Francesco Bolzetta, 1645.
Subject: Mismanagement of disease, illness, medicine, materia medica.
Q/.569

Bk. 1: "Errori quali si commettono contro la medicina"; bk. 2: "Contro il medico"; bk. 3: "Contro gli amalati in letto"; bk. 4: "Contro gl' infermi in piazza"; bk. 5: "Contro le donne grauide, e parturienti"; bk. 6: "Contro fanciulli occorrono"; bk. 7: "Tutti gli errori, che occorrono nelle cause delle malattie, cioè nell' aere, moto, equiete, mangiare, bere, dormire, veghiare, euacuationi del corpo, & dolori, con gli suoi rimedij."

## [Merrifield, John]

*Catastrophe Galliae: or The French King's fatal downfal, predicted to happen in or about the years 1691 or 92. Together with his nativity calculated according to the rules of astrology. . . . by J. P., student in astrology.*
London: printed for Rowland Reynolds, 1691.
STC II M1845A
Subject: Astrology.
Case/B/8635/.57

## [Merret, Christopher] 1614–1695.

*A letter concerning the present state of physick, and the regulation of the practice of it in this kingdom. Written to a doctor here in London.*
London: printed for Jo. Martyn and Ja. Allestry, 1665.
STC II M1837
Subject: Medical practice.
Case/I/045/.88/no. 1

Binders title: *Tracts on physick, 1665–70.*

## Merrett, Christopher.

*Self-conviction; or an enumeration of the absurdities, railings, &c. against the college, and physicians in general; but more especially, the writers against the apothecaries non-sense, irrational conclusions, falsities in matters of fact, and in quotations, concessions, &c. of a nameless person. And also an answer to the rest of "Lex talionis."*
London: printed for James Allestry, 1670.
STC II M1841
Subject: Apothecaries, medicine.
Case/I/045/.88/no. 4

Binders title: *Tracts on physick, 1665–70.*

## Mersenne, Marin, 1588–1648.

*Cogitata physico-mathematica. In quibus tam naturae quàm artis effectus admirandi certissimis demonstrationibus explicantur.*
Paris: Antonius Bertier, 1644.
Subject: Ballistics, hydrostatics, mathematics, mechanics, nautical sciences.
Case/4A/1018

In 6 tracts, including "De hydraulico-pneumaticis phaenomenis";"De arte nautica, seu histiodromia, & hydrostatica"; "De mechanicis phaenomenis"; "De ballisticis, seu acontismologicis phaenomenis."

## Mersenne, Marin.

*Harmonicorvm libri: in qvibvs agitvr de sonorvm natvra, cavsis et effectibvs: de consonantiis, dissonantiis, rationibus, generibus, modis, cantibus, compositione, orbísque totius harmonicis instrumentis. Opus vtile grammaticis, oratoribus, philosophis, iurisconsultis, medicis, mathematicis, atque theologis.*

Paris: Gvillielmus Bavdry, 1636.
Subject: Acoustics, harmonics, mechanics, music, music and medicine.
Case/fV/201/.572

In 8 books, e.g., bk. 3: De fidibvs, nervis et chordis atque metallis, ex quibi fieri solent.

## Mersenne, Marin.

*Harmonie vniverselle, contenant la theorie et la pratiqve de la mvsiqve. Où il est traite de la nature des sons, & des mouuemens, des consonances, des dissonances, des genres, des modes, de la composition, de la voix, des chants, & de toutes sortes d'instrumens harmoniques.*
Paris: Sebastien Cramoisy, 1636, 1637.
Subject: Acoustics, instruments, music.
Case/6A/141

2 vols. Some MS marginalia. "Table des propositions des 19 liures de l'harmonie vniuerselle" at end of pt. 2. The tracts in this copy are assembled in the order given in the table, an order seldom followed. Among other tracts are "De l'vtilité de l'harmonie, & d'autres parties des mathematiques" (pt. 2, bk. 7) and "Novvelles observations physiqves et mathematiqves" (pt. 2, bk. 8).

## Mersenne, Marin.

*Les prelvdes de l'harmonie vniverselle, ov qvestions cvrievses: vtiles aux predicateurs, aux theologiens, aux astrologues, aux medecins & aux philosophes.*
Paris: Henry Gvenon, 1634.
Subject: Judicial astrology.
Case/3A/742

Horoscope of a perfect musician; asks why people reject astrology; discusses the humors.

## Mersenne, Marin.

*Qvaestiones celeberrimae in Genesim cvm accvrata textvs explicatione... Graceorum & Hebraeorum musica instauratur. . . . Opvs theologis, philosophis, medicis, iurisconsultis, mathematicis, musicis verò, & catoptricis praesertim vtile.*
Paris: Sebastian Cramoisy, 1623.
Subject: Astrology.
Case/fV/501/.57

Among the questions: Apparitiones angelorum in mathematicas refundi non posse (no. 7). De speculo vstorio parabolico (no. 13). Astra non esse causas miraculosarum curationum (no. 22). Quantum in corde, pulmonibus, & alijs partibus stomachi diuina prouidentia reluceat (p. 1130). BOUND WITH *Observationes, et emendationes ad Francisci Georgii veneti problemata: Hoc opere cabala evertitvr.*

## Merula, Gaudenzio, 1500–1555

*Nvova selva di varia lettione tradotta di Latino in lingua italiana.*
Venice: Gio. Andrea Valuassori [called] Guadagnino, 1561.
Subject: Encyclopedia of science, scientific miscellany.
Case/L/0114/.572

Among subjects are heavens, elements, man, marvels and wonders (pyramids, earthquakes), trees, stones, cosmography, transmutation of metals.

## Merula, Paulus, 1558–1607.

*Cosmographiae generalis libri tres: Item geographiae particvlaris libri qvatvor: quibus Evropa in genere; speciatim Hispania, Gallia, Italia, describuntur.*
[Leyden] Plantiniana Raphelengij, 1605.
Subject: Cosmography
Ayer/*7/M/45/1605

In 2 parts. Pt. 1: De mundo, de sphaeram (coelestem, terrestrem); pt. 2 focuses on Europe, esp. France, Spain, Italy.

## Metellus, Johannes Matalius, Sequanus.

*Germania superior. 38. Inferior qvae etiam Belgivm dicitvr.*
Cologne: Ioannes Christophorus, 1598.
Subject: Cosmography.
Case/5A/547/no. 2

BOUND WITH Acosta, *Geographische vnd historische*, and 2 works by Botero, *Geographische landtaffel* and *Theatrvm principvm*, qq.v.

## [Michaelis, Sébastien], 1543–1618.

"The admirable history of the possession and conversion of a penitent woman. Seduced by a magician that made her to become a witch."
[London: printed for William Aspley, 1613.]
STC 17854
Subject: Witchcraft.
Case/B/88/.57

Translated by W. B. Title in MS, taken from title on opening page of work: "The admirable history of a penitent woman conuerted, who was seduced by a magician in the country of Prouince in France: and of the end of the said magician." 2 vols. in 1. Vol. 2: *A discovrse of spirits.*

## Michaelis, Sebastian.

*Histoire admirable de la possession et conversion d'vne penitente, seduite par vn magicien, la faisant sorciere & princesse des sorciers au païs de Prouence, conduite à la Ste. Baume pour y etre exorcizee l'an M. DC. X. Au mois de Nouembre, souz l'authorité du R. P. F. Sebastien Michaelis. . . . Ensemble la pneumologie ou discours des esprits du sudit P. Michaelis reueu, corrigé & augmenté.*
Paris: Charles Chastellain, 1613.
Subject: Witchcraft.
Case/3A/2772
2d ed. BOUND WITH his *Discovrs des esprits en tant qv'il est de besoin, pour entendre & resoudre la matiere difficile des sorciers.*

## Michelot, Henry.

*Le portulan de la mer Mediterranée, ou, le vray guide des pilotes costiers. Dans lequel on verra la veritable maniere de naviguer le long des côtes d'Espagne, Catalogne, Provence, Italie, les isles d'Yvice, Mayorque, Minorque, Corse, Sicile, & autres.*

Amsterdam: Pierre Mortier, 1709.
Subject: Navigation.
Ayer/8.9/N2/M62/1709

With "L'usage des tables de la declinaison du soleil" (p. 157).

## Middendorp, Jacob, 1538?–1611.

*Academiarvm celebrivm vniversi terrarvm orbis libri VIII.*
Cologne: Gosuinum Cholinum, 1602.
Subject: Science teaching, academies and learned societies.
I/7030/.574

Bk. 1.: learned academies in general, what they are, with description of subjects taught. Following books describe learned academies in Judaea, Italy, Germany, Bohemia, Poland, Denmark, France, Spain, and England.

## Middleton, John.

*Practical Astrology.*
London: printed by J. C. for Richard Preston, 1679.
STC II M1978
Subject: Astrology.
Case/B/8635/.574

In 2 parts. 1: "An easie introduction to the whole art of astrologie"; 2: "The resolution of all manner of horary questions" (e.g., "how to find the nature and kinde of disease").

* * *

*Mirabilis annvs, or the year of prodigies and wonders, being a faithful and impartial collection of several signs that have been seen in the heavens, in the earth, and in the waters; together with many remarkable accidents and judgments befalling divers persons, according as they have been testified by very credible hands.*
[London] Printed in the year 1661.
STC II E3127
Subject: Meteors, prodigies, wonders.
Case/B/863/.26

BOUND WITH *Mirabilis annvs secundus*, q.v.

* * *

*Mirabilis annus secundus: or, the second part of the second years prodigies. Being a true additional collection of many strange signs and apparitions, which have this last year been seen in the heavens, and in the earth, and in the waters.*
[London] Printed in the year 1662.
STC II M2204
Subject: Meteors, prodigies, wonders.
Case/B/863/.26

["Printed by stealth."]

* * *

*Mirabilis annus secundus, or, the second year of prodigies. Being a true and impartial collection of many strange signes and apparitions, which have this last year been seen in the heavens, and in the earth, and in the waters. Together with many remarkable accidents and judgements befalling divers persons.*
[London] Printed in the year 1662.
STC II M2205
Subject: Meteors, prodigies, wonders.
Case/B/863/.261

## Mizauld, Antoine, d. 1578.

*Centuriae IX memorabilivm, vtilivm, ac jvcvndorvm in aphorismos arcanorum omnis generis locupletes, perpulcre digestae; Accessit his appendix nonnullorum secretorum, experimentorum, antidotorumque contra varios morbos, tam ex libris manuscriptis quam typis excusis, collecta. Seorsvm excvsa harmonia caelestivm corporum & humanorum, dialogus vndecim astronomice & medice. . . . Item memorabilium aliquot naturae arcanorum syluula, rerum variarum sympathias & antipathias.*
Frankfurt: Ioannes Saurius, 1599.
Subject: Astrological medicine, astronomy, sympathy and antipathy.
Case/3A/1060

## Mizauld, Antoine.

*Ephemerides aeris perpetvae: sev popularis & rustica tempestarum astrologia, vbique terrarum & vera, & certa.*
Antwerp: Gerard Spelmann, 1555.
Subject: Earthquakes, relation between the stars and storms.
Case/-B/868/.58

## Mizauld, Antoine.

*Harmonia svperioris natvrae mvndi et inferioris; vnà cum admirabili foedere & sympatheia rerum vtriusque; quibus annectuntur paradoxa doctrina coelesti accommoda.*
Paris: Federicus Morell, 1577.
Subject: Astronomy, harmony between man and heavens, sympathy and antipathy.
Case/L/900/.584

BOUND WITH *Paradoxa rervm coeli.*

## Mizauld, Antoine.

*Memorabilium utilium, ac ivcvndorvm centvriae novem, in aphorismos arcanorum omnis generis locupletes, perpulchrè digestae.*
Paris Federicus Morell, 1567.
Subject: Astrological medicine, astronomy, sympathy and antipathy.
Case/Y/682/.M7672

Remedium doloris capitis (p. 13).

## [Moerman, Jan]

*Apologi creatvrarvm.*
[Antwerp: C. Plantin, 1584?]
Subject: Religion and science.
Case/Y/682/.M7675

Engraved title with "G. de Jode excu 1584" mounted on 1st sheet. In MS on this title sheet: "Auctore Johanne Moërmanno Johanni Aviario D." Dedication signed Iohannes Moerman. No colophon leaf.

## Molitor, Ulrich, d. 1492.

*De lamiis et phitonicis mulieribus teutonice vnholden vel hexen.*
[Reutlingen: Johann Otmar, not before January 1489.]
H-C *11536^ BMC (XV)II:587^ Pr. 720^ Still. M683^ Goff M-795
Subject: Demons, incantations, transformation of witches to women and vice versa, witchcraft.
Inc./2716/.5

## Monardes, Nicolàs, 1512–1588.

*Delle cose, che vengono portate dall' Indie Occidentali pertinenti all' vso della medicina. Parte seconda, distinta in due libri.*
Venice: heirs of Francesco Ziletti, 1589.
Subject: Botany, herbals, materia medica.
Ayer/109.9/B6/07/1589

In 2 bks. Bk. 1: "Del tabaco & sue grandi virtù" (pp. 8–24), sassafras (pp. 24–46), "Del zarzapariglia di Guaiquil" (chap. 13, p. 87), "Del ambragriso" (chap. 15, pp. 93–101); bk. 2: "Del sangre di drago" (p. 72). BOUND WITH Orta, q.v.

## [Monardes, Nicolàs]

*Ioyfvll nevvs ovt of the newe founde worlde wherein is declared the rare and singuler vertues of diuerse and sundrie hearbes, trees, oyles, plantes, and stones, with their applications, as well for phisicke as chirurgerie, the saied beyng well applied bryngeth suche present remedie for all deseases, as maie seme altogether incredible: notwithstandyng by practize founde out, to bee true: Also the portrature of the saied hearbes, very aptly described.*
London: William Norton, 1577.
STC 18005a
Subject: Herbal medicine, New World.
Ayer/109.9/M4/M7/1577

Translated by John Frampton. "The thynges that are brought from the occidentall Indias serueth for the use of medicine." In 3 parts.

## Monardes, Nicolàs.

*Ioyfvll newes out of the new found world, wherein are declared the rare and singular vertues of diuers and sundrie herbs, trees, oyles, plants & stones. . . . Newly corrected. . . . Whereunto are added three other bookes treating of the bezaar stone, the herbe escuerçonera, the properties of yron and steele, in medicine and the benefite of snowe.*
London: William Norton, 1580.
STC 18006
Subject: Herbal medicine, metallurgy, New World.
Ayer/*109.9/M4/M7/1580

The three added tracts are: 1. *The dialogve of Yron, which treateth of the greatnesse thereof, and howe it is the most excellent metall of all others, and the thing most necessary for the seruice of man; and the greate medicinall vertues which it hath.* 2. *A booke which treateth of two medicines most excellent agaynst all venome, which are the bezaar stone, and the hearbe escuerconera.* 3. *The boke of which treateth of the snow, and of the properties and vertues thereof; and of the manner that should be vsed to make the drink cold therewith: & of the other wayes wherewith drinke is to be made colde: whereof is shewed partly, in the latter parte of the seconde dialogve of yron.*

## Monardes, Nicolàs.

*Primera parte del libro qve trata de las cosas que se traen delas Indias Occidentales, que siruen al vso de medicina, y de la orden que se ha de tener en tomar la rayz del Mechoacan. Do se descubren grandes secretos de naturaleza y grandes experiencias. Hecho y copilado por el doctor Monardes, medico de Seuilla. Parte segunda. De las cosas que traen del arbol que traen del Florida llamado sassafras. Parte tercera. De la historia de las cosas que se traen de nuestras Indias Occidentales, que siruen al vso de medecina.*
Seville: Alonso Escriuano, 1574.
Subject: Herbal medicine, metallurgy, New World.
Ayer/109.9/M4/M7/1574

Contains also *Libro qve trata de dos medicinas excellentissimas contra todo veneno: que son la piedra bezaar, y la yerua escuerçonera; Dialogo del hierro, y de svs grandezas, y como es el mas excelente metal de todos, y la cosa mas necessaria para seruicio del hombre, y de las grandes virtudes medicinales que tiene;* and *Libro qve trata de la nieve y el modo qve se ha de tener, en el beuer enfriado con ella: y delos otros modos que ay de enfriar.*

## Monardes, Nicolàs.

*De simplicibvs medicamentis ex occidentali India delatis, qvorvm in medicina vsvs est. Auctore D. Nicolao Monardes Hispalensi medico; interprete Carolo Clvsio Atrebate.*
Antwerp: Christopher Plantin, 1574.
Subject: Herbal medicine.
Ayer/*109.9/M4/M7/1574a

Printed note on front pastedown describes this as 1st Latin ed. of the earliest American herbal as issued by Carolus Clusius, the eminent botanist. See "Monardes tells us of various herbs and gums brought from the New World and of what the herbalist had been able to learn of their medicinal virtues. The second part contains the first written account and illustration of tobacco and its use as a wound-herb and in the religious ceremonies of the Red Indians" (Rohde, *Old English Herbals*, pp. 121–24).

## Mondino dei Luzzi, d. 1326.

*Mundinus de omnibus humani corporis interioribus menbris [sic] anathomia.*
Strassburg: Martinus Flach, 1513.
Subject: Anatomy.
Case/oQM/21/.M66/1513

Ed. by Johannes Adelphus.

## Montagnana, Bartolomeo, d. ca. 1460.

*Consilia Bartholomei Montagnane. Tractatus tres de balneis patauinis. De compositione & dosi medicinarum. Antidotarium eiusdem.*
[Venice:] Bonetus Locatellus for Octavianus Scotus, 1497.
H-C(Add) 11552*^ BMC (XV)V:448^ Pr. 5081^ Still. M702^ Goff M-815

Subject: Balneology, consilia, medicine.
Inc./f5081

### Montaigne, Michel Eyquem de, 1533–1592.

*Apologia di Raimondo di Sebonda . . . nel qvale si tratta della debolezza, & incertitudine del discorso humano.*
Venice: Marco Ginammi, 1634.
Subject: Natural theology, physiognomy, prognostication.
Case/Y/762/.M7697

In *Saggi di Michel Sig. di Montagna ouero discorsi natvrali, politici, e morale, trasportati della lingua Francese nell' Italiana per opera di Marco Ginammi.* Apology of R. Sebonde has separate title page.

### Montaigne, Michel Eyquem de.

*Essais de Messire Michel seignevr de Montaigne . . . livre premier & second.*
Bordeaux: S. Millanges, 1580.
Tchemerzine VIII:402^ Le Petit, pp. 99–100
Subject: Natural theology, physiognomy, prognostication.
Case/3A/584

Bk. 2, chap. 12: *Apologie de Raimond Sebond.*

### Montaigne, Michel Eyquem de.

*Les essais de Michel Seignevr de Montaigne.*
Paris: Abel L'Angelier, 1600.
Subject: Natural theology, physiognomy, prognostication.
Case/Y/762/.M7547

In 3 vols. "Edition novvelle, prise sur l'exemplaire trouué apres le deceds de l'autheur, reueu & augmenté d'vn tiers outre les precedentes impressions." Included in vol. 1: on prognostications, on cannibals; vol. 2: Apology of R. Sebonde; vol. 3: on physiognomy.

### Montaigne, Michel Eyquem de.

*Les essais de Michel, Seignevr de Montaigne.*
Paris: Iean Camvsat, 1635.
Subject: Natural theology, physiognomy, prognostication.
Case/fY/762/.M7548

"Edition novvelle." In 3 books. Included in bk. 1: "Des prognostications" (p.26); bk. 2: "Apologie de Raimond de Sebonde" (p.329), and "D'un enfant monstrueux" (p. 557); and 3: "De la physiognomie" (p. 807).

### Montaigne, Michel Eyquem de.

*Les essais de Michel seignevr de Montaigne.*
Paris: chez Michel Blageart, 1640.
Subject: Natural theology, physiognomy, prognostication.
Case/Y/762/.M755

"Edition novvelle."

### Montaigne, Michel Eyquem de.

*Les essais de Michel, Seigneur de Montaigne.*
Brussels: François Foppens, 1659.
Subject: Natural theology, physiognomy, prognostication.
Case/*Y/762/.M756

3 vols. "Nouvelle edition exactemen pvrgée des defavts des precedentes, selon le vray original." Each vol. has separate title page.

### Montaigne, Michel Eyquem de.

*Les essais de Michel de Montaigne.*
Paris: Laurent Rondet [etc.], 1669.
Subject: Natural theology, physiognomy, prognostication.
Case/Y/762/.M7566

3 vols. "Nouuelle edition." See Sayce & Maskell, *A descriptive bibliography of Montaigne's* Essais (London, 1983).

### Monte, Guido Ubaldo, Marchese del, 1545–1607.

*Mechanicorvm liber.*
Pesaro: Hieronymus Concordia, 1577.
Subject: Mechanics, technology.
Wing/fZP/535/.C74

De libra, de vecte, de trochlea, de axe in peritrochio, de cuneo, de cochlea.

### Monteiro de Carvalho, José.

*Noticia astronomica, ou discurso do cometa, que na noute de vinte e outo de Dezembro, se vio sobre esta cidade de Lisboa.*
Lisbon [1744]
Subject: Comets.
Greenlee/4504/P855

### Moore, Sir Jonas, 1617–1679.

*Bibliotheca mathematica optimis libris diversarum linguarum refertissima.*
London: Edward Millington, 1684.
Subject: Auction catalogues, geography, mathematics, medicine.
STC II M2567
Case/Z/4911/.054

Binder's title: *Auction catalogues.* A collection of book auction catalogues bound together. Most catalogues include works on geography, mathematics, medicine. First catalogue in collection, *Bibliotheca Angleseiana,* has MS table of contents and lists (in MS) prices paid for books.

### Moore, Sir Jonas.

*England's interest, or, the gentleman and farmer's friend, shewing how to make wines, cider, brew malt liquor, instructions for breeding horses, husbandry of bees, profitable ordering of fish ponds, & for breeding of fish.*
London: J. How, 1707.
Subject: Agriculture, household medicine.
Case/R/5145/.592

4th ed. Medicinal virtues of meath and metheglin (p. 145).

### Moore, Sir Jonas.

*Modern fortification: or, elements of military architecture. . . . with the use of a joint-ruler or sector, for the speedy description of any fortification.*

London: Obadiah Blagrave 1689.
Subject: Instrumentation, surveying.
Case/U/26/.59

Chap. 1: Basic surveying amd measuring principles; chap. 2: Use of a joint-rule or sector.

### Morato, Fulvio Pellegrino, d. 1547.

*Significato de i colori e de' mazzolli.*
Venice: Lucio Spineda, 1599.
Subject: Color symbolism.
Case/W/72/.74

BOUND WITH Rinaldi, Giovanni de', q.v.

### Mordecai Ben Judah Löb Ashkenazi, fl. 1701.

*Sepher Ha-Yashar* [Book of Jasher].
Furth: Model of Ansbach, 1701.
Subject: Cabala.
Case/C/13/.19

By J. L. Sarahya (formerly ascribed to Jac. Tam). According to the card catalogue entry, this work is "an introduction to Eshel Abraham by Abraham Rovigo."

### More, Henry, 1614–1687.

*An antidote against atheism, or an appeal to the naturall faculties of the minde of man, whether there be not a god.*
London: printed by J. Flesher, 1655.
STC II M2640
Subject: Religion and science.
Case/B/73/.592

2d ed. corrected and enlarged. With an appendix. Binders title: *H. More on atheism, enthusiasme.* With this are bound the following, each having separate pagination but some lacking title pages: (1) *A short discourse of the nature, causes, kindes, and cure of enthusiasme;* (2) *Observations upon* [Thomas Vaughan's] *"Anthroposophia theomagica" and "Anima magica abscondita" by Alazonomastix Philalethes.* London: printed by J. Flesher, 1655. Subject: Alchemy. "But tell me, Mr. Alchymist! in all your skill and observation in your Experiments, if you have hit on any thing that will settle any considerable point controverted amongst Philosophers which may not be done as effectually at lesse charges. Nay, whether you may not lose Nature sooner then find her by your industrious *vexing* of her, and make her appear something else then what she really is" (p. 100); (3) *The second lash of Alazonomastix* [same imprint]; (4) *The contents of Mastix his letter to a private friend concerning his reply.* Subject: Mysticism (Jacob Behmen). BOUND WITH More's *Conjectura cabbalistica,* q.v.

### More, Henry.

*A collection of several philosophical writings of Dr Henry More.*
London: printed by James Flesher for William Morden, 1662.
STC II M2646
Subject: Cabala, natural theology.
Case/fB/245/.59

Includes *Enthusiasmus triumphatus or, a brief discourse of the nature, causes, kinds, and cure of enthusiasm; Letters to Des-Cartes, &c.; Conjectura cabbalistica. Or, a conjectural essay of interpreting the mind of Moses, in the first three chapters of Genesis, according to a threefold cabbala: viz. literal, philosophical, mystical or divinely moral.*

### More, Henry.

*Conjectura cabbalistica. Or, a conjectural essay of interpreting the minde of Moses, according to a threefold cabbala: viz. literal, philosophical, mystical, or divinely moral.*
London: printed by James Flesher, 1653.
STC II M2647
Subject: Cabala.
Case/B/73/.592

Followed by *The defence of the threefold cabbala* (continuously paginated).

### More, Henry.

*Tetractys anti-astrologica, or, the four chapters in the explanation of the grand mystery of godliness, which contain a brief but solid confutation of judiciary astrology.*
London: printed by J. M. for Walter Kettilby, 1681.
STC II M2679
Subject: Astrology.
Case/B/8635/.594

Chap. 15: The general plausibilities of the art of astrology propounded. Of the twelve celestial houses. Chap. 16: That the stars and planets are not useless tho there be no truth in astrology. Chap. 17: Their [the astrologers'] fallacious allegation of events answering to predictions. An answer to that evasion of theirs, that the error is in the artist, not in the art.

### Morelli, Gregorio.

*Scala di tvtte le scienze, et arti del l'eccellente medico et filosofo messer Gregorio Morelli, divisa in qvattro settioni.*
Venice: Gabriel Giolito de' Ferrari, 1568.
Subject: Experimental science, instruments, medicine.
Case/3A/349

In dialogue form: Tomitano et Morello.

### Moretti, Tomaso, d. 1675.

*A general treatise of artillery with an appendix of artificial fire-works for war and delight by Sir Abraham Dager.*
London: printed by A. G. & J. P. for Obadiah Blagrave, 1683.
STC II M2726
Subject: Geometry, pyrotechnics.
Case/3A/1734

Moretti's treatise trans. Sir Jonas Moore. The appendix of artificial fire-works contains formulas and recipes for the making of fireworks.

### Morgagni, Joannes Baptista, 1682–1771.

*Epistolae anatomicae duode viginti ad scripta pertinentes celeberrimi viri Antonii Mariae Valsalvae.*
Venice: Franciscus Pitter, 1741.

Subject: Anatomy.
UNCATALOGUED

2 vols. Valsalva's tract (ends p. 96), followed by 13 letters of Morgagni in vol. 1; letters 14–20 of Morgagni constitute vol. 2.

## Morhof, Daniel Georg, 1639–91.

*Polyhistor, literarius, philosophicus et practicus. . . . Brevis notitia alphabetica, ephemeridum literariarum, et aliorvm qvorvndam scriptorvm eivsmodi divrnorvm, hebdomadariorvm, menstrvorvm, anniversariorvmqve, avcta et ad annvm 1747 continvata a Io. Ioachimo Schwabio.*
Lübeck: P. Boeckmann, 1747.
Subject: Encyclopedia of logic and metaphysics, magic, mathematics, medicine, natural history, physics.
Y/011/.595

3 vols. in 2. 4th ed. List of learned journals at front. Includes "De polymathia" (vol. 1, bk. 1, chap. 1); "De arte Lulliana similibusque inventis" (bk. 2, chap. 5); "Polyhistoris practici liber VII medicvs, de scriptoribvs medicis" (p. 691); "De Paracelso" (p. 249); "De qualitatibvs occvltis" (vol. 2, bk. 2, pt. 2, p. 302).

## Moro, Anton-Lazzaro, 1687–1764.

*De' crostacei e degli altri marini corpi che si truovano su' monti libri due.*
Venice: printed by Stefano Monti, 1740.
Subject: Earth sciences, fossils.
QL/435/.M67/1740

Discusses Woodward and Burnet. "Come avvenuto sia, che certe porzioni di monti siano tanto piene di crostacei, che pajono de essi composte ed impastate: e come in certi monti ritrovinsi ossa di balene, o d'altri mostri marini, e crostacei che sembrano di mari stranieri" (bk. 2, chap. 22).

* * *

*A most rare and true report of such great tempests, straunge sightes, and wonderfull accidents, which happened by the prouidence of God, in Hereford shire, at a place called the Hay, and there abouts, besides the sightes of strannge [sic] fowles, which there were seene, most fearefull to beholde, with their horrible cryes & strangenes, with the great hurt was done by them.*
London: printed for Thomas Law, 1585.
STC 20889.5
Subject: Supernatural.
Case/3A/618

## Moxon, Joseph, 1627–1700.

*Ein Kurtzer discours von der schiff-fahrt bey dem Nord-pol nach Japan, China, und so weiter. Durch drey erfahrunge dargethan und erwiesen nebenst beantwortungen aller ein würffe welche wieder die fahrt auff diesen weg sönnen eingewendet worden. . . . Sampt einer land-charte so aller länder nechst dem polo anweiset. Aus dem Englischen ins hochdeutsche übersetzet.*
Hamburg: Johan Nauman and G. Wolff, 1676.
Subject: Geography, navigation.
Ayer/*7/M78/1676

Translation of Moxon's *Brief Discourse of a passage by the North Pole to Japan.*

## Moxon, Joseph.

*Mathematicks made easie: or, A mathematical dictionary explaining the terms of art, and difficult phrases used in arithmetick, geometry, astronomy, astrology & other mathematical sciences.*
London: printed for Joseph Moxon, 1679.
STC II M3006
Subject: Mathematical dictionary.
Wing/ZP/645/.M8762

Appendix with weights and measures. 2 flyleaves with MS notes. Verso of title page has partial table of contents in MS.

## Moxon, Joseph.

*Mechanick excercises, or the doctrine of handy-works. Began Jan. 1. 1677. And intended to be continued. By Joseph Moxon, member of the Royal Society, and hydrographer to the King's most excellent majesty.*
London: printed for Joseph Moxon, 1683.
STC II M3014
Subject: Carpentry, forging, printing technology, smithing, turning.
Case/Wing/Z/4029/.6

2 vols. First edition with woodcuts. Printed with separate title pages—as a serial—e.g., from Jan. 1, 1677, to Feb. 1, 1677, and intended to be monthly continued, but issues appeared at irregular intervals: no. 3 is Feb.–Mar. 1677; no. 4, Apr. 1–May 1, 1678; no. 7 is 1679; no. 10 is 1680. Vol. 1 goes to no. 14 (1680). Vol. 2: *Mechanick exercises: or, the doctrine of handy-works. Applied to the art of printing* [same imprint]. Concentrates on printing technology: letter cutting, mold making, composition, sinking matrices, pressman's trade, warehouseman's office.

## Moxon, Joseph.

*Regulae trium ordinum literarum typographicarum or the rules of the three orders of print letters: viz the Roman, Italick and English capitals and small. Shewing how they are compounded of geometrick figvres, and mostly made by rule and compass. Useful for writing masters, painters, carvers, masons, and others that are lovers of curiosity.*
London: printed for Joseph Moxon, 1676.
STC II M3019
Subject: Geometry of lettering, type design, typography.
Case/Wing/Z/4035/.598

## Moxon, Joseph.

*A tutor to astronomie and geographie. Or an easie and speedy way to know the use of both the globes, coelestial & terrestrial. In six books. The first teaching the rudiments of astronomie and geographie. The 2, 3, 4, 5, 6 shewing by the globe the solution of astronomical and geographical prob., problemes in navigation, astrological problemes. Gnomonical problemes. Trigonometrical problemes. More fully and amply than hath yet been set forth either by Gemma Frisius, Metius, Hues,*

*Wright, Blaew, or any others that have taught the use of the globes: and that so plainly and methodically that the meanest capacity may at first reading apprehend it, and with a little practice grow expert in these divine sciences.*
London: printed by Joseph Moxon, 1670.
STC II M3023
Subject: Astronomy, geography.
Wing/ZP/645/.M876

2d ed. corrected and enlarged.

## Moxon, Joseph.

*A tutor to astronomy and geography. Or an easie and speedy way to know the use of both the globes, coelestial and terrestrial.*
London: printed by S. Roycroft for Joseph Moxon, 1686.
STC II M3025
Subject: Astronomy, geography.
Case/L/900/.598

4th ed. corrected and enlarged.

## Moxon, Joseph.

*A tutor to astronomy and geography. Or an easie and speedy way to know the use of both the globes coelestial and terrestial. In 6 books. Teaching the rudiments of astronomy and geography, shewing by the globes the solution of astronomical and geographical problems, problems in navigation, astrological problems, gnomonical problems, trigonometrical problems more fully than hath been set forth, either by Gemma Frisius, Metius, Hues, Wright, Blaew, or any others that have taught the use of the globes: And that so plainly and methodically, that the meanest capacity may at first reading apprehend it, and with a little practice grow expert in the divine sciences. Whereunto is added the antient poetical stories of the stars; shewing reasons why the several shapes and forms are pictured on the coelestial globe. As also a discourse of the antiquity, progress and augmentation of astronomy.*
London: printed for W. Hawes, 1699.
Subject: Astronomy, geometry.
Case/QB/41/.M68/1699

5th ed. corrected and enlarged.

## Müller, Johannes, Regiomontanus, 1436–1476.

*Calendario.*
Venice: Bernhard Maler, Peter Löslein, & Erhard Ratdolt, 1476.
H 13789^ BMC (XV)V:243–44^ Pr. 4366^ Still. R99
Subject: Astrology, astronomy, calendar.
Inc./f4366

In Italian. "Qvesta opra da ogni parte e un libro doro non fu piu preciosa gemma mai dil kalendario: che tratta cose asai con gran facilita: ma gran lauoro qui numero aureo: e tutti i segni suoro." Bernardus pictor de Augusta, Petrus Loslein de Langencen, Erhardus Ratdolt de Augusta.

## Müller, Johannes, Regiomontanus.

*Calendario a Joanne de Monte Regio Germanorum decoris nostre etatis astronomorum principis editi Jacobi Sentini Ricinensis.*
Venice: Erhard Ratdolt, 1483.
H *13778^ Still. R90^ Goff R-95
Subject: Astrology, astronomy, calendar.
Inc./4390.5

## Müller, Johannes, Regiomontanus.

*Ephemerides. Aetatis nostrae astronomorum principis ephemerides.*
Venice: Erhard Ratdolt, 1481.
Reichling 1046^ Still. R101^ Goff R-105
Subject: Almanacs, astronomy.
Inc./4381.5

Ephemerides, 1482–1506.

## [Mueller, Johannes, Regiomontanus]

*Ephemerides, solis & lune planetarumque perpetue.*
Venice: Petrus Liechtenstein, 1498.
H-C *13798^ BMC (XV)V:578^ Pr. 5641^ Still. 105^ Goff R-110
Subject: Almanacs, astronomy.
Inc./5641

## Müller, Johannes, Regiomontanus.

*Epitoma in almagestum Ptolomei.*
Venice: Johannes Hamman, 1496.
H-C *13806^ BMC (XV)V:427^ Pr. 5197^ Still. R-106^ Goff R-111
Subject: Astronomy.
Inc./f5197

[Incipit] "Epytoma Joannis de Monte Regio in almagestum Ptolomei. Cl. Ptolemei alexandrini astronomorum principis [Greek text] id est in magnam constructionem: Georgij Purbachij. Eiusque discipuli Johannis de Regio Monte astronomicon epitoma."

## Münster, Sebastian, 1489–1552.

[*Cosmographia universalis*]
*Cosmographie uniuersalis lib. VI, in quibus iuxta certioris fidei scriptorum traditionem describuntur, omnium habitabilis orbis partium situs propriaeque dotes. Regionum topographicae effigies. Terrae ingenia, quibus fit ut tam differentes & uarias specie res, & animatas & inanimatas, ferat. Animalium peregrinorum naturae & picturae.*
Basel: Henricvs Petri, 1550.
Subject: Cosmography, descriptive geography.
Ayer/7/M8/1550

## Münster, Sebastian.

[*Cosmographia universalis*]
*Cosmographie uniuersalis lib. VI.* [Another ed.]
Basel: Henricvs Petri, 1552.
Ayer/7/M8/1552

**Münster, Sebastian.**
[*Cosmographia universalis*]
*Cosmographiae uniuersalis. lib. VI.*
[Basel: Henricus Petri, 1559.]
Case/fG/117/.6

Imperfect. Over 30 leaves missing from scattered locations throughout vol.

**Münster, Sebastian.**
[*Cosmographia universalis*]
*Cosmographia vniversale, nella quale secondo che n'hanno parlato ipiu veraci scrittori, son designati siti di tutti gli paësi, i siti de tutte le parti del mondo habitabile, & le proprie doti: le tauole topographice delle regioni. Le naturali qualita del terreno, onde nascono tante diferenze, & varietà di cose, & animate & non animate. Le nature, & le dipinture degli animali pellegrini. . . . Raccolta primo da diuersi autori per Sebastiano Munstero, & dappoi corretta & repurgata, per gli censori ecclesiastici, & quei del re Catholico nelli paesi bassi, & per l'inquisitore di Venetia.*
Cologne: printed by heirs of Arnold Birckman, 1575.
Subject: Cosmography, descriptive geography.
Ayer/7/M8/1575

**Münster, Sebastian.**
[*Cosmographia universalis*]
*La cosmographie vniverselle de tovt le monde. En laquelle, suiuant les auteurs plus dignes de foy, sont au vray descriptes toutes les parties habitables, & non habitables de la terre, & de la mer, leurs assiettes & choses qu'elles produisent: puis la description & peincture topographique des regions, la difference de l'air de chacun pays, d'où aduient la diuersité tant de la complexion des hommes que des figures des bestes brutes. . . . Auteur en partie Mvnster, mais beaucou plus augmentée, ornée & enrichie, par Francois de Belle-Forest.*
Paris: Michel Sonnius, 1575.
Subject: Cosmography, descriptive geography.
Case/fG/117/.601

In 2 vols.

**Münster, Sebastian.**
[*Cosmographia universalis*]
*La cosmographie vniverselle de tovt le monde . . . Des terres descovuertes de nostre temps ausquelles on a donné le nom de nouueau monde & d'Indes Occidentales, ou Amerique.*
Paris: Michel Sonnius, 1575.
Subject: Cosmography, descriptive geography.
Ayer/7/M8/1575a

Imperfect: Book 7 and last only (cols. 2034–2235).

**Münster, Sebastian.**
*Weldt vnd Indianischen Königreichs newe vnnd wahrhaffte history, von allen geschichten, handlungen, thaten, strengem vnnd ernstlichem regiment der Spanier gegen den Indianern, ungläublichem grossem gut von goldt, sylber, edelgestein, peerlin, schmaragdt, vnnd andern reichtumb, so die Spanier darinn erobert: sambt von den forglichen schiffarthen kriegen, schlachten vnnd streit, eroberung vnd verhergung vieler prouinz, landtschafften, vnd Königreich, so sich bey vnser gedächtnuss haben darinn verlossen vnd zugetragen.*
Basel: Henricpetri, [1579].
Subject: Exploration, voyages.
Ayer/108/B4/1579a

BOUND WITH a map of The Netherlands and a work by Luigi Guicciardini.

**Münster, Sebastian.**
*Cosmographia, das ist: Beschreibung der gantzen welt darinnen aller monarchien keyserthumben. königreichen Eürstenthumben, graff-und herzschafften, ländern, stätten und gemeinden.*
Basel: Henricpetri, 1628.
Subject: Cosmography.
Case/fG/117/.602

"Das erste buch der cosmographey Sebastiani Münster, auss Ptolomeo, Strabone, Solino, Pomponio auch andren alten vnd newen erfahrneit cosmographis, trewlich zusammen gezogen und verteutschet."

**Münster, Sebastian.**
*Germaniae atqve aliarvm regionvm. . . . ex historicis atque cosmographis, pro tabula Nicolai Cusae intelligenda excerpta. item eiusdem tabulae canon.*
[Basel: A. Cratander, 1531.]
Subject: Cosmography.
Case/G/004/.602

**Münster, Sebastian.**
*Kalendarium hebraicum.*
Basel: Ioan. Froben, 1527.
Subject: Astrology, astronomy, calendar.
Wing/ZP/538/.F92093

Names and signs of planets that influence the days of the week. Good and evil aspects of the celestial signs. Eclipses. In Hebrew and Latin on facing pages, book format conforming to Hebrew style.

**Mulcaster, Richard, 1530?–1611.**
*Positions vvherein those primitive circumstances be examined which are necessarie for the training vp of children, either for skill in their booke, or health in their bodie.*
London: T. Vautrollier, 1581.
STC 18253a.
Subject: Child development, exercise, fitness.
Case/I/40165/.608

Chap. 6: "Of the exercises and training of the body. . . . What health is , and how it is maintained: what sicknesse is, how it cometh, and how it is prevented"; chap. 7: "The braunching order, and methode kept in this discourse of exercises."

**Muratori, Lodovico Antonio, 1672–1750.**
*Del governo della peste, e delle maniere di guardarsene.*
Modena: Bartolomeo Soliani, 1722.

Subject: Disease prevention, medicine, pest tracts.
Q/.604

2d ed. In 3 parts: governo politico, medico, ecclesiastico.

## Murner, Thomas, 1475–1537.

*Logica memoratiua chartiludium logice, siue totius dialectice memoria: & nouus Petri Hyspani textus emendatus: Cum incundo pictasmatis exercitio.*
Strassburg: Ioannes Gruninger, 1509.
Subject: Logic, mnemonics, symbolism.
Wing/ZP/547/.G915

First printed in Cracow, 1507.

## [Mussard, Pierre].

*Historia deorum fatidicorum.*
Geneva: P. Chouët, 1675.
Subject: Prognostication, prophesying, soothsaying.
Case/B/863/.608

In addition to a section, "De rervmn fvtvrarvm praecognitione, vaticinatione, & oraculis," there is also a section of portraits and short biographies of seers and soothsayers, including, e.g., Hermes Trismegistus, Tiresias, Iamblichus, Pythagoras, Calchas, and Mopsus.

## [Mylius, Martinus], 1542–1611.

*Chronologia scriptorum Philippi Melanchthonis.*
Görlitz: Ambrosius Fritschi, 1582.
Subject: Astrology, physics.
Case/C/515/.567

Chronological bibliography of Melanchthon's writings, including *Tractatus physica* and *De dignitate astrologiae.*

## Myritius, Joannes.

*Opvscvlvm geographicvm rarvm, totivs eivs negotii rationem, mira indvstria et brevitate complectens, iam recens ex diversorvm libris ad chartis, summa cura ac diligentia collectum & publicatum.*
Ingolstadt: Wolfgang Eder, 1590.
Subject: Astronomy, cosmography, geography.
Ayer/7/M9/1590

In 2 parts. Pt. 1: "De principiis astronomiae et geographiae"; pt. 2: "De particvlari orbis descriptione."

# N

## Najera, Antonio de, 17th cent.

*Svmma astrologica y arte para ensenar hazer pronosticos de los tiempos, y por ellos conocer la fertilidad, o esterilidad del año, y las alteraciones del aire, por el juyzio de los eclypses del sol, y luna, por la reuolucion del año, y mas en particular por las conjunciones, opposiciones, y quartos que haze la luna con el sol todos los meses, y semanas.*
Lisbon: Antonio Aluarez, 1632.
Subject: Astrology.
Case/B/8635/.61

* * *

*A natural history of Ireland, in three parts.*
Dublin: printed by & for George Grierson, 1726.
Subject: Archaeology, geography.
Case/G/42003/.6

"By several hands." Pt. 1: physical description; pt. 2: "A collection of such papers as were communicated to the Royal Society referring to some curiosities in Ireland"; pt. 3: about the Danish mounts, forts and towers in Ireland (archaeology).

## Naudé, Gabriel, 1600–1653.

*Apologie pour tous les grands personnages qui ont esté faussement soupçonnez de magie.*
The Hague: Adrian Vlac, 1653.
Subject: Defense of magic.
Case/B/88/.61

Mathematics mistaken for magic. Zoroaster neither author nor fomenter of forbidden magic. Defense of Pythagoras, Alkindi, Geber, Iamblichus, Scaliger, Cardan, Lull, Paracelsus, Agrippa, Merlin, Savonarola, Saint Thomas, Nostradamus, Roger Bacon, Michael Scot, Albertus Magnus, Robert of Lincoln.

## Naudé, Gabriel.

*The history of magick by way of apology, for all the wise men who have unjustly been reputed magicians, from the creation, to the present age.*
London: printed for John Streater, 1657.
STC II N246
Subject: Magic.
Case/B/88/.614

Englished by J. Davies.

## Naudé, Gabriel.

*Pentas quaestionum iatro-philologicarum.*
[N.p.: Samuel Chouët] 1647.
Subject: Medicine.
Y/682/.N227

1. "An magnum homini venenus periculum?" 2. "An vita hominvm hodie quàm olim breuior?" 3. "An matutina studia vespertinis salubriora?" 4. "An liceat medico fallere aegrotum?" 5. "De fato & fatali vitae termino."

## Naudé, Gabriel.

*Syntagma de stvdio liberali.*
Urbino: Mazzantinus & Aloysius Ghisonus, 1632.
Subject: Education, humanism, quadrivium.

"Adiunxit postea Galenus methodum, ac ordinem, exemplo nature" (p. 23).

## Navarro, Gaspar.

*Tribunal de supersticion ladina, explorador del saber, astucia, y poder del demonio; en que se condena lo que fuele correr por bueno en hechizos, agueros, ensalmos, vanos saludadores, maelficios, conjuros, arte notoria, caualista, y paulina, y semejantes acciones vulgares.*
Huesca: Pedro Bluson, 1631.
Subject: Demonology, witchcraft.
B/88/.616

## Nazari, Giovanni Battista.

*Della tramvtatione metallica sogni tre, di Gio. Battista Nazari Bresciano; nel primo de quali si tratta della falsa tramutatione sofistica: nel secondo della vtile tramutatione detta reale vsuale: nel terzo della diuina tramutatione detta reale filosofica. Aggiontovi di nvovo la concordanza de filosofi, & loro prattica; nellaquale, si vede i gradi, & termini di esso diuino magistero, & della verissima compositione della filosofia naturale, con laquale ogni cosa diminuta si riduce al vero solificio, & lunificio.*
Brescia: Pietro Maria Marchetti, 1599.
Subject: Alchemy.
Case/B/8633/.614

## Neander, Michael, 1525–1595.

*Chronicon sive synopsis historiarvm.*
Leipzig, 1586.
Subject: Descriptive geography.
Case/F/09/.614

Pp. 188–209: "Partivm orbis terrae veteris et recentis svccincta enumeratio simplexque & plana explicatio."

## Neander, Michael.

*Orbis terrae partivm svccincta explicatio, sev simplex envmeratio, distributa in singularum partum regiones: vbi singvlis regionibvs svae vrbes, elogia, et praeconia, personae, siue illustres, siue infames, fontes, merces, singularia & propria singulis, & caetera quacunque; ratione insignia, admiranda & noua adtribuuntur.*

Leipzig: Abraham Lamberg, 1589.
Subject: Descriptive geography.
Case/G/117/.61

3d ed. See, e.g., leaves 56v ff.: medicine in "regiones germaniae meridionales."

### Nedelius, Johannes, 1554–1612.

*Copulatio animae & corporis in homine.*
[Leipzig: Iohannes Rhamba] 1576.
Subject: Body and soul.
Case/B/511/.616

### Nemesius, bp. of Emesa, late 4th cent.

*De natura hominis.*
Antwerp: Christophorus Plantin, 1565.
Subject: Study of man, physiology, psychology.
Wing/ZP/5465/.P7

Printed in Greek and Latin, in succession. Includes material on apprehension, anxiety, fear, cognition, memory, the senses, physiology (breathing).

### Nesius, Ioannes, b. 1456.

*Oraculum de novo saeculo.*
Florence: Lorenzo Morgiani, [1496].
H *11693^ BMC (XV)VI:686^ Pr. 6373^ Still. N10^ Goff N-12
Subject: Prognostication.
Inc./6373

Date as supplied by Goff: 1497.

### Neudörffer, Anton, d. 1628

*Künstliche vnd ordentliche anweyssung der gantzen practic vff den jetzigen schlag vnd der selben herlichen geschwinden exempel vffs kurtzt zusammen getzogen &c neuen lieben discipeln zu sonderlich em nuzen gestelt durch mich Anthonium Newdorffer.*
Nuremberg: [Paulus Kauffmann,] 1599.
Subject: Arithmetic.
Wing/ZW/547/.N385

Part 2, "Schreibkunst," Nuremberg: Author, 1601.

### Neuhaus, Heinrich, fl. 1618.

*Pia & utilissima admonitio, de fratribvs Rosae-crvcis nimirum, an sint? Qvales sint? Vnde nomen illud sibi asciuerint? & quo fine eiusmodi famam sparserint?*
[Danzig?] [for sale by] Christoph Vetter, 1618.
Subject: Rosicrucians.
K/975/.619

### Newcastle, Margaret (Lucas) Cavendish, duchess of, 1624?–1674.

*Grounds of natural philosophy.*
London: A. Maxwell, 1668.
Subject: Medicine, natural history, natural philosophy.
STC II N851
fB/245/.622

2d ed. In 13 parts, with a 5-part appendix. Includes sections on matter, motion, vacuum, animals, man, motions of the mind, humors, diseases, elements.

### Newcastle, Margaret (Lucas) Cavendish, duchess of.

*Observations upon experimental philosophy: to which is added, the description of a new blazing world.*
London: printed by A. Maxwell, 1668.
STC II N858
Subject: Experimental science, natural history, natural philosophy.
Case/fY/1565/.N43

2d ed. 2 vols. in 1. Includes "Of the eyes of flyes; of the walking motions of flyes, and other creatures." Its principal divisions include "Further observations upon experimental philosophy; reflecting withall, upon some principal subjects in contemplative philosophy."

### Newcastle, Margaret (Lucas) Cavendish, duchess of.

*The philosophical and physical opinions.*
London: printed by J. Martin & J. Allestrye, 1655.
STC II N863
Subject: Medicine, natural history, natural philosophy.
Case/fB/245/.62

Earlier edition of *Grounds of Natural Philosophy*. Includes "Of matter and motion" (pt. 1); "Of metals, wetness, oyle, loadstone" (pt. 3); "The natural wars in animal figvres" (pt. 5); knowledge of diseases (chap. 208). From "An epistle to the unbelieving readers in natural philosophy" (p. 53): "I had rather live in a general remembrance than in a particular life."

### Newcastle, Margaret (Lucas) Cavendish, duchess of.

*Philosophical and physical opinions.*
London: printed by William Wilson, 1663.
STC II N864
Subject: Medicne, natural history, natural philosphy.
Case/6A/119

From "Epistle to the reader": "Thus, noble readers, you will find, that this present vvork contains pure natural philosophy, without any mixture of theology. . . . Neither can theology and natural philosophy agree, for philosophy is built upon all human sense, reason, and observation, whereas theology is onely built upon an implicit faith. . . . [With respect to religion] it is better to pray than to preach [because] disputations and arguments do rather deform religion than inform the creature. . . Natural philosophy doth not onely instruct men to know the course of the planets and the seasons of the year, but it instructs men in husbandry, architecture, and navigation, as also combination andassociation, but above all it instructs men in the rules and arts of physick." In 7 parts whose topics include physics, epistemology, medicine.

## Newcastle, Margaret (Lucas) Cavendish, duchess of.

*Poems and fancies: Written by the right honorable the Lady Newcastle.*
London: printed by T. R. for J. Martin & J. Allestrye, 1653.
STC II N869
Subject: Natural history, natural philosophy.
Case/fY/185/.N46

Scientific metaphors. Poems on motion, vacuum, the motion of the blood (p. 42), the ant, "A world made by atomes" (pp. 5–17), beasts, fish, birds.

## Newton, Isaac, 1642–1727.

*Philosophia naturalis principia mathematica. Imprimatur S. Pepys.*
London: J. Streater for the Royal Society, 1687.
STC II N1048^ Gray 5
Subject: Astronomy, mathematics, motion, physics.
Case/5A/162

1st ed., 1st issue. Chatsworth copy. Greek marginalia. Preface not signed in this copy.

## Nicander, of Colophon, fl. 2d cent. B.C.

*Les oevvres de Nicandre medecin et poete grec, tradvictes en vers François. Ensemble, deux liures des venins, ausquels il est amplement discouru des bestes venimeuses, theriaques, poisons & contrepoisons par Iaques Gréuin.*
Antwerp: Christophle Plantin, 1567.
Subject: Medicine, poisons and antidotes.
Wing/ZP/5465/.P701

Contains "Les theriaques de Nicandre" and "Les contrepoisons de Nicandre." BOUND WITH Grévin, q.v.

## Nicander, of Colophon.

*Nicandri Theriaca. Eiusdem Alexipharmaca. Interpretatio innominati authoris in theriaca. Commentarii diuersorum authorum in Alexipharmaca. Expositiò ponderum, signorum, & characterum.*
[Venice: Aldus & Andreas Asulanus associates, 1522.]
Subject: Medicine.
Case/*Y/642/.N2

MS notes in *Theriaca.*

## Nicephorus Gregoras [Blemmidas], d. 1274.

*Logica.*
Venice: Simon Bevilaqua, 1498.
H-C 11748*^ BMC (XV)V:523^ Pr. 5408^ Goff N-44
Subject: Astrolabe, astronomy, mathematics, medicine, music, plague.
Inc./f 5408

21 tracts, trans. Giorgio Valla, of which the following are on relevant subjects: 3. "Hypsiclis, Indeputatum Euclidi uolumen"; 4. "Euclidis, "Elementorum quartusdecimus liber"; 5. "Nicephorus Astrolabi"; 6. "Procli Diadochi, De fabrica usuque astrolabi"; 7. "Aristarchus Samii, De magnitudinibus & distantiis solis & lunae"; 8. "Timae Locri de vniversitatis natura. Timaevs de mundo"; 11. "Cleomedes, De mundo"; 12. "Cleomedes, Circularis inspectionis meteorum"; 14. "Aristotelis, De coelo volvminvm quatvor"; 16. "Galeni, De optima corporis nostri confirmatione"; 17. "Galeni, De inaeqvali distemperanti"; 18. "Galeni, De praesagitura"; 19. "Galeni, Succidanea" (list of substances and their uses); 20. "Alexander Aphrodiseivs, De febribvs"; 21. "Rhazes, De pestilentia." [Newberry Library copy catalogued "Blemmidas, Nicephorus."]

## Nicolas, Augustin, 1622–1695.

*Si la torture est un moyen seur a verifier les crimes secrets; dissertation morale et juridique, par laquelle il est amplement traitté des abus qui se commettent par tout en l'instruction des procés criminels, & particulierement en la recherche du sortilege.*
Amsterdam: Abraham Wolfgang, 1682.
Subject: Witchcraft.
I/2685/.624

"Artifices des demons pour entretenir l'abus de la torture" (p. 90).

## Nicolay, Nicolas de, Sieur d'Arfeuille, 1517–1583.

*Les navigations, peregrinations et voyages, faicts en la Turquie.*
Antwerp: Guillaume Silvius, 1576.
Subject: Geography, travel.
Wing/ZP/5465/.S587

In 4 books. "Des medecins de Constantinople" (bk. 3, chap. 12, p. 168). Nicolay was "valet de chambre & geographe ordinaire du roy de France."

## Nieremberg, Juan Eusebio, 1595–1658.

*Historia naturae, maxime peregrinae libris XVI, distincta. In quibus rarissima naturae arcana, etiam astronomica, & ignota Indiarum animalia, quadrupedes, aues, pisces, reptilia, insecta, zoophyta, plantae, metalla, lapides, & alia mineralia, fluuiorumqûe & elementorum conditiones, etiam cum proprietatibus medicinalibus, describuntur; nouae & curiosissima quaestiones disputantur, ac plura sacrae scripturae loca eruditè enodantur. Accedunt de miris & miraculosis naturis in Europâ libri duo: item de iisdem in terrâ Hebraeis promisâ liber vnus.*
Antwerp: Plantiniana, Balthasar Moreti, 1635.
Subject: Natural history, New World.
Ayer/*8.9/B7/N67/1635

1st ed. Illustrated by Christoffel Jeger. (No illustrations in added sections about Europe and the Promised Land.)

## Nieuwentijt, Bernard, 1654–1718.

*The religious philosopher: or, the right use of contemplating the works of the creator. I. In the wonderful structures of animal bodies, and in particular man. II. In the no less wonderful and wise formation of the elements. and their various effects upon animal and vegetable bodies. And, III. In the most amazing structure of the heavens, with all its furniture. Design'd for the conviction of atheists and infidels. Throughout which, all*

*the late discoveries in anatomy, philosophy, and astronomy, together with the various experiments made use of to illustrate the same, are most copiously handled by that learned mathematician Dr. Nieuwentyt.*
London: printed for Tho. Longman, 1724.
Subject: Anatomy, astronomy, natural science.
B/72/.625

2 vols. 3d ed. Translated by J. T. Desaguliers.

## Nifo, Agostino, 1473?–1538.

*De auguriis lib. II. His adiecimus Ori Apollinis Niliaci de hieroglyphicis notis lib. II à Bernardino Vicentino latinitate donatos.*
Basel: Iohannes Hervagivs, 1534.
Subject: Astrology, hieroglyphics.
Case/B/8635/.624

Application of zodiac signs to parts of the body. P. 90: Tabella de saltibvs et sternutamentis e.g., capvt; collvm & cervix; hvmeri et brachia; pectvs; latera et stomachvs; venter; venes et vertebras; pudenda; coxas; genva; crvra; pedes.

## Nifo, Agostino.

*De nostrarum calamitatum.*
Venice: Bonetus Locatellus for heirs of Octavianus Scotus, 1505.
Subject: Astrology, astronomy.
Case/fY/682/.N/528

Relation of stars and planets to disastrous events on earth.

## Nifo, Agostino.

*Avgvstini Niphi medicis libri dvo, de pvlchro, primvs. De amore secvndvs.*
Lyon: Godefridus & Marcellus Beringos (brothers), 1549.
Subject: Psychology.
Case/B/529/.623

## Nifo, Agostino.

*De pulchro et amore.*
[N.p.], 1614.
Subject: Psychology.
Case/B/529/.625

## Nifo, Agostino.

*In libros Aristotelis de coelo & mundo.*
Naples: Sigismund Mayr Alemanus, 1517.
Subject: Motion, natural philosophy, physics.
Case/fY/642/.A/83924

Aristotelis de coelo & mundo liber [secundus et] tertius interprete & expositore Augustino Nipho. In 4 parts: the universal world, parts of the world (stars), generation and corruption (elements), and gravity.

## Nifo, Agostino.

*Super postrema cum tabula Eustychi Augustini Niphi medices philosophi Suessani commentaria in libris posteriorum Aristotelis.*
[Venice: heirs of Octavianus Scotus, 1526].
Subject: Metaphysics, philosophy.
Case/fY/642/.A/86246

MS marginalia in bk. 2. Title page in MS. Bound in vellum MS.

## Niger, Dominicus Marius.

*Dominici Marii Nigri Veneti geographiae commentariorum libri XI, nunc primum in lvcem magno stvdio editi, qvibus non solum orbis totius habitabilis loca regiones prouinciae, urbes, montes, insulae, maria, flumina, & caetera, ut nostro tempore sunt sita & denominata, uerum etiam omnium ferè populorum & uariarum gentium mores, leges, ac ritus tam sacri quam prophani exactè describuntur, ita uel ipso Strabone utilior nostris temporibus, autor hic doctorum quorundam iudicio meritò habetur.*
Basel [H. Petri, 1557].
Subject: Geography.
Case/fG/117/.62

Geography of Africa, Asia. Includes geography of Laurentius Corvinus and Strabo, the latter with an epitome translated by Hieronymus Gemusaeus.

## Norden, John, 1548–1625?

*Specvlvm Britanniae. The first parte an historicall, & chorographicall discription of Middlesex. Wherein are also alphabeticallie sett downe, the names of the cyties, townes, parishes, hamlets, howses of name &c. With direction speedelie to finde anie place desired in the mappe & the distance betwene place and place without compasses. By the travaile and view of Iohn Norden, 1593.*
[London: Eliot's Court Press] 1593.
STC 18635
Subject: Chorography.
Case/G/45004/.6

With folding plates. Title page engrapved by Pieter vanden Keere.

## Norden, John.

The surveiors dialogve, very profitable for all men to pervse, but especially for gentlemen, farmers, and husbandmen, that shall either haue occasion, or be willing to buy, hire, or sell lands: as in the ready and perfect surveying of them diuided into six bookes by I. N.
London: printed by Thomas Snodham, 1618.
STC 18641
Subject: Surveying.
Case/H/9445/.63

"The third time imprinted."

## Norfolk, Henry Howard, 6th duke of, 1628–1684.

[Royal Society of London. Library.] *Bibliotheca Norfolciana: sive catalogus libb. manuscriptorum & impressorum in omni arte & lingua, quos illustriss. princeps Henricus, dux Norfolciae &c. Regiae societati londinensi pro scientia naturali promovenda donavit.*
London: Ric. Chiswel, 1681.
STC II N1230
Subject: Natural science.
Case/Z/79/.R812

Also contains a catalogue of books bequeathed to the Royal Society by Georgius Entius Armig.

## [Northampton, Henry Howard, Earl of], 1540–1614.

*A defensatiue against the poyson of supposed prophesies: Not hitherto confuted by the penne of any man, which being grounded, eyther vppon the warrant and authority of olde paynted bookes, expositions of dreames, oracles, reuelations, inuocations of damned spirites, judicialles of astrologie, or any other kinde of pretended knowledge whatsoeuer . . . haue been causes of disorder in the common wealth, and cheefely among the simple and vnlearned people: very needfull to be published at this time, considering the late offence which grew by most palpable and gross errours in astrology.*
London: Iohn Charlewood, 1583.
STC 13858.
Subject: Astrology, falseness of prophesies.
Case/B/863/.628

* * *

*Nürnbergische observation dess neuen cometems, das ist: kurze historische erzehlung; dess im nechsten monat decembris vergangenen 1664. jahrs, erscheinenen erschröcklichen comet/ sterns wunderbaren weiten lauffs veränderung und ende. Wie solcher anzusehen und zu betrachten; auch was von solchen zu förchten und zu hoffen seyn möchte. . . . Nach möglichkeit observirt und vorgestellet, von einem cultore der christlichen astrologiae.*
Nuremberg: Johann Andreas Endter, and Wolffgang the younger, heirs, 1665.
Subject: Comets.
Case/B/863/.635

## Nunes, Pedro, fl. 1537.

[*Opera*]
*Petri Nonii Salaciensis opera, qvae complectvntvr, primvm, dvos libros, in qvorvm priore tractantvr pvlcherrima problemata. In altero traduntur ex mathematicis disciplinis regulae & instrumenta artis nauigandi, quibus uaria rerum astronomicarum* [in Greek: phainomena] *circa coelestium corporum motus explorare possumus. Deinde annotationes in Aristotelis problema mechanicum de motu nauigij ex remis. Postremo, annotationes in planetarum theoricas Georgii Pvrbachii, quibus multa hactenus perperam intellecta, ab alijsque praeterita exponuntur.*
Basel: Henricpetrina, [1566].
Subject: Astronomy, geometry, instrumentation, navigation.
Ayer/*8.9/N2/N97/1566

### Obsequens, Julius, n.d.

*Iules Obsequent des prodiges. Plvs trois liures de Polydore Vergile sur la mesme matiere. Traduis de Latin en François par George de la Bouthiere Autunois.*
Lyon: Ian de Tovrnes, 1555.
Subject: Prodigies, prognostication.
Case/Y/672/.O/069

Three ways of making prognostications: by means of the Holy Spirit, Nature, or Art.

### Obsequens, Julius.

*Giulio Ossequente de' prodigii. Polidoro Vergilio de' prodigii lib. III. per Damiano Marassi, fatti Toscani & Di Giovacchino Camerario, a' l Cl. V. Andrea Fusso, la Norica ò vero de gl'ostenti.*
Lyon: Giovan. di Tovrnes, 1554.
Subject: Prodigies, prognostication.
Wing/ZP/539/.T6463

Description of catastrophic natural events during the reigns of various rulers of antiquity.

### Obsequens, Julius.

*[Iulius Obsequens de prodigiis liber cum notis J. Schefferj.]*
Amsterdam: Henricus & Theodor Boom, 1679.
Subject: Prodigies, prognostication.
Y/672/.O/067

### Obsequens, Julius.

*Ivlii Obseqventis prodigiorvm liber, ab vrbe condita usque; ad Augustum Caesarem, cuius tantum extabat fragmentum, nunc demum historiarum beneficio, per Conradum Lycostenem . . . Polydori Vergilij Vrbinatis de prodigijs libri III Ioachimi Camerarij Paberg, de ostentis libri II.*
Lyon: Ioan. Tornaesivs and G. Gazeius, 1553.
Subject: Prodigies, prognostication.
Case/–Y/672/.O/065

### Olearius, Adam (Oelschlaeger, called), d. 1671.

*Des welt-berühmten Adami Olearii colligirte und viel vermehrte reise beschreibungen bestehend in der nach Musskau und Persien wie auch Johann Albrechts von Mandelslo morgenländischen und Jürg[en] Anderersens und Volq[uard] Yversens orientalischen reise: mit angehängter Chinesischen revolution.*
Hamburg: Zacharias Herteln and Thomas von Wiering, 1696.
Subject: Voyages.
fG/131/.638

### Olearius, Adam.

*The voyages and travels of the ambassadors sent by Frederick Duke of Holstein, to the great duke of Muscovy, and the king of Persia. Begun in the year 1633 and finish'd in 1639. Containing a compleat history of Muscovy, Tartary, Persia, and other adjacent countries . . . in seven books whereto are added the travels of John Albert de Mandelslo, (a gentleman belonging to the embassy) from Persia, into the East-Indies, in three books. Written originally by Adam Olearius, secretary to the embassy.*
London: printed for Thomas Dring, and John Starkey, 1662.
STC II O269
Subject: Voyages.
Case/5A/506

"Faithfully rendred into English by John Davies, of Kidwelly."

### Olearius, Adam.

*The voyages and travells of the ambassadors sent by Frederick Duke of Holstein, . . . containing a compleat history of Muscovy, Tartary, Persia. And other adjacent countries. . . . In VII books. Whereto are added the travels of John Albert de Mandelslo, (a gentleman belonging to the embassy) from Persia to the East-Indies . . . in III books . . . written originally by Adam Olearius.*
London: printed for John Starkey, and Thomas Basset, 1669.
STC II O270
Subject: Voyages.
Case/G/6003/.64

2 vols. in 1. 2d ed. corrected.

### Olearius, Adam.

*Voyages très-curieux & très renommez faits en Moscovie, Tartare et Perse, par le Sr. Adam Olearius, . . . dans lesquels on trouve une description curieuse & la situation exacte des pays & etats, par où il a passé, tels que sont la Livonie, la Moscovie, la Tartarie, la Medie, & la Perse; et où il est parlé du naturel, des manieres de vivre, des moeurs, & des coutumes de leurs habitans; . . . des raretz qui se trouvent dans ce pays.*
Amsterdam: Michel Charles Le Céne, 1727.
Subject: Voyages.
Greenlee/5000/045/1727

2 vols. in 1. "Traduits de l'original & augmentez par le Sr. de Wicquefort."

### Olympiodorus, 5th cent., A.D.

*Olympiodori in meteora Aristotelis commentarii. Ioannes Grammatici Philoponi scholia in primvm meteorvm Aristotelis.*
Venice: [Aldus sons], 1551.

Subject: Meteorology.
Case/*Y/642/.025

2 vols. Title in Greek and Latin, text in Greek. Colophon: Venetiis apud Aldi filios, expensis nobilis uiri Federici de Turrisanis eorum auunculi.

* * *

*Opera Nvova. Intitolata dificio de ricette, nella quale si contengono tre utilissimi ricettari. Nel primo si tratta di molte & diuerse uirtu, nel secondo se insegna a comporre uarie sorti de soaui & utili odori, nel terzo & ultimo si tratta di alcuni rimedi secreti medicinali necessari in risanar li corpi humani, come nella tauola qui sequente si puo uedere.*
[Venice: Giovanni Antonio and the brothers da Sabbio] 1528.
Subject: Materia medica, medicine.
Wing/ZP/535/.N54

* * *

*[Oracula Sibyllina] Sibyllinorum oraculorum lib. VIII. Addita Sebastiani Castalionis interpretatione latina, quae Graeco eregionè respondeat. Cum annotationib. Xysti Betuleij in Greca.*
Basel, Ioannes Oporinus [1555].
Subject: Dream interpretation, occult.
Case/Y/642/.0705

2d ed. Greek and Latin on facing pages.

* * *

*Oracula Sibyllina.*
[Greek title] *hoc est, Sibyllina oracvla ex uett. codd. aucta, renouata, et notis illustrata à D. Iohanne Opsopaeo Brettano, cum interpretatione Latina Sebastiani Castalionis.*
Paris, 1599.
Subject: Dream interpretation, occult.
Case/Y/642/.0706

4 pts. in 1 vol. Portraits and brief descriptions of 12 sybils; 8 bks. of oracles. In Greek and Latin on facing pages.

* * *

*Oracula sibyllina. Hoc est, Sibyllina oracula ex veteribus codicibus emendata, ac restituta et commentariis diversorum illustrata, operâ & studio Servatii Gallaei accedunt etiam oracula magica Zoroastris, Jovis, Apollonis, &c.*
Amsterdam: Henry & widow of Theodore Boom, [1689].
Subject: Dream interpretation, occult.
Case/Y/642/.O/7068

Ed. Georgios Gemistos Plethon. In Greek and Latin, with commentary beneath. Separate half title for "Oracula metrica Jovis, Apollonis, Hecates, Serapidis, et aliorum deorum ac vatum tam virorum ac foeminarum a Joanne Opsopoeo collecta. Item Astrampsychi oneirocriticon, a Jos. Scaligero."

## Orbellis, Nicolaus de, d. 1475.

*Cursus librorum philosphiae naturalis [secundum] viam doctoris Scoti.*
Basel: Michael Furter, 1494.
H-C 5864^ BMC (XV)III:783^ Pr. 7726^ Goff O-74
Subject: Generation and corruption, natural philosophy, natural science.
Inc./7726

Consists of De physica; De coelo et mundo; de generatione et corruptione; libri metherorum; de anima; de metaphysica; de ethica; de mathematica.
BOUND WITH his *Summulae philosophie*, q.v.

## Orbellis, Nicolaus de.

*Eximii doctoris magistri Nicolai Orbellis super sententias compendium perutile: elegantiora doctoris subtilis dicta summatim complectens: quod dudum multis viciatum erroribus: castigatissime fuit recognitum: ac noue impressioni in Hagenarv commendatum.*
[Hagenau: H. Gran, J. Rynman, 1503.]
Subject: Motion, philosophy, supernatural, theology.
Inc./2765.5

Mentions Aristotle and Avicenna.

## Orbellis, Nicolaus de.

*Expositio logicae cum textu Petri Hispani.*
Venice: Albertinus Rubeus, Vercellensus, 1500.
H-C 12052^ BMC (XV)V:422^ Goff O-79
Subject: Logic, specualtive sciences.
Inc./5151

Action, passion, position, "De motu."

## Orbellis, Nicolaus de.

*Expositio super textu Petri Hispani.*
Parma: Damianus de Moyllis & Joannes Antonius de Montalli, 1482.
H-C *12043^ BMC (XV)VII:940–41^ Pr. 6861^ Goff O-75
Subject: Logic, speculative sciences.
Inc./6861

MS marginalia.

## Orbellis, Nicolaus de.

*Summulae philosophiae rationalis seu logica secundum doctrinam doctoris Scoti.*
Basel: Michael Furter, 1494.
Hain 12044^ BMC (XV)III:782^ Goff O-80
Subject: Generation and corruption, natural philosophy, natural science.
Inc./7725

MS marginalia. BOUND WITH his *Cursus librorum*, q.v.

## Origanus, David, 1588–1628.

*Novae motuum coelestium, ephemerides Brandenbvrgicae, annorvm LX, incipientes ab anno 1595, & desinentes in annum 1655, calculo duplici luminarium, Tychonico & Copernicaeo, reliquorum planetarum posteriore elaboratae, & varijs diversarum nationum calendarijs accommodatae, cum introductione hac pleniore, in qua chronologica, astronomica & astrologica ex fundamentis ipsis tractantur. . . . Opus medicis, mathematicis, historicis, & universis ferè qui literas & tempora tractant, utilissimum.*
Frankfurt: Ioannes Eichorn for David Reichard, 1609.

Subject: Astrology, astronomy.
Greenlee/5000/069/1609

## Orsini, Latino, 16th cent.

*Trattato del radio latino istrumento giustissimo & facile più d'ogni altro per prendere qual si voglia misura, & positione di luogo, tanto in cielo come in terra: Il quale, oltre alle operationi proprie sue, fà anco tutte quelle della gran regola di C. Tolomeo & del antico radioastronomico, inuentato, dall' illustrissimo & excellentissimo signor Latino Orsini.*
Rome: Marc' Antonio Moretti and Iacomo Brianzi, 1586.
Subject: Astronomy, instrumentation, surveying.
Ayer/*7/B26/1585

Pt. 1: "Della fabrica"; pts. 2 and 3: "Del vso del radio latino." BOUND WITH Barozzi, F., *Cosmographia*, 1585, q.v.

## Orta, Garcia de, 16th cent.

*Aromatvm, et simplicivm aliqvot medicamentorvm apvd indos nascentivm historia: primùm quidem Lusitanica lingua per diagolos conscripta, D. Garcia ab Horto.... Nunc verò Latino sermone in epitomen contracta, & iconibus ad viuum expressis, locupletioribusque annotatiunculis illustrata à Carolo Clvsio Atrebate.*
Antwerp: Christophorus Plantin, 1574.
Subject: Botany, herbals, materia medica.
Case/3A/1018

## Orta, Garcia de.

*Dell' historia de i semplici aromati, et altre cose; che vengono portate dall' Indie Orientali, pertinenti all'vso della medicina. Di Don Garzia dall' Horto, medico Portughese; con alcune breui annotationi di Caro Clvsio. Et dve altri libri parimente di quelle cose che si portano dall' Indie Occidentali; di Nicolò Monardes, medico di Siuiglia.*
Venice: heirs of Francesco Ziletti, 1589.
Subject: Botany, herbals, materia medica.
Ayer/109.9/B6/O7/1589

Translated into Italian by Annibale Brigante. BOUND WITH Monardes, *Delle cose . . . dall' Indie Occidentali, parte seconda*, q.v.

## Orta, Garcia de.

*Dell' historia de i semplici aromati et altre cose, che vengono portate dall' Indie Orientali pertinenti all' uso della medicina. Di Don Garzia da l'Horto . . . et dve libri . . . di Nicolò Monardes.*
Venice: heirs of Girolamo Scotto, 1597.
Subject: Botany, herbals, materia medica.
Ayer/109.9/B6/O7/1597

## Orta, Garcia de.

*Dve libri dell' historia de i semplici, aromati, et altre cose, che vengono portate dall' Indie orientali, pertinenti alla medicina, . . . et dve altri libri . . . di Nicolò Monardes.*
Venice, 1576.
Subject: Botany, herbals, medicine, materia medica.
Case/Q/.64

Trans. Annibale Brigante.

## Orta, Garcia de.

[*Histoire de drogves*]
*Garcie de Iardin Histoire de drogves espisceries, et de certains medicamens simples, qvi naissent es Indes & en l'Amerique. Ceste matiere comprise en six liures: dont il y en a cinq tirés du Latin de Charles de l'Escluse: & l'histoire du baulme adioustee de noueau: ou il est proué, que nous auons le varay* [*sic*] *baulme d'Arabie; contre l'opinion des anciens & modernes.*
Lyon: Iean Pillehotte, 1619.
Subject: Botany, herbals, materia medica.
Case/N/980/.57

4 vol. set, 2d ed., "reueuë & augmentée," trans. Antoine Colin; binder's title: *Médicaments de l'Amérique, I–IV*. Vol. 1 is Orta; vol. 2: *Traicté de Christophle de la Coste medecin et chirvrgien. Des drogues & medicamens qui naissent aux Indes. Seruant beaucoup pour l'esclaircissement & intelligence de ce que Garcie du Iardin a escrit sur ce subject*; vol. 3: *Histoire du bavlme. Ov il est provvé qve novs avons vraye cognoissance de la plante qui produict le baulme, & par consequent de son fruict, & de son bois. Contre l'opinion commvne de plusieurs medecins, & apoticaires anciens & modernes.* [P. 79: "La raison povrqvoy novs avons icy adiovsté les obseruations faictes par Pierre Bellon en Egypte, touchant la plante du baulme."] *Version françoise tirée de Prosper Alpin par Antoine Colin*; vol. 4: *Histoire des simples medicamens apportés de l'Ameriqve, desqvels on se sert en la medicine.* [Original text by Monardes, Latin trans. C. de l'Ecluse.]

## Osio, Teodato, fl. 1637.

*L'armonia del nvdo parlare con ragione di nvmeri pitagorici, discoperta da Teodato Osio.*
Milan: Carlo Ferrandi [1637].
Subject: Mathematics and music, proportions of harmonic numbers.
Case/3A/862

Letterpress title page: *L'armonia del nvdo parlare overo la mvsica ragione della voce continva nella qvale a forza di aritmetiche et di mvsiche specvlationi si pongono all prova le regole sino al presente stabilite da gl'osservatori della prosa et del verso.*

## Oviedo y Valdés, Gonzálo Fernández de, 1478–1557.

*Coronica delas Indias. La hystoria general delas Indias agora nueuamente impressa corregida y emendada. Y la conquista del Peru.*
Salamanca: Juan de Junta, 1547.
Subject: Natural history, New World.
Ayer/*108/09/1547

## Oviedo y Valdés, Gonzálo Fernández de.

*L'histoire natvrelle et generalle des Indes, isles, et terre ferme de la grand mer oceane. Tradvicte de Castillan en François.*
Paris: Michel de Vascosan, 1555.
Subject: Natural history, New World.

Ayer/108/09/1555

Translated by Jean Poleur. MS marginalia in book 8.

### Oviedo y Valdés, Gonzálo Fernández de.

*La historia general delas Indias yslas & tierra firme del mar oceano: escripta por el capitan Gonçalo Hernandez de Oviedo & Valdes. . . . Por cuyo mandado el auctor escriuio las cosas marauillosas que ay en dido & supo en veynte & dos años.*
Seville: Iuam Cromberger, 1535.
Church 71
Subject: Natural history, New World.
Ayer/*108/09/1535

Referred to as "part 1" in colophon.

### Oviedo y Valdés, Gonzálo Fernández de.

*Oviedo dela natural hystoria delas Indias.*
[Toledo: Remo de Petras, 1526.]
Church 59^ Harrisse 139
Subject: Natural history, New World.
Ayer/*108/09/1526

# P

## Paccioli, Luca, d. 1514.

*Summa di arithmetica geometria proportioni & proportionalita.*
Venice: Paganinus de Paganinis, 1494.
H-C 4105^ BMC (XV) V:457-58^ Pr. 5168^ Still. L282^ Goff L-315
Subject: Practical arithmetic, geometry.
Inc./f5168

In 5 parts: 1. practical aspects of numbers (useful or necessary in their applications to, e.g., perspective [architects] and legislative negotiations [lawyers]) to music, astrology, cosmography; 2. business arithmetic, conversion of money, speculation, algebra; 3. trade; 4. weights and measures; 5. geometry as applied to surveying. Woodcut showing hand signals for numbers up to 9,000 (leaf 36v).

## Paccioli, Luca.

*Diuina proportione; opera a tutti glingegni perspicaci e curiosi necessaria oue ciascun studioso di philosophia: prospectiua pictura sculptura: architectura: musica: e altre mathematice: suauissima: sottile: e admirabile doctrina consequira: e de lectarassi: con varie questione de secretissima scientia.*
[Venice] A. Paganius Paganinus, [1509].
Subject: Mathematics, proportion, solid geometry.
Wing/fZW/14/.P123

Chap. 1 includes "Medici e astronomi supremi de sua D. celsitudine." Has a list of figures named in Latin, Greek, and then transliterated. BOUND WITH *Libellus in tres partiales tractatus diuisus quibus corporum regularium & dependentium actiue perscrutationis.* [In 3 tracts: triangles, squares, and circles and other geometric shapes, e.g., polyhedrons.] Also BOUND WITH a geometrically analyzed alphabet showing proportions, and a section on solid geometry illustrated with figures. Finally, "arbor proportio et proportionalitas."

## Pagan, Blaise François de, comte de Merveilles, 1604–1665.

*L'astrologie natvrelle dv comte de Pagan. Premire partie. Contenant les principes ou les fondemens de la science.*
Paris: Antoine de Sommaville, 1659 [i.e., 1660].
Subject: Astrology.
B/8635/.653

In 3 books: 1. history and background; 2. interaction between the stars and man's and nature's actions; 3. zodiac, aspects of the 12 houses.

## Pagan, Blaise François de, comte de Merveilles.

*Les fortifications de monsievr le comte de Pagan. Avec ses theorêmes sur la fortification.*
Paris: Cardin Besongne . . . & A. Besongne, 1669.
Subject: Geometry, military architecture, trigonometry.
U/26/.6543

3d ed.

## Pagan, Blaise François de, comte de Merveilles.

*An historical and geographical description of the great country & river of the Amazones in America. Drawn out of divers authors and reduced into a better forme; with a mapp of the river, and of its provinces, being the place where Sr Walter Rawleigh intended to conquer and plant, when he made his voyage to Guiana.*
London: printed for John Starkey, 1661.
STC II P162
Subject: Anthropology, descriptive geography, New World.
Ayer/*1269/B8/P12/1661

Trans. William Hamilton. "Of the qualities of the air & grounds of the Great Amazone" (chap. 50).

## Pagan, Blaise François de, comte de Merveilles.

*Relation historiqve et geographiqve, de la grande riviere des Amazones dans l'Ameriqve par le Comte de Pagan.*
Paris: Cardin Besongne, 1655.
Subject: Anthropology, descriptive geography, New World.
Case/G/989/.65/1655

## Palatino, Giovanni Battista, fl. 1540.

*Compendio de gran volvme de l'arte del bene et leggiadramente scrivere tvtte le sorti di lettere et caratteri con le lor regole misure, & essempi, di M. Gioanbattista Palatino cittadino Romano . . . con un breue discorso del cifre.*
Rome: heirs of Valerio & Luigi Dorici, 1566.
Subject: Numbers, secret writing.
Wing/ZW/535/.P171

Has "on numbers," rebus, writing instruments.

## Palatino, Giovanni Battista.

*Compendio del gran volvme dell' arte del bene & leggiadramente scriuere tutte le sorti di lettere e caratteri. Con le lor regole, misure, & essempi. . . . Con vn nuouo, breue, & vtil discorso delle cifre.*
Venice: Aluise Sessa, 1588.
Subject: Numbers. proportion, secret writing.
Wing/ZW/535/.P172

Has "Delle cifre"; "Cifre qvadrate, et sonetto figvrato"; "Degli instrvmenti." Colophon: In Venetia per Gio. Antonio Rampazetto, 1588.

**Palatino, Giovanni Battista.**

*Libro di M. Giovanbattista Palatino cittadino Romano, nel qual s'insegna à scriuere ogni sorte lettera, antica, et moderna, di qualunque natione, con le sue regole, et misure, et essempi: Et con vn breve et vtil discorso de le cifre.*
[Rome: Antonio Blado Asolano, 1547.]
Subject: Numbers, secret writing.
Wing/ZW/535/.P1737

Includes "Delle cifre"; "Cifre qvadrate, et sonetto figvrato"; "De gli instrvmenti."

**Palatino, Giovanni Battista.**

*Libro di M. Giovanbattista Palatino.* [Another copy]
Wing/ZW/535/.P1738

**Palatino, Giovanni Battista.**

*Libro di M. Giovanbattista Palatino.* [Another ed.]
[Rome: Antonio Blado Asolano, 1548.]
Wing/ZW/535/.P1739

Includes "De gli instvmenti." [*sic*] "Tutti i modi di scriuer secreto, che veggiano vsati cosi da gliantichi, come da moderni . . . sono di due sorte in genere, cioè visibili & inuisibili."

**Palatino, Giovanni Battista.**

*Libro di M. Giovanbattista Palatino.* [Another ed.]
[Rome: Valerio Dorico, for m. Gíouan della Gatta, 1561.]
Wing/ZW/535/.P174

**Palatino, Giouanni Battista.**

*Libro nvovo d'imparare a scrivere tvtte sorte lettere antiche et moderne di tvtte nationi con nvove regole misvre et essempi, con vn breue & vtile trattato de le cifere composto per Giouambattista Palatino cittadino Romano.*
[Rome, 1540.]
Subject: Numbers, secret writing.
Wing/ZW/535/.P173

**Palingenius, Marcellus, Stellatus, fl. 1528.**

*The firste syxe bokes of the mooste Christian poet Marcellus Palingenius, called the Zodiake of life. Newly translated out of Latin into English by Barnaby Googe.*
London: printed by Jhon Tisdale for Rafe Newbery, 1561.
STC 19149.
Subject: Astrology.
Case/Y/682/.P154

With "a table brefelye declaryng the signification and meanynge of all such poetical wordes as are conteined wythin this boke, for the better understanding thereof." (See Robert Schuler, *English magical and scientific poems to 1700* [New York: Garland, 1979]).

**Palingenius, Marcellus, Stellatus.**

*Zodiacus uitae, hoc est, De hominis vita, studio, ac moribus optimè instituendis libri duodecim. . . . opus mirè eruditum, planeque philosophicum: nunc tertio longè quàm antea cum emendatius, tum diligentius excusum.*
Basel [R. Winter], 1543.
Subject: Astrology.
Case/3A/1044

**Palingenius, Marcellus, Stellatus.**

*Zodiacus Vitae, hoc est, de hominis vita, studio, ac moribus optimè instituendis libri XII.*
Lyon: Ioannes Tornaesius & Gul. Gazius, 1556.
Subject: Astrology.
Wing/-ZP/539/.T6468

**Palingenius, Marcells, Stellatus.**

*Zodiacvs Vitae, hoc est, de hominis uita, studio, ac moribus optimè instituendis libri XII.*
Basel: Nicolaus Brylin, 1563.
Subject: Astrology.
Case/3A/867

**Palingenius, Marcells, Stellatus.**

*Zodiacvs Vitae, hoc est, de hominis vita, stvdio, ac moribvs optimè instituendis libri XII.*
Paris: Hieronymus de Marnef, 1564.
Subject: Astrology.
Case/3A/866

**Palingenius, Marcellus, Stellatus.**

*Zodiacvs vitae: hoc est, de hominis vita, studio, ac moribus optimè instituendis, libri XII.*
Lyon: Antonius de Harsy, 1606.
Subject: Astrology.
Case/Y/682/.P156

**Palingenius, Marcellus, Stellatus.**

*Marcelli Palingenii Stellati poetae, Zodiacus vitae, id est de hominis vita, studio, ac moribus optimè instituendis libri XII.*
Rotterdam: Joannes Hofhout, 1722.
Subject: Astrology.
Y/682/.P15

**Palingenius, Marcellus, Stellatus.**

*Zodiacvs vitae pvlcherrimvm opvs atqve vtillissimvm, Marcelli Palingenii stellati.*
[Venice: Bernardinus Vitalis Venetus, 1531.]
Brunet, p. 318
Subject: Astrology.
Case/Y/682/.P146

**Palingenius, Marcellus, Stellatus.**

*The Zodiake of life written by the godly and zealous poet Marcellus Pallingenius stellatus, wherein are conteyned twelve bookes disclosing the haynous crymes & wicked vices of our corrupt nature: and plainly declaring the pleasant and perfit pathway vnto eternall lyfe, besides a number of digressions both pleasaunt & profitable, newly translated into Englishe verse by Barnabae Googe.*
London: Henry Denham for Rafe Newberye, 1565.
STC 19150
Subject: Astrology.

Case/Y/682/.P1542

Listed in Schuler (see above) as encyclopedia of astrology, morality, etc.

## Palingenius, Marcellus, Stellatus.

*Le zodiaque de la vie humaine, ou préceptes pour diriger la conduite & les moeurs des hommes. Nouvell edition, revûë, corrigée, & augmentée de notes historiques, critiques, politiques, morales, & sur autres grands sçiences, par Mr J. B. C. de la Monnerie.*
London: Le Prevost, & company, 1733.
Subject: Astrology.
Y/682/.P158

2 vols. Vol. 2 has folding plate, "Table de l'écoulement des estres, et le principe des arts et des sçiences." Notes, p. 135: "Science ecrite de tout l'art Hermetique, qui n'a pas été puisée dans les livres d'autrui; mais qui a été justifiée & prouvée par l'expérience même." "Sommaire de livre XI le verseau: Ce livre donne des préceptes astronomiques."

## Palingenius, Marcellus, Stellatus.

*Zwölff bücher zu Latein zodiacus vitae, das ist gürtel dess lebens genannt gründtlich verteutscht vnd in reimen verfasst . . . durch M. Johan Spreng von Augspurg.*
Frankfurt [Sigmund Feyerabend], 1564.
Subject: Astrology.
Case/3A/361

Colophon: Gedruckt zu Franckfurt am mayn durch Georg Raben, Sigmund Feyerabendt, und Weygand [with heirs] 1564.

## Pancirolli, Guido, 1523–1599.

*Rerum memorabilium libri duo. Quorum prior deperditarum: posterior noviter inventarum est.*
[vol. 1] Hamburg: Michael Forster, 1612; [vol. 2] Frankfurt: Christoph Vetter, 1617.
Subject: Metallurgy, technology.
Case/R/2/.658

3d ed. Vol. 1, clepsydra, precious stones; trans. from Italian by Henricus Salmuth; vol. 2, *Nova reperta sive rerum memorabilium recens inventarvm & veteribvs incognitarum.* Subjects include "De alchymia, de horlogiis, de typographia, de quadratum circuli, de aucupio quodcum accipitre, niso, falcone, & aliis auibus peragitur."

## Paradin, Claude, 16th cent.

*Devises heroiqves et emblemes de M. Claude Paradin. Reueuës & augmentées de moitié par Messire François d'Amboise.*
Paris: Rolet Bovtonné, 1622.
Subject: Emblems.
Case/W/1025/.6541

## Paré, Ambroise, 1517?–1590.

*The works of Ambrose Parey, chyrurgeon to Henry II. Francis II. Charles IX. and Henry III. kings of France. Wherein are contained an introduction to chirurgery in general: A discourse of animals, and of the excellency of man. The anatomy of man's body. A treatise of praeternatural tumors in general. Of their cure in particular. Of wounds in general. Of wounds with their cure in the particular parts of man's body. Of wounds made by gun-shot, and other destructive engines. Of contusions and gangrenes. Of ulcers and fistulaes. Of bandage. Of fractures and dislocations. Of almost all preternatural affections, whose cure is performed by manual operation. Of the gout. Of the French pox, with all its symptoms. Of the small pox and measles. Of worms and leprosie. Of poisons, with the cure of wounds, made by the biting of a mad dog, and other venomous creatures. Of the plague. Of artificial supplying those things which by nature or accident are wanting. Of monsters and prodigies. The way of making reports, and embalming dead bodies. . . . Recommended by the University of Paris to all students in physick and chirurgery, particularly such as practised in camps and the sea.*
London: printed by Jos. Hindworth, 1691.
STC II P352
Subject: Medicine.
UNCATALOGUED

Trans. Thomas Johnson.

## Parker, George.

*Mercurius Anglicanus, or the English Mercury: being a double ephemeris for the year of our Lord, 1691* [–93] *heliocentrical & geocentrical, or the planets places and aspects referred both to the sun and earth; exactly calculated from Astronomia Carolina for the meridian of London . . . but generally useful to England, Scotland, and Ireland &c. with general and monthly predictions thereon. . . . The like not extant in any other. By George Parker, a lover of the coelestial sciences.*
London: printed for the Company of Stationers, 1691 [–1699].
STC II A2004, II A2006-2009, II A2013
Subject: Almanacs.
Case/A/1/.0175

A collection of almanacs by various writers. Binder's title: *Almanacs, 1691–1700.* Includes eds. of Parker's *Ephemeris* (London: printed for the author) for the years 1695–99. The 1698 and 1699 eds. contain, respectively, attacks on John Partridge's villainy and his ignorance.(Also bound into this collection *Ephemeris or a diary astronomical, astrological, meteorological for the year of our Lord 1694*[1698–99, 1700] by John Gadbury (STC II A1773, II A1777-79). See also "Almanacs."
London: printed by J. R. for the Company of Stationers.

## Parkinson, John, 1567–1650.

*Paradisi in sole paradisvs terrestris. Or, a choise garden of all sorts of rarest flowers, with their nature, place of birth, time of flowring, names and vertues to each plant, useful in physick, or admired for beauty. To which is annext a kitchin-garden furnished with all manner of herbs, roots, and fruits . . . with the art of planting an orchard . . . collected by John Parkinson, apothecary of London, and the King's herbalist.*
London: printed by R. N., 1656.
STC II P495
Subject: Gardening, herbals.

Case/fR/57/.653

2d impression much corrected and enlarged.

## Parkinson, John.

*Theatrvm botanicvm: the theater of plants, or an herball of a large extent: containing therein a more ample and exact history and declaration of the physicall herbs and plants that are in other authours, encreased by the accesse of many hundreds of new, rare, and strange plants from all the parts of the world, with sundry gummes, and other physicall materials, than hath beene hitherto published by any before; and a most large demonstration of their natures and vertues. . . . Collected by the many years travaile, industry, and experience in this subject by John Parkinson apothecary of London, and the kings herbarist.*
London: Thomas Cotes, 1640.
Subject: Herbals.
Case/N/405/.66

MS notes. Divided into 17 classes or tribes, including "purging plants, venemous, sleepy, and hurtfull plants, and their counterpoysons, wound herbes, cooling herbes, hot and sharpe biting plants." With physical description of each plant and its "vertues," summary of its habitat, and its vernacular and Latin names.

## Parran, Antoine, 1582–1650.

*Traité de la mvsique théorique.*
Paris, 1646.
Subject: Consonance and disonance, music, proportion.
Case/4A/1019

"Figure de Ptolomée nommée helicon, comprenant toute sorte de consonances" (p. 142).

## Parsons, Robert, 1546–1610.

*The second parte of the booke of Christian exercises.*
London: printed by Iohn Charlwood for Simon Waterson, 1592.
STC 19382.
Subject: Natural philosophy, philosophy of science.
Case/oBV/4500/.P27/B6/1592

"The first argument in naturall philosophy, Aristotle, lib. 7 & 8 (Primum mobile)" (p. 45).

* * *

*Partitio magnitvdinis vniversi orbis in ivgera terrae, et significatio, qvanta eorum pars colatur habiteturque; nempe 13 miliartia 134. milliones 278. millia iugerum, quae in omnes quatuor orbes partes diuiduntur: quo fine recensentur omnes Europae Asie Africe et Americe prouincie indicaturque: quot iugera vnaquaque earum.*
Cologne: Lambertus Andrea, 1596.
Subject: Cosmography.
Case/5A/547/no./3

BOUND WITH Botero, *Theatrvm* and *Geographische landtaffel*; Metellus, *Germania*; and Acosta, *Geographische und historische beschreibung*, qq.v.

* * *

*Parvvlvs philosophiae natvralis: Ivvenilibvs ingeniis phisicen desiderantibvs oppido qvam necfssarivs.* [sic]
[Vienna: Hieronymus Vietor Philovallis] 1510.
Subject: Natural science.
Case/Y/642/.A8678

Manuscript notes. A compendium of the Physica abridged by unknown scholar. Edited by Joachim Vadianus. Heavily annotated with MS marginalia. In 3 tracts: motion, elements (qualities), the soul.

## Pascal, Blaise.

[*Pensées*]
Amsterdam: A. Wolfgang, 1688.
Subject: Philosophy.
Case/3A/2449

See "Pensées diverses," no. 2 (pp. 213 ff.): "Il y a beaucoup de difference entre l'esprit de geometrie & l'esprit de finesse."

## Pasch, Georg, 1661–1707.

*De novis inventis, quorum accuratiori cultui facem praetulit antiquitas, tractatus, secundum ductum disciplinarum, facultatum atque artium in gratiam curiosi lectoris concinnatus, editio secunda, priori quarta parte auctior.*
Leipzig: heirs of Joh. Gross, 1700.
Subject: Inventions, mechanical arts and sciences, natural philosophy.
A/91/.66

Chap. 6: de inventis medicis; chap. 7: de inventis physico-mathematico-mechanicis.

## Pastrengo, William, 14th cent.

*De originibvs rervm libellvs avthore Gvlielmo Pastregico Veronese. in qvo agitvr de scriptvris virorum illustrium. De fundatoribus vrbium. De primis rerum hominibus. De inuentoribus rerum. De primis dignitatibus. Deque magnificis institutionibus.*
[Venice: Nicolaus de Bascarinis, 1547.]
Subject: Astrology, biographical dictionary, inventions and inventors.
Case/E/3/.658

"Illustrium virorum gentilium philosophi & astrologi" (p. 16).

## Patin, Guy, 1601–1672.

*Lettres choisies de feu mr. Guy Patin.*
Paris: Jean Petit, 1692.
Subject: Medicine.
E/5/.P2739

2 vols. See esp. letters 119-29 ff., "The King fell ill at Mardik," and, in vol. 2, letter 334.

## Patin, Guy.

*Naudaeana et Patiniana; ou singularitez remarquables, prises des conuersations de Mess. Naudé & Patin.*
Paris: Florentin & Pierre Delaulne, 1701.
Subject: Medicine.

Y/762/.N219

With "Catalogus omnium operum Gabrielis Naudaei." First part (Naudaeana) is a miscellany, e.g., p. 69: "Les scorpions en Italie ne sont point venimeux: je me souviens que sous un degré qu'on abbatit pour le rétablir, on trouva dans une fosse plus de trois grands tombereaux des scorpions: on les jetta dans une riviere voisine. Les poissons les mangent & s'engraissent; les courtesans en Italie en ont dans leurs lits l'été pour se rafraîchir." P. 107: "M. Patin a beau dire le Quina-quina est un bon febrifuge: c'est l'ecorce d'un arbre qu'on trouve dans le province de Quito en Amerique."

## Patin, Guy.

*Naudaeana et Patiniana.* [Another ed.]
Amsterdam: François Vander Plaats, 1703.
Subject: Medicine.
Y/762/.N22

"Seconde edition revuë, corrigée & augmentée d'additions au Naudaeana qui ne sont point dans l'edition de Paris." "Je n'ai jamais vû en Italie ni ailleurs aucun hermaphrodite parfait" (p. 76); "Divinatio morientium" (those near death have the sudden power of prophecy [p. 86]). The "Catalogus omnium operum Gabrielis Naudaei" is by Cardinal Mazarin.

## Patrizi, Francesco, 1529–1597.

[*Francisci Patricii Panvrgia. Vniuersae lvcis tractatio nova, et acvtissima.*]
[Ferrara: Benedictus Mammarelli, 1591.]
Subject: Cosmology, occult science.
Case/fB/42/.664

In 8 bks. A collection of works, each with special title page; entire vol. has general title page. Works include *Vniversae lvcis tractatio nova, et acvtissima; Novae philosophiae—an mvndvs sit animatvs; De aethere, ac, rebvs coelestibvs lib. XIIII; Zoroaster et eivs CCCXX oracvla chaldaica; Hermetis Trismegisti libelli integri XX et fragmenta. Asclepii eivs discipvli libelli III a Francisco Patricio locis plusquam mille emendati.*

## Paulus Aegineta.

*Praecepta salvbria Gvilielmo Copo Basileiensi interprete.*
Paris: Simon Colinaeus, 1527.
Subject: Anatomy, biology, diet, medicine.
Case/Y/6809/.66

Dietetics, obstetrics. BOUND WITH Chappusius, Nicolaus, *De mente & memoria libellus vtilissimus.* Paris: Ascensianus, 1513. Subject: Mnemonics. Also BOUND WITH Archithrenii Fictv *ab effectu vocabulo, Joannis nomine & Neustrii seu Normanni natione.* Ascensianus, 1517.

## Paulus Middelburgensis, bp., d. 1534.

*De recta Paschae.*
[Fossombrona: Octavianus Petrutius, 1513.]
Subject: Astrology.
Ayer/107/.58/P3/1513

Leaf CCiiii: "De stella magorum" (planetary predictions of Christ's Passion); leaf FFiiii: discoveries made by Columbus (referring to the Antipodes), and by Vespucci (called Almarici).

## Peacham, Henry, 1576?–1643?

*The compleat gentleman fashioning him absolute in the most necessary & commendable qualities concerning minde or bodie that may be required in a noble gentleman.*
London: Francis Constable, 1622.
Subject: Cosmography, education, geography.
STC 19502
Case/B/69/.659

1st ed.

## Peacham, Henry.

*The compleat gentleman. Whereunto is annexed a description of the order of a maine battaile . . . as also certaine necessarie instructions concerning the art of fishing.*
London: printed for Francis Constable, 1627.
STC 19503
Subject: Cosmography, education, geography.
Case/B/69/.66

2d impression.

## Peacham, Henry.

*The compleat gentleman. Fashioning him absolut, in the most necessary and commendable qualities concerning minde or body, that may be required in a noble gentleman.*
London: printed for Francis Constable, 1634.
STC 19504
Subject: Cosmography, education, geometry.
Case/B/69/.662

Includes chapters on "the dutie of parents in their childrens education"; chap. 7: "of cosmography"; chap. 8: "of memorable observation in survey of the earth"; chap. 9: of geometry.

## Peacham, Henry.

*The compleat gentleman To which is added the gentlemans exercise or an exquisite practise, as well for drawing all manner of beasts, as for making colours, to be used in painting limming. &c.*
London: printed by E. Tyler for Richard Thrale, 1661.
STC II P943
Subject: Cosmography, education, geometry.
Case/B/69/.663

3d impression. Pp. 249–50, 255–56 missing and replaced by photocopies. Penciled note on front flyleaf: "From this edition Dr. Johnson drew all the heraldic definitions in his dictionary."

## Peckham, John, abp. of Canterbury, ca. 1230[or 35]–1292.

*Jo. Archiepiscopi Cantuariensis Perspectiua communis.*
Venice: Io. Baptista Sessa [1504].
Subject: Optics, vision.

Case/W/5922/.655

Most widely used of all optical texts of the 14th–16th centuries. Deals with anatomy and physiology of the eye, propagation of light and color. "Per L. Gauricvm Neapolitanvm emendata." MS marginalia.

## Peckham, John, apb. of Canterbury.

*Prospectiva communis d. Johannis archepiscopi Cantuariensis fratris ordinis minorum decim psaurum unguem castigata per eximium artium et medicine ac iuris utriusque doctorem ac mathematicum peritissimum. D. Facium Cardanum Mediolanensem in uenerabili colegio iuris peritorum Mediolani residentem.*
[Milan:] Petrus de Corneno [1482/83?]
H*9425^ BMC (XV)VI:759^ Pr. 5974^ Still. J 355^ Goff J-394
Subject: Optics, vision.
Inc./5974

Ed. by Facius Cardanus, father of Hieronymus Cardanus. With geometrical diagrams in margins passim.

## Peletier, Jacques, 1517–1582.

*L'algebre de Iaqves Peletier dv Mans, departie an deus liures.*
Lyon: Ian de Tournes, 1554.
Subject: Algebra, algorithms, binomials, equations, irrational numbers.
Wing/ZP/539/.T64632

## Pellegrini, Antonio, 16th Cent.

*I segni de la natvra ne l'hvomo.*
[Venice: Giouanni de Farri & bros., 1545]
Subject: Sciences of man, psychology.
Case/B/475/.67

## Pellegrini, Antonio.

*I segni de la natvra ne l'hvomo.*
Venice: Gio. Grifo, 1569.
Subject: Sciences of man, psychology.
Case/B/475/.672

"Ma nondimeno, essendo commune opinione di tutti i medici, & di Galeno massimamente, che quatro siano le membra principali, onde parimente habbiano origine le quatro nostre maggiori affettioni; come è dal ceruello l'ingegno, l'ira dal core, dal fegato la letitia, & la libidine al fine da i genitali; mi pare che per gli occhi maggiormente si debbano conoscere quegli affetti che regnano in noi, non in potentia, ma in atto" (p.104).

## Pellison-Fontanier, Paul, 1624–1693.

*The history of the French Academy, erected at Paris by the late famous Cardinal de Richelieu, and consisting of the most refined wits of the nation. Wherein is set down its original and establishment, its statutes, daies, places, and manner of assemblies &c. with the names of its members, a character of their persons, and a catalogue of their works. Written in French by Mr. Paul Pellison, counseller and secretary to the King of France.*
London: printed by J. Streater for Thomas Johnson, 1657.
STC II P1110
Subject: Academies and learned societies.
Case/F/3200/.47

"Le Sieur de Taneur, having published in the year 1650 a treatise of the incommensurable quantities, with a translation of the tenth book of Euclide, added thereto a very excellent discourse to the gentlemen of the French Academy, concerning a way to explain the sciences in French" (p. 131).

## [Pellison-Fontanier, Paul]

*Relation contenant l'histoire de l'Academie Françoise par M. P. jouxte la copie imprimé a Paris.*
[Amsterdam], 1671.
Subject: Academies and learned societies.
Case/*A/9/.6307

2d ed. 5 pts., including (3) De ce qu'ell a fait depuis son institution (p. 67); (5) Des academiciens en particulier (p.147). Appears to deal principally with grammar, rhetoric, etc.

## [Pemberton, Henry] 1694–1771.

*A view of Sir Isaac Newton's philosophy.*
London: printed by S. Palmer, 1728.
Subject: Astronomy, mechanics, motion, optics, physics.
fB/455/.623

In 3 books: Motion, Planets, Optics, followed by a list of subscribers.

* * *

*Pennas, que cahiram de huma das azas ao celebrado fenix das tempestades, que poder à servir de segunda parte escolhidas no taboleiro das escadas do hospital real entre espedaçados livros, en hum terça feira, e offerecidas ao fenix Arabio por Cosme Fragozo de Matos.*
Lisbon: Bernardo da Costa de Carvalho, 1733.
Subject: Astrology, comets, plagues.
Greenlee/4504/P855

List of disastrous events, beginning with a comet in A.D. 78.

## Pereira, Benito, 1535 (ca.)–1610.

*Aduersus fallaces & superstitiosas artes. Id est, de magia, de obseruatione somniorum, & de diuinatione astrologica.*
Lyon: ex officina Iuntarvm, 1592.
Subject: Astrology, dreams, natural magic.
Case/B/85/.669

In 3 books: (1) De arte cabalistica; (2) De somniis; (3) De diuinatione astrologica.

## Pereira, Benito.

*Adversvs fallaces & superstitiosas artes, id est, de magia, de observatione somniorum, & de diuinatione astrologica libri tres.*
Lyon: Horativs Cardon, 1603.
Subject: Astrology, dreams, natural magic.

Case/oBF/1600/.P47/A38/1603

### Pereira, Benito.

*De magia, de observatione somniorvm, et de divinatione astrologica, libri tres. Adversuvs fallaces, et svperstitiosas artes.*
Cologne: Ioannes Gymnicus, 1598.
Subject: Astrology, dreams, natural magic.
Case/B/85/.668

### Pereira,Benito.

*De magia, de observatione somniorvm et de divinatione astrologica libri tres. Adversvs fallaces, et superstitiosas artes.*
Cologne: Ioannes Gymnicus, 1612.
Subject: Astrology, dreams, natural magic.
Case/B/85/.67

Bk. 1: "Quotuplex sit magia & primo de magia astronomica; de arte cabalistica; de necromantica, an per artem magicam possit verum aurum fieri ab alchimistis, de magis pharaonis [mentioned in Exodus].

### Perkins, William, 1558–1602.

*A discovrse of the damned art of witchcraft, so farre forth as it is reuealed in the Scriptures, and manifest by true experience.*
[Cambridge]: printed by Cantrel Legge, 1610.
STC 19698
Subject: Witchcraft.
Case/B/88/.674

The nature, parts, and working of witchcraft; punishment of witches; casting out of devils.

### Perrault, Charles, 1628–1703.

*Paralelle des anciens et des modernes.*
Paris: widow of Jean Baptiste Coignard and Jean Baptiste Coignard, son, 1692–1696.
Subject: Ancients vs. Moderns, astronomy, geography, medicine.
Y/0102/.675

2d ed. 4 vols., all but vol. 3 referred to as 2d ed. Vol. 1: arts and sciences; vol. 2: eloquence; vol. 3: poetry; vol. 4: astronomy, geography, navigation, war, philosophy, medicine.

### Perrault, Claude, 1613–88.

*Memoires pour servir a l'histoire naturelle des animaux.*
Paris: Imprimerie royale, 1671–76.
Subject: Natural history.
Wing/+ZP/7391/.1

Pt. 1 of made-up vol. in 2 pts., neither title page nor text in this pt., which consists of engravings of, e.g., animal anatomy, leather tanning, papermaking, typefounding, printing.

### Perrault, François, 1572–1657.

*L'antidemon de Mascon, ov la relation pvre et simple des principales choses qui ont esté faites & dites par vn demon il y a quelques années dans la ville de Mascon en la maison du Sr. Perravd. Opposees a plvsievrs favssetez qui en ont couru ci deuant sou le nom du diable de Mascon comme celles qui ont couru dés plusieurs siecles en ça contre le Comte de Mascon, dont l'apologie est aussi inseree ci apres. Ensemble la demonologie ov discovrs en general tovchant l'existence, puissance, & impuissance des demons & des sorciers, & des vrais exorcismes & remedes contre iceux.*
Geneva: Pierre Chouët, 1656.
Subject: Demonology, witchcraft.
Case/3A/232

2d ed.

### Perrault, François.

*Demonologie ov traitte des demons et sorciers: de leur puissance & impiussance. Ensemble l'antidemon de Mascon, ou histoire veritable de ce qu'un demon a fait & dit, il y a quelques années, en la maison dudit Sr. Perreaud à Mascon.*
Geneva: Pierre Aubert, 1653.
Subject: Demonology, witchcraft.
Case/3A/50

### Perrault, François.

*The divell of Mascon, or, a true relation of the chief things which an vnclean spirit did and said at Mascon in Burgundy, in the house of one Mr. Francis Pereaud, minister of the Reformed church in the same town. Published in French lately by himself.*
Oxford: Henry Hall for Ric. Davis, 1679.
STC II P1588
Subject: Demonology, witchcraft.
Case/B/891/.68

5th ed. Translated by A. D.

### Petau, Denis, 1583–1652.

*Dionysii Petavii de doctrina temporum . . . juxta editionem Antuerpiensem anno MDCCIV.*
Verona: Petrus Antonius Berno, & Venice: Jo. Baptista Recurti, 1734–36.
Subject: Astronomy, calendar, chronology.
+F/017/.676

3 vols. Vol. 1: "Et Joannis Harduini . . . praefatio ac dissertatio de LXX hebdomadibus"; vol. 2: "In quo temporum [Greek word] disputantur, tum doctrina usus atque fructus chronico libro traditur" [Verona, 1735]; vol. 3: "In quo uranologium et alia ipsius Petavii, aliorumque varia opuscula"[Verona, 1736].

### Petau, Denis.

*The history of the vvorld: or, an account of time. Compiled by the learned Dionisius Petavius. And continued by others to the year of our Lord 1659. Together with a geographicall description of Europe, Asia, Africa, and America.*
London: printed by J. Streater, 1659.
STC II P1677C
Subject: Chronology, descriptive geography.
Case/F/09/.67

With separate title page: *A geographicall description of the world.*

### Petau, Denis.

*The history of the vvorld: or, an account of time.* [Another copy]
London: printed by J. Streater, 1659.
STC II P1677D
Ayer/*7/P4/1659

Includes folding plate map of the world.

### Petau, Denis.

*Rationarium temporum . . . duo opuscula Jacobi Usserii.*
Verona: Petrus Antonius Bern, 1741.
Subject: Astronomy, calendar, chronology.
+F/09/.668

Pt. 1: De chronologia sacra veteris testamenti. Pt. 2: De Macedonum, & Asianorum anno solari. Usserius's work is on astronomy and the calendar.

### Petrarca, Francesco, 1304–1374.

*Francisci Petrarchae Florentini, philosophi, oratoris, et poetae clarissimi, reflorescentis literatvrae, Latinaeqve lingvae, aliqvot secvlis horrenda barbarie inquinatae, ac penè sepultae, assertoris & instauratoris, opera, quae extant omnia. In quibus praeter theologica, naturalis, moralisque philosophiae praecepta, liberalium quoque artium encyclopediam, historiarum thesaurum, & poësis diuinam quandam uim, pari cum sermonis maiestate, coniuncta inuenies.*
Basel: Sebastianvs Henricpetri, 1581.
Subject: Medicine.
Case/fY/712/.P/40551

Pp. 1087–1117: "Invectiva in medicum quendam" (in 4 books).

### Petrucci, Gioseffo, d.1680.

*Prodomo [sic] apologetico alli studi Chircheriani nella quale con un' apparato di saggi diversi, si dà prova dell' esquisito studio ha tenuto il celebratissimo padre Atanasio Chircher, circa il credere all' opinioni degli scrittori, sì de' tempi andati, come de' presenti, e particolarmente intorno a quelle cose naturali dell' India, chegli furon portate, ò referte da' quei, che abitarono quelle parti.*
Amsterdam: Janssonio-Waesberg, 1677.
Subject: Anthropology, natural history of India.
Case/L/0114/.676

### Petto, Samuel, 1624?–1711.

*A faithful narrative of the wonderful and extraordinary fits which Mr. Tho. Spatchet . . . was under by witchcraft: or, a mysterious providence in his even unparallel'd fits. With an account of his first falling into, behaviour under, and (in part) deliverance out of them.*
London: printed for John Harris, 1693.
STC II P1897
Subject: Witchcraft.
Case/B/88/.678

### Peucer, Kaspar, 1525–1602.

*Commentarivs de praecipvis generibvs divinationvm, in qvo a prophetiis avtoritate diuina traditis, & à physicis coniecturis, discernuntur artes & imposturae diabolicae, atque obseruationes natae ex superstitione & cum hac coniunctae: Et monstrantur fontes ac causae physicarum praedictionum: Diabolicae verò ac superstitiosae confutae damnantur ea serie, quam tabella praefixa ostendit.*
Wittenberg: Iohan Schvvertel [1576].
Subject: Divination, witchcraft.
Case/B/863/.676

"Includes De diuinationum generibus; de oraculis; de magia; de teratoscopia."

### Peucer, Kaspar.

*Commentarivs, de praecipvis divinationvm generibvs, in qvo prophetiis, authoritate diuina traditis, & à physicis coniecturis, discernuntur artes & imposturae diabolicae, atque obseruationes natae ex superstitione, & cum hac coniunctae: Et monstrantur fontes ac cause physicarum praedictionum: diabolicae vero ac superstitiosae confutatae damnantur, ea serie, quam tabella praefixa ostendit.*
Frankfurt: printed by Wecheliana for Claudius Marnius & heirs of Joan. Aubri, 1607.
Subject: Divination, witchcraft.
Case/B/863/.678

Includes "De incantationibvs"; "De divinatione extispicvm"; "De avgvriis et arvspicina"; "De sortiis"; "De somniis"; "De praesagiis medicorvm"; "De astrologia."

### Peucer, Kaspar.

*De dimensione terrae et geometrice nvmerandis locorum particularium interuallis ex doctrina triangulorum sphaericorum & canone substensarum liber, denuo editus, sed auctius multò & correctius quàm antea.*
Wittenberg [Iohannes Crato], 1554.
Subject: Cosmography.
Ayer/*7/.P41/1554

MS marginalia. Also contains Brocardo Monacho, *Descriptio locorvm terrae sanctae exactissima* and *Aliquot insignivm locorum terrae sanctae explicatio & historiae per Philippvm Melanthonem.*

### Peucer, Kaspar.

*Les devins, ov commentaire des principales sortes de devinations: distingué en quinze liures, esquels les ruses & impostures de Satan sont descouuertes, solidement refutees, & separees d'auec les saincte propheties & d'auec les predictions naturelles. Escrit en Latin par M. Gaspar Pevcer.*
Antwerp: Hevdrik Connix, 1584.
Subject: Divination, witchcraft.
Case/B/863/.68

Divination by animal entrails (hieroscopie); by flight and movement of birds (arvspicine); by dreams (oneiropolie); and by prodigies, monstrous, or multiple births (teratoscopie).

### Peurbach, Georg von, 1423–1461.

*Theoricae novae planetarvm Georgii Pvrbachii astronomi celebratissimi.*

[Nuremberg: Johann Müller of Königsberg, before 1475.]
H-C 13593^ BMC (XV)II:456^ Pr. 2208^ Goff P-1134
Subject: Astronomy.
Inc./f2208

## Pexenfelder, Michael, b. 1613.

*Apparatus eruditionis tam rerum quam verborum per omnes artes et scientias.*
Sultzbach: Michael & Joh. Friderich Endter, 1680.
Subject: Dictionary encyclopedia of science.
A/911/.678

Includes anatomy, astronomy, biology, medicine & surgery, natural history, "vocabulorum Germanico-Latinorum."

## Physiologus.

[Title in Greek] *Sancti patris nostri epiphanii episcopi Constantiae Cypri ad physiologum.*
Rome: Zannettus & Ruffinillus, 1587.
Subject: Emblems.
Case/Y/652/.P522

In Greek and Latin. Allegorical method of interpreting natural history. Includes lion, magnet, ostrich, turtle dove, etc., total of 51 subjects.

## Pianero, Giovanni, 1480 ca.–1570.

*Ioannis Planerii qvintiani Brixiensis artivm et medicine doctoris varia opuscula.*
Venice: Francisco Ziletti, 1584.
Subject: Astronomy, medicine.
Case/Y/682/.P57

Included is "De comete" (1577); "De lacte" (antidote to "febris hectica").

## Piccolomini, Alessandro, 1508–1578.

*De le stelle fisse libro vno; doue di tutte le XLVIII. imagin celesti minutissimamente si tratta, & non solo le fauole loro ordinatamente si narra, ma ancora le figure di ciascheduna n'apparon cosi manifeste, & distinamente disposte, & formate, come à punto per il ciel si distendono.*
Venice: heirs of Giouanni Varisco [n.d.]
Subject: Astronomy.
Ayer/7/P5/1595

BOUND WITH *La sfera del mondo di M. Alessandro Piccolomini.* Venice: heirs of Giovanni Varisco, 1595.

## Piccolomini, Alessandro.

*Della filosofia natvrale.*
Venice: Giovanmaria Bonello, 1552.
Subject: Natural philosophy.
Case/B/4/.685

2 vols. in 1. Vol. 1: *La prima parte della filosofia natvrale* (operations of nature, time [bk. 4: transmutation, motion, generation, 4 universal causes]). Printer's name "Bonelli" in colophon. Vol. 2: *L'instrvmento della filosofia* (logic, syllogisms, 5 parts of knowledge).

## Piccolomini, Alessandro.

*Della filosofia natvrale.*
Venice: Francesco Lorenzini da Torino, 1560.
Subject: Natural philosophy.
Case/B/4/.6853

Four books of vol. 1 of the natural philosophy.

## Piccolomini, Alessandro.

*Della filosofia natvrale di M. Alessandro Piccolomini, distinta in dve parti, con vn trattato intitolato instrumento. Et di nvovo aggiunta a queste, la terza parte, di Portio Piccolomini suo nipote.*
Venice: Francesco di Franceschi Senese, 1585.
Subject: Natural philosophy.
Case/B/4/.6849

## Piccolomini, Alessandro.

*Della grandezza della terra et dell' acqva.*
Venice: Giordano Ziletti, 1561.
Subject: Geography.
Case/M/55/.688

Of which is there a larger proportion on earth: land or water?

## [Piccolomini, Alessandro]

*Della sfera del mondo libri quattro in lingua Toscana, i quali non per uia di traduttione, ne à qual si voglia particolare scrittore obligatima parte da i migliori raccogliendo, e parte di nuouo producendo, contengono in se tutto quel, ch'intorno à tal materia si possa desiderara, ridotti à tanta ageuolezza, & à cosi facil modo di mostrare, che qual si uoglia poco essercitato ne gli studii di mathematica potrà ageuolissimamente, & con prestezza intenderne il fatto. Delle stelle fisse. Libro vno con le figure, e con le sue tauole; doue con marauigliosa ageuolezza potrà ciascheduno conoscere qualunque stella de le XLVIII imagini del cielo stellato, e le fauole loro integramente: & sapere in ogni tempo de l'anno, à qual si voglia hora di notte, in che parte del cielo si trouino, non solo le dette imagini, ma qualunque stella di quelle.*
Venice: Nicolò Beuilacqua, 1561.
Subject: Astronomy.
Case/L/900/.686

## Piccolomini, Alessandro.

*La prima parte delle theoriche overo speculationi de i pianeti.*
Venice: Giouann Varisco & associates, 1568.
Subject: Astronomy.
Case/L/900/.688

## Piccolomini, Alessandro.

*La seconda parte de la filosofia natvrale.*
Venice: Vincenzo Valgrisio, 1554.
Subject: Articficial, celestial, and natural bodies, mathematics, motion.
Case/B/4/.686

In 4 books. Bk. 1: "Quali sieno li corpi naturali."

## Piccolomini, Francesco di Niccolo, 1520–1604.

*Vniversa philosophia de moribvs.*
Venice: Franciscus de Franciscis, 1583.
Subject: Mind, psychology, soul.
Case/f/B6/.688

Chap. 3, p. 146: "De instinctu naturae."

## Picinelli, Filippo, 1604–ca. 1667.

*Mondo simbolico ò sia vniversitá d'imprese scelte, spiegate, ed ilvstrate con sentenze, ed eruditioni sacre, e profane.*
Milan: archiepiscopal press, 1653.
Subject: Emblems, encyclopedias, natural history.
Case/W/1025/.689

Divided by subject, e.g., celestial bodies, elements, heroes of antiquity, birds, quadrupeds, fish, serpents, plants, stones, metals, various types of instruments, etc.

## Picinelli, Filippo.

*Mundus symbolicus, in emblematum universitate formatus, explicatus, et tam sacris, quàm profanis eruditionibus ac sententiis illustratus: Subminstrans oratoribus, praedicatoribus, academicis, poetis &c. innumera conceptum argumenta.*
Cologne: Hermann Demen, 1681.
Subject: Emblems, encyclopedias, natural history.
Case/W/1025/.69

2 vols. in 1. Translated from Italian into Latin by R. D. Augustino Erath. Index in 6 languages: Latin, German, Italian, French, Spanish, and Flemish. In 2 parts, "corpora naturalia" (in 13 bks., e.g., astronomy, zoology, botany, geology, metallurgy) and "corpora artificialia" (in 12 bks., e.g., domestic, mechanical, nautical, mathematical, and rustic instruments).

## Pico della Mirandola, Giovanni, 1463–1494.

*Opera.*
Bologna: Benedetto d'Ettore Faelli, 1496.
H-C (add.)12992*^ BMC (XV)VI:843^ Pr. 6630–31^ Goff P-632
Subject: Astrology, mysticism.
Inc./f6630–6631

Editio princeps. Ed. Giovanni Francesco Pico della Mirandola. This copy has bookplate engraved by Albrecht Dürer: "Liber Bilibaldi Pirckheimer." Contents include "Apologia tredecim quaestionum," "Tractatus de ente & uno cum obiectionibus quibusdam & responsionibus"; BOUND WITH *DispvtationesIoannis Pici Mirandulae litterarum principis aduersus astrologiam diuinatricem quibus penitus subneruata corruit.* [Inc. f 6631].

## Pico della Mirandola, Giovanni.

*Opera.*
Venice: Bernardinus de Vitalibus, 1498.
H-C *12993^ BMC (XV)V:548^ Pr. 5526^ Goff P-634
Subject: Astrology, mysticim.
Inc./f5526

Reprint of 1496 ed.

## Pico della Mirandola, Giovanni.

*Opera. Nouissime accurate reuisa (addito generali super ominbus memoratu dignis regesto) quarumcunque facultatum professoribus tam iucunda que proficua.*
[Strassburg: Ioannes Prüss, 1504.]
Subject: Astrology, mysticism.
Case/fY/682/.P578

## Pico della Mirandola, Giovanni.

*Ioannis Pici Mirandvlae omnia opera.*
[Venice: Gulielmus de Fontaneto de Monteferrato, 1519.]
Subject: Astrology, mysticism.
Case/Y/682/.P58

MS marginalia in Latin. No title page.

## Pico della Mirandola, Giovanni.

*Opera Omnia.*
Basel: ofc. Henricpetrina, [1572–73].
Subject: Astrology, mysticism.
Case/fY/682/.P5801

Included in contents of tomo 1: "Heptaplus"; "De ente & vno"; "De astrologia disputationum." Tomo II (Ioannes Francisco Pico). Contents include "Defensio de vno & ente"; "Physici, libri duo"; "De praenotione libri nouem," with "Aduersus astrologiam argumentatur"; "Contra chiromantes, geomantes, augures."

## Pico della Mirandola, Giovanni.

*Dispvtationvm adversvs astrologos.*
[Venice: Bernardus Venetus, 1498.]
Subject: Astrology, mysticism.
Inc./f5526a

Duplicate of last part of *Opera* (Inc. f5526). BOUND WITH Bellanti, Lucio, q.v. MS marginalia passim.

## Pico della Mirandola, Giovanni Francesco, 1470–1533. [Nephew of G. Pico]

*Dialogo intitolato la strega ouero de gli inganni de demoni.*
Pescia [Lorenzo Torrentino], 1555.
Subject: Demonology, witchcraft.
Case/B/88/.688

## Pico della Mirandola, Giovanni Francesco.

*Dialogus in tres libros diuisus titulus est strix, siue de ludificatione daemonum, de reformandis moribus, ad excitatum genus humanum a uitae huius somno ad futurae uigiliam carmen. Eiusdem passerendis a calumnia libris Dionysii Areopagitae epistola. Eiusdem ad uigiliam carmen.*
[Bologna: Hieronymo de Benedictis Bonon(iensis), 1523.]
Subject: Demonology, witchcraft.
Case/Y/682/.P583

### Pico della Mirandola, Giovanni Francesco.

*De phantasia avreolvs sane . . . in quo qvae imaginationis facvltas, natvra, quaeque ei erroris caussae sit, quibusque, remediorum praesidijs occurri possit.*
Wittenberg: Mattheus VVelack, 1598.
Subject: Imagination, psychology.
Case/B/5283/.689

Chap. 9: "Qvomodo imaginationis morbus falsitasqve de corporis temperatura deque obiectis sensuum proueniens corrigi, curarique possit."

### Pictorius, Georg, ca. 1500–1569.

*Opera noua, in quibus mirifica, iocos salesque, poetica, historica & medica lib. V complectitur.*
Basel, Sixtus Henricpetri [1569].
Subject: Geography, medicine, natural history.
Case/Y/682/.P584

"Praeterea, in Marsilii Ficini de tuenda studiosorum sanitate librum scholia"; "Item, in Plinij naturalis historiae septimum librum annotationes." Book 3 includes "Onomatologos [Greek transliteration], id est nomenclator montivm, silvarum, fontium, lacuum, fluminum, stagnorum & marium."

### Pierce, Robert, 1622–1710.

*Bath Memoirs: or, observations in three and forty years practice at the Bath, what cures have been there wrought (both by bathing and drinking these waters) by God's blessing, on the directions of Robert Pierce . . . a constant inhabitant in Bath, from the year 1653 to this present year 1697.*
Bristol: printed for H. Hammond, 1697.
Subject: Balneology, medicine.
Case/F/490195/.688

MS notes and marginalia passim. Treatments for palpitation of the heart, asthma, dropsy, etc.

### [Pinder, Ulrich], d. 1510 or 1519.

*Speculum intellectualis foelicitas humanae: atque breuis compendii de bone valitudinis cura.*
Nuremberg[?]: printer for the Sodalitas Celtica[?], 1510.
Graesse VI, 462 and V, 289^ Brunet V, 483^ Pr. 11033
Subject: Hygiene, medicine.
Case/*C/55/.695

BOUND WITH *Speculum phlebothomye* (in 2 pts.): 1: "phlebothomye"; 2: "simplicium medicinarum."

### Pino, Bernardino, d. 1601.

*Gli affetti ragionamenti famigliari . . . nel quale sotto varie persone, si scoprono con piaceuole modi varie passione humane, & si mostra il modo di regolarle.*
Venice: Gio. Battista & Gio. Bernardo Sessa, 1597.
Subject: Psychology
Case/Y/712/.P6466

In dialogue form.

### Pinzio, Paolo.

*Fisionomia con grandissima breuità raccolta da i libri de antichi filosofi, nuouamente fattà volgare per Paolo Pinzio.*
Lyon: Giovan di Tournes, 1550.
Subject: Natural history, physiognomy.
Case/B/56/.692

### Pirckheimer, Wilibald, 1470–1530.

*Dissertationvm lvdicrarvm et amoenitatvm scriptores varij.*
Lyon: Francisco Hegerus & Hackius, 1638.
Subject: Gout, medicine.
Case/Y/6894/.23

Miscellaneous tracts, including Pirckheimer, "Apologia, sive lavs podagrae"; Nicolavs Wynmann, "Colymbetes, sive de arte natandi"; Philip Melanchthon, "Lavs formicae"; Jvstvs Lipsivs, "Lavs elephantis"; Gul. Menapio Insulano, "Encomivm febris quartanae."

### Pirckeheimer, Wilibald.

*The praise of the govt, or the govts apologie. A paradox both pleasant profitable. Englished by William Est.*
London: printed by G. P. for I. Budge, 1617.
STC 19947
Subject: Gout, medicine.
Case/Y/682/.P67

### Pisan, Christine de, 1363–1431.

*The fayt of armes and chyvalrye.*
[Westminster: William Caxton, 1489.]
H-C 4988 (=15918)^ Duff 96^ Pr. 9677^ GW VI:6648 (this copy lacks the variants listed in GW)^ Goff C-472^ STC 7269^ Copy 36 of Item 28 in Ricci, Census of Caxtons (Vaughan-Crocker copy)
Subject: Military technology.
Inc./f9677

Most of this has to do with law and custom on and around the field of battle (e.g., can an English priest be imprisoned in France?). Sig. C8v: "Here speketh of the passage of ostis ouer flodes and ryuers" (making temporary bridges). Pt. 2, chap. 2: "What powdres longen to gonnes and other engyns."

### Pisanelli, Baldassare, fl. 1577–1587.

*Dell' anima libri dve . . . ne' quali s'espone la natura tutta dell' anima, le sue parti, le potenze, gli habiti, l'operationi, e gli oggetti.*
Venice: Roberto Meietti, 1594.
Subject: Psychology, soul.
Case/B/51/.694

### Pisanelli, Baldassare.

*Trattato della natvra de' cibi e del bere . . . nel quale non solo tutte le virtv, & i vitij di quelli minutamenti si palesano; ma anco i rimedij per correggere i loro difetti copiosamente s'insegnano . . . tutto ripieno della dottrina de' piu celebrati medici, e filosofi: con molte belle historie naturali.*
Bergamo: Comino Ventura and associates, 1587.

Subject: Diet.
Case/R/999/.688

An encyclopedia of foods and beverages, giving for each, "elettione, gioumanti, nocumenti, rimedio, gradi, tempi, etadi, complessioni, et historia natvrali."

## Pisis, H. de [n.d.]

*Opvs geomantiae completvm in libros tres divisvm, qvorvm I. vniversam geomanticam theoriam, II. praxim, III. varias à diuersis authorib. decerptas questiones continet.*
Lyon: Ioan. Ant. Hvgvetan, 1638.
Subject: Divination, geomancy.
B/863/.694

## Piso, Willem, 1611–1678.

*Historia natvralis Brasiliae, . . . in qua non tantum plantae et animalia, sed indigenarum morbi, ingenia et mores describuntur et iconibus supra quingentas illustrantur.*
Leyden: Franciscus Hackius, and Amsterdam: Lud. Elzevir, 1648.
Subject: Anthropology, diseases, medicine.
Ayer/1269/B8/P67/1648

Bk. 1: "De Brasiliae aëre, aquis & locis"; bk. 2: "De morbis endemiis"; bk. 3: "De venenis eorumque antidotis"; bk. 4: "De facultatibus simplicium." Contains Georgi Marcgravi de Liebstad, *Historia rervm natvralivm Brasiliae, libri octo.* First 3 books: plants; bk. 4: fish; bk. 5: birds; bk. 6: quadrupeds and serpents; bk. 7: insects; bk. 8: of the region and its inhabitants. Has half title, separate pagination but catchwords carry over as though both parts were meant to be bound together.

## Piso, Willem.

*Gulielmi Pisonis medici amstelaedamensis de Indiae utriusque re naturali et medica libri qvatvordecim.*
Amsterdam: Ludovic and Daniel Elzevir, 1658.
Subject: Anthropology, biology, descriptive geography, medicine.
Ayer/*109/.9/M4P6/1658

Contains the works of Piso, including "De natura & cura morborum occidentali Indiae imprimis Brasiliae, familiarum," "de arboribus fructibus & herbis medicis atque alementariis," "de noxiis et venenatis," and "mantissa aromatica &c." Vol. also contains Georgius Marcgravius de Liebstad, "Tractatvs topographicus & meteorologus Brasiliae, cum eclipsi solari; quibus additi sunt illius & aliorum commentarium de Brasilensium & Chilensium indole & lingua." Also Jacob Bondt, "Historiae naturalis & medicae Indiae Orientalis libri sex." Entire contents listed on letterpress half-title page following engraved general title page. Binders title: *Gvlielmi Pisonis medica.*

## Piso, Willem.

*De Indiae utriusque.* [Another copy]
Case/fM/069/.7

Binders title: *Pisonis de India hist nat. et medica.*

## Pistorius of Nidda, Johann, comp., 1546–1608.

*Artis cabalisticae: hoc est, reconditae theologiae et philosophiae scriptorvm.*
Basel: Sebastianvs Henricpetri, 1587.
Subject: Cabala.
Case/fC/13/.695

"In quo praeter Pavli Ricii . . . sunt Latini penè omnes & Hebraei . . . opvs omnibvs theologis, et occvltae abstrvsaeqve philosophiae stvdiosis pernecessarium." Contains also Reuchlin's *De arte cabalistica* (3 bks.) and *De verbo mirifico*; Archangeli Bvrgonovensis, *Interpretationes in selectiora obscuriaque cabalistarum dogmata*; Leo Hebraeus, *Dialogi de amore*, and *Abrahami de creatione cabalistenis.*

## Pitt, Rob[ert], 1653–1713.

*The craft and frauds of physick expos'd. The very low prices of the best medicines discover'd. The costly medicines, now in greatest esteem, such as bezoar, pearl. &c. As also the distill'd waters, censur'd. And the too frequent use of physick prov'd destructive to health. With instructions to prevent being cheated and destroy'd by the prevailing practice.*
London: printed for Tim. Childe, 1703.
Subject: Materia medica, medicine.
oRS/78/.P67/1703

Includes "Prescriptions which may be made up by the family, viz. a cephalick draught" (p. 127). "A cardiac haustus [baum water and barly, cinnamon and plague-water, make a draught to raise the motion of the blood]" (p. 127).

## Pius II, Pope 1405–1464 [Piccolomini, Aeneas Sylvius].

*Asiae Europaeque elegantissima descriptio. . . . Accessit Henrici Glareani, Heluetii, poetae laureati compendiaria Asiae, Africae, Europaeque descriptio.*
Paris: Galeotus a Prato, 1534.
Subject: Descriptive geography.
Case/G/117/.692

## Pius II, Pope [Piccolomini, Aeneas Sylvius].

*Cosmographia Pii Papae in Asiae & Europae eleganti descriptione Asia. Historias rerum vbique gestarum cum locorum descriptione complectitur. Europa temporum authoris varias continet historias.*
Paris: printed by Henricus Stephanus for himself & Jean Hongonti, 1509.
Subject: Geography.
Case/G/117/.03

Edited by Geoffroy Tory. Numerous MS notes and marginalia passim. BOUND WITH Berosvs Babilonicvs, *De his quae praecesserunt inundationem terrarum.*

## Pius II, Pope [Piccolomini, Aeneas Sylvius].

*De educatione puerorum.*
Cologne: Ulrich Zell, ca. 1470.
H *205^ BMC (XV)I:191^ Still. P627^ Goff P-691

Subject: Education and educational reform.
Inc./859

"De cura corporum & qualite nutriendi sunt pueri" (leaf [5]).

## Plat, Hugh, 1552–1611?

*The jewell house of art and nature, containing divers rare and profitable inventions, together with sundry new experiments in the art of husbandry. With divers chemical conclusions concerning the art of distillation, and the rare practises and uses thereof. . . . Whereunto is added, a rare and excellent discourse of minerals, stones, gums, and rosins, wtih the vertues and use thereof, by D. B. gent.*
London: Elizabeth Alsop, 1653.
Subject: Agriculture, household remedies.
Case/A/15/.687

P. 70, no. 86: Sweet and delicate dentifrices or rubbers for the teeth; p. 140, no. 105: The manner of drawing or extracting the oyles out of herbs or spices.

## Platina (Bartolomeo de' Sacchi di Piadena), 1421–1481.

*De honesta voluptate et valetudine.*
Cividale: Gerardus de Lisa de Flandria, 1480.
H-C *13052^ BMC (XV)VII:1094^ Pr. 7266^ Still. P694^ Goff P-763
Subject: Diet, hygiene.
Inc./7266

First book printed at Cividale. MS marginalia passim.

## Platina (Bartolomeo de' Sacchi di Piadena).

*De honesta uoluptate et ualetudine.*
Venice: Bernardinus Benalius, 1494.
H-C-R 13058^ BMC (XV)V:375^ Pr. 4890^ Still. P698^ Goff P-767
Subject: Diet, hygiene.
Inc./4890

In Italian. Sig. "a" misbound (following sig. "b").

## Platina (Bartolomeo de' Sacchi di Piadena).

*De honesta voluptate. De ratione victus, & modo viuendi. De natura rerum & arte coquendi libri X.*
Paris: Ioannes Parui, 1530.
Subject: Cookery, hygiene.
Case/B/692/.696

BOUND WITH *De falso & vero bono dialogi*, etc. (n.a.)

## Plato.

*Opera.*
Florence: Laurentius Venetus, 1484 [before April 1485].
H-C 13602*=H 13068?^ Pr. 6405^ BMC (XV)VI:666^ Goff P-771
Subject: Creation, natural history, philosophy.
Inc./f. 6405

"Tabula librorum platonis a Marsilio Ficino florentino traductorum. Item insunt partim argumenta partim autem commentaria. Phedon de anima, Timeus de generatione mundi." [Printer according to Goff: Laurentius (Francisci) de Alopa, Venetus.]

## Plato.

*Opera.*
Venice: Bernardinus de Choris, de Cremona & Simon de Luere, for Andreas Torresanus, 1491.
H-C 13063*^ Pr. 5216^ BMC (XV)V:465^ Goff P-772
Subject: Creation, natural history, philosophy.
Inc./f.5216

## Plato.

*Omnia divini Platonis opera tralatione Marsilii Ficini, emendatione et ad Graecvm codicem collatione Simonis Grynaei, summa diligentia repurgata, quibus subiectus est index quam copiosissimus.*
Basel: Frobeniana, 1546.
Subject: Creation, natural history, philosophy.
Case/f*Y/642/.P525

Colophon: Basel: Hier. Frobenivs et Nic. Episcopivm.

## Plato.

*Divini Platonis opera omnia Marsilio Ficino interprete.*
Lyon: Antonivs Vincentivs, 1567.
Subject: Creation, natural history, philosophy.
Case/+Y/642/.P5255
At end of index: Lyon: excudebat Joannes Marcorelius.

## Plato.

*Divini Platonis opera omnia qvae exstant. Marsilio Ficino interprete.*
Lyon: Guillelmus Laemarius, 1590.
Subject: Creation, natural history, philosophy.
Case/fY/642/.P5059

Platonis vita auctore Diogene Laertio (double cols., Greek/Latin). P. 754: "Marsilii Ficini in Platonis libros argvmenta et commentaria."

## Plato.

*Il dialogo di Platone, intitolato 'Il Timeo, overo della natvra del mondo,' tradotto di lingva Greca in Italiana.*
Venice: Comin da Trino, 1558.
Subject: Creation, natural history, philosophy.
Case/B/49/.272

Edited by Girolamo Ruscelli. Discusses universal principles of Pythagoras, elements, chain of being: "Corpo, anima, menta, vita, esser vno." BOUND WITH Erizzo, Sebastiano, q.v.

## Plato.

*Le timée de Platon traittant de la natvre du monde, & de l'homme, & de ce qui concerne vniuersellement tant l'ame, que le corps des deux: translaté de grec en françois auec l'exposition des lieux plus obscurs & difficiles. Par Loys le Roy dit Regius.*
Paris: Abel l'Angelier, 1582.

Subject: Creation, natural history, philosophy.
Case/4A/673

P. 137: Plvtarqve de la creation de l'ame, qve Platon descrit en son liure du Timée. BOUND WITH "Le sympose" and "Le Phedon."

## [Plautius, Caspar, abbot of Seitenstetten] fl. 1621.

*Nova typis transacta navigatio, novi orbis Indiae occidentalis . . . e varijs scriptoribus in vnum collecta, & figuris ornata.*
[n.p.] 1621.
Subject: Discovery, exploration, New World.
Ayer/*107.5/P7/1621

Includes "Observatio de magnete navtico," "Navigatio in novvm mvndvm, sive Americam."

## [Plautius, Caspar, abbot of Seitenstetten]

*Nova typis transacta navigatio.* [Another copy]
Ayer/*107.5/P7/1621a

Compare Sabin and *Catalogue of the Huth Library.* With folding plates.

## [Plautius, Caspar, abbot of Seitenstetten]

*Nova typis transacta navigatio.* [Another copy]
Case/G/801/.691

Description from sales catalogue pasted onto front flyleaf attributes plates to W. Kilian and states, further, that Honorius Philoponus may possibly have been assumed name of Caspar Plautius.

## [Plautius, Caspar, abbot of Seitenstetten]

*Nova typis transacta navigatio.* [Another copy]
Case/fG/801/.69

## Plemp, Vopiscus Fortunatus, 1601–1671.

*Vopisci Fortvnati Plempii Amstelredamensis fvndamenta medicinae ad scholiae acribiologiam aptata. Accedit Danielis Vermostij breve apologema pro authore.*
Louvain: Hieronymus Nempaeus, 1654.
Subject: Medicine.
UNCATALOGUED

3d ed.

## Plethon, Georgios Gemistos, 15th cent.

*De Platonicae atque Aristotelicae philosophiae differentia, libellus ex Greca lingua in Latinam conuersus.*
Basel: Petrus Perna, 1574.
Subject: Astronomy, philosophy.
Case/Y/6809/.643

One of a group of seven treatises, bound together, most having to do with law and politics. No. 7, Plethon, chap. 11: metaphysics; chap. 14: astronomy; 5th body heaven and earth; Plato and elements; chap. 15: stars; chap. 16: sun.

## Plinius Secundus, C., 23–79.

[*Historia naturalis*]
*Naturalis historiae.*
[Venice: Johann von Speyer, 1469.]
H-C-R 13087^ Pr. 4018^ BMC (XV)V:153^ Goff P-786
Subject: Natural history.
Inc./+4018

Bk. 1: index. With illuminated capitals.

## Plinius Secundus, C.

[*Historia naturalis*]
*Historia naturalis.*
Venice: Nicolas Jenson, 1472.
H-C *13089^ BMC (XV)V:172^ Pr. 4087^ Still. P718^ Goff P-788
Subject: Natural history.
Inc./4087

## Plinius Secundus, C.

[*Historia naturalis*]
*Historia naturalis.*
Parma: Andreas Portilia, 1481.
H-C *13094^ BMC (XV)VII:937^ Pr. 6851^ Still. P722^ Goff P-793
Subject: Natural history.
Inc./f*/6851

Edited by Filippo Beroaldo. Illuminated capitals.

## Plinius Secundus, C.

[*Historia naturalis*]
[Colophon: *Caii Plynii Secundi Naturalis hystoriae liber trigesimus septimus & ultimis finit.*]
Venice: Rainaldi de Nouimagio Alamani, 1483.
H-C *13095^ BMC (XV)V:257–58^ Pr. 4445^ Still. P723^ Goff P-794
Subject: Natural history.
Inc./f4445

## Plinius Secundus, C.

[*Historia naturalis*]
*Caii Plinii secvndi natvralis hystoriae [liber primuvs].*
Venice: Marinus Saracenus, 1487.
H-C 13096^ BMC (XV)V:413–14^ Pr. 5157^ Still. P724^ Goff P-795
Subject: Natural history.
Inc./f5157

## Plinius Secundus, C.

[*Historia Naturalis*]
*C. Plinius Secundus de naturali hystoria diligentissime castigatus.*
Brescia: Angelus & Jacobus Britannicus, brothers, 1496.
H-C 13098^ BMC (XV) VII: 977–78^ Pr. 6992^ Still. P726^ Goff P-797
Subject: Natural History.
Inc./+6992

Edited by Mattheus Rufus.

**Plinius Secundus, C.**

[*Historia naturalis*]

*C. Plinii secundi naturae historiarum libri XXXVII, e castigationibus Hermolai Barbari quam emendatissime editi.*

Venice: Joannes Alvisius, 1499.

H-C 13104^ BMC (XV)V:572^ Pr. 5636^ Still. P729^ Goff P-800

Subject: Natural history.

Inc./f5636

**Plinius Secundus, C.**

[*Historia naturalis*]

*C. Plinii secundi Veronensis historia naturalis libri XXXVII ab Alexandro Benedicto physico emendatiores redditi.*

[Venice: Joannes & Bernardinus Rubeus Vercellences, brothers, 1507]

Subject: Natural history.

Greenlee/5100/P729/1507

Edited by Alessandro Benedetti. Introduction to table: "Hic simplicia litterarum ordine explicabuntur: in quibus Dioscoridis summi medici imaginem uidebis. Haec ex tota uniuersitate simplicium aceruata: & in quendam cumulum coniecta ostendimus pro medicorum utilitate: uel quibusuis de medicinae facultatibus iudicium asserra uolentibus. Reliqua si quis addere uoluerit bona uena subjungat."

**Plinius Secundus, C.**

[*Historia naturalis*]

*Historia natvrale di C. Plinio Secondo di Latino in uolgare tradotta per Christophoro Landino ... et con somma diligenza corretta per Antonio Brucioli.*

Venice: Gabriel Iolito di Ferrarii, 1543.

Subject: Natural History.

Case/Y/672/.P624

**Plinius Secundus, C.**

[*Historia naturalis*]

*C. Plinii Secundi historiae mundi libri xxxvij ex postrema ad vetvstos codices collatione cvm annotationibvs, et indice.*

Paris: Andreas Berthelin & G. Roland, 1543.

Subject: Natural history.

Wing/fZP/539/.F434

At end of text: "Parisiis excudebat Michaël Fezandat M. XLIII."

**Plinius Secundus, C.**

[*Historia naturalis*]

*C. Plinii Secvndi historiae mvndi libri XXXVII denvo ad vetvstos codices collati, et plurimis locis emendati, ut patet ex adiunctis annotationibus.*

Basel: Frobeniana, 1545.

Subject: Natural history.

Case/6A/157

Historic binding (painted strap work, gauffered edges), very fine wood engravings (initial letters depicting Socrates, Democritus, etc.).

**Plinius Secundus, C.**

[*Historia naturalis*]

*Historia natvrale di C. Plinio Secondo nvovamente tradotta di Latino in vvlgare Toscano per Antonio Brvcioli.*

Venice: Alessandro Brucioli & brothers, 1548.

Subject: Natural History.

Case/*Y/672/.P628

**Plinius Secundus, C.**

[*Historia naturalis*]

*C. Plinii secvndi natvralis historiae mundi libri trigintaseptem a Paulo Manutio multis in locis emendati. Castgationes Sigismvndi Gelenii.*

Venice: Paulus Manutius, Aldi, F., 1559.

Subject: Natural history.

Case/*Y/672/.P6055

**Plinius Secundus, C.**

[*Historia naturalis*]

*Sommaire des singularitez de Pline, extrait des seize premiers liures de de sa naturelle histoire par P. de Changy.*

Paris: Richard Breton, 1559.

Subject: Natural History.

Wing/ZP/539/.B74

**Plinius Secundus, C.**

[*Historia naturalis*]

*The historie of the world commonly called the natvrall historie of C. Plinivs Secvndvs. Tr. into English by Philemon Holland, doctor in physicke.*

London: Printed by Adam Islip, 1601.

STC 20029

Subject: Natural history.

Case/*fY/672/.P616

2 vols. in 1. Has "A briefe catalogue of the words of art, with the explanation thereof."

**Plinius Secundus, C.**

[*Historia naturalis*]

*Historia natural de Cayo Plinio segvndo, tradvcida por el licenciado Geronimo de Huerta, medico y familiar del santo oficio de la inqvisicion. Y ampliada por el mismo, con escolios y anotaciones, en que aclara lo escuro y dudoso, y añade lo no sabido hasta estos tiempo.*

Madrid: Luis Sanchez, 1624.

Subject: Natural history.

Ayer/*8.9/B7/P72/1624

In 2 vols. Imprint in vol. 2 reads, Madrid: Juan Gonçalez, 1629. With "tabla de las efigies de gentes monstrvosas, animales, pescados, aves, y insectos, contenidos en este volvmen."

**Plinius Secundus, C.**

[*Historia naturalis*]

*C. Plinij Secvndi historiae mvndi libri XXXVII.*

Geneva: Iacob Crispin, 1631.

Subject: Natural history.

Case/fY/672/.P606

With castigationes of Gelenius. Under supervision of the late Jacobi Dalecampii [Jacques Dalechamps].

### Plinius Secundus, C.

[*Historia naturalis*]
*The historie of the world commonly called, the natvrall historie of C. Plinivs Secvndvs. Translated into English by Philemon Holland, doctor of physicke.*
London: Adam Islip, 1634–35.
STC 20030a
Subject: Natural History.
Case/fY/672/.P6163

2 vols in 1. "The second tombe" [*sic*] dated 1634. 2d tome "To the reader: Forasmuch as this second tome treateth most of physicke, and the tearms belonging thereto be for the most part borowed from the Greek . . . I could not content my selfe to let them pass without some explanation. . . In regard whereof I thought good to prefix a briefe catalogue of such words of art, as euer and anon shall offer themselues in these discourses that insue, with the explanation thereto annexed."

### Plinius Secundus, C.

[*Historia naturalis*]
*C. Plinii Secvndi historiae naturalis libri XXXVII.*
Leyden: Elzeviriana, 1635.
Subject: Natural history.
Case/*Y/672/.P6063

3 vols. Engraved title page in vol. 1 only.

### Plinius Secundus, C.

[*Historia naturalis*]
[Another copy]
Wing/ZP/646/.E5232

MS marginalia passim.

### Plinius Secundus, C.

[*Historia naturalis*]
*C. Plinii Secundi naturalis historiae, tomus primus [-tertius] cum commentariis & adnotationibus Hermolai Barbari, Pintiani, Rhenani, Gelenii, Dalechampii, Scaligeri, Salmasii, Is. Vossii & variorum.*
Leyden, Rotterdam: Hackios, ex officina Hackiana, [1668–]1669.
Subject: Natural history.
Y/672/.P6066

3 vols., with vols. 2 and 3 dated 1668.

### [Plot, Robert] 1640–1696.

*The natural history of Oxford-shire, being an essay toward the natural history of England.*
Oxford: printed at the theater, 1677.
Subject: Natural history.
Case/F/024565/.7

In 10 chapters, including "Of the heavens," "Of men and women," "Of arts" (with some geometry and physics). MS marginalia.

### Plotinus, 205?–270.

*Opera.*
[Florence: Antonivs Miscominvs, 1492]
H-C 13121^ BMC (XV) VI: 640^ Pr. 6156^ Goff P-815
Subject: Natural philosophy, neoplatonism.
Inc./f6156

Translated by Marsilio Ficino. Includes "De tribus principalibus substantiis"; "De duabus materiis"; "De circulari motu caeli"; "De numeris."

### Plotinus.

*Plotini Platonicorvm facile coryphaei opervm philosophicorvm omnivm libri LIV in sex enneades distribvti. Ex antiquiss. codicum fide nunc primùm graecè editi, cum Latina Marsilii Ficini interpretatione & commentatione.*
Basel: Perneas Lecythvs, 1580.
Subject: Natural philosophy, neoplatonism.
Case/Y/642/.P59

Double columns, Greek and Latin.

### Pluche, Noël Antoine, 1688–1761.

*The history of the heavens, considered according to the notions of the poets & philosophers, compared with the doctrines of Moses.*
London: printed for J. Osborn, 1740.
Subject: Alchemy, cosmogony.
B/42/.7

In 2 vols. Translated by J. B. De Freval. From "The plan of the work," p. i: "The history is . . . the collection and examination of what those who lived before us have thought or learnt from their fathers concerning the origine of the heavens, and their relations to the earth." Bk. 2 contains principles of the alchemists, the world of Descartes and of Newton; bk. 3, "The physicks of Moses."

### Plutarchus, 46?–?120.

*De natura et effectionibus daemonum libelli duo.*
[Leipzig: Vögelianis for I. Steinman, 1576.]
Subject: Demonology, oracles, superstition.
Case/Y/642/.P6487

"In libellos duos Plvtarchi Cheronei, vnvm de oraculis quae defecerint, alterum de consecrata figura EI. Delphis proemium." [MS marginalia passim in proemium] Also "De defectv oraculorum liber. Adriano Tvrnebo interprete."

### Plutarchus.

*Plvtarchi Cheronei de placitis philosophorvm natvralibvs libri quinq.*
[Rome: Iacobus Mazochius, 1510]
Subject: Astronomy, divination, earth, elements, medicine, natural history.
Case/*Y/642/.P66

BOUND WITH other works by Plutarch. MS marginalia.

### Plutarchus.

*Dialogo di Plvtarco circa l'avertire de gl'animali quali sieno piu accorti, ò li terrestri ò li marini.*
Venice: by Bortolamiu, called L'Imperatore, 1545.
Subject: Aquatic vs. terrestrial fauna.
Case/Y/642/.P6895

### Plutarchus.

*Plutarchi Cheronei diui traiani praeceptoris Graecorum clarissimi historici: ac philosophi problemata emendatissima.*
Venice: Dominicus de Siliprandis, 1477.
H-C 13137^ Pr. 4454^ BMC (XV) V:263^ Goff P-828
Subject: Philosophy.
Inc./4454

Ed. Joannes Calphurnius; trans. Joannes Petri Lucensis.

### Plutarchus.

*Traitté de la superstition, composé par Plvtarqve, & traduit par Mr. Le Fèvre.*
Savmvr: Iean Lesnier, 1666.
Subject: Superstition.
Y/642/.P/689

### Polemo, Antonius, 88–ca. 145.

*Fisonomia di Polemone tradotta di Greco in Latino dall' illustrissimo signor Co: Carlo Montecvccoli, con annotationi del medemo; et poscia di Latino fatta volgare dal Co. Francesco suo fratello.*
Padua: Pietro Paolo Tozzi, 1623.
Subject: Physiognomy.
B/56/.714

BOUND WITH Porta, Giovanni Battista della; and Ingegneri, Giovanni, qq.v.

### Pollux, Julius, of Naucratis, 2d cent. A.D.

*Pollucis uocabularii index in Latinum tralatus, ut uel Graece nescientibus nota sint, quae a Polluce tractantur.*
Venice: Aldus, 1502.
Subject: Encyclopedias.
Case/*Y/642/.P694

Bk. 1 contains, in addition to general cosmography and history (including building of campsites), some natural history: "de partibus arborum"; bk. 2: parts of man; bk. 3: of generation; bk. 4: knowledge of arts and sciences; bk. 7: de arte forensitus; bk. 10: instruments and implements used in various sciences.

### Pomponazzi, Pietro, 1462–1525.

*Tractatus acutissimi, vtillimi,* [sic] *& mere peripatetici.*
[Venice: heirs of Octavianus Scotus, 1525.]
Subject: Philosophy.
Case/fB/235/.704

Includes De intensione & remissione formarum ac de paruitate & magnitudine; De reactione; De immortalitate anime; De nutritione & augmentatione.

### [Pontano, Giovanni Gioviano], 1426–1503.

[*Opera*]
Venice: Aldi Ro., 1505
Subject: Astrology, astronomy.
Case/*Y/682/.P77

In verse. Includes "Vrania, siue de stellis"; "Meteorum" (see Mauro di Nichilo, *I poemi astrologici di Giovanni Pontano*).

### Pontano, Giovanni Gioviano.

*Opera.*
Venice: heirs of Aldus Manutius & Andreas Socerus, 1533.
Subject: Astrology, astronomy.
Case/3A/2681

### Pontano, Giovanni Gioviano.

*Io. Ioviani Pontani Commentariorvm in centum Claudij Ptolemaei sententias, libri duo.*
[Basel]: And. Cratander, 1531.
Subject: Aphorisms, natural history.
Case/Y/642/.P91

### Pontano, Giovanni Gioviano.

*De rebus coelestibus lib. XIIII.*
[Florence: heirs of Philip Iunta, 1520.]
Subject: Astronomy.
Case/B/8635/.702

Vol. 6 of Pontano's works as listed in *Catalogo delle edizioni e traduzioni a stampa delle opere di Gio. Gioviano Pontano*, ed. Michele Tafuri (Naples, 1827). With this is bound his *Centvm Ptolemaei sententiae.*

### Pontano, Ioh. Isacius, 1571–1640?

*Discvssionvm historicarvm libri duo quibus praecipuè quatenus & quodnam mare liberum vel non liberum clausumque accipiendum dispicitur expenditurque. Accedit praeter alia Casparis Varrerii Lvsitani de Ophyra regione & ad eam navigatione commentarivs.*
Harderwijk: Nicolaus à Wieringem, 1637.
Subject: Navigation.
L/995/.702

Contains also Bart. Keckermann "Problemata nautica," pp. 331–93.

### Pontus, Antonius.

*Antonii Ponti Consentini rhomitypion . . . que totius terrae, situs, aeris, & superiorum omnium cognitio. Quicquid est ab ultima circumferentia ad centrum uniuersi. Et demum breuissima cosmographia.*
[Rome: Antonio Blado, d' Assola, 1524.]
Subject: Cosmography.
Ayer/*108/P79/1524

### [Popowitsch, Johann Sigmund Valentin] 1705–1774.

*Unterschungen vom meere, die auf veranlassung einer schrift de Colvmnis Hercvlis.*
Frankfurt & Leipzig, 1750.
Subject: Geography, natural history.
Bonaparte/Coll./No./9568

Added entry: Linné.

### Porcacchi, Tommaso, ca. 1530–1585.

*L'isole piv̀ famose del mondo.*
Venice: printed by Girolamo Porro for Simon Galignani, 1572.
Subject: Geography.
Ayer/*f7/P8/1572

1st ed. "Prohemio" contains a glossary defining such terms as continent, peninsula, zodiac. With engraved map of each island as headpiece.

### Porcacchi, Tommaso.

*L'isole piv̀ famose del mondo. . . . con l'aggiunta di molte isole all' ill're S. Conte Georgio Trivltio.*
Venice: printed by Girolamo Porro for Simon Galignani, 1576.
Subject: Geography.
Case/G/117/.7

Compare to Ayer copy; p. 144 (p. no. misplaced) and p. 155 (misnumbered).

### Porcacchi, Tommaso.

*L'isole piv̀ famose del mondo.* [Another copy]
Ayer/7/P8/1576

No errors on pp. 144, 155.

### Porcacchi, Tommaso.

*L'isole piv̀ famose del mondo.*
Venice: heirs of Simon Galignani, 1590.
Subject: Geography.
Ayer/7/P8/1590

With engraved maps, glossary.

### Porcacchi, Tommaso.

*L'isole piv̀ famose del mondo.*
Venice: heirs of Simon Galignani, 1604.
Subject: Geography.
Case/6A/245

### Porcacchi, Tommaso.

*L'isole piv̀ famose del mondo. . . . [con nova aggionta]*
Padua: Paolo & Francesco Galignani, brothers, 1620.
Subject: Geography.
Ayer/7/P8/1620

### Porcacchi, Tommaso.

*L'isole piv̀ famose del mondo, descritte da Tomaso Porcacchi da Castiglione Arretino, et intagliate da Girolamo Porro Padovano. Di nuovo corrette, & illustrate con l'aggiunta dell' Istria, & altre isole scogli, e nuove curiosità.*
Venice: Pietr' Antonio Brigonci, 1686.
Subject: Geography.
Ayer/7/P8/1686

In 3 books. Begins with a description of Iceland.

### Porta, Giovanni Battista della, 1538?–1615.

*Della celeste fisonomia libri sei. Nei qvali ribvttata la vanità dell' astrologia givdiciaria, si dà maniera di essattamente conoscere per via delle cause naturali tutto quello, che l'aspetto, la presenza, & le fattezze de gl' huomini possono fisicamente significare, e promettere.*
Padua: Pietro Paolo Tozzi, 1616.
Subject: Astrology, astronomy, physiognomy.
Case/B/8635/.706

Influence of humors (rather than zodiac) on individual's temperament, makeup.

### Porta, Giovanni Battista della.

*Della celeste fisonomia.* [Another edition]
Padua: Pietro Paolo Tozzi, 1623.
Subject: Astrology, astronomy, physiognomy.
B/56/.714

BOUND WITH Ingegneri, Giovanni; Polemo, Antonius; and Porta, Giovanni Battista, *Della fisonomia*, qq.v.

### Porta, Giovanni Battista della.

*Della fisonomia dell' hvomo . . . libri sei. Tradotta da Latino in volgare, e dall' istesso autore accresciuta di figure, & di luoghi necessarij à diuerse parte dell' opera.*
Naples: Gio. Giacomo Carlino, and Constantino Vitale, 1610.
Subject: Physiognomy, physiology.
Wing/fZP/635/.C2

### Porta, Giovanni Battista della.

*Della fisonomia dell'hvomo di Gio. Batt. della Porta.*
[Padua: Pietro Paolo Tozzi, 1613]
Subject: Physiognomy, physiology.
B/56/.713

Discusses various parts of the body: hair, skin, body types. Bk. 6: "Si tratta di cosi più mirabilissima, e degnissima di questa; noua, e degna d'esser amata, e desiderata, cio è che conosciuti i tuoi, ò gli altrui vitij, possi leuarli via, e scancellati del tutto."

### Porta, Giovanni Battista della.

*Della fisonomia dell'hvomo del signor Gio Battista dalla Porta Napolitano libri sei. . . . Et hora in questa terza & vltima editione migliorati in più di mille luoghi, che nella stampa di Napoli si leggeuano scorrettissimi, & aggiontaui la fisonomia naturale di Monsignor Giovanni Ingegneri.*
Padua: Pietro Paolo Tozzi, 1622.
Subject: Physiognomy, physiology.
B/56/.714

BOUND WITH Ingegneri, Giovanni; Polemo, Antonius; and Porta, *Della celeste fisonomia*, qq.v.

**Porta, Giovanni Battista della.**
*De fvrtivis literarvm notis vvlgo, de ziferis libri IIII.*
Naples: Ioa. Maria Scotus, 1563.
Subject: Ciphers, secret writing.
Case/Z/216/.7

**Porta, Giovanni Battista della.**
*De fvrtivis literarvm notis, vvlgo de ziferis libri quinque.*
Naples: Ioannes Baptista Subtiles, 1602.
Subject: Ciphers, secret writing.
Wing/fZP/635/.S941

**Porta, Giovanni Battista della.**
*De hvmana physiognomonia . . . libri IV qvi ab extimis, quae in hominum corporibus conspiciuntur signis, ita eorum naturas, mores, & consilia (egregiis ad viuum expressis iconibvs) demonstrant, vt intimos animi recessus penetrare videantur.*
Vrsell: Cornelius Sutor, 1601.
Subject: Physiognomy, physiology.
Case/B/56/.71

Note on front flyleaf: "With portrait and 83 curious woodcuts." Features of humans and animals compared.

**Porta, Giovanni Battista della.**
*De hvmana physiognomia li VI in qvibvs docetvr qvomodo animi propentibvs naturalibvs remediis compesci possint.*
Naples: Tarquinius Longus, 1602 [colophon: 1601].
Subject: Physiognomy, physiology.
Case/B/56/.711

Appears to be large format and expanded version of 1601 ed. above. Illustrated with copperplate engravings rather than woodcuts.

**Porta, Giovanni Battista della.**
*De humana physiognomonia Joannis Baptistae Portae Neapolitani libri IV.*
Rouen: Ioannis Berthelin, 1650.
Subject: Physiognomy, physiology.
B/56/.7105

**Porta, Giovanni Battista della.**
[*La physionomie hvmaine de Iean Baptiste Porta Neapolitain.*
Rouen: Clavde Grivet, 1654.]
Subject: Physiognomy, physiology.
B/56/.707

Title page wanting [title taken from table of contents].

**Porta, Giovanni Battista della.**
*Magiae natvralis, sive de miracvlis rervm natvralivm libri IIII.*
Antwerp: Christophorus Plantin, 1561.
Subject: Alchemy, chemistry, natural magic, sympathy and antipathy.
Case/L/0114/.702

Originally published 1558. Written when Porta was 16. Bk. III, chap. 2: "De sublimatione, in calcem reductione, & reliquis in hoc necessarijs."

**Porta, Giovanni Battista della.**
*Magiae natvralis libri xx. Ab ipso authore expurgati, & superaucti, in quibus scientiarum naturalium diuitiae, & delitiae demonstrantur.*
Naples: Horatius Saluianus, 1589.
Subject: Alchemy, chemistry, natural magic, sympathy and antipathy.
Case/fL/0114/.703

Bk. 2: "De varijs animalibus gignendis"; bk. 3: "De nouis plantis producendis"; bk. 5: "De metallorum transmutatione"; bk. 9: "De mulierum cosmetice"; bk. 18: "De staticis experimentis"; bk. 20: "Chaos."

**Porta, Giovanni Battista della.**
*La magie natvrelle diuisée en quatre liures, par Jean Baptiste Porta, contenant les secrets, & miracles de nature, et nouuellement, l'introdvction à la belle magie.*
Lyon: André Olier, 1669.
Subject: Alchemy, chemistry, natural magic, sympathy and antipathy.
Case/Q/155/.P67/1669

"Par Lazare Meyssonnier, medecin du Roy." Magic of the ancients; agricultural magic ("Comme nous pourrons faire produire des fruicts hastifs & tardifs"); chemical magic; light, optics.

**Porta, Giovanni Battista della.**
*De i miracoli et maravigliosi effeti dalla natvra prodotti, libri IIII.*
Venice: heirs of Iacomo Simbeni, 1588.
Subject: Natural magic, natural science.
Case/M/O/.699

"Della sapienza naturale"; botany, cookery; alchemy, alchemical medicine; optics.

**Porta, Giovanni Battista della.**
*De occvltis literarvm notis sev artis animi sensa occulte alijs significandi, aut ab alijs significata expiscandi enodandique libri IIII.*
Montbeliard: Iacobvs Foillet for Lazarus Zetzner, 1593.
Subject: Cyphers, cryptography, occult writing.
Case/Z/216/.702

* * *

Portugal. Laws, statutes etc. 1557–78 (Sebastian).
*Legge del serenissimo e molto potente re di Portogallo sopra la tratta del pepe drogherie e mercantie dell' Indie del suo gran regno.*
Florence: Royal printing house, 1571.
Subject: Botany, materia medica.
Greenlee/4612/P85/1571

Laws relating to importation of herbs and drugs.

## Portuguese pamphlets, Lisbon, etc., 1541–1721 (?)

*Dialogvs noctis Atticae.*
Coimbra: Nicolaus Carualho, printer to the University, 1619.

*Pronostico geral, e lvnario perpetvo, assidas luas nouas, & cheas, como quartos crescentes, & minguantes. Composto por Guaspar Cardozo de Sequeira.*
Lisbon: Nicolao Carualho, printer to the University, 1614.
Subject: Perpetual calendar.

*Relaçam de mais extraordinaria admiravel e lastimosa tormenta de vento, que entre as memoraueis do mundo socedeo na India oriental na cidade de Baçaim & seu destricto na era de 1618 aos 17, do mes de Mayo.*
Lisbon: Pedro Craesbeeck, 1619.
Subject: Storms, weather.

*Relaçam do lamentavel, e horroroso terremoto . . .* [1748]
Subject: Earthquakes.

*Relaçam dos danos causa dos pelos terremotos, que houve no reyno de Sicilia no mez de Janeiro deste anno de 1693.*
Printed in Rome, published in the Court of Lisbon.
Subject: Earthquakes.

*Relaçam dos terremotos socedidos em la cidade de Traina, no reyno de Sicilia es anno de 1643, & dos effeitos, que causaram em as circunvisinhas. Mandouao o Doutor Silvestre Randeli.* [1644]
Subject: Earthquakes.

*Relaçaõ do formidavel e lastimoso terremoto succedido no reino de Valença no dia 23 Março deste presente anno de 1748.*
Lisbon: Francisco Luiz Ameno. [1748]
Subject: Earthquakes.

*Relacion del lastimoso sucesso que nuestro señor seruido sucediesse en la Isla de la Tercera Cabeça de las siete islas de los Azores, de la corona del reyno de Portugal, en veynte y quatro de Mayo Sabado de Santa Iuliana deste año de 1614 a las tres horas de la tarde, con tres temblores que duraron por espacio de dos Credos.* [1614]
Subject: Earthquakes.
Greenlee/4504/P855

3,000 pamphlets, filed chronologically, in boxes.

## Porzio, Simone.

*De dolore.*
Florence: Lorentius Torrentinus, 1551.
Subject: Medicine, pain.
Case/B/656/.706

Sententiae of Galen, opinions of Avicenna, Averroës, and Aristotle with reference to pain.

## Porzio, Simone.

*De rervm naturalivm principiis libri dvo: qvibvs plvrimae eaqué; haud contemnendae quaestiones naturales explicantur.*
Naples: Gio. Maria Scotvs, 1561.
Subject: Form and substance, generation and corruption, matter, natural philosophy.
Case/L/0114/.708

Chap. 8: "An reperiatvr alivd corpvs, praeter naturale & mathematicum"; chap. 9: "An materia appetitv natvrali expetat formam."

## Postel, Guillaume, 1510–1581.

*Alcorani seu legis Mahometi et evangelistarum concordiae liber, in quo de calamitatibus orbi Christiano imminentibus tractatur. Additus est libellus de vniersalis conuersionis, iudicii ve tempore, & intra quot annos sit expectandum, coniectatio, ex diuinis ducta authoribus veroque proxima.*
Paris: Petrus Gromorsus, 1543.
Subject: Prognostication.
Case/C/5238/.706

## Postel, Guillaume.

*Cosmographicae disciplinae compendium, in suum finem, hoc est ad diuinae prouidentiae certissimam demonstrationem conductum.*
Basel: Ioannes Oporinus, 1561.
Subject: Cosmography.
Ayer/7/P85/1561

## Postel, Guillaume.

*De cosmographica disciplina et signorum coelestium vera configuratione: libri II. Ex museo Joan Balesdens.*
Leyden: Joan. Mairs [*sic*], 1636.
Subject: Astrology, cosmography.
Case/-G/117/.708

## Postel, Guillaume.

*De Etrvriae regionis, qvae prima in orbe Evropaeo habitata est, originibus, institutis, religione & moribus, & imprimis de avrei saecvli doctrina et uita praestantissima quae in diuinationis sacrae usu posita est, Guilielmi Postellis commentatio.*
Florence: [L. Torrentinus?] 1551.
Subject: Anthropology, origin of divination.
Case/F/0235/.706

## Postel, Guillaume.

*De Foenicvm literis, sev de prisco Latinae & Graecae linguae charactere, eiusque antiquissima origine & vsu.*
Paris: Vivantius Gaultherot, 1552.
Subject: Comparative linguistics, philology.
Case/–Z/11/.708

With folding plate table of alphabets in, e.g., Samaritan, Gallic, Latin, Greek.

## Postel, Guillaume.

*Liber de cavsis sev de principiis & originibus naturae vtriusque, in quo ita de eterna rerum veritate agitur vt & authoritate & ratione non tantùm vbiuis particularis Dei prouidentia, sed & animorum & corporum immortalitas ex ipsius Aristotelis verbis rectè intellectis & non detortis demonstretur clarissimè.*
Paris: Sebastianus Niuellius, 1552.
Subject: Natural philosophy.
Case/–B/239/.712

### Postel, Guillaume.

*De orbis terrae concordia libri quatuor, mvlti ivga ervditione ac pietate referti; quibus nihil hoc tam perturbato rerum statu uel utilius, uel accommodatius potuisse in publicum edi, quiuis aequus lector indicabit.*
[Basel, 1544]
Subject: Religion and science.
Case/fC/523/.706

### Postel, Guillaume.

*De vniuersitate liber, in quo astronomiae doctrinaeúe coelestis compendium terrae aptatum, & secundum coelestis influxus ordinem precipuarúmque originum rationem totus orbis terrae quatenus innotuit, cum regnorum temporibus exponitur.*
Paris: Martinus Iuuenes, 1563.
Subject: Astronomy, cosmography.
Case/B/117/.71

2d ed. 2 vols. in 1. Vol. 1: "De vniversitate sev de cosmographia compendium"; vol. 2: "Alterius siue secundae partis operis Gvilelmi Postelli de vniversitate expositio, cui nomen imposuit Ptolemeolus."

### Potts, Thomas, fl. 1612–1618.

*The vvonderfvll discoverie of witches in the covntie of Lancaster. With the arraignement and triall of nineteene notorious witches . . . together with the arraignement and triall of Iennet Preston.*
London: printed by W. Stansby for I. Barnes, 1613.
STC 20138
Subject: Witchcraft.
Case/B/8845/.708

### [Powell, Thomas] 1608–1666.

*Humane industry: or, a history of most manual arts, deducing the original, progress, and improvement of them. Furnished with a variety of instances and examples, shewing forth excellency of humane wit.*
London: printed for Henry Herringman, 1661.
STC II P3072
Subject: Instrumentation, technology.
Case/A/15/.71

Sections on, e.g., use of wild animals (elephant, lion) as beasts of burden or to draw vehicles; dials, clocks, mariner's compass; glassmaking; "Of sundry machins [*sic*] and artificial motions, by water and air."

### Prieur, Claude.

*Dialogve de la lycanthrophie, ov transformation d'hommes en lovps, vulgairement dits loups-garous, & si telle peut faire. Auquel en discourant est traicté de la maniere de se contregarder des enchantemens & sorcelleries, ensemble de plusieurs abus & superstitions, lesquelles se commettent en ce temps.*
Louvain: Iehan Maes and Philippe Zangre, 1596.
Subject: Magic, witchcraft.
Case/B/862/.706

### Prior, Thomas, 1682?–1751.

*An authentick narrative of the success of tar-water, in curing a great number and variety of distempers, with remarks and occasional papers relative to the subject. To which are subjoined, two letters from the author of* Siris, *shewing the medicinal properties of tar-water, and the best manner of making it.*
Dublin printed, London reprinted for W. Innys et al., 1746.
Subject: Medicine, materia medica.
Q/.7165

Testimonials to the efficacy of tar-water.

### Proclus Diadochus, 410?–485.

*Platonis ex Timeo de animorvm generatione, cvm explicatione et digressione Proclilytii tradvctio.*
[Venice: Bernardinus Vitalis Venetus, 1525.]
Subject: Constitution and generation of the soul.
Case/Y/642/.P8096

### Proclus Diadochus.

*La sfera di Proclo, nvovamente tradotta dal Greco essemplare in idioma Italiano.*
Venice: Gabriel Giolito de Ferrari & brothers, 1556.
Subject: Astronomy.
Wing/ZP/535/.G421

BOUND WITH Scandianese, Tito, q.v.

* * *

*Prognosticon: Opusculum reportorii prognosticon in mutationes aeris. Hippocratis libellus de medicorum astrologia.*
Venice: Erhard Ratdolt, 1485.
H-C 13393^ BMC (XV)V:291^ Pr. 4401^ Still. P920^ Goff P-1006
Subject: Astrological medicine.
Inc./4401

Translated by Petrus de Abbano. "Opusculum reportorii pronosticon in mutationes aeris tam via astrologica que metheorologica vt sapientes experientia comperientes voluerunt proque vtilissime ordinatum."

### Psellus, Michael, 11th cent.

*Pselli . . . Liber de quatuor mathematicis scientijs arithmetica, musica, geometria, & astronomia: Graecè & Latinè nunc primùm editus. Gvilielmo Xylandro Augustano interprete.*
Basel: Ioannes Oporinus, [1556].
Subject: Astronomy, geometry, mathematics.
Case/3A/731

Latin text followed by Greek text.

### Ptolemy, Claudius, 2d cent. A.D.

[*Cosmographia*]
[Vicenza: H. Liechtenstein, 1475.]
H 13536^ Sabin 66469^ Phillips 351^ Goff P-1081
Subject: Cosmography, geography.
Ayer/*6/P9/1475

Translated by J. d'Angelo. Edited by Angelus Vadius and Barnabas Picardus Vicentinus. No plates. Colophon: "En tibi lector cosmographia Ptolemaei ab Hermano

Leuilapide Coloniensi Vicenciae accuratissime impressa. Benedicto Triuisano & Angelo Michaele praesidibus." With 4 MS leaves in an early hand tipped in at end (a table of contents). A 2-leaf MS history and description of the Stevens collection, written by Henry Newton Stevens, son of the original collector and dated 18 May 1898.

### Ptolemy, Claudius.

[*Cosmographia*]
[Rome: Arnoldvs Bvckinck, 1478.]
H 13537^ Sabin 66470^ Goff P-1083
Subject: Cosmography, geography.
Ayer/*6/P9/1478

Trans. by Jacobus Angelus; ed., with emendations of Georgius Gemistus, by Domitius Calderinus of Verona. 1st ed. with maps. Only known work bearing imprint of Buckinck. Some MS marginalia, leaves 8–10, 14–16.

### Ptolemy, Claudius.

[*Cosmographia*]
[Vlm: Leonardvs Hol (Leinhart Holle), 1482.]
H 13538^ Sabin 66472^ Phillips 353^ Goff P-1084
Subject: Cosmography, geography.
Ayer/*6/P9/1482a

In 8 books. Trans. by Jacobus Angelus; ed. by Donnus Nicolaus Germanus, who redrew, corrected, and improved maps, adding 5 new ones, incl. first printed representation of Greenland. Recto of last leaf is xylographic.

### Ptolemy, Claudius

[*Cosmographia*] *Clavdii Ptholomei viri Alexandrini cosmographie.*
[Vlm: Leonardvs Hol, 1482]
H 13538^ Sabin 66472^ Phillips 353^ Goff P-1084
Subject: Cosmography, geography.
Ayer/*6/P9/1482b

In 8 books. (2d issue?) Recto of last leaf is typeset rather than xylographic.

### Ptolemy, Claudius.

[*Cosmographia*]
[Ulm: printed by Johann Reger for Justus de Albano de Venetiis, 1486.]
H 13540^ Sabin 66473^ Phillips 354^ Goff P-1085
Subject: Cosmography, geography.
Ayer/*6/P9/1486

Has added index. Woodcut maps are followe by a newly included treatise, "De locis ac mirabilibvs mvndi et primvs de tribvs orbis partibvs."

### [Ptolemy] Stobniczy, Jan ze (Ioannes de Stobnicza), d. 1530.

[*Cosmographia*] *Introductio in Ptolomei cosmographiam cum longitudinibus & latitudinibus regionum & ciuitatum celebriorum.*
Cracow: Hieronymus Vietor, 1519.
Harrisse 95
Subject: Cosmography, geography.
Ayer/*6/P9 S8/1519

2d ed.

### Ptolemy, Claudius.

[*Geographia*]
Florence: Nicolaus Todescho, 1480.
Sabin 66500
Subject: Cosmography, geography.
Ayer/*6/P9 B5/1480a

1st issue. Recto of 1st leaf and entire last leaf blank. Italian terza rima paraphrase by Francesco Berlinghieri. Verso of 1st leaf: "In qvesto volvmne si contengono septe giornate della geographia di Francesco Berlingeri Fiorentino allo illvstrissimo Federigo Dvca Dvrbino."

### Ptolemy, Claudius.

[*Geographia*] *Geographia di Francesco Berlinghieri Fiorentino in terza rima et lingva Toscana distincta con le sve tavole in varii siti et provincie secondo la geographia et distinctione delle tauole di Ptolomeo.*
Florence: Nicolaus Todescho, 1480.
Sabin 66501
Subject: Cosmography, geography.
Ayer/*6/P9 B5/1480b

Title page printed in red, probably added, according to statement on bookplate, after 1500. Register and colophon probably added after 1st ed. but before title page. Bk. 1 has hand-colored initials and some illumination; bks. 2 ff. have scattered MS marginalia and notes. Maps are colored and have notes or indexes on verso.

### Ptolemy, Claudius.

[*Geographia*]
Rome: Petrus de Tvrre, 1490.
H *13541^ Pr. 3966 (2 Rome 35–1)^ BMC (XV)IV:133–34^ C 19313^ Goff P-1086
Subject: Cosmography, geography.
Ayer/*6/P9/1490

2d Rome ed.

### Ptolemy, Claudius.

[*Geographia*] *In hoc operae haec continentvr geographia Cl. Ptholemaei a plurimis uiris utriusque linguae doctiss. emendati: & cum archetypo graeco ab ipsis collata. Schemata cum demonstrationibus suis correcta a Marco Monacho Caelestino Beneuentano; & Ioanne Cota Veronënsi uiris mathematicis consultissimis.*
Rome: Bernardinus Venetus de Vitalibus, 1507.
Sabin 66476^ Phillips 356
Subject: Cosmography, geography.
Ayer/*6/P9/1507

New ed. of Jacobus Angelus.

### Ptolemy, Claudius.

[*Geographia*] *Clavdii Ptolemaei Alexandrini liber geographiae cvm tabvlis et vniversali figvra et cvm additione locorvm qvae*

*a recentioribvs reperta svnt diligenti cvra emendatvs et impressvs.*
Venice: Jacob Pentius de Leucho, 1511.
Subject: Cosmography, geography.
Ayer/*6/P9/1511a

[From description on bookplate:] "This edition is principally esteemed for the New World map drawn on a heart-shaped projection, which contains the first printed delineation of any part of the North American continent."

## Ptolemy, Claudius.

[*Geographia*] *Clavdii Ptholemaei Alexandrini liber geographiae cvm tabvlis et vniversali figvra et cvm additione locorvm qvae a recentioribvs reperta svnt diligenti cvra emendatvs impressvs.*
Venice: Jacob Pentius de Leucho, 1511.
Subject: Cosmography, geography.
Ayer/*6/P9/1511b

## Ptolemy, Claudius.

[*Geographia*] *Claudii Ptolemei viri Alexandrini mathematice disciplinae philosophi doctissimi geographiae opus nouissima traductione e Graecorum archetypis castigatissime pressum: caeteris ante lucubratorum multo praestantius.*
[Strassburg: Ioannes Schott, 1513.]
Sabin 66478^ Phillips 359
Subject: Cosmography, geography.
Ayer/*6/P9/1513

Ed. and published by Jacobus Essler and Georgius Ubelin,.

## Ptolemy, Claudius.

[*Geographia*] *In hoc opera continentur noua translatio primi libri geographiae Cl. Ptolomaei.*
Nuremberg: Ioannes Stuchs, 1514.
Sabin 66479
Subject: Cosmography, geography.
Ayer/*6/P9/1514

Trans. by Ioannes Vernero [Werner?]. Includes "De his quae geographiae debent adesse: Georgii Amirucii Constantinopolitani opusculum; Ioannis de Regiomonte epistola . . . de compositione & vsu cuiusdam meteoroscopii." No maps.

## Ptolemy, Claudius

[*Geographia*] *Ptolemaevs avctvs restitvtvs. Emacvlatvs. cvm tabvlis veteribvs ac novis.*
[Strassburg: Ioannes Schott, 1520.]
Sabin 66480^ Phillips 360
Subject: Cosmography, Geography.
Ayer/*6/P9/1520

Ed. by Georgius Ubelin. 2d Strassburg ed. Has some scattered hand coloring on maps.

## Ptolemy, Claudius.

[*Geographia*] *Clavdii Ptolemaei Alexandrini . . . opus geographiae nouiter castigatum & emaculatum additionibus. . . . Hec bona mento Laurentius Phrisius artis Appollineae doctor & mathematicarum artium clientulus, in lucem iussit prodire.*
[Strassburg: Ioannes Grieninger, 1522.]
Sabin 66481^ Phillips 361
Subject: Cosmography, geography.
Ayer/*6/P9/1522

3d Strassburg ed.

## Ptolemy, Claudius.

[*Geographia*] *Clavdii Ptolemaei geographicae enarrationis libri octo Bilibaldo Pircheymhero interprete Annotationes de Regio Monte in errores commissos a Iacobo Angelo in translatione sua.*
[Strassburg: Iohannes Grieninger for Iohannis Koberger, 1525.]
Sabin 66482^ Phillips 362
Subject: Cosmography, geography.
Ayer/*6/P9/1525

4th Strassburg ed.

## Ptolemy, Claudius.

[*Geographia*] [Greek title.] *Clavdii Ptolemaei Alexandrini philosophi cum primus eruditi, de geographia libri octo, summa cum uigilantia excusi.*
Basel: Froben, 1533.
Sabin v. 16, p. 86^ Phillips 363
Subject: Cosmography, geography.
Ayer/*6/P9/1533

1st complete Greek edition. Dedication [to Theobald Fettichi] signed by Desiderius Erasmus. First printing of Greek text. No maps.

## Ptolemy, Claudius.

[*Geographia*] *Clavdii Ptolemaei Alexandrini geographicae enarrationis libri octo . . . à Michaële Villanouano iam primum recogniti. Adiecta insuper ab eodem scholia, quibus exoleta urbium nomina ad nostri seculi morem exponuntur.*
Lyon: Melchior and Gaspar Treschel, brothers, 1535.
Sabin 66483^ Phillips 364
Subject: Cosmography, geography.
Ayer/*6/P9/1535

Translated by Wilibald Pirckheimer. Woodcut borders and ornaments said to be the work of Hans Holbein and Graf of Basel. Many copies of this ed. said to have been burned on Calvin's orders when the editor, Michael Villanovanus (better known as Servetus) was executed in 1553.

## Ptolemy, Claudius

[*Geographia*] *Cl. Ptolemaei Alexandrini philosophi et mathematici praestantissimi libri VIII de geographia è Graeco denuo traducti . . . redactis numquam antea uisa commoditate simili: Ioannis Noviomagi opera.*
Cologne: Joannes Ruremundanus, 1540.
Subject: Cosmography, geography.
Ayer/*6/P9/1540a

New Latin trans. (from Greek) by Joannes Noviomagus (Johann Bronchorst).

**Ptolemy, Claudius.**

[*Geographia*] *Geographia vniversalis, vetvs et nova, complectens Clavdii Ptolemaei Alexandrini enarrationis libros VIII. Quorum primus noua translatione Pirckheimheri et accessione commentarioli illustrior quàm hactenus fuerit, redditus est.*
Basel: Henricus Petri, 1540.
Sabin 66484; Phillips 365
Subject: Cosmography, geography.
Ayer/*6/P9/1540b

New ed., revised and ed. Sebastian Münster, who also redesigned the maps and added a geographical appendix.

**Ptolemy, Claudius**

[*Geographia*] [Another copy]
Basel: Henricus Petri, 1540.
Case/fY/642/.P895

**Ptolemy, Claudius.**

[*Geographia*] *Clavdii Ptolemaei Alexandrini geographicae enarrationis libri octo. Ex Bilibaldi Pirckeymheri tralatione, sed ad Graeca & prisca exemplaria à Michaële Villanouano secundò recogniti, & locis innumeris denuò castigati. Adiecta insuper ab eodem scholia, quibus & difficilis ille primus liber nunc primum explicatur, & exoleta vrbium nomina ad nostra seculi morem exponuntur. Quinquaginta illae quoque cum ueterum tum recentium tabulae adnectantur, uarijque incolentium ritus & mores explicantur.*
Lyon: Hugo Porta. [printed: Vienna: Gaspar Treschel, 1541.]
Sabin 66485^ Phillips 366
Subject: Cosmography, geography.
Ayer/*6/P9/1541

2d ed. of Villanovanus (Servetus), newly revised and corrected.

**Ptolemy, Claudius.**

[*Geographia*] [Another copy]
Lyon: Hugo Porta. [printed: Vienna: Gaspar Treschel, 1541.]
Case/fY/642/.P8951

**Ptolemy, Claudius.**

[*Geographia*] *Geographia vniversalis, vetvs et nova, complectens Clavdio Ptolemaei Alexandrini enarrationis libros VIII. Quorum primus noua translatione Pirckheimeri et accessione commentarioli illustrior quàm hactenus fuerit, redditus est. Reliqui cum Graeco & alijs uetustis exemplaribus collati, in infinitis ferè locis castigationes facti sunt. Addita sunt insuper scholia . . . succedunt tabulae Ptolemaicae, opera Sebastiani Munsteri nouo paratae modo. His adiectae sunt plurimae nouae tabulae, modernam orbis faciem literis & pictura explicantes, inter quas quaedam antehàc Ptolemaeo non fuerunt additae. Vltimo annexum est compendium geographicae descriptionis.*
Basel: Henricus Petri, 1542.
Sabin 66486^ Phillips 367
Subject: Cosmography, geography.
Ayer/*6/P9/1542

2d ed. of Münster, reprinted without addition or alteration from the 1540 ed.

**Ptolemy, Claudius.**

[*Geographia*] *Geographia vniversalis, vetvs et nova, complectens Clavdii Ptolemaei Alexandrini enarrationis libros VIII. . . . Adiectae sunt huic posteriori editioni nouae quaedam tabulae, quae hactenus apud nullam Ptolemaicam impressuram uisae sunt.*
Basel: Henrichvs Petri, 1545.
Sabin 66487^ Phillips 368
Subject: Cosmography, geography.
Ayer/*6/P9/1545

3d ed. of Münster; with 6 new maps.

**Ptolemy, Claudius.**

[*Geographia*] *Ptolemeo la geografia di Clavdio Ptolemeo Alessandrino, con alcuni comenti & aggiunte fatteui da Sebastiano Munstero Alamanno, con le tauole non solamente antiche & moderne solite di stamparsi, ma altre noue aggiunteui di messer Iacopo Gastaldo Piamontese cosmographo, ridotta in uolgare Italiano da M. Pietro Andrea Mattiolo Senese medico eccellentissimo.*
Venice: Giovanni Baptista Pedrezano, 1548.
Sabin 66502^ Phillips 369^ Harrisse 285
Subject: Cosmography, geography.
Ayer/*6/P9/1548

1st Italian ed.

**Ptolemy, Claudius**

[*Geographia*] *Geographiae Clavdii Ptolemaei Alexandrini, philosophi ac mathematici praestantissimi, libri VIII, partim à Bilibaldo Pirckheymero translati ac commentario illustrati, partim etiam Graecorum antiquissimorumque exemplariorum collatione emendati atque in integrum restituti.*
[Basel: Henrichus Petri, 1552.]
Sabin 66488^ Phillips 370
Subject: Cosmography, geography.

4th ed. of Münster with additional treatise, enlarged indexes, and some alterations in text and lettering in maps.

**Ptolemy, Claudius.**

[*Geographia*] *La geografia di Clavdio Tolomeo Alessandrino, nuouamente tradotta di Greco in Italiano da Girolamo Rvscelli, con espositioni del medesimo . . . et con nuoue . . .figure. . . . Aggiuntoui vn pieno discorso di M. Gioseppe Moleto matematico.*
Venice: Vincenzo Valgrisi, 1561.
Sabin 66503^ Phillips 371
Subject: Cosmography, geography.
Ayer/*6/P9/1561

New Italian ed. [Ruscelli = Allesio of Piedmont, pseud.]

**Ptolemy, Claudius.**

[*Geographia*] *La geografia.* [Another copy]
Case/Y/642/.P894

**Ptolemy, Claudius.**

[*Geographia*] *Geographia Cl. Ptolemaei Alexandrini olim a Bilibaldo Pirckheimherio traslata, at nunc multis codicibus Graecis collata, pluribusque in locis ad pristinam ueritatem redacta a Iosepho Moletio mathematico.*
Venice: Vincentius Valgrisius, 1562.
Sabin 66489^ Phillips 372
Subject: Cosmography, geography.
Ayer/*6/P9/1562

"Addita sunt in primum & septimum librum amplissima eiusdem commentaria."

**Ptolemy, Claudius.**

[*Geographia*] *La geografia di Clavdio Tolomeo Alessandrino, nuouamente tradotta di Greco in Italiano da Ieronimo Rvscelli, con espositioni del medesimo . . . et con nvove . . . figure. . . . Aggivntovi un pieno discorso di M. Gioseppe Moleto.*
Venice: Giordan [*sic*] Ziletti, 1564.
Sabin 66504
Subject: Cosmography, geography.
Ayer/*6/P9/1564b

2d ed. of Ruscelli. BOUND WITH Moletius's "Discorso" (separate title page and separate pagination).

**Ptolemy, Claudius.**

[*Geographia*] *Geographia Cl. Ptolemaei Alexandrini olim ab alijs translata, at nunc multis codicibus Graecis collata, pluribusqué in locis ad pristinam ueritatem mira accuratione redacta. . . . Adsunt LXIIII tabulae, XXVII nempe antiquae, & XXXVII nouae.*
Venice: Giordano Zileti, [*sic*] 1564.
Sabin 66504
Subject: Cosmography, geography.
Ayer/*6/P9/1564a

Reissue (by different publisher) of 1562 Latin ed. of Moletius, whose name does not appear on the reprinted title page but is retained on the (reprinted) dedication page.

**Ptolemy, Claudius.**

[*Geographia*] *La geografia di Clavdio Tolomeo Alessandrino, già tradotto di Greco in Italiano da M. Giero. Rvscelli & hora in questa nuoua ed. da M. Gio. Malombra ricorretta, & purgata d'infiniti errori. . . . con l'espositioni del Rvscelli.*
Venice: Giordano Ziletti, 1574.
Sabin 66505^ Phillips 380
Subject: Cosmography, geography.
Ayer/*6/P9/1574

3d ed. of Ruscelli, revised and corrected by Malombra.

**Ptolemy, Claudius.**

[*Geographia*] *La geografia.* [Another copy]
Case/Y/642/.P8941

**Ptolemy, Claudius.**

[*Geographia*] *Tabvlae geographicae Cl: Ptolemaei ad mentem autoris restitutae & emendatae.*
[Cologne: Godefridus Kempen, 1578.]
Sabin 66490^ Phillips 384
Subject: Cosmography, geography.
Ayer/*6/P9/1578

1st ed. of Mercator's maps for Ptolemy's geography but without text. Each map accompanied by brief description.

**Ptolemy, Claudius.**

[*Geographia*] *Tabvlae geographicae Cl: Ptolemaei ad mentem autoris restitutae & emendatae per Gerardum Mercatorum illustriss. ducis Cliui &c. Cosmographum.*
Cologne: G. Kempen, 1578.
Subject: Cosmography, geography.
Ayer/*6/P9/1578a

MS notes passim in Spanish. Engraved title page hand colored. Also hand-drawn diagrams on verso of folding plates and elsewhere throughout.

**Ptolemy, Claudius.**

[*Geographia*] *Cl. Ptolemaei Alexandrini geographiae libri octo. recogniti iam & diligenter emendati cum tabulis geographicis ad mentem autoris restitutis ac emendatis, per Gerardvm Mercatorem, illustriss. Ducis Cliuensis etc. Cosmographum.*
Cologne: Godefridus Kempen, 1584.
Sabin 66491^ Phillips 390
Subject: Cosmography, geography.
Ayer/*6/P9/1584

Text is version of Pirckheimer, ed. by Arnaldus Mylius.

**Ptolemy, Claudius.**

[*Geographia*] *Geographiae vniversae tvm veteris tvm novae absolvtissimvm opvs dvobvs volvminibvs distinctvm, in quorum priore habentur Cl. Ptolemaei Pelvsiensis geographicae enarrationis libri octo. . . . In secund* [*sic*] *volumine insunt Cl. Ptolemaei antiquae orbis tabulae XXVII . . . et tabulae XXXVII recentiores.*
Venice: heirs of Simon Galignani de Karera, 1596.
Subject: Cosmography, geography.
Ayer/*6/P9/1596

Pt. 2 has separate letterpress title page. Plates by Hieronymus Porro, ed. Giovanni Antonio Magini of Padua.

**Ptolemy, Claudius.**

[*Geographia*] *Geographiae vniversae tvm veteris, tvm novae absolvtissimvm opus, duobus voluminibus distinctum.*
Cologne: Petrvs Keschedt, 1597.
Sabin 66493^ Phillips 404
Subject: Cosmography, geography.
Greenlee/4890/P97/1597

2d ed. of Magini. No colophon.

**Ptolemy, Claudius.**

[*Geographia*] *Geographiae vniversae.* [Another copy]
Ayer/*6/P9/1597b

Colophon: Arnhemii, apud Ioannem Iansonium Bibliopolam, Anno M.D.XCVII.

**Ptolemy, Claudius.**

[*Geographia*] *Geographiae vniversae.* [Another copy]
Ayer/*6/P9/1597a

No colophon.

**Ptolemy, Claudius.**

[*Geographia*] *Geografia cioè descrittione vniversale della terra partita in dve volumi, nel primo de' quali si contengono gli otto libri della geografia di Cl. Tolomeo, nuouamente con singolare studio rincontrati, & corretti dall' eccell.mo sig. Gio. Ant. Magini Padovano. . . . Nel secondo vi sono poste XXVII tauole antichi di Tolomeo . . . illustrate da ricchissimi commentarij di detto Sig. Magini.*
Venice: Gio. Battista & Giorgio Galignani, brothers, 1598.
Sabin 66506^ Phillips 405
Subject: Cosmography, geography.
Ayer/*6/P9/1598b

"La seconda parte della geografia di Cl. Tolomeo, la qual, oltra l'antiche tauole d'esso Tolomeo, contiene le moderne ancora . . . intagliate da Girolamo Porro."

**Ptolemy, Claudius.**

[*Geographia*] *Geografia di Clavdio Tolomeo Alessandrino.*
Venice: heirs of M. Sessa, 1598.
Sabin 66507 (date 1598)^ Phillips 409 (date 1599)
Subject: Cosmography, geography.
Ayer/*6/P9/1598a

4th ed. of Ruscelli trans. Title page: 1598. Untrimmed.

**Ptolemy, Claudius.**

[*Geographia*] *Geografia di Clavdio Tolomeo Alessandrino, tradotta di Greco nell' idioma volgare Italiano da Girolamo Rvscelli, et hora nuouamente ampliata da Gioseffo Rosaccio.*
Venice: heirs of Melchior Sessa, 1599.
Sabin 66507^ Phillips 409
Subject: Cosmography, geography.
Ayer/*6/P9/1599

4th ed. of Ruscelli trans.

**Ptolemy, Claudius.**

[*Geographia*] *Claudii Ptolemaei Alexandrini Geographiae libri octo Graeco-Latini Latinè primùm recogniti & emendati cum tabulis geographicis ad mentem auctoris restitutis per Gerardum Mercatorem: Iam verò ad Graeca & Latina exemplaria à Petro Montano iterum recogniti, et pluribus locis castigati. Adiecta insuper ab eodem nomina recentia et aequipollentia ex variis auctoribus veteribus et recentiorib. magna cura collecta; in gratiam et usum geographiae studiosorum.*
Amsterdam: Cornelius Nicolaus, Iodocus Hondius, 1605.
Subject: Cosmography, geography.
Ayer/*6/P9/1605a

1st ed. of Greek and Latin texts together. Ed. Petrus Montanus. Preface by Hondius contains account of the origin of this edition. Hand-colored plates.

**Ptolemy, Claudius.**

[*Geographia*] *Claudii Ptolemaei Alexandrini Geographiae libri octo Graeco-Latini.*
Frankfurt: printed by Iodocus Hondius, sold by Cornelius Nicolaus, 1605.
Subject: Cosmography, geography.
Ayer/*6/P9/1605b

**Ptolemy, Claudius.**

[*Geographia*] *Claudii Ptolemaei Alexandrini Geographiae libri octo Graeco-Latini.*
Amsterdam: printed by Iodocus Hondius, sold by Cornelius Nicolaus, 1605.
Subject: Cosmography, geography.
Ayer/*6/P9/1605c

Imprint on letterpress title page: "Sumtibus Cornelii Nicolai & Iudoci Hondii, Amsterodami, 1605."

**Ptolemy, Claudius.**

[*Geographia*] *Geographiae vniversae tvm veteris tvm novae, absolvtissimvm opvs, dvobvs volvminibvs distinctvm, in quorum priorum habentur Cl. Ptolemaei Pelvsiensis geographicae enarrationis libri octo. . . . In secundo volumine insunt Cl. Ptolemaei antiquae orbis tabulae XXVII.*
Cologne: Petrus Keschedt, 1608.
Sabin 66495
Subject: Cosmography, geography.
Ayer/*6/P9/1608

3d ed. of Giovanni Antonio Magini.

**Ptolemy, Claudius.**

[*Geographia*] *Geographiae, tum veteris, tum novae, volumina duo. In quorum priore, Cl. Ptol. Peluensis geographicae enarrationis libri octo. . . . In posteriore, eiusdem Ptol. antiqui orbis tabulae 27.*
Arnheim: Ioannes Ianssonius, 1617.
Sabin 66497^ Phillips 433
Subject: Cosmography, geography.
Ayer/*6/P9/1617

4th ed. of Giovanni Antonio Magini of Padua. Ed. Gaspar Ens. Maps by Hieronymus Porro. 2d part has separate title page: *Geographiae Cl. Ptolemaei pars secvnda, continens praeter antiqvas ipsius Ptol. recentiores etiam tabulas, quae vniversae terrae faciem nostro aevo cognitam exhibent.*

**Ptolemy, Claudius.**

[*Geographia*] *Theatri geographiae veteris tomus prior in quo Cl. Ptol. Alexandrini geographiae libri VIII Graecé et Latiné Graeca ad codices Palatinos collata aucta et emendata sunt Latina infinitis locis correcta opera P. Bertii.*
Amsterdam: Iodocus Hondius, 1618. [Printed at Leyden by Isaac Elzevir.]
Sabin 66497^ Phillips 433
Subject: Cosmography, geography.
Ayer/*6/P9/1618a

Maps by Mercator and Ortelius. Engraved title page of pt. 2: *Geographia ocvlvs historiarvm theatri geographiae veteris tomvs posterior in quo itinerarivm Antonini imperatoris terrestre & maritimum provinciarvm Romanarvm libellus civitates provinciarvm Gallicarvm itinerarivm a Burdigala Hierosolymam vsque tabvla Pevtingeriana cum notis Marci Velseri ad tabulae eius partem parergi Orteliani tabulae aliquot edente P. Bertio.* Amsterdam: Iudoci Hondii, 1619.

### Ptolemy, Claudius.

[*Geographia*] *Theatri geographiae veteris, duobus tomis distinctum.*
Leyden: printed by Isaacvs Elzevirivs for Ivdocus Hondius, 1618.
Subject: Cosmography, geography.
Ayer/*6/P9/1618b

BOUND WITH *Petri Kaerii Germania inferior id est, XVII provinciarum ejus novae et exacte tabulae geographicae, cum luculentis singularum descriptionibus additis. A Petro Montano.* Amsterdam: printed for Pet. Kaerii, 1622. Customs, history of ancient Belgium and the Netherlands. With quotations from Strabo, Diodorus, Mela, Pliny, and others. Both vols. have hand-colored engraved title pages.

### Ptolemy, Claudius.

[*Geographia*] *Geografia cioè descrittione vniversale della terra partita in due volumi, nel primo de' quali si contengono gli otto libri della geografia di Cl. Tolomeo, nuouamente con singolare studio rincontrari, & corretti dall' eccell.mo Sig. Gio. Antonio Magini Padovano.... Nel secondo visono poste XXVII tauole antichi di Tolomeo... Opera vtilissima & specialmente necessaria allo studio dell' historie, dal Latino nell' Italiano tradotta dal R. D. Leonardo Cernoti Vinitiano.*
Padua: Galignani Brothers, 1621.
Phillips 436
Subject: Cosmography, geography.
Ayer/*6/P9/1621

2d ed. of Cernoti's translation of of Magini. Untrimmed copy.

### Ptolemy, Claudius.

*Claudii Ptolemaei harmonicorum libri tres.*
Oxford: Theatro Sheldoniano, 1682.
Subject: Acoustics, harmonics.
STC II P4169
Case/V/331/.717

John Wallis, 1616–1703 (Savilian Professor of geometry at Oxford and member of the Royal Society) ed. and trans. In double cols., Greek and Latin. Bk. 3, chap. 8: "De similitudine perfecti systematis, & zodiaci circuli"; chap. 10: "Quomodo stellarum motui in longitudinem, assimilatur, continuus in sonis motus"; chap. 13: "De analogia quae est inter tetrachorda, & aspectus ad solem."

### Ptolemy, Claudius.

*Liber de analemmate, a Federico Commandino Vrbinate instauratu & commentariis illustratus,... eiusdem Federici Commandini liber de horologiorum descriptione.*
Rome: Paulus Manutius Aldus brothers, 1562.
Subject: Cosmography, gnomonics, mathematical astronomy.
Case/L/900/.716

Some MS marginalia.

### Ptolemy, Claudius.

*C. Ptolemaei mathematicae constrvctionis liber primus. Additae explicationes aliquot locorum ab Erasmus Reinholt Salueldensi.*
Paris: Gulielmus Cauellat, 1557.
Subject: Astronomy, geometry, mathematics.
Case/3A/734

In 2 books. Some MS marginalia. Bk. 1, chap. 3: "Qvod terra sit sphaerica ad sensum secundum vniuersas partes"; chap. 6: "Qvod terra non moueatur locali motu, seu mutatione loci"; bk. 2 title page: *Liber secundus Latina interpretatione recēns donatus.* BOUND WITH *Le liure de la musique d'Euclide traduit par P. Forcadel* [attributed to Cleonides]. Paris: Charles Perier, 1566. Subject: mathematics and harmony.

### Ptolemy, Claudius.

*Ptolemaei planisphaerivm. Iordani planisphaerivm. Federici Commandini Vrbinatis in Ptolemaei planisphaerivm commentarivs. In quo uinuersa scenographices ratio quambreuissime traditur, ac demonstrationibus confirmatur.*
Venice: Aldus, 1558.
Subject: Astronomy.
Ayer/*6/P9/1558

MS notes in early hand on front flyleaf and scattered in Commandino's commentary.

### Ptolemy, Claudius.

[*Quadripartitum centiloquium cum commento Hali*] *Liber quadripartiti Ptolemaei id est quatuor tractatum: in radicanti discretione per stellas de futuris & in hoc mundo constructionis & destructionis contingentibus.*
Venice: Erhard Ratdolt, 1484.
H-C *13543^ BMC (XV)V:288^ Pr. 4394^ Still. P994^ Goff P-1088
Subject: Astrology.
Inc./4394

### Ptolemy, Claudius.

*Tractatus de judicandi facultate et de animi principatu, e Graeco in Latinum versus ab Ismaele Bullialdo: cujus accedunt notae breves ad Renati Cartesii opiniones ... et solutiones aliquot quibus primae philosophiae principia astruuntur: per Franciscus DuLaurens.*
The Hague: Adrian Vlacq, 1663.
Subject: Astrology, elements, humors.
Ayer/6/P9/1663

Descartes: 1: "De anime"; 2: "Corpus non sentire sed animam"; 3: "De speciebus ab objectis emissis." Tract of Du Laurens has separate signatures and separate pagination.

**Purchas, Samuel, 1577?–1626.**

*Purchas his pilgrim. Microcosmvs, or the historie of man. Relating to the wonders of his generation, vanities in his degeneration, necessity of his regeneration.*
London: printed by W. S. for Henry Fetherstone, 1619.
STC 20503
Subject: Macrocosm-Microcosm.
Ayer/3A/269

Chap. 4: "Man a little world: the correspondence betwixt him and the greater world"; chap. 63: the diseases of physicians.

**Purchas, Samuel.**

*Pvrchas his pilgrimage, or relations of the vvorld and the religions observed in all ages and places discouered, from the Creation vnto this present.*
London: printed by William Stansby for Henrie Fetherstone, 1613.
STC 20505
Subject: Geography.
Ayer/*110/P9/1613

In 9 bks. Bk. 1, chap. 13: "The Chaldean and Assyrian chronicle, or computation of times"; bk. 5, chap. 12: "Of the creatures, plants, and fruits in India."

**Purchas, Samuel.**

*Pvrchas his pilgrimage, or relations of the vvorld and the religions observed in all ages and places discouered, from the Creation vnto this present.*
London: printed by William Stansby for Henrie Fetherstone, 1614.
STC 20506
Subject: Geography.
Ayer/*110/P9/1614

2d ed. "much enlarged with additions through the whole worke."

**Purchas, Samuel.**

*Pvrchas his pilgrimage.* [Another edition]
London: printed by William Stansby for Henrie Fetherstone, 1617.
STC 20507
Subject: Geography.
Ayer/*110/P9/1617

3d ed. "much enlarged with additions through the whole worke."

**Purchas, Samuel.**

*Pvrchas his pilgrimage.* [Another edition]
London: printed by William Stansby for Henrie Fetherstone,1626.
STC 20508.5
Subject: Geography.
Ayer/*110/P9/1626

4th ed. "much enlarged with additions, and illustrated with mappes through the whole worke; and three whole treatises annexed, one of Russia and other northeasterne regions by Sr. Ierome Horsey; the second of the Gulf of Bengala by Master William Methold; the third of the Saracenicall empire translated out of Arabike by T. Erpenivs." Binders title: *Purchas's pilgrimes vol. 5*

**Purchas, Samuel.**

[*Purchas his pilgrimes*]
*Haklytvs posthumus or Purchas his pilgrimes. Contayning a history of the world, in sea voyages & lande trauells by Englishmen & others wherein Gods wonders in nature & prouidence, the actes, arts, varieties, & vanities of men, with a world of the world's rarities, are by a world of eyewitness-authors, related to the world Some left written by Mr. Hakluyt at his death more since added. His also perused & perfected. All examined, abbreuiated, illustrated with notes. Enlarged with discourses. Adorned with pictures, and expressed in mapps. In fower parts, each containing fiue bookes.*
London: printed by William Stansby for Henrie Fetherstone, 1625.
STC 20509
Subject: Discovery, exploration, navigation, travel.
Case/fG/12/.71

In 4 vols. Fifth vol. in the set is *Pvrchas his pilgrimage, or relations of the vvorld and the religions obserued in all ages and places discouered, from the creation vnto this present. Contayning a theologicall and geographicall historie of Asia, Africa, and America, with the ilands adiacent.* [4th ed.]

**Purchas, Samuel.**

[*Purchas his pilgrimes*]
*Haklvytvs posthumus.* [Another copy]
Ayer/*110/P9/1625

In 4 vols. with *Pvrchas his pilgrimage*, 4th ed., q.v., added to the set as vol. 5. Penciled note on front flyleaf: "The only copy known (besides that in the Grenville Library) which has the right map on p. 65, book 1 of vol. 1. All other copies have on that page a duplicate of the inferior world-map on page 115 of the same book."

**Purchas, Samuel.**

*A theatre of politicall flying-insects, wherein especially the nature, the vvorth, the vvork, the wonder, and the manner, of right-ordering of the bee is discovered and described.*
London: printed by R. I. for T. Parkhurst, 1657.
STC II P4224
Subject: Apiculture, entomology.
Case/R/63/.71

In 2 parts. Pt. 1: authorities of antiquity on bees and their value; pt. 2: "Meditations and observations, theologicall and morall . . . upon the nature of bees."

## Quarles, Francis, 1592–1644.

*Emblemes.*
London: printed by G. M[iller], 1635.
STC 20540
Subject: Emblems.
Case/W/1025/.715

To the reader (sig. A3): "An embleme is but a silent parable." Sundial (p. 177), and a few other mystical and scientific symbols.

## Quercetanus (Duchesne), Josephus, 1544?–1609.

*Diaeteticon polyhistoricon. Opvs vtiqve varivm, magnae vtilitatis ac delectationis, quod multa historica, philosophica, & medica, tàm conseruandae sanitati, quàm varijs curandis morbis necessaria continent.*
Geneva: Petrus Chouët, 1626.
Subject: Diet, medicine.
R/82/.244

In 3 sections: (1) perturbations of the soul, emotions; (2) preparation of food and drink; (3) various herbs, foods, and drinks and their effects on man.

* * *

*Querela geometrica, or geometry's complaint of the injuries lately received from Mr. Thomas VVhite in his late tract, entituled, Tutela geometrica. In the end you have some places at large out of Mr. White's Tutela, and Gulden's Centrobaryca, reprinted, and faithfully translated into English.*
London: printed by R. W., 1660.
STC II Q162.
Subject: Geometry, mathematics.
Case/3A/2019

Translated from *De centro gravitatis* (p. 73), *The center of gravity* appears on p. 81.

* * *

*Quinta scienza cava' pr forza d'inzegn, e d'vrdign dà i influss d' tutt quant l'dsuers strell ch' in tal ciel, e lambichè cunform alla vera art, e al me solit, cumpusition salutifra pr sauer acgnosr al ben dal mal, e pr sauer ab mod, e al temp d'guardars dai incontr cattiu, ch' mn azza l'strell in tutta la carriera d' st' ann MDCLV.*
Bologna: Iacm Mont [1654?].
Subject: Almanacs, astrology.
Bonaparte/Coll./No./5417

Dedication signed Giacomo Monti. "Discors astrologigh in dialg sovra l'ann MDCLV."

# R

## R. accademia delle scienze dell' Istituto di Bologne.

*De Bononiensi scientiarum et artium instituto atque academia commentarii.*
Bologna: Laelii a Vulpe, 1731.
Subject: Academies and learned societies
fA/9/.103

Among scientific subjects: "de crystallo montana" (p.88); "chymica" (p. 113); "anatomica" (p. 123); "medica" (p. 145); "de observationibus astronomicis variis" (p. 265); "mechanica" (p. 213); "analytica—de geometria principiis" (p. 241); Dominici Gulielmini, "Epistola hydrostatica" (p. 545).

## Raimondo, Annibale.

*Opera dell' antica et honorata scientia de nomandia, specchio d'infiniti beni, & mali, che sotto il cherchio della luna possono alli viuenti interuenire, per l'eccellentiss. astrologo, geomante, clairomante, & fisionomo M. Annibale Raimondo Veronese, ridotta insieme, & castigata.*
Venice: Iouita Rapirio & associates, 1549.
Subject: Divination, physiognomy.
Case/B/863/.73

Colophon: Venice: Pietro & Gioan. Maria brothers of Nicolini de Sabio, for Iouita Rapirio, & associates, 1550. "Aggiontoui la fisonomia" (p. 184).

## Raleigh, Sir Walter, 1522–1618.

*Abridgement of Sir Walter Raleigh's history of the world, in five books.... Wherein the particular chapters and paragraphs are succinctly abridg'd according to his own method in the larger volume. To which is added, his premonition to princes with some genuine remains of that learned knight, viz I. Of the first invention of shipping; II. A relation of the action at Cadiz; III. A dialogue between a Jesuit and a Recusant; IV. An apology for his unlucky voyage to Guiana.*
London: printed for M. Gillyflower, 1700.
STC II R152
Subject: Creation, instruments (compass), navigation.
Case/F/09/.7332

## Raleigh, Sir Walter.

*The discoverie of the large, rich, and bevvtifvl empyre of Gviana, with a relation of the great and golden citie of Manoa . . . performed in the yeare 1595 by Sir W. Ralegh knight.*
London: Robert Robinson, 1596.
STC 20636^ Sabin 67554
Subject: Discovery, exploration, voyages, New World.
Ayer/*118/R2/1596

Imperfect copy: [p. 103] leaf torn (possibly a damaged sheet from the printer).

## Raleigh, Sir Walter.

*Judicious and selecte observations. By the renowned and learned knight, Sir Walter Raleigh. Upon the first invention of shipping. The misery of invasive warre. The navy royall and sea-service. Wtih his apologie for his voyage to Guiana.*
London: printed by T. W. for Humphrey Moseley, 1650.
STC II R170
Subject: Inventions and technology.
Case/Y/145/.R13

In 3 sections; 1: "A discourse of the invention of ships, anchors, compasse, &c." 2: on invasive war; 3: "Excellent observations and notes concerning the Royal Navy and sea-service."

## Rameau, Jean Philippe 1683–1764.

*Démonstration du principe de l'harmonie, servant de base à tout l'art musical théorique & pratique. Approuvée par messieurs de l'Académie Royale des sciences.*
Paris: Durand, 1750.
Subject: Acoustics, consonance and dissonance, harmonics.
Case/V/22/.733

## Rameau, Jean Philippe.

*Nouveau systême de musique theorique, où l'on découvre le principe de toutes les regles necessaires à la pratique, pour servir d'introduction au traité de l'harmonie.*
Paris: printed by Jean-Baptiste Christophe Ballard, 1726.
Subject: Acoustics, harmony.

Has "Catalogue des autres livres de musique théorique, imprimez en France, dont on peut trouver des exemplaires."

## Ramelli, Agostino, 1531–ca. 1600.

*Le diverse et artificiose machine del capitano Agostino Ramelli dal ponte della tresia ingegniero del Christianissimo re di Francia et Pollonia. Nellequali si contengono uarij et industriosi mouimenti, degni digrandissima speculatione, per cauarne beneficio infinito in ogni sorte d'operatione; composte in lingua Italiana et Francese.*
Paris: by the author, 1588.
Subject: Inventions, technology.
Wing/fZP/539/.R143

Fig. 188 (p. 317): reading machine.

## Ramelli, Agostino.

[*Le diverse et artificiose machine nel Capitano Agostino Ramelli . . . composte in lingua Italiana et Francesa.*
Paris: by the author, 1588.]
Subject: Inventions, technology.
Greenlee/Misc./5000/R17/1588

Imperfect copy: lacking title page (inserted in MS) and scattered pages following fol. 222. A list of missing folios is laid in. Italian and French on facing pages.

## Ramusio, Giovanni Battista, 1485–1557.

*Delle navigationi et viaggi raccolte da M. Gio. Battista Ramvsio, nelle quali con relatione fedelissima si descriuono tutti quei paesi, che da già 300. anni sin'hora sono stati scoperti, così di verso Leuante, & Ponente, come di verso mezzo dì, & tramontana; . . . con discorsi à suoi luoghi, & imprese diuerse d'imperatori di Tartari . . . & tauole di geografia secondo le carte di nauicare. . . . et net fine con aggiunta nella presente quinta impressione del viaggio di. M. Cesare de' Federici, nell' India orientale, nel quale si descriue le spetierie, droghe, gioie, & perle che in dette paesi si trouano.*
Venice: Giunti, 1606–13.
Subject: Natural science, navigation, voyages.
Case/fG/12/.73

In 3 vols. Vol. 1: 6th ed, 1613; vol. 2: 4th ed., 1606; vol. 3: 3d ed., 1606. Contains descriptions (by various authors) of voyages and navigation with commentaries by Ramusio; e.g., vol. 1: "Viaggio fatto nell' India per Giovanni da Empoli"; vol. 2: "Dichiartione d'alcuni luoghi ne libri di M. Marco Polo con l'historia del rheubarbaro," "Parte del trattato dell' aere, dell' acqva, e de lvoghi d'Ippocrate nella qvale si ragiona delli Scithi"; vol. 3: "Nel quale si contiene le nauigationi al mondo nuouo, à gli antich incognito, fatte da Don Christoforo Colombo Genouese" (and numerous others). See Sabin 67735, 67739, 67742.

## Ramusio, Giovanni Battista.

*Delle navigationi et viaggi raccolte da M. Gio. Battista Ramvsio, in tre volvmini divise.*
Venice: Giunti, 1559–1606.
Subject: Natural science, navigation, voyages.
Greenlee/4583/R18/1606

Another set (see previous entry). Vol. 1 (5th ed. [reprint of 4th], 1606); vol. 2 (1st ed., 1559); vol. 3 (2d ed. [reprint of 1st], 1565).

## Rantzau, Henrik, 1526–1598.

*Catalogvs imperatorvm, regvm, ac principvm qvi astrologicam artem amarunt, ornarunt, & exercuerunt: quibus additae sunt astrologica quaedam praedictiones vera ac mirabiles omnium temporum, desumptae ex Iosepho, Suetonio, Tacito, Dione, Xiphilino, Cuspiniano, & aliis, ex quibus certitudo ac veritas harum disciplinarum colligi potest. Adiectus est praeterea tractatus de annis climactericis, vnà cum variis exemplis illustrium virorum qui annis ijsdem & praesertim anno 49, 56, & 63 periere: versus insuper nonnulli de planetis ac signis, mensiumque laboribus, quae omnia tam lectu, quam scitu necessaria videntur: collecta ab Henrico Rantzovio, ac edita à Theophilo Siluio.*
Antwerp: Christophorus Plantin, 1580.
Subject: Astrology.
Case/B/8635/.732

## Rantzau, Henrik.

*De conservanda valetvdine liber, in privatvm liberorvm svorvm vsum ab ipso conscriptus, ac editus. A. Dethlevo Silvio Holsato. In qvo de diaeta, itinere, annis climactericis, & antidotis praestantissimis, breuia & vtilia praecepta continentur.*
Frankfurt: Nicolas Hoffmann for Ioan. Rhodius, 1604.
Subject: Disease, medicine, symptomology.
Case/3A/357

Followed (continuous pagination) by Gulielmo Grataroli, "De memoria reparanda, avgenda, conservandaqve, ac de reminiscentia; tutiora omnimoda remedia, praeceptionesque optimae."

## Rapin, René, 1621–1687.

*The whole critical works of Monsieur Rapin.*
London: printed for H. Bonwicke [etc.], 1706.
Subject: Metaphysics, philosophy, physics.
Y/682/.R17

In 2 vols. Vol. 2 contains his reflections on philosophy in general, on physics (chap. 7), and on metaphysis (chap. 8).

## Rathborne, Aaron, 1572–1618.

*The surveyor in foure books.*
London: printed by W. Stansby for W. Burre, 1616.
STC 20748
Subject: Surveying.
Case/5A/481

Bk. 1: Geometry; bk. 2: problems and practical application; bk. 3: instruments, surveying, including latitude and longitude; bk. 4: legal aspect of surveying.

## Rattray, Sylvester, fl. 1650.

*Aditus nouus ad occultas sympathiae et antipathiae cavsas inveniendas: per principia philosophiae naturalis ex fermentorum artificiosa anatomia hausta, patefactus, à Sylvestro Rattray, med. doct. Glasguensi Scoto.*
Tübingen: Johannes Henricus Reisius, 1660.
Subject: Sympathetic medicine, sympathy and antipathy.
B/529/.231

Man's antipathy to certain animal parts. BOUND WITH *Theatrum Sympatheticum*, q.v.

## Raunce, John.

*Astrologica accusata pariter & condemnata. Or the diabolical art of judicial astrologie, receiving the definitive sentence of final condemnation: being delivered in this following discourse, where the said art is briefly and manifestly opened, justly arraigned, diligently examined, and experimentally condemned by him, who was a student in the same. By John Raunce, sometime a practitioner of astrologie, and student in the magick art.*
London: printed by J. Clowes for W. Learner, 1650.
STC II R317
Subject: Astrology.
Case/B/8635/.88

Binder's title: *Tracts on astrology, 1642–1652.* BOUND WITH Gell, Geree, and E. R. [Edmund Reeve], qq.v.

### Ray, John, 1627?–1705.

A collection of English words . . . with catalogues of English birds and fishes: and an account of the preparing and refining such metals and minerals as are gotten in England.
London: printed by H. Bruges for Tho. Burrel, 1674.
STC II R388
Subject: Metallurgy, zoology.
Bonaparte/Coll./No./11,702

Catalogue of English birds, pp. 81–112; metallurgy, pp. 113–78.

### Ray, John.

*A collection of English words . . . with an account of the preparing and refining such metals and minerals as are gotten in England.*
London: printed for Christopher Wilkinson, 1691.
STC II R389
Subject: Metallurgy.
Bonaparte/Coll./No./11,703.

2d ed. Metallurgy begins p. 174.

### Ray, John.

*Observations topographical, moral, & physiological made in a journey through part of the low-countries, Germany, Italy, and France: with a catalogue of plants not native of England, found spontaneously growing in those parts, and their virtues. Whereunto is added a brief account of Francis Willughby.*
London: printed for J. Martyn, 1673.
STC II R399
Subject: Botany, diet, natural history, travel.
Case/G/307/.736

The baths at Aken (p.64); "Several sorts of meats, fruits, sallets, &c. used in Italy" (p. 403); "A relation of a voyage made through a great part of Spain by Francis Willughby esq. containing the chief observables he met with there, collected out of his notes" (p. 466). BOUND WITH *Catalogus stirpium in exteris regionibus a nobis observatarum; suae vel non omnino vel parcè admodum in Anglia sponte proveniunt.* London: printed by Andrea Clark for J. Martyn, 1673.

### Ray, John.

*Philosophical letters between the late Mr. Ray and several of his ingenious correspondents, natives and foreigners. To which are added those of Francis Willughby esq.; the whole consisting of many curious discoveries in the history of quadrupeds, birds, fishes, insects, plants, fossils, fountains, &c.*
London: printed by William & John Innys, 1718.
Subject: Natural history.
Bonaparte/Coll./No./11,361

"Mr. Ray of the number of plants [in the world]" (p. 344).

### Ray, John.

*Three physico-theological discourses concerning I. The primitive chaos and creation of the world. II. The general deluge, its causes & effects. III. The dissolution of the world, and future conflagration. Wherein are largely discussed the production and use of mountains; the original of fountains, of formed stones, and sea-fishes bones and shells found in the earth; the effects of particular floods and inundations of the sea; the eruptions of volcano's; the nature and causes of earthquakes.*
London: printed for Sam. Smith, 1693.
STC II R409
Subject: Cosmogony, earth sciences, fossils.
Case/C/257/.729

2d ed. corrected and very much enlarged.

### Ray, John.

*Three physico-theological discourses.* [Another ed.]
London: printed for William Innys, 1713.
Subject: Cosmogony, earth sciences, fossils.
C/257/.73

3d ed. enlarged. From the author's own MSS.

### Ray, John.

*The wisdom of God manifested in the works of the creation in two parts viz. the heavenly bodies, elements, meteors, fossils, vegetables, animals (beasts, birds, fishes, and insects); more particularly in the body of the earth, its figure, motion, and consistency; and the admirable structure of the bodies of man, and other animals; as also in their generation &c. with answers to some objections.*
London: printed by R. Harbin for William Innys, 1717.
Subject: Natural history, natural theology.
B/72/.738

Argument from natural history for the existence of God.

### Raymundus de Sabunde d. ?1437.

*De natura hominis, Raemundi Sabundii dialogi: Viola animi ab ipso autore inscripti.*
Lyon: Seb. Gryphius, 1544.
Subject: Natural theology.
Case/C/57/.734

In 7 dialogues.

### Raymundus de Sabunde.

*Theologia naturalis, siue liber creaturarum, specialiter de homine, & de natura eius inquantum homo. & de his que sunt necessaria ad cognoscendum seipsum & deum. et omne debitum ad quo homo tenetur. et obligationum tam deo que primo.*
Strassburg: Martin Flach, 1496.
H-C 14069^ BMC (XV)I:154^ Pr. 703^ Still. R 32^ Goff R-33
Subject: Natural theology.
Inc./703

* * *

*Recherche du contentement et bien souverain de l'homme. Par I. H. P.*
Paris, 1584.
Subject: Psychology.
Case/B/475/.736

## Recorde, Robert, 1510?–1558.

*The castle of knowledge. Containing the explication of the sphere bothe celestiall and materiall, and diuers other thinges incident thereto.*
[London: R. Wolfe, 1556.]
STC 20796
Subject: Astronomy, instrumentation, sphere.
Case/5A/164

In 4 treatises: 1) introduction to the sphere and its parts; 2) making of the sphere; 3) use of the sphere; 4) other uses, tables (proofs, etc.). In dialogue form (Master-Schollar).

## Recorde, Robert.

*The castle of knowledge.* [Another copy]
Case/fL/900/.742

MS marginalia passim.

## Recorde, Robert.

[*The grovnd of arts. . . .*]
[London: by M. F. for John Harison? 1646?]
STC II R650 (?)
Subject: Arithmetic.
Case/L/130/.73

Title page lacking. Referred to in preface as "Book of the art of numbering; Book of arithmetick" (cf. David Eugene Smith, *Rara arithmetica*, pp. 219–20.) In 3 pts., in dialogue form (Master-Scholar [cf. *Castle of knowledge*]). Pt. 2 (p. 261): "The second part of arithmetick; touching fractions, briefly set forth"; pt. 3 (p. 409): "The third part, or, addition to this book: entreateth of brief rules, called the rules of practise, of rare, pleasant, and commodious effects abridged into a briefer method then hitherto hath been published. . . . Set forth by John Mellis, school-master." Tables by R[obert] H[artwell].

## Recorde, Robert.

*Recorde's arithmetick: or, the ground of arts; teaching the perfect work and practice of arithmetick, both in whole numbers and fractions, after a more easie and exact form than in former time hath been set forth. Made by Mr. Robert Record, Dr. in physick. Afterwards augmented by Mr. John Dee, and since enlarged with a third part . . . with divers necessary rules incident to the trade of merchandise . . . by John Mellis. And now diligently perused . . . with an appendix of figurative numbers . . . with tables of board and timber-measure, and new tables of interest . . . the first calculated by R. C. but corrected by Ro. Hartwell.*
London: printed by E. Flesher, 1673.
STC II R648
Subject: Arithmetic.
Case/L/130/.74

Includes bartering, weights and measures, sports and pastimes (e.g., dice).

## Recorde, Robert.

*The Whetstone of witte, which is the seconde parte of arithmetike: containyng the extraction of rootes: the cossike practise, with the rule of equation: and the woorkes of surde nombers.*
[London]: to be sold by Jhon Kyngstone, [1557].
STC 20820
Subject: Algebra, mathematics.
Case/3A/660

First appearance of plus and minus signs in an English work on algebra; Recorde is said to have devised the equals sign (according to extract from booksellers catalogue pasted onto front flyleaf).

## Redi, Francesco, 1626–1689.

*Opere.*
Venice: Gio. Gabbriello Ertz, 1712–1729.
Subject: Consilia, medicine, natural history.
Y/712/.R241

7 vols. in 6. Vols. 1 and 2 (1712): general science and biology, e.g., "generazion degl' insetti; osservazioni intorno a' pellicelli del corpo umano; osservazioni intorno quelle gocciole e fili di vetro, che rotte in qualsisia parte, tutte quante si stritolano; lettera intorno all invenzione degli occhiali; esperienze intorno a' sali fattizj"; [vols. 3–5 n.a.] vol.6, *Consulti lettere e rime*, Florence: Giuseppe Manni, 1729; vol. 7, *Consulti medici opuscoli di Francesco Redi appartenenti alla medicina ed alla storia naturale*, Florence: Manni, 1726.

## Redi, Francesco.

*Esperienze intorno a diverse cose natvrali, e, particolarmente a qvelle, checi son portate dall' Indie, fatte da Francesco Redi, . . . e scritte in vna lettera al reverendissimo padre Atanasio Chircher.*
Florence: Piero Matini, 1686.
Subject: Medicine, psychology, natural history.
Ayer/*108/R31/1686

Medicinal stones to aid in childbirth (pp. 59 ff.).

## Redi, Francesco.

*Esperienze intorno a diverse cose naturali.* [Another ed.]
Naples: Giacomo Raillard, 1687.
Ayer/108/R31/1687

## [Reeve, Edmund]

*The new Jerusalem: the perfection of beauty: the joy of the whole earth. Described in the booke of the revelation, . . . in a sermon composed for the learned society of astrologers, at their generall meeting Aug. 14 anno 1651.*
London: printed by J. G. for Nath. Brooks, 1652.
STC II R668
Subject: Astrology.
Case/B/8635/.88

Binder's title: *Tracts on Astrology, 1642–1652.* BOUND WITH Gell, Geree, and Raunce, qq.v.

* * *

[*Regimen sanitatis Salernitanum*]
*L'escole de medecins de Salerne, qvi enseigne comme il favt sainement & longuement viure. Par la connoissance qv'elle donne des facultez, de tous les aliments qui entrent au corps humain. Enrichie de plusieurs beaux & doctes discours, sur*

*les choses naturelles, non naturelles, & contre nature. Et sur les proprietez des medicaments qui seruent pour la guerison des plus facheuses maladies. Augmenté de l'epistre que Diocle Carystien medecin, enuoye a Antigon roy d'Asie, pour le preseruer ou guerir de toutes les indispositions qui luy pouuoient arriuer. Oeuvres necessaires à toutes sortes de personnes. Et qui donnera vn agreable diuertissement aux lecteurs, & beaucoup de profi aux amateurs de leur santé. Tradvit dv Grec en Francois.*
Rouen: Antoine Ferrand, 1660.
Subject: Diagnosis, diet, hygiene, medicine.
oRA/775/.R34/1660

Diagnoisis and treatment of ailments; "remedes contra le rheume"; "de la garison des fistules" (texte 88).

* * *

*Regimen sanitatis Salernitanum.*
Paris: André Bocard, 1493.
C 5069^ Pr. 8154^ Still. R60^ Goff R-66
Subject: Hygiene, medicine.
Inc./8154

Probably written by Joannes de Mediolano. Commentary by Arnold of Villanova. Two MS paragraphs on verso of back flyleaf.

* * *

[*Regimen sanitatis Salernitanum*] *Regimen sanitatis Salerni. This booke teachyng all people to gouerne them in healthe, is translated out of the Latyne tonge into Englyshe by Thomas Paynell, which boke is amended, augmented, and diligently imprynted.*
[London: Abraham Uele, 1557.]
STC 21600
Subject: Hygiene, medicine.
Case/B/8615/.74

Expatiation on 7 doctrines relating to health and hygiene.

## Reinking, Dietrich, 1590–1664.

*Tractatus synopticus de retractu consanguinitatis—Huic accessit ejusdem auctoris responsum juris, de processu contra sagas et maleficos.*
Cologne: Sebastian Ketteler, 1708.
Subject: Legal sanctions against witchcraft.
K/74/.7017

4th ed. BOUND WITH his *Responsum juris in ardua et gravi quadam causa, concernente processum quendam, contra sagam, nulliter institutum, & inde exortam diffamationem. Ubi quaestiones quaedam, de nocturnis sagarum conventiculis, saltationibus, usurpatione suppellectilium piorum in illis, transmutationibus personarum in alia animalia, confessionibus, assertionibus & denuntiationibus, exactè examinantur, & requisita totius processus criminalis propununtur.*

## Reinzer, Franz, 1661–1708.

*Meteorologia philosophico-politica, das ist philosophische und politische beschreib- und erklärung der meteorischen oder in der obern lufft erzeugten dinge; in zwölff zerschiednen.*
Augsburg: J. Wolff, 1709.
Subject: Astrology, meteorology.
Case/fB/8635/.735

Thunder and lightning, rainbows, winds and clouds, earthquakes, metals & fossils. "Conclusio politica": De auro . . . sine labor nihil.

## Reinzer, Franz.

*Meteorologia philosophico-politica.* [Another ed.]
Augsburg: printed by Peter Detleffsen for Jeremia[h] Wolff, 1712.
Case/fB/8635/.7351

## [Reisch, Gregorius] d.1525.

[*Margarita philosophica*] *Margarita philosophica nova.*
[Strassburg: Joannes Grüninger, 1508.]
Subject: Encyclopedias.
Case/A/251/.737

Geometry (bk. 6); astronomy/astrology (bk. 7). MS marginalia in bk. 7.

## [Reisch, Gregorius]

[*Margarita philosophica*] *Margarita philosophica cum additionibus nouis: ab auctore suo studiosissima reuisione quarto super additis.*
[Basel: Michael Furter, 1517.]
Subject: Encyclopedias.
Wing/ZP/538/.F985

Rational philosophy in 12 books: bk. 4: arithmetic; bk. 6: geometry; bk. 7: astronomy/astrology; bk. 8: natural philosophy.

## [Reisch, Gregorius]

*Margarita filosofica del R. P. F. Gregorio Reisch, nella quale si trattano tutte le dottrine comprese nelle ciclopedia. Accrescivta di molte belle dottrine da Oratio Fineo matematico regio.*
Venice: Barezzo Barezzi & others, 1599.
Subject: Encyclopedias.
Ayer/7/R2/1599

Translated into Italian by Paolo Gallvcci. "Aggionta al libro settimo" (p. 976): "Due quadrature del cerchio; quadratura della spera; "Diuerse compositione dei quadrante, & l'uso dechiaratione dell' astrolabio da Messalat" (p. 1016); "principii di cosmografia di Gioanni Hontero Coronense" (p. 1114). "Aggionta al decimo libro cioe nel secondo trattato. Introdottione nella scienza prospettiua di Carlo Bouillo di Cambrai."

* * *

*Relaçam do lamentavel, e horroroso terremoto que sentio, na noute do ultimo dia do mez de Março para o primeiro de Abril de 1748 a ilha da Madeira.*
[Madeira, 1748]
Subject: Earthquakes.
Greenlee/4504/P855

* * *

*Relaçam dos danos causados pelos terremotos que houve no reyno de Sicilia no mez de Janeiro deste anno de 1693.*

Lisbon: Manoel Lopes Ferreyra, 1693.
Subject: Earthquakes.
Greenlee/4504/P855

* * *

*Relaçam de mais . . . extraordinaria, admiravel, e lastimosa tormenta de vento, que entre as memoraueis do mundo socedeo na India Oriental na cidade de Baçaim, & seu destricto, na era de 1618 aos 17 do mes de Mayo.*
Lisbon: Pedro Craesbeeck, 1619.
Subject: Storms, wind.
Greenlee/4504/P855

* * *

*Relaçam dos terremotos socedidos em a cidade de Traina, no reyno de Sicilia, este anno de 1643 & dos effeitos, que causaram em as circunvisinhas.*
Lisbon: Domingos Lopez Roza, 1644.
Subject: Earthquakes.
Greenlee/4504/P855

* * *

*Relaçaõ do formidavel e lastimoso terremoto succedido no reino de Valença no dia 23 de Março deste presente anno de 1748.*
Lisbon: Francisco Luiz Ameno, 1748.
Subject: Earthquakes.
Greenlee/4504/P855

* * *

*Relação do novo incendio do monte Etna. Chamado vvlgarmente Mongibello, na ilha de Sicilia, com as ruinas de muitos lugares circumvesinhos â cidade de Catania . . . com a destrvic, ão de ovtras terras da mesma ilha, causada dos tremores de terra.*
Lisbon: Antonio Craesbeeck, 1669.
Subject: Earthquakes, volcanoes.
Greenlee/4504/P855

* * *

*Relacion del lastimo sucesso que nuestra Señor fue seruido sucediesse en la isla de la Tercera, cabeça de las siete islas de los Azores . . . as las tres horas de la tarde con tres temblores que duraron por espacios de dos Credos.*
Saragossa: Iuan de Larumbe, 1614.
Subject: Earthquakes.
Greenlee/4504/P855

## Remi, Nicolas, 1554–1600.

*Nicolai Remigii sereniss. dvcis Lotharingiae a consiliis interioribus, et in eivs ditione Lotharingica cognitioris publici daemonolatreiae libri tres. Ex ivdiciis capitalibvs non gentorum plus minus hominum, qui sortilegij crimen intra annos quindecem in Lotharingia capite luerunt.*
Cologne: Henricus Falckenburg, 1596.
Subject: Demon worship, witchcraft.
Case/B/88/.738

## [Remmelin, Johann] 1583–1632.

*Pinax microcosmographicvs hoc est, admirandae partium hominis creaturarum divinarum praestantissimi universarum fabricae, historica brevis at perspicua enarratio.*
[n.p.] Stephan Michelspacher, 1615.
Subject: Anatomy, Macrocosm-Microcosm.
UNCATALOGUED

With "Elvcidarivs, tabulis synopticis microcosmici."

## [Renaudot, Eusebe] d. 1679.

*A general collection of discourses of the virtuosi of France, upon questions of philosophy, and other natural knowledg.*
London: printed for Thomas Dring and John Starkey, 1664.
STC II R1034
Subject: Natural philosophy, scientific miscellany.
Case/Y/7644/.33

Translated by G. Havers. 100 "conferences" (or tracts) on miscellaneous scientific and philosophical topics, including Of fire, air, water, earth, motion, sympathy and antipathy, Cabala, lycanthropy, judicial astrology, disease, chiromancy, occult writing, hermaphrodites.

* * *

*Repertorivm dictorvm: Aristotelis: Averoys: aliorvmqve philosophorvm.*
Bologna: Bazalerius de Bazaleriis for Benedictus Hectoris, 1491.
H 1934^ BMC (XV)VI:834^ Goff A-1204
Subject: Medicine, natural science.
Inc./6579.5

Ed. Andreas Victorius.

## Reuchlin, Johann, 1455–1522.

*De arte cabalistica libri tres.*
[Hagenau: Thomas Anshelm, 1517.]
Subject: Cabala, mysticism.
Case/fC/13/.738

## Reuchlin, Johann.

*De arte cabalistica.* See Columna, Pietro, for 1550 ed.

## Reuchlin, Johann.

*De verbo mirifico libri tres.*
Cologne: Eucharius, 1532.
Subject: Cabala, philology.
Case/C/13/.74

## Reusner, Nikolaus von, 1545–1602.

*Emblemata Nicolai Revsneri IC. Partim ethica, et physica, partim verò historica & hieroglyphica, sed ad virtutis, morumque; doctrinam omnia ingeniose traducta: & in quatuor libros digesta, cum symbolis & inscriptionibus illustrium & clarorum virorum. Ex recensione Ieremiae Reusneri Leorini.*
Frankfurt: Ioannes Feyerabendt, 1581.
Subject: Emblems.
Case/*W/1025/.745

## Reynolds, Edward, bp. of Norwich, 1599–1676.

*A treatise of the passions and faculties of the soule of man; with the severall dignities and corruptions thereunto belonging.*

London: printed for Henry Cripps and Edward Farnham, 1658.
STC II R1298
Subject: Psychology, theory of the passions.
Case/B/529/.736

### Riccioli, Giovanni Battista, 1598–1671.

*Almagestvm novvm astronomiam veterem novamqve complectens observationibvs aliorvm, et propriis nouisque theorematibus, problematibus, ac tabulis promotam, in tres tomos distribvtam. . . . Avctore P. Ioanne Baptista Ricciolo.*
Bologna: Heirs of Victor Benatius, 1651.
Subject: Astrology, astronomy.
Ayer/8.9/A8/R49/1651

Vol. 1, pt. 1 only. Includes 2 indexes: chronological list of astronomers and astrologers from Zoroaster, 1990 b.c., to Riccioli, 1651; followed by alphabetical list and brief biography of each individual.

### Riccioli, Giovanni Battista.

*Geographiae et hydrographiae reformatae libri dvodecim quorum argumentum sequens pagina explicabit.*
Bologna: heirs of Victor Benatius, 1661.
Subject: Geography, hydrography, navigation, surveying.
Ayer/7/R69/1661

A loose sheet laid in: The gift of Mr. Thomas Heatley, citizen and iron-monger of London, to the mathematical school in Christ's Hospital, anno do. 1700."

### [Richer, Edmond] 1560–1631.

*Obstetrix animorum. Hoc est brevis et expedita ratio docendi, studendi, conuersandi, imitandi, iudicandi, componendi.*
Paris: Ambrosivs Drovart, 1600.
Subject: Education, mnemonics, pedagogy.
Case/I/40187/.742

Chap. 10, sec. 3: "Summa memoriae & locorum communium in omnibus scientijs necessitas"; sec. 4: "Ac de facilitate artes memoriae comparandae."

### Richter, Christoph, d. 1680.

*Berichtendes send-schreiben vom cometen so in Christmonat des 1664. Christen jahres is erscheinen: darinnen derselbige astronomicè, physicè, astrologicè, theologicè, betrachtet und erkläret wird von Christophoro Richtern Görlicensi, pfarrern zu gnandstein in weissen. Anno, qvo seraph cvm gladio vibrante tibi astat.*
Subject: Comets.
Case/B/863/.743

### Richter, Matthaeus, [Mathäus Iudex] 1528–1564.

*De typographiae inventione, et de praelorvm legitima inspectione, libellvs brevis et vtilis.*
Copenhagen: Johannes Zimmerman, 1566.
Subject: History of technology, typography.
Wing/ZP/550/.Z65

Invention (discovery?) of papyrus.

### [Richter, Samuel] fl. 18th cent.

*Die Warhaffte und vollkommene bereitung des philosophischen steins, der brüderschafft aus dem orden des gülden- und Rosen-Creutzes. Darinne die materie zu diesem geheimniss mit seinem nahmen genennet. Auch die bereitung & von anfang biss zum ende mit allen hand-griffen gezeiget ist.*
Breslau: Esaiae Fellgiebels Wittwe, 1714.
Subject: Alchemy, Rosicrucians.
K/975/.744

With rules under which the Brotherhood operated.

### Ricius, Paulus, 16th cent.

*De novem doctrinarvm dorinibvs: et totivs perypatetici dogmatis nexv compendivm, conclvsiones atqve oratio.*
Augsburg: Ioann. Miller, 1515.
Subject: Philosophy.
Case/C/13/.754

4th part of his "De sexcentum et tredecim Mosaice sanctionis edictis." See "Conclvsiones," p. 22: "Motum vna cum multitudine & magnitudine naturalis, multitudinem cum magnitudine mathematicus, multitudinem absquemagnitudine motu metaphysicus in ente admirantur."

### Rinaldi, Giovanni de'.

*Il mostrvosissimo mostro . . . diuiso in due trattati. Nel primo de' qvali si ragiona del significato de' colori. Nel secondo si tratta dell' herbe, & fiori.*
Venice: Lucio Spineda, 1602.
Subject: Floral emblems, herbals.
Case/W/72/.74

Tract 2 is an alphabetical list of plants, herbs, and flowers and their symbolism. BOUND WITH Morato, Fulvio Pellegrino, q.v.

### Rinaldi, Oratio, 16th cent.

*Speccio di scienze, et compendio delle cose.*
Venice: Francesco Ziletti, 1583.
Subject: Dictionary encyclopedia of psychology.
Case/A/15/.741

With alphabetical table listing the qualities described in the volume, e.g., jealousy, medicine, continence.

### Ringelbergh, Joachim Sterck van, 1499?–1536?

*Institutiones astronomicae ternis libris contentae. Quorum primus sphaerae ac mundi naturam declarat. Secundus orbium. Tertius circulorum.*
[Venice: Io. Antonius de Nicolini de Sabio for D. Melchior Sessa, 1535.]
Subject: Astrology, astronomy.
Case/L/900/.746

1. "Quid inter astronomiam astrologiam intersit"; 2. "Duplex mundi dissectio"; 3. "Circulorum in signa, gradus, siue parteis, & minuta dissectio."

### Ringhieri, Innocentio, fl. 1550.

*Il sole di M. Innocentio Ringhieri gentilhvomo Bolognese.*
Rome: [A. Blado], 1543.
Subject: Natural philosophy, relation of man and heavens.
Case/L/974/.746

Dialogue between Diligenza and Otio. "Il tutto quasi in duomondi si diuide, elementale, et celeste, quello alterabile corruttibile, generabile" (sig. B1).

### Ripa, Cesare, fl. 1600.

[*Iconologia*] *Della novissima iconologia de Cesare Ripa Pervgino.*
Padua: Pietro Paolo Tozzi, 1624.
Subject: Emblems.
Case/W/1025/.7477

Indexes include "De imagini principali"; "tauola de gesti, moti, & positure del corpo humano"; "De ordigni diuersi, & altre cose artificiali"; "de gli animali"; "delle piante"; "di pesci"; "di colori"; "de gli autori citati."

### Ripa, Cesare.

[*Iconologia*] *Della piv che novissima iconologia de Cesare Ripa Pervgino. Parte prima nella quale si esprimono varie imagini di virtù vitij, affetti, passioni humane, arte, discipline, humori, elementi, corpi celesti, prouincie d'Italia, fiumi, & altre materie infinite vtili ad ogni stato di persone.*
Padua: Donato Pasquardi, 1630.
Subject: Emblems.
Case/W/1025/.7478

### Ripa, Cesare.

*Iconologia: or, moral emblems by Caesar Ripa wherein are express'd, various images of virtues, vices, passions, arts, humours, elements and celestial bodies; as design'd by the ancient Egyptians, Greeks, Romans, and modern Italians.... Illustrated with three hundred twenty-six humane figures, with their explanations.*
London: Benj. Motte, 1709.
Subject: Emblems.
Case/W/1025/.748

Arranged alphabetically. With description of the figure followed by explanation of symbolic meanings of various devices.

### Ripa, Cesare.

[*Iconologia*] *Iconologie, ou nouvelle explication de plusieurs images, emblemes, & autres figures hyerogliphiques des vertus, des vices, des arts, des sciences, des causes naturelles, des deux parties.*
Paris: Louis Billaine, 1677.
Subject: Emblems.
Case/W/1025/.7482

"Tirée des recherches & des figures de Cesar Ripa, moralisées par J. Baudouin.

### Ripa, Cesare.

[*Iconologia*] *Iconologie ou la science des emblmes devises &c.*
Amsterdam: Adrian Braakman, 1698.
Subject: Emblems.
Case/W/1025/.7484

### Ritter, Stephan.

*Cosmographia prosometrica. Hoc est, universi terrarum orbis, regionum, populorum, insularum, urbium, fluviorum, montium, marium, aliarumque rerum cosmographicarum, tum ex probatis autoribus, variorumque locorum antiquitatibus, oratione soluta; tum ex variis variorum poetarum monumentis oratione ligata descriptio.*
Marburg: Paulus Egenolph, 1619.
Cosmography, descriptive geography.
Case/G/117/.745

In 8 bks. 1.: "Cosmographiam universalem"; 2.: "Nesographiam, sive insularum descriptionem"; 3.: "Chorographiam, sive regionum & populorum descriptionem"; 4.: "poligraphiam, sive urbium descriptionem"; 5.: "Potamographiam, sive fluviorum"; 6.: "Orographiam, sive montium"; 7.: "Thalassographiam, sive marium"; 8.: "Rerum cosmographicarum promiscuarum descriptionem" [7 wonders of the world].

### Rivière, Lazare, 1589–1655.

*Lazari Rivierii consiliarii ed medici regii: atque in Monspeliensi Universitate medicinae professoris . . . praxis medica.*
The Hague: Adrian Vlacq, 1664.
Subject: Consilia, diagnosis, practice of medicine.
oR/128/.R58/1664

10th ed.

### Rivinus, Augustus Quirinus, 1652–1723.

*Bibliotheca Riviniana sive catalogvs librorvm philologico-philosophico-historicorvm, itinerariorvm, inprimis avtem medicorvm, botanicorvm et historiae naturalis scriptorvm &c. rariorvm, qvam magno stvdio et svmptv sibi comparavit D. Avg. Qvir. Rivinvs.*
Leipzig: Immanuel Titius, 1727.
Subject: Catalogue of scientific books and writers.
Z/491/.R528

Divided (roughly) by subject (e.g., Thomas Bartholin, de ponderibus & mensuris is s.v. "antiqvarii"). Includes "Scriptores physici, scriptores mathematici, lexica medica," and "scriptores medici & alchemici," among others.

### Rivius, Walter.

Von rechtem verstandt wag vnd gewicht, etliche büchlein zu sonderlichem verstandt, künstlicher mechanischer inuention, der geometrischen messung angehenckt.
[Nuremberg: Gabriel Heyn, n.d.]
Subject: Geometry, measurement.
Wing/fZP/.H152

BOUND WITH Fronsperger, Leonhard, q.v.

### Roberts, Alexander, d. 1620.

*A treatise of witchcraft. Wherein sundry propositions are laid downe, plainely discouering the wickednesse of that damnable*

*art. . . . with a true narration of the witchcrafts which Mary Smith, wife of Henry Smith glouer did practise.*
London: printed by N. O. for Samvel Man, [16]16.
STC 21075
Subject: Witchcraft.
Case/B/88/.75

## Rochefort, Charles de, 1605–1690?

*Beschreibung von Tabago: einer insul von denen Antilles, oder Caribischen eylanden in America. Auss dem Frankösischen tableau des Monsr. de Rochefort übersetzt.*
Hamburg: Thomas von Wiering, [17–?].
Subject: Anthropology, natural history.
Ayer/1000.5/T62/R67/17–

## [Rochefort, Charles de]

*Histoire naturelle et morale des iles Antilles de l'Amerique.*
Rotterdam: Arnould Leers, 1658.
Subject: Anthropology, natural history.
Ayer/1000/R6/1658

Goods, occupations and recreations, marriage, birth and education of children, illnesses, remedies.

## [Rochefort, Charles de]

*Histoire naturelle et morale des iles Antilles de l'Amerique.*
Rotterdam: Arnout Leers, 1665.
Subject: Anthropology, natural history.
Ayer/*1000/R67/1665

2d ed. 2 vols. in 1: vol. 1 is natural history, vol. 2, moral history.

## [Rochefort, Charles de]

*Histoire natvrelle des iles Antilles de l'Ameriqve.*
Lyon: Christofle Fovrmy, 1667.
Subject: Anthropology, natural history.
Ayer/1000/R6/1667

2 vols. Vol. 2, chap. 4: "Du trafic & des occupations des habitans etrangers du pais: & premierement de la culture & de la preparation du tabac"; chap. 5: "De la maniere de faire le sucre, de preparer l gingembre, l'indigo & le cotton."

## [Rochefort, Charles de]

*Histoire naturelle et morale des iles Antilles de l'Amerique.*
Rotterdam: Reinier Leers, 1681.
Subject: Anthropology, natural history.
Case/G/976/.75

"Derniere edition." With "Recit de l'estat present des celebres colonies."

## [Rochefort, Charles de]

*Histoire naturelle et morale des iles Antilles de l'Amerique, . . . Reveüe & augmentée par l'autheur d'un recit de l'estat present des celebres colonies de la Virginie, de Marie-Land, de la Caroline, du nouveau duché d'York, de Penn-Sylvania, & de la Nouvelle Angleterre, situées dans l'Amerique septentrionale . . . Tiré fidelement des memoires des habitans des mêmes colonies.*
Rotterdam: Renier Leers, 1681 (1690?).
Subject: Anthropology, natural history.
Ayer/1000/R6/1681

"Derniere edition."

## [Rochefort, Charles de]

*The history of Barbados, St. Christophers, Mevis, St. Vincents, Antego, Martinico, Monserrat, and the rest of the Carriby-Islands, in all XXVIII. In 2 books. Englished by J. Davies of Kidwelly.*
London: printed for John Starkey and Thomas Dring junr., 1666.
STC II R1739
Subject: Anthropology, natural history.
Ayer/1000/R6/1666

## [Rochefort, Charles de]

*The history of the Caribby-islands, viz. Barbados, St. Christophers, St. Vincents, Martinico, Dominico, Barbouthos, Monserrat, Mevis, Antigo, &c. in all XXVIII rendered into English by John Davies.*
London: printed by J. M. for Thomas Dring and John Starkey, 1666.
STC II R1740
Subject: Anthropology, natural history.
Ayer/1000/1666a

## Rochefort, Charles de.

*Relation de l'isle de Tabago, ou de la novvelle ovalere, l'vne des isles Antilles de l'Ameriqve.*
Paris: Lovys Billaine, 1666.
Subject: Natural history.
Ayer/1000.5/T62/R6/1666

## [Rochefort, Charles de]

*Le tableau de l'isle de Tabago, ou de la nouvelle Oüalchre, l'une des isles Antilles de l'Amerique.*
Leyden: Jean le Carpentier, 1665.
Subject: Natural history.
Ayer/*1000.5/T62/R67/1665

## Rodler, Hieronymus.

*Perspectiua. Eyn schön nützlich büchlin vnd vnderweisung der kunst des messens, mit dem zirckel, richtscheidt oder linial. Zu nutz allen kunstlieb habern fürnemlich den malern, bildhawern, goldschmiden, seidensticktern, steinmetzen, schreinern auch allen andern so sich der kunst des messens (perspectiua zu Latein genant) zu gebrauchen lust haben.*
Frankfurt: Cyriacus Jacob, 1546.
Subject: Measurement, perspective.
Wing/fZP/547/.H152

BOUND WITH Rivius, [Ryff] Walter, *De architectür,* and Krönsperger, Leonhart, *Bauw Ordnung,* qq.v.

## Rogers, Woodes, d. 1732.

*A cruising voyage around the world:... containing a journal of all the remarkable transactions... and a brief description of several countries in our course noted for trade, especially in the South-Sea.*
London: Printed for A. Bell... & B. Lintot, 1712.
Subject: Voyages.
Ayer/118/R67/1712

Account of the River Amazons, esp. p. 73.

## Rogers, Woodes.

*Voyage fait autour du monde, par le capitaine Woodes Rogers. Traduit de l'Anglois.... Avec la relation de la grande riviere des Amazones & de la Guyane dans le nouveau monde, traduit de l'Espagnol, par feu Monsieur de Gomberville... où il a joint la description des côtes, rades, havres, rochers, bas-fonds, isles, caps, aiguades, criques, anses, aspects, gisemens, distances, &c.... avec des observations sur les animaux que chaque contrée produit.*
Amsterdam: L'Honoré & Chatelain, 1725.
Subject: Descriptive geography, discovery, voyages.
Ayer/118/R67/1725

3 vols., 1709, 1709–11, 1725.

## [Rolevinck, Werner] 1425–1502.

*Fasciculus temporum omnes antiquorum cronicas complectens.*
[Strassburg: Johann Pruss, not before 1490.]
HC* 6915^ BMC (XV)I:127^ Pr. 571^ Still. R 267^ Goff R-275
Subject: Chronology.
Inc./f571

Some MS notes.

## Rondelet, Guillaume, 1507–1566.

*Gvlielmi Rondeletii doctoris medici et medicinae in schola Monspeliensi professoris regii Libri de piscibus marinis, in quibus verae piscium effigies expressae sunt.*
Lyon: Matthias Bonhomme, 1554–55.
Subject: Marine fauna, zoology.
Case/P/1/.75

1st ed. In 2 parts. Pt. 2: "De testaceis vniversae aquatilium historiae pars altera" (1555).

## Rorario, Girolamo, 1485–1556.

*Quòd animalia bruta ratione utantur meliùs homine.*
Amsterdam: Joannes Ravestein, 1654.
Subject: Comparative psychology, reason, man vs. beasts.
Case/B/58/.756

Edited by Gabriel Naudé. In 2 books.

## Rosaccio, Giuseppe, 1530–ca.1620.

*Mondo elementare, et celeste di Gioseppe Rosaccio cosmografo, & dottore in filosfia, & medicina. Nelquale si tratta de' moti, & ordini delle sfere; della grandezza della terra; dell' Europa, Africa, Asia & America.*
Treviso: Evangelista Deuchino for Gio. Battista Ciotti, 1604.
Subject: Astronomy, cosmography.
Ayer/7/R7/1604

Leaf 237v: "Del cielo cristalino, con la sua grandezza, e quanto sia lontano da noi."

## Roscius, L. Vitruvius.

*De docendi stvdendíque modo ac de claris puerorum moribus, libellus planè aureus. L. Vitrvvio Roscio Parmensi autore.*
Basel: Robert VVinter, 1541.
Subject: Education.
Case/I/409/.756

Tracts on aspects of education, e.g., Iacobi Comitis Pvrliliarvm, de generosa liberorum educatione opusculum; Maphei Vegii Lavdensis, de edvcatione liberorum & claris eorum moribus; Petri Pavli Vergerii Ivstinopolitani, de ingenuis moribus ac liberalibus studijs libellus; Joachim Camerarius Pabergensi, praecepta vitae pverilis.

## [Rosnel, Pierre de]

*Le mercvre indien, ov le tresor des Indes. Dans laquelle est traité des pierres precieuses & des perles, ensemble de leur origine, de leur formation, de leur vsage, & de leur valeur.*
Paris: [Robert Chevillion], 1672.
Subject: Earth sciences, gemology, philosophers stone.
Ayer/8.9*/M6/R82/1672

## Ross, Alexander, 1590–1654.

*Arcana microcosmi; or, the hid secrets of man's body discovered in an anatomical duel between Aristotle & Galen concerning the parts thereof: As also, by a discovery of the strange and marvelous diseases, symptomes & accidents of man's body. With a refutation of Doctor Brown's vulgar errors, the Lord Bacon's natural history, and Doctor Harvy's book De generatione, Comenius, and others.*
London: printed by Tho. Newcomb, 1652.
STC II R1947
Subject: Anatomy, disease, medicine.
Case/B/8615/.76

## Ross, Alexander.

*Medicus medicatus; or, the physicians religion cured by a lenitive or gentle potion. With some animadversions vpon Sir Kenelm Digbie's Observations on religio medici.*
London: printed by James Young, 1645.
STC II R1961
Subject: Chiromancy, medicine, natural theology, sympathy and antipathy.
Case/C/53/.122

In 2 parts: pt. 1: natural theology; pt. 2: "Of physiognomie and palmestry."

## Ross, Alexander.

*The philosophicall touch-stone: or observations upon Sir Kenelm Digbie's discourses of the nature of bodies, and of the reasonable soule. In which his erroneous paradoxes are refuted, the truth, and Aristotelian philosophy vindicated, the immortality of man's soule briefly, but sufficiently proved.*

London: printed for James Young, 1645.
STC II R1979
Subject: Magic, metaphysics, natural philosophy, sympathy and antipathy.
Case/B/79/.243

Has material on sense of touch, refutation of natural, mathemtical, anddiabolical magic (sec. 37), on weapon-salve (sec. 38), and on the lodestone (sec. 40). BOUND WITH Digby, Sir Kenelm, *Two treatises*, q.v.

## Ross, Alexander.

*The philosophical touchstone.* [Another copy]
Case/B/79/.244

## Rosselius, Cosmas, d. 1578.

*Thesavrvs artificiosae memoriae, concionatoribus, philosophis, medicis, iuristis, oratoribus, procuratoribus, caeterisque; bonarum litterarum amatoribus.*
Venice: Antonius Paduanius, 1579.
Subject: Astrology, encyclopedias.
Case/*B/528/.76

In 2 parts: pt. 1: natural history (cosmos, including heaven and hell); pt. 2: "De figuris naturalibus minoribus" (alphabetical list of stones and minerals); correspondence of Hebrew and Turkish alphabets to parts of the human body; astrological medicine.

## Rothmann, Johann, fl. 1595.

[*Chryomantia*]: *or, the art of divining by the lines and signatures engraven in the hand of man, by the hand of nature, theorically, practically. Wherein you have the secret concordance, and harmony betwixt it, and astrology, made evident in 19. genitures.... A matchlesse piece. Written originally in Latine by Io: Rothmanne, ... and now faithfully Englished by Geo: Wharton.*
London: printed by J. G. for Nathaniel Brooke, 1652.
STC II R2001
Subject: Astrology, chiromancy.
Case/B/864/.758

"I have nothing to do with the ignorant and malevolent" [closing line].

## Rowning, John, 1701?–1777.

*A compendious system of natural philosophy: with notes containing the mathematical demonstrations, and some occasional remarks.*
London: printed for Sam. Harding, 1744.
Subject: Astronomy, mathematics, physics.
L/0/.763

In 4 parts: 1: properties of bodies, laws of motion, 4th ed. (1745); 2: hydrostatics and pneumatics (1745), 5 dissertations, 2d ed. (1738); 3: optics, 2d ed., and catoptrics (1738); 4: astronomy (1742), globes, orrery (1743).

## Ruano, Ferdinando, d. 1560.

*Sette alphabeti di varie lettere. Formati con ragion geometrica.*
Rome: Valerio Dorico & Luigi Bressani, brothers, 1554.
Subject: Geometry, lettering.
Wing/fZW/14/.R82

"Scrittor della biblioteca Vaticana."

## Rueff, Jacob, 1500–1558.

*De conceptu et generatione hominis: de matrice et eivs partibvs, nec non de conditione infantis in vtero, & grauidarum cura & officio: De partu & parturientium, infantumque cura omnifaria: De differentijs non naturalis partus & earundem curis: De mola alijsque falsis vteri tumoribus, simulque de abortibus & monstris diuersis, nec non de conceptus signis varijs: De sterilitatis causis diuersis, & de praecipuis matricis aegritudinibus omniumque horum curis varijs, libri sex, opera clarissimi viri Iacobi Rveffi, chirurgi Tigurini quondam congesti.*
Frankfurt: Petrus Fabricius for Sigismund Feyerabend, 1587.
Subject: Medicine, obstetrics.
Case/*Q/.76

Illustrated by Jobst Amman.

## Ruysch, Frederick, 1638–1731.

*Opera omnia anatomico-medico-chirurgica.*
Amsterdam: Janssonio-Waesberg, 1721.
Subject: Anatomy, medicine.
UNCATALOGUED

Includes *Epistola anatomica problematica octava. Authore Johanne Henrico Graetz.* Amsterdam: Wolters, 1692; *Epistola anatomica, problematica quarta & decima. Authore Mauritio à Reverhorst.* Amsterdam: Wolters, 1701; Godfridus Bidloo, *Vindiciae quarundam delineationum anatomicarum contra ineptas animadversiones Fred: Ruyschii.* Leyden: Jordanus Luchtmans, 1697; and various rejoinders.

## Ruysch, Frederick.

*Museum anatomicum Ruyschianum.*
Amsterdam: Henry & the widow of Theodor Boom, 1691.
Subject: Anatomy, medicine.
oQM/21/.R89/1691

BOUND WITH Ruysch, *Observationum*, q.v.

## Ruysch, Frederick.

*Observationum anatomica-chirvrgicarum centuria. Accedit catalogus rariorum quae in musaeo Ruyschiano asservantur.*
Amsterdam: Henry & the widow of Theodor Boom, 1691.
Subject: Anatomy, medicine.
oQM/21/.R89/1691

BOUND WITH his *Museum*, q.v.

## Ruysch, Frederick.

*Thesaurus animalium.*
Amsterdam: Joannes Wolters, 1710.
Subject: Anatomy, medicine.
UNCATALOGUED

In 2 vols. In Latin and Dutch. Anatomical tracts and "adversariora."

## Ryff, Walther Hermann, 16th cent.

*Der architectur fürnembsten, notwendigsten, angehörigen mathematischen vnd mechanischen kunst, eygentlicher bericht, vnd verstendliche vnterrichtung, zu rechtem verstandt der lehr Vitruuij, in drey fürneme bücher abgetheilet. Als der newen perspectiua das I. buch. Vom rechten gewissen geometrischen grund, alle regulierte vnd anregulirte cörperliche ding dessgleichen ein yeden baw vnd desselbigen angehörige glider, vnd was vns im gesicht fürkomen mag künstlichen durch mancherley vortheil vnd gerechtigkeit zirckels vnnd richtscheidts auff zureissen, in grundt zu legen, vnd nach perspectiuischer art auff zu ziehen, mit weiterem bericht des grundts der abkürtzung, oder vermerung aller ding nach verendrung der distanz mit erklerung der fürnembsten puncten, künstlichs vnd perspectiuischen reissens vnd malens verstandt der farben.*

Nuremberg: Gabriel Heyn, 1558.

Subject: Architecture, engineering, geometry, perspective.

Wing/fZP/547/.H152

BOUND WITH Rodler and with Krönsperger, qq.v.

# S

## Sacro Bosco, Johannes de, fl. 1230.

*Opusculum Johannis de Sacrobusto spericum cum notabili commento.*
[Leipzig: M. Landsberg, ca. 1510?]
Subject: Astronomy, geometry.
Case/L/93/.768

MS notes. Binders title: *De Sacro Bosco opusculum sphericum*. BOUND WITH 8 tracts, (n.a.) by Heinrich Bebel, Hermann von dem Busche, Andreas Crappus, Agostino Dati, Philipp Engelbrecht, Johannes, Hess, Chilianus Reuther, and a group of poems by Beroaldi, Fortunati, Lactantius, and others.

## Sacro Bosco, Joannes de.

*Sphaera mundi.*
Venice: Franciscus Renner, de Heilbron, 1478.
H-C *14108^ BMC (XV)V:195^ Pr. 4175^ Goff J-402
Subject: Astronomy.
Inc./4175

MS notes. In 4 chaps.: 1. Definition of sphere; 2. circle from which material sphere is composed; 3. rising of signs; 4. orbits and motions of planets, causes of eclipses. Includes Gerard of Cremona [Gerard of Sabbioneta, fl. 1255–1259], "Theorica planetarum."

## Sacro Bosco, Joannes de.

[*Sphaera mundi*] *Ioannis de Sacro busto sphericum opusculum Georgiique Purbachii in motibus planetarum accuratiss. theorice necnon contra Cremonensia in eorundem planetarum theoricas deliramenta Ioannis de Monte Regio disputationes tam accuratiss. quam utiliss.*
[Venice: Erhard Ratdolt, before Nov. 1485.]
H *1411^ BMC (XV)V:290^ Pr. 4402^ Redgrave 57^ Goff J-406
Subject: Astronomy.
Inc./4402

Bookplate of Gilbert Redgrave. Penciled note states that this is 2d ed. and the "first time more than two coloured inks [were] employed simultaneously." MS note (by Redgrave) on front flyleaf states that this was printed at Venice, not Augsburg (Ratdolt was not yet at Augsburg), and that between 1st ed. (1482), whose blocks were hand colored, and this one, Ratdolt introduced the use of colored blocks. With table of signs above colophon not present in 1st ed.

## Sacro Bosco, Joannes de.

*Johannis de Sacro Busto sphere mundi opusculum, vna cum additionibus per opportune interfertis. ac familiarissima textus expositione Petri C. D. felici sidere incohat.*
Paris: Jehan Petit, 1498–99.
Hain 14/20 = 5363
Subject: Astronomy.
Inc./f8015

Title page in facsimile: *Uberrimum sphere mundi commentum interfertis etiam questionibus divini Petri de Aliaco.* Some MS marginalia at preface: "Petri C. D. [Ciruelli Darocensis] in astronomicum sphere mundi opusculum prefatio." In 4 bks.: 1. "De forma mundi. Habet tres partes"; 2. "De circulis decem ex quibus sphera materialis componitur et de illis quos in sphera celesti ymaginamur" (3 pts.); 3. "De ortu et occasu signorum" (4 pts); 4. "De circulis eccentricis & epiciclis planetarum" (3 pts.), followed by "Disputatorius dialogus" [participants are Darocensis and Burgensis].

## Sacro Bosco, Joannes de.

*Sphaera mundi.*
Venice: [Bonetus Locatellus for] Octavianus Scotus, 1490.
H-C *14113^ BMC (XV)V:438^ Pr. 5023^ Goff J-409
Subject: Astronomy.
Inc./5023

Contains also *Disputationum Ioannes de Monteregio contra Cremonensia in planetarum theoricas deliramenta praefatio; Theoricae nouae planetarum Georgii Purbachii astronomi celebratiss.*

## Sacro Bosco, Joannes de.

[*Sphaera mundi*] *Sphera mundi cum tribus commentis nuper editis vz. Cicchi Esculani; Francisci Capuani de Manfredonia; Jacobi Fabri Stapulensis.*
[Venice: Simon Beuilaqua, 1499]
H *14125^ BMC (XV)V:524^ Pr. 5414^ Goff J-419
Subject: Astronomy.
Inc./5414

## Sacro Bosco, Joannes de.

[*Sphaera mundi*] *Textus de sphera Johannis de Sacrobosco cum additione (quantum necessarium est) adiecta: Nouo commentario nuper edito ad vtilitatem studentium philosophice Parisiensis. Academie illustratus cum compositione anuli astronomici Boni Latensis et geometria Euclidis Megarensis.*
[Paris: Johann Petit, 1515.]
Subject: Astronomy.
Ayer/*6/S2/1507

## Sacro Bosco, Joannes de.

[*Sphaera Mundi*] *Sphaera Iani de Sacrobvsto astronomiae ac cosmographiae candidatis scitu apprime necessaria per Petrum Apianum accuratissima diligentia denuo recognita ac emendata.*
Ingolstadt: Apianus, 1526.

Subject: Astronomy.
Ayer/6/S2/1526

**Sacro Bosco, Joannes de.**

[*Sphaera mundi*] [*Liber Joannis de Sacrobusto de sphaera*]
Venice: Ioan. Anto. & brothers de Sabio, for D. Melchior Sessa, 1532.
Subject: Astronomy.
Ayer/6/S2/1532

Title page lacking. Replaced by MS transcription on front flyleaf.

**Sacro Bosco, Joannes de.**

[*Sphaera mundi*] *Sphera volgare novamente tradotta con molte notande additioni di geometria, cosmographia, arte, navicatoria, et stereometria, proportioni, et qvantita delli elementi, distanze, grandeze, et movimenti di tvtti li corpi celesti, cose certamente rade et maravigliose. Avtore M. Mavro Fiorentino phonasco et philopanareto.*
Venice: printed by Bartholomeo Zanetti for M. Giouann' Orthega de Carion Burgense Hyspano, 1537.
Subject: Astrology, astronomy, cosmography, geography, geometry.
Ayer/6/S2/1537

**Sacro Bosco, Joannes de.**

[*Sphaera mundi*] *Trattato della sphera, nel qual si dimonstrano, & insegnano i principii della astrologia raccolto da Giouanni de Sacrobusto, & altri astronomi, & tradotto in lingua Italiana.*
Venice: Antonio Brucioli & brothers, 1543.
Subject: Astrology, astronomy.
Ayer/6/S2/1543

Translated by Antonio Brucioli.

**Sacro Bosco, Joannes de.**

[*Sphaera mundi*] *La sfera di messer Giovanni Sacrobosco tradotta emendata & distinta in capitoli da Pieruincentio Dante de Rinaldi con molte et utili annotazioni del medesimo.*
Florence: Giunti, 1571.
Ayer/6/S2/1571

[Colophon: 1572]

**Sacro Bosco, Joannes de.**

[*Sphaera mundi*] *Sphaera Ioannis de Sacrobvsto. Addita svnt qvaedam ad explanationem eorum quae in sphaera dicuntur facientia.*
Venice: heirs of Melchior Sessa, 1580.
Subject: Astronomy.
Greenlee/4890/S12/1580

Bound in vellum music MS.

**Sacro Bosco, Johannes de.**

[*Sphaera mundi*] *Sphaera Johannis de Sacro Bosco.*
Leyden: Bonaventvra & Abraham Elzevir, 1647.
Subject: Astronomy.
Case/L/93/.77

**Sagri, Nicolò.**

*Ragionamenti sopre la varietà de i flvssi et riflvssi del mare oceano occidentale, fatti da Andrea di Noblisia, pedotto biscaino, & Vicenzo Sabici, nocchiero, & Ambrosio di Goze, raguesi; raccolti da Nicolo Sagri & in vn dialogo dall' istesso ridotti, diuiso in due parti, ad vtilità di ciascuno nauigante.*
Venice: Domenico & Gio. Battista Guerra, brothers, 1574.
Subject: Navigation, tides.
Case/Y/712/.S128

**Sambucus, Johannes, 1531–1584.**

*Emblemata, cvm aliqvot nvmmis antiqvi operis.*
Antwerp: Christophorvs Plantin, 1564.
Subject: Emblems.
Case/W/1025/.77

1st ed. "Alchimia vanitas," p. 185.

**Sánches de Arévalo, Rodrigo, bp. [Rodericus Zamorensis] 1404–1470.**

*Speculum vitae humanae.*
Augsburg: Günther Zainer, 1471.
H-C *13940^ BMC (XV)II:316^ Pr. 1525^ Still. R209^ Goff R-215
Subject: Epistemology, soul.
Inc./f1525

MS marginalia. Incipit: "... ad recte viuendi documenta: editus a Rodorico Zamorensi."

**Sánches de Arévalo, Rodrigo.**

*Speculum vite humanae. Quia in eo cuncti mortales in quouis fuint statu vel officio spirituali.*
[Cologne: Ulrich Zell, 1472.]
H-C *13933^ BMC (XV)I:187^ Pr. 852^ Still. R212^ Goff R-218
Subject: Epistemology, soul.
Inc./852

**Sánches de Arévalo, Rodrigo.**

*Speculum vite humanae qua in eocuncti mortales in quouis fuerint statu nel officio spirituali aut temporali speculabuntur cuius liber artis et vite prospera et aduersa ac recta uiuendi documenta.*
Lyon: Guillaume Le Roy with Barthélemy Buyer, 1477.
H-C 13946^ BMC (XV)VIII:234^ Pr. 8500^ Still. R220^ Goff R-225
Subject: Epistemology, soul.
Inc./f * 8500

"Editus a Rodorico episcopo Zamorensi postea calgorritano hispano." In 2 parts: temporal life; spiritual life. Mechanical arts (chaps. 23–31). medicine (chaps 32–37), mathematics, geometry (chaps. 37–40).

**Sánchez, Francisco, 1550ca.–1623.**

*Tractatus philosophici, quod nihil scitur. De divinatione per somnum ad Aristotelem.*
Rotterdam: Arnold Leers, 1649.
Subject: Divination by dreams, philosophy.

Case/-B/4/.395

"In libr: Aristotelis physiognomicon commentarius"; "De longitudine & brevitate vitae."

## Sanson, Nicolas, 1600–1667.

*Description de tout l'univers, en plusieurs cartes, & en divers traitez de geographie et d'histoire où sont décrits succinctement & avec une methode belle & facile ses empires, ses peuples, ses colonies, leurs moeurs, langues, religions, richesses, &c. Et ce qu'il y a de plus beau et de plus rare dans toutes ses parties & dans ses isles.*
Amsterdam: François Halma, 1705.
Subject: Cosmography, descriptive geography.
Ayer/135/S19/1700

In 4 parts: Europe, Asia, Africa, America; hand-colored maps.

## Sarazinus, Joannes.

*Horographvm catholicvm sev vniversale, quo omnia cuiuscunque generis horologia sciotherica in quacunque superficie data compendio ac facilitate incredibili describuntur.*
Paris: Sebastian Cramoisy, 1630.
Subject: Clocks, horology, instrumentation.
Wing/ZP/639/.C84

"Dicat et consecrat inventor Ioannes Sarazinus." Description and use of clocks and astronomical instruments.

## Saumaise, Claude de, 1588–1653?

*Cl. Salmasii de annis climactericis et antiqua astrologia diat ribae.*
Leyden: Elzevir, 1648.
Subject: Astrological medicine.
Case/B/8635/.78

## Savigny, Christofle de, 1530–1608.

*Tableaux accomplis de tovs les arts liberaux, contenans brievement et clerement par singvliere methode de doctrine, vne generale et sommaire partition des dicts arts, amassez et redvicts en ordre povr le soulagement et profit de la ievnesse.*
Paris: Iean & François de Gourmont, brothers, 1587.
French STC 715.1.8 (vol. 2, p. 395)
Subject: Education, quadrivium.
Wing/+ZP/539/.G74104

Ed. Nicholas Bergeron. Has a diagram of each liberal art followed by descriptive text. Includes astrology "contemplatrice" (also called "astronomie judiciaire") medicine, physics, and others.

## Savonarola, Giovanni Michele, 1384?–1462?

[*Practica medicinae*] *Practica Joannis Michaelis Sauonarole tractatus quartus de simplicibus & compostis cum suis virtutibus secundis, tertijs & quartis vsualibis magis.*
Venice: Bonetus Locatellus for Octavianus Scotus, 1497.
H-C *14484^ BMC (XV)V:448^ Pr. 5080^ Still. S273^ Goff S-298
Subject: Astrological medicine, disease, iatrochemistry.
Inc./f5080

## Savonarola, Girolamo Maria Francesco Matteo, 1452–1498.

*Opere singolare del Reuerendo Padre F. Hieronimo Sauonarola contra l'astrologia diuinatrice in corraboratione delle refutatione astrologice del S. Conte Ioan. Pico de la Mirandola. Con alcune cose dil medemo di nuouo aggionte.*
Vinegia: [M. Bernardino Stagnino], 1536.
Subject: Astrology.
Case/3A/1303

## Savonarola, Girolamo Maria, Francesco Matteo.

*Hieronymi Savonarolae ... opvs eximivm, aduersvs diuinatricem astronomiam, in confirmationem confutationis eiusdem astronomicae praedictionis, Ioan. Pici Mirandulae comitis, ex italico in latinum translatum.*
Florence: Georgius Marescotus, 1582.
Subject: Astrology.
Case/B/8635/.786

Translated by Thoma Boninsignio. In 3 tracts.

## Savonarola, Raffaelo, 1546–1730.

*Universus terrarum orbis scriptorum calamo delineatus, hoc est auctorum feré omnium, qui de Europae, Asiae, Africae, & Americae regnis, provinciis, populis, civitatibus, oppidis, arcibus, maribus, insulis, montibus, fluminibus, fodinis, balneis, publicis hortis, & de aliis tam super, quam subtus terram locis.*
Padua: formerly Frambotti, now Jo. Baptista Conzatti, 1713.
Subject: Geographical and historical encyclopedia.
Ayer/*7/S24/1713

In 2 vols. Name on title page an anagram of Raphaelis Savonarola, "Alphonsi Lasor a Varea."

## [Savonarola, Raffaello]

[*Universus terrarum orbis scriptorum calamo delineatus*] *Regnorum provinciarum, civitatumque, ac quorumcumque locorum orbis terrarum nomina Latina, tàm juxtà antiquos quàm recentes geographos, duobus tomis exposita quibus congruuni* [sic] *Italica, ut facilis inveniantur illa.*
[n.p.?] P. Coronelli, 1716.
Subject: Geographical and historical encyclopedia.
Case/6A/168

In 2 vols. A pirated ed. with canceled title page. Title page information is taken from the Ayer collection copy listed above. With author index, subject index, bibliographic index, and list of abbreviations.

## Scaliger, Joseph Juste, 1540–1609.

*Iosephi Scaligeri Ivl. Caesaris F. opvs novvm de emendatione temporvm in octo libros tribvtvm.*
Paris: Mamertus Patisson [for] Robert Stephan, 1583.
Subject: Calendar.
Case/fF/017/.788

Includes "De anno lvnari"; "De anno solari"; a chronology beginning with the Flood. Comparison of Greek, Hebrew, Arabic, and other calendars. Book 8: "De anno coelesti & tropico castigatio methodi Ivliani. De anno Liliano et eivs tetracosieteride."

## Scaliger, Joseph Juste.

*Ios. Ivsti Scaligeri Ivlii Caesaris a Bvrden filii opvscvla varia ante hac non edita.*
Frankfurt: Jacob Fischer, 1612.
Subject: Botany, engineering.
Case/Y/682/.S275

Includes "Iosephi Ivsti Scaligeri animadversiones in Melchioris Gvilandini commentarivm in tria C. Plinii de papyro capita libri XIII" and "Discours de la ionction des mers, du dessechemens des marais & de la reparation des riuieres, pour les rendre nauigeables" (p. 467).

## Scaliger, Joseph Juste.

*Cyclometrica elementa dvo.*
Leyden: Plantiniana, Franciscus Raphelengius, 1594.
Subject: Geometry.
Wing/fZP/5465/.P7055

In Greek and Latin. Includes "Ambitv circvli" and "De potentia circvli." BOUND WITH Iosephi Scaligeri Ivl. Caes. F. *Mesolabium.*

## Scaliger, Julius Caesar, 1484–1558.

*Exotericarvm exercitationvm liber qvintvs decimvs, De svbtilitate, ad Hieronymvm Cardanvm.*
Paris: Michel Vascosan, 1557.
Subject: Encyclopedia of natural history.
Case/A/911/.1538

Astronomy, biology, origins of various plants and animals, etc.

## Scaliger, Julius Caesar.

*Exotericarvm exercitationvm liber XV. De svbtilitate, ad Hieronymvm Cardanvm.*
Hanover: Wecheliana by Daniel & David Aubrios, and Clement Schleichius, 1620.
Subject: Encyclopedia of natural history.
A/911/.154

## Scandianese, Tito Giovanni, 1517 or 1518–1582.

*I quattro libri della caccia con la dimonstratione de lvochi de Greci et Latini scrittori, & con la tradottione della sfera di Proclo Greco.*
Venice: Gabriel Giolito de Ferrari and brothers, 1556.
Subject: Astronomy, hunting, zoology.
Wing/ZP/535/.G421

In verse. Includes cosmology, seasons (p. 23); discussion of animals in bk. 4. BOUND WITH Proclus, q.v.

## Schedel, Hartmann, 1475–1537.

*[Liber chronicarum] Registrum huius operis libri cronicarum cum figuris et ymaginibus ab initio mundi.*
Nuremberg: Anton Koberger, 1493.
H-C *14508^ BMC (XV)II:437^ Goff S-307
Subject: Chronology.
Inc./+*2084

## Schedel, Hartmann.

[*Liber chronicarum*] *Register des buchs der croniken vnd geschictens mit figuren vnd pildnussen von anbegin der welt bis auf dise vnsere zeit.*
Nuremberg: Anton Koberger, 1493.
H-C *14510^ BMC (XV)II:437–38^ Pr. 2086^ Still. S283^ Goff S-309
Subject: Chronology.
Inc./+2086

## Schedel, Hartmann.

[*Liber chronicarum*] [Another copy]
Inc./+2086a

## Schedel, Hartmann.

[*Liber chronicarum*] *Buch der croniken unnd geschichten mit figuren und pildnussen von anbeginn der welt biss auff diese unsere zeyt.*
Augsburg: Johann Schönsperger, 1500.
H-C*14512^BMC (XV)II:375^ Pr. 1807A^ Still. S285^ Goff S-311
Subject: Chronology.
Inc./f1807/.5

## Scheffer, Johannes, 1621–1679.

*Histoire de la Laponie, sa description, l'origine, les moeurs, la maniere de vivre de ses habitans, leur religion, leur magie, & les choses rares du païs.*
Paris: Widow of Olivier de Varennes, 1678.
Subject: Anthropology, geography, natural history.
Bonaparte/Coll./No./3240

## Scheffer, Johannes.

*Joannis Schefferi Laponia id est, regionis Lapponum et gentis nova et verissima descriptio. In qua multa de origine, superstitione, sacris magicis victu, cultu, negotiis Lapponum, item animalium, metallorumque indole, quae in terris eorum proveniunt, hactenus incognita produntur, & eiconibus adjectis cum cura illustrantur.*
Frankfurt: Joannis Andreae for Christian Wolff, 1673.
Subject: Anthropology, geography, natural history.
Bonaparte/Coll./No./2688

Includes chapters on diseases of the Lapps, their sacred and profane magic, wild and domestic animals, birds, fish etc.; trees, plants, metals, stones, rivers, mountains.

## Scheffer, Johannes.

*The history of Lapland wherein are shewed the original, manners, habits, marriages, conjurations, &c. of that people.*

Oxford: at the Theater, 1674.
STC II S851A
Subject: Anthropology, geography, natural history.
fG/536/.788

### Scheffer, Johannes.

*Joannis Schefferi Argentoratensis de natura & constitutione philosophiae italicae seu Pythagorciae liber singularis.*
Upsala: Henricus Curio, 1664.
Subject: Pythagorean philosophy.
Case/B/122/.792

### Scherer, Heinrich, 1628–1704.

*Atlas novus exhibens orbem terraqueum per natura opera, historiae novae ac veteris monumenta, artisque geographicae leges et praecepta. Hoc est, geographia universa in septem partes contracta, et instructa ducentis fere chartis geographicis, ac figuris.*
Monaco: printed by Maria Magdalena Rauchin, widow, for Joannes Caspar Bencard, 1710.
Subject: Geography.
Ayer/135/S326/1710

Vol. contains 1st 3 parts only: 1. geographia naturalis; 2. geographia hierarchica; 3. geographia Mariana. Each part has individual title page.

### Scheuchzer, Johann Jacob, 1672–1733.

*Bibliotheca scriptorum historiae naturali omnium terrae regionum inservientium. Historiae naturalis Helvetiae prodromus. Accessit celeberrimi viri Jacobi Le Long, . . . De scriptoribus historiae naturalis Galliae. Collegit Joh. Jacobus Scheuchzer.*
Zurich: Heinrich Bodmer, 1716.
Subject: Natural history.
m̄/.8

Biobibliographical list (by Scheuchzer) is arranged by country, alphabetically within each country. Includes Spain and Portugal, France, Germany, Belgium, Switzerland, Italy, Hungary, Poland, Britain, Denmark and Sweden, Asia, Africa, America. Le Long's list (p. 213) is an appendix.

### Schneider, Conrad Victor, 1614–1680.

*Liber de catarrhis.*
Wittenberg: printed by Michael Wendt for heirs of D. Tobias Mevius & Elerdus Schumacher, 1660.
Subject: Medicine.
UNCATALOGUED

4 bks. in 3 vols.

### Schneuber, Johann Matthias, fl. 1650–1665.

*Umständliche beschreibung dess grossen cometen. Welchen in anfang dess Christmonats 1664.*
Strassburg: J. Pastorius, 1665.
Subject: Astronomy, comets.
Case/B/863/.795

A series of 13 observations of a comet, 8 December–24 January, describing distance, declination, altitude, followed by epigrams or poetic allusions, etc.

### Schonefeld, Stephen von, fl. 1573–1624.

*Icthyologia et nomenclatvrae animalium marinorvm, fluviatilium, lacustrium quae in florentissimus, ducatibus slesvici et Holsatiae & celeberrimo emporio Hamburgo occurrunt trivialis. . . . Auctore Stephano à Schonevelde doctore medico, cive Hamburg.*
Hamburg: Hering, 1624.
Subject: Dictionary encyclopedias, natural history.
Bonaparte/Coll./No./10,067

Author's presentation copy. Has index of names in Latin and German.

### Schöner, Johann, 1477–1547.

*Luculentissima quaedam terrae totius descriptio cum multis vtilissimis cosmographiae iniciis. Nouaquae & quae ante fuit verior Europae nostrae formatio. Praeterea, fluuiorum: montium: prouintiarum: vrbium: & gentium quaeplurimorum vetustissima nomina recentioribus admixta vocabulis. Multa etiam quae diligens lector noua vsuique futura inueniet.*
Nuremberg: Ioannis Stuchssen, 1515.
Subject: Descriptive geography, topography.
Ayer/*7/S3/1515

### Schöner, Johann.

*Opera mathematica Ioannis Schoneri Carolostadii in vnvm volvmen congesta, et pvblicae vtilitati studiosorum omnium, ac celebri famae Norici nominis dicata.*
Nuremberg: Ioannis Montanus and Vlric Neuber, 1551.
Subject: Astrology, astronomy, instrumentation.
Ayer/7/S3/1551

### [Schonheintz, Jacobus]

*Apologia astrologie.*
[Nuremberg: Georgius Schenck, 1502.]
Subject: Astrology.
Case/B/8635/.812

### Schopper, Hartmann, b. 1542.

*De omnibvs illiberalibvs sive mechanicis artibvs, hvmani ingenii sagacitate atque industria iam inde ab exordio nascentis mundi vsque ad nostram aetatem adinuentis, luculentus atque succintus liber.*
Frankfurt: [C. Feyerabent] 1574.
Subject: Liberal and mechanic arts.
Wing/ZP/547/.F427

Illustrated by Jost Amman (1539–91). Illustrations of such occupations as astronomus (de sternbeseher), apotecarius (der apotecker), medicinae (der doc. der artzney), book arts, goldsmith.

### Schorer, Christoph, 1618–1671.

*Bedencken von dem cometen dess 1652.*
Basel: Emanuel König, 1653.

Subject: Comets.
Case/B/863/.796

## Schorer, Christoph.

*Kurtzer relation und discurs von dem cometen dess 1664 jahrs. Auff vieler begehren in truck gegeben.*
Vienna: Balthasar Kühnen, 1665.
Subject: Comets.
Case/B/863/.7962

## Schott, Franciscus Sebastianus.

*Cosmvs in micro-cosmo. Hoc est, mundus opere sex dierum creatus, in homine velut parvo in magno micro-cosmo consummatus.*
Vienna: printed by Andreas Heyinger, 1701.
Subject: Creation, macrocosm-microcosm.
B/76/.79

## Schott, Gaspar, 1608–1666.

*Magia universalis naturae et artis, sive recondita naturalium & artificialium rerum scientia, cujus ope per variam applicationem activorum cum passivis, admirandorum effectum spectacula, abditarumque inventionum miracula, ad varios humanae vitae usus, eruuntur.*
Würzburg: Henricus Pigrin for the heirs of Joannes G. Schönwetter, 1657–59.
Subject: Acoustics, mathematics, optics, physics.
L/0114/.797

4 vols. in 2.

## Schott, Gaspar.

*Mathesis Caesarea, sive amussis Ferdinandea in lucem publicam, & usum eruditae posteritatis, . . . atque problemata universae matheseos, praesertim verò architecturae militaris explicata.*
Würzburg: for the widow & heirs of J. Godefridus Schönwetter by Jobus Hertz, 1662.
Subject: Arithmetical computation, geometry, gnomonics, instrumentation.
L/011/.798

Description of a sector invented by Schott and its uses.

## Schott, Gaspar.

*Mechanica hydraulico-pneumatica, qua praeterquàm quòd aquei elementi natura, proprietas, vis motrix, atque occultus cum aëre conflictus, à primis fundamentis demonstratur.*
[Frankfurt]: printed by Henricus Pigrin for heirs of Joannis Godefridus Schönwetter, 1657.
Subject: Machines, water power, technology.
S/4/.797

In 2 parts: theory, praxis.

## Schott, Gaspar.

*Technica curiosa, sive mirabilia artis, libris XII comprehensa; Quibus varia experimenta variaque technasmata pnevmatica, hydraulica, hydrotechnica, mechanica, graphica, cyclometrica, chronometrica, automatica, cabalistica, aliaque artis arcana ac miracula, rara, curiosa, ingeniosa, magnamque partem nova & antehac inaudita, eruditi, orbis utilitate, delectationi, disceptationique proponuntur.*
Würzburg: printed by Jobus Hertz for Wolfgang Mauritius Endter, 1687.
Subject: Artificial magic, astrology, occult, technology.
L/0114/.798

In 2 vols.Engraved title page: "Sumptibus Io. Andreae Endtori et Wolffgangi junioris haeredum." Arranged by locality: Mirabilia Anglicana, mirabilia Magdeburgica, etc. With an appendix by Athanasius Kircher.

## Schwalenberg, Henricus.

*Aphorismi hieroglyphici, qvibvs vetervm philosophorvm mysteria quaedam declaratur: ex commentariis Ioannis Pierii Valeriani & Caelij Augustini Curionis collecti . . . editi per Henricvm Schvalenberg.*
Leipzig: Valentin Vögelin, 1592.
Subject: Animal emblems, hieroglyphics, mnemonics, symbols.
Case/B/835/.798

Animals used as mnemonic devices on ancient monuments. "Explicatio nominis: . . . Primi, inquiens, Aegyptij per figuras animalium sensus mentis effingebant, & antiquissima monumenta memoriae humanae saxis impressa cernuntur, & literarum inuentores perhibentur."

## [Schwenter, Daniel] 1585–1636.

*Deliciae physico-mathematicae. Mathemat. und philosophische erquickstunden sechshundert drey vnd sechsig schöne, liebliche vnd annehmliche kunst stücklein auffgaben, vnd fragen auf der rechenkunst, landt, messen, perspectiv, naturkündigung, vnd andern wissenschaften genommen begriffen seindt.*
Nuremberg: Wolffgang Moritz Endter, & Johann Andrea Enders, 1636–92.
Subject: Arithmetic, astronomy, chemistry, motion, natural philosophy, optics.
Case/V/22/.802

Composed of 16 books on mathematical subjects.

* * *

*La science des hieroglyphes, ou l'art d'exprimer par des figures symboliques, les vertus, les vices, les passions, & les moeurs; &c. avec differentes devises historiques.*
The Hague: Jaques van den Kieboom, 1736.
Subject: Emblems.
Case/W/1025/.802

Note on front flyleaf (dated 1860) states that this work, by one "Daniel de la Feuille," had appeared in 1700 in Amsterdam, apparently a plagiarism. It attributes translation to Cesare Ripa. "Mais le plus intéressant de ce qu'on tresure dans ce volume, ce sons les 36 emblémes es devises composées pour former l'abrégé de la vie de Marie IIe du nom, reine d'Angleterre, morte en 1695."

## Scot, Reginald, 1538?–1599.

*The discouerie of witchcraft, wherein the lewde dealing of witches and witchmongers is notablie detected, the knauerie of coniurors, the impietie of inchantors, the follie of soothsayers,*

*the impudent falshood of cousenors, the infidelitie of atheists, the pestilent practises of pythonists, the curiositie of figurecasters. . . . the abomination of idolatrie, the horrible art of poisoning, the vertue and power of naturall magike and all the conueiances of legierdemaine and iuggling are deciphered: and many other things opened, which haue long lien hidden, howbeit verie necessarie to be knowne. Heerevnto is added a treatise vpon the nature and substance of spirits and diuels &c.*
London: William Brome, 1584.
STC 21864
Subject: Witchcraft.
Case/*B/88/.799

1st ed. Contains the two starred leaves between pp. 336 and 337 that are often missing.

## Scot, Reginald.

*The discovery of witchcraft: proving that the compacts and contracts of witches with devils and all infernal spirits and familiars, are but erroneous novelties and imaginary conceptions. Also discovering how far their power extendeth in killing, tormenting, consuming, or curing the bodies of men, women, children, or animals, by charms, philtres, periapts, pentacles, curses, and conjurations. Wherein likewise the unchristian practices . . . and the knavery of juglers, conjurers, charmers, soothsayers, figure-casters, dreamers, alchymists and philterers; . . . in sixteen books. Whereunto is added an excellent discourse of the nature and substance of devils and spirits, in two books; the first by the aforesaid author: the second now added in this third edition.*
London: printed for A. Clark, 1665.
STC II S945.
Subject: Witchcraft.
Case/fB/88/.7992

## Scot, Reginald.

*Scot's discovery of vvitchcraft: proving the common opinions of witches contracting with divels, spirits, or familiars; and their power to kill, torment, and consume the bodies of men, women, and children, or other creatures by diseases or otherwise; wherein also, the lewde unchristian practises of witchmongers . . . with many other things opened that have long lain hidden.*
[London] Printed by R. C., 1651.
STC II S943
Subject: Witchcraft.
Case/B/88/.7991

With facsimile title page. Bound in at the end: "Select elements from Cornelius Agrippa" (36 pp. in manuscript).

## Scott, Michael, 1175?–?1234.

*Physiognomia.*
[Venice: Jacob de Fivizano] 1477.
H-C *14550^ BMC (XV)V:242^ Pr. 4364^ Goff M-551
Subject: Astrology, physiognomy.
Inc./4364

"Incipit liber phisionomiae: quem compilauit magister Michael Scotus ad preces. D. Federici romanorum imperatoris scientia cuius est multum tenenda in secreto: eo quae est magnae efficaciae: continens secreta artis naturae: que sufficiunt omni astrologo. Et cum haec pars libri phisionomiae constet ex tribus partibus."

## Scott, Michael.

[*Physiognomia*] *Phisonomia Michaelis Scoti.*
[Passau: Johann Petri, ca. 1487–88.]
H-C *14547^ BMC (XV)II:616^ Goff M-559
Subject: Astrology, physiognomy.
Inc./2833

## Scribonius, Willem Adolf, 16th cent.

*De sagarvm natvra et potestate, deqve his recte cognoscendis et pvniendis physiologia . . . vbi de pvrgatione earvm per aqvam frigidam.*
Marburg: Paul Egenolph, 1588.
Subject: Witchcraft.
Case/B/88/.805

## Scultetus, Johannes, 1595–1645.

*Johannis Sculteti physici Ulmensis . . . armamentarium chirurgicum olim auctum triginta novem tabulis, tam veteres quàm recenter excogitatas machinas & operationes exhibentibus; una cum observationum medico-chirurgicarum centuria ex praecipuis hujus seculi practicis collecta. Nunc verò observationibus quibusdam curiosissimis denuo locupletatum, & ab innumeris mendis expurgatum.*
Leyden: Cornelius Boutesteyn, Jordanus Luchtmans, 1693.
Subject: Medicine, surgery.
Case/oRD/30/.s3/1693

BOUND WITH his *Auctarium ad armamentarium chirurgicum, opera defuncti haeredum editum.* Leyden: Cornelius Boutesteyn, Jordanus Luchtmans, 1692.

## Seba, Albert, 1665–1736.

*Locupletissimi rerum naturalium thesauri accurata descriptio et iconibus artificiocissimus expressio, per universam physices historiam.*
Amsterdam: Janssonio-Waesbergios & J. Wetstenius & Gul. Smith, 1734–65.
Subject: Natural curiosities and natural history.
Wing/+M/O/.804

In 4 vols., in Latin and French. I (1734): anatomy of plants, rare and unusual quadrupeds and birds; II (1735): serpentology; vols. III and IV (posthumous, 1753, 1765): creatures of the sea and insects, respectively.

## Seidel, Bruno, 1530–1591.

*Liber, morborum incurabilium causas, mira brevitate, summa lectionis jucunditate eruditè explicans. . . . Accessit Fabritii de Paduanis tractatus de morbis.*
Leyden: Petrus Hackius, 1662.
Subject: Medicine, pathology.
oR/128.6/.S45/1662

## Seixas y Lover, Francisco de, fl. 1688–1690.

*Theatro naval hydrographico, de los flvxos, y reflvxos, y de las corrientes de los mares, estrechos, archipielagos, y passages aquales del mundo, y de las differencias de las uariaciones de la aguja de marear, y effectos de la luna, con los uientos generales, y particvlares qve reynan en las quatro regiones maritimas del orbe.*
Madrid: printed by Antonio de Zafra, 1688.
Subject: Astronomy, hydrography, navigation.
Ayer/*1262/S46/1688

## Selenus, G. pseud. (August II, Duke of Braunschweig-Lüneburg).

*Cryptomenytices et cryptographiae libri ix. In quibus & planissima steganographiae à Johanne Trithemio . . . enodatio traditur.*
Lüneberg: Johannes & Henry Stern, 1624.
Subject: Secret writing.
Bonaparte/Coll./No./2691

Includes "de natura cryptographiae"; "de mediis, seu modis occultandi"; "de scriptionis instrumentis &c."

## Seller, Jeremiah.

*The English pilot.*
London: printed by George Larkin, 1690.
STC II S2472A
Subject: Navigation.
Ayer/*135/E55/1690

Pt. 2: Description of south coast of England, Ireland, etc."The fourth part of the general English pilot" [separate title page and pagination] includes description of east coast of America from Greenland to the River Amazones. London: printed by J. Dawks, for Jer. Seller and Cha. Price, 1703. Pt. 1 text wanting; title page and map only.

## Seller, John, fl. 1700.

*A new system of geography, designed in a most plain and easie method, for the better understanding that science. Accomodated with new maps . . . with geographical tables.*
London: printed for Jer. Seller and Char. Price, . . . and John Senex, 1703.
Subject: Geography.
Case/3A/1704

3d ed. Primarily maps, small amount of text.

## Seller, John.

*A pocket book, containing severall choice collections in arithmetick, astronomy, geometry, surveying, dialling, navigation, astrology, geography, measuring, gageing. By John Seller, hydrographer to the king.*
London, [1677? 1682?]
STC II S2480A
Subject: Almanacs, arithmetic, astronomy, navigation.
Ayer/8.9*/S46/1677

## Seller, John.

*Practical navigation: or, an introduction to the whole art. Containing the doctrine of plain and spherical triangles. Plain, Mercator, great-circle sailing; and astronomical problems. The use of divers instruments; as also of the plain-chart, Mercator's chart, and both globes. Sundry useful tables in navigation: and a table of 1000 logarithms, and of the logarithm-sines, tangent, and secants.*
London: printed for Richard Mount . . . and Jeremiah Seller, 1699.
STC II S2484
Subject: Mathematics, navigation.
Ayer/8.9*/N2/S46/1699

On front pastedown: "Richard Thomas his book 1700. Price 6s. 5d. 8–6 sterling." A MS provenance written by a great-grandson of original owner is laid in.

* * *

*Sendtschreiben oder einfeltige antwort an die hocherleuchte brüderschafft des hochlöblichen ordens dess Rosen Creutzes. auff die von ihnen aussgefertige famam und confessionem der fraternitet: durch einen waren liebhaber der vollkommenen weissheit gestellet und aussgesandt.*
Frankfurt: Johann Bringer, 1615.
Subject: Rosicrucians.
K/975/.807

Signed C. H. C.

## Seneca, Lucius Annaeus, ca. 4 B.C.–A.D. 65.

*L. Annai Senecae natvralivm qvaestionvm libri VII. Matthaei Fortunati in eosdem libros annotationes.*
[Venice: Aldus & Andreus Asulanus associates, 1522]
Subject: Comets, earthquakes, meteorology.
Case/*Y/672/.S2565

See *DNB* s.v. Seneca. Sections include "De ignibvs coelestibvs et de arcv praefatio"; "De tonitribvs, fvlgvrationibvs et fvlminibvs, fvlgvrvmqve generibvs"; "In qvo de nive, grandine, et plvvia, atqve in primis de natvra nili"; "De ventis et aeris motv agitvr."

## Sennault, Jean François, 1601–1672.

*De l'vsage des passions.*
Paris: widow of Iean Cammsat, 1641.
Subject: Passions of the mind.
Case/B/529/.802

In 2 parts: (1) the passions in general (nature of the passions, power of passions over will); (2) the passions in particular (love and hate, hope and despair, anger, courage and fear).

## Sennault, Jean François.

*De l'vsage des passions. Suivant la copie imprimée a Paris.*
[Leyden: Elsevier] 1643.
Subject: Passions of the mind.
Case/–B/529/.803

"Dernière edition."

### Sennault, Jean François.

*The vse of passions written in French by J. F. Senault and put into English by Henry Earle of Monmouth.*
London: printed by W. G. for John Sims, 1671.
STC II S2505
Subject: Passions of the mind.
Case/B/529/.805

### Sennert, Daniel, 1572–1637.

*Danielis Sennerti Vratislaviensis, doctoris et medicinae professoris in academia Wittebergensi, operum in sex tomos divisorum.*
Lyon: printed for Joannis Antonii Huguetan, 1676.
Subject: Medicine, natural science.
UNCATALOGUED

6 vols. in 3. Editio novissima. Includes "Hypomnemata physica" (vol. 1); "Epitome febrium . . . fasciculus medicamentorum contra pestem" (vol. 6); "Hypomnematus II de occultis qualitatibus."

### Sennert, Daniel.

*Institvtionvm medicinae Danielis Sennerti D. medici, tomvs secvndvs.*
Geneva: Philipp Gamonet, n.d. [1646?]
Subject: Materia medica, medicine.
UNCATALOGUED

Vol. 2 only (begins with bk. 4).

### Sensi, Lodovico.

*La historia del l'hvomo composta da messer Lodouico Sensi giureconsulto Perugino. Divisa in tre libri.*
Perugia: Baldo Saluiani, 1577.
Subject: Natural theology, psychology.
Case/B/475/.796

*Nel primo de' qvali si ragiona di quello che ha l'huomo per natura dentro, & fuora di se. Nel secondo di quello, che può sopranaturalmente hauer per gratia. Nel terzo si parla dello stato della innocentia, del primo peccato, & de i disordini ne' quali l'huomo incorse per lo primo peccato.* 4 interior senses: "senso commune; fantasia; della stimatiua, ouero cogitatiua; memoria."

### Sequeira, Gaspar Cardoso de, fl. 1605–1631.

*Pronostico geral, e lvnario perpetvo, assidas luas nouas, & cheas, como quartos crescentes, & minguantes. Composto por Guaspar Cardozo de Sequeira, mathematico.*
[Coimbra] Nicolao Carualho, 1614.
Subject: Astrology, prognostication.
Greenlee/4504/P855

### Serlio, Sebastiano, 1475–1552.

*Tvtte l'opere d'architettvra, et prospetiva di Sebastiano Serlio.*
Venice: heirs of Francesco di Franceschi, 1600.
Subject: Geometry, perspective as elements of architecture.
Wing/ZP/635/.F/848

2d ed. Bk. 1: geometry, bk. 2: perspective, bks. 3 and 4: architecture.

### Serrão Pimentel, Luiz, 1613–1679.

*Methodo Lvsitanico de desenhar as fortificaçoens das praças regulares, & irregulares, fortes de campanha, e ovtras obras pertencentes a architectura militar distribuido em duas partes operativa, e qvalificativa.*
Lisbon: Antonio Craesbeeck de Mello, 1680.
Subject: Geometrical, mathematical, trigonometrical approach to military architecture.
Greenlee/f4547/S 487/1680

"Practicada arithmetica decimal ov dizima" (p. 548).

* * *

[*Seven planets.*] *In dieser figur vindt mann die sieben planeten.*
[Germany, not before 1460.]
Subject: Almanacs, astrology.
Inc./Wing/+31/.6

Broadside epitome of the seven planets: Sol, Merkurius, Venus, Mars, Luna, Saturnus, Jupiter.

### Seymour, Robert [pseud.].

*An accurate survey of the cities of London and Westminster, and borough of Southwark: . . . the whole being an improvement of Mr. Stow's, and other surveys.*
London: printed by the booksellers, 1736.
Subject: Natural science, topography.
Wing/G/455/.848

"Of schools, and other houses of learning" (esp. pp. 231 ff.) "The curiosities of the Royal Society," with a list of the Royal Society in 1733 (p. 241). See Stow, John. [Seymour is said to be pseud. for John Mottley, 1692–1750.]

### Shaw, Thomas.

*Travels, or observations relating to several parts of Barbary and the Levant.*
Oxford: printed at the Theatre, 1738.
Subject: Descriptive geography, natural history, weather.
fG/79/.806

Descriptions of flora and fauna in Algiers and Tunis, Syria, Egypt. Includes (pt. 6) a catalogue of some of the rare plants of Barbary, Egypt, and Arabia. With an appendix, "De coralliis & eorum affinibus." Other catalogues of stones, fossils, fish, weather, etc.

### Shaw, Thomas.

*Travels.* [Another copy]
Wing/fG/79/.807

With *A supplement to a book entituled Travels or observations, &c. wherein some objections, lately made against it, are fully considered and answered: with several additional remarks and dissertations.* [1746]

### Sherley, Thomas, 1638–1678.

*A philosophical essay: declaring the probable causes whence stones are produced in the greater world. From which occasion is taken to search into the origin of all bodies, discovering them to proceed from water and seeds. Being a prodromus to a*

*medicinal tract concerning the causes and cure of the stone in the kidneys and bladders of men.*
London: printed for William Cademan, 1672.
Subject: Macrocosm-microcosm, medicine, petrifaction.
Case/Q/809

## Sibbald, Sir Robert, 1641–1722.

*Scotia illustrata, sive prodromus historiae naturalis in quo regionis natura, incolarum ingenia & mores, morbi iisque medendi methodus, & medicina indigena accuratè explicantur.*
Edinburgh: Jacob Knible, Joshua Solingen, & Johannes Colmar for the author, [1683–]1684.
Subject: Medicine, natural history.
Case/+G/43003/.81

3 vols. in 1.

## Silva Fernandes, José da.

*Discurso apologetico cirurgico-medico, escrito em estylo epistolar.*
Lisbon: Miguel Rodrigues, 1729.
Subject: Medicine.
Greenlee/4532.1/M48/S583/.729

## [Simon, Jean Baptiste]

*Le gouvernement admirable, ou la republique des abeilles. Avec les moyens d'en tirer une grande utilite.*
The Hague: Pierre de Hondt, 1740.
Subject: Apiculture, bees.
O/975/.82

A detailed treatise on the life cycle of the honey bee.

## Simplicius of Cilicia, 6th cent. A.D.

*Simplicii commentarii in octo Aristotelis Physicae avscvltationis libros cvm ipso Aristotelis textv.*
Venice: Aldus [Manutius, the elder] and Andreas Asulanus Socerus, 1526.
Subject: Physics.
Case/*Y/642/.S32

Aristotle's text in Greek.

## Siria, Pedro de.

*Arte de la verdadera navegacion. En que se trata de la machina del mundo, es a saber, cielos, y elementos: de las mareas, y señales de tempestades: del agua de marear: del modo de hazer cartas de nauegar. del vso dellas; de la declinacion ni rodeo: el modo come se sabra el camino, y leguas que ha nauegado el piloto, por qualquier rumbo; y vltimamente el saber tomar el altura del polo. Compvesta por Pedro de Syria.*
Valencia: Iuan Chrysostomo Garriz, 1602.
Sabin 94133
Subject: Navigation.
Ayer/8.9/N2/S619/1602

## Sirigatti, Lorenzo, fl. 1596.

*La pratica di prospettiva del cavaliere Lorenzo Sirigatti.*
Venice: Girolamo Franceschi Sanese, 1596.
Subject: Perspective.
Case/Wing/+ZA/3/.82

In 2 books: bk. 1: 2-dimensional renderings; bk. 2: renderings of 3-dimensional objects—solids, hollow shapes, architectural components. Bk. 1 has text and plates; bk. 2 consists of plates only.

## Skinner, Stephen, 1623–1667.

*Etymologicon linguae anglicanae, . . . Accedit etymologicon botanicum, seu explicatio nominum omnium vegetabilium, praesertim solo nostro assuetorum, aut quae, licèt peregrina sint, vulgò nota sunt; omissis interim quae manifestè vel à Latino vel à Graeco fonte promanant. . . . Tandem ultimo etymologicum onomasticon, seu explicatio etymologica nominum fluviorum, regionum, urbium, oppidorum.*
London: printed by T. Roycroft, 1671.
STC II S3947
Subject: Etymology of plant names and medicinal uses of plants.
Case/fX/993/.823

Ed. Thomas Henshaw, 1618–1700. Entries include descriptions and medical value (e.g., "`much-good,' oreoselinum, q.d. herba polychresta, quà sc. ad omnes obstructiones & movendam urinam utilis est"; "neck-weed, cannabis sic dicta quia funes quibus malefici suspenduntur ex hac materia texuntur").

## Skinner, Stephen.

*Etymologicon linguae anglicanae.* [Another copy]
Bonaparte/Coll./No./11416

## Sloane, Sir Hans, bart., 1660–1753.

*A voyage to the islands Madera, Barbados, Nieves, S. Christophers and Jamaica, with the natural history of the herbs and trees, four-footed beasts, fishes, birds, insects, reptiles, &c. of the last of those islands; to which is prefix'd an introduction, wherein is an account of the inhabitants, air, waters, diseases, trade, &c. of that place, with some relations concerning the neighbouring continent, and islands of America.*
London: printed by B. M. for the author, 1707, 1725.
Subject: Natural history, voyages.
Case/M/0974/.81

In 2 vols. Vol. 1 includes natural history of Jamaica and illustrations of plants and fishes. Text of vol. 2 (1725) includes material on trees, insects, crustaceans, fish, birds, quadrupeds, serpents, minerals.

## [Smith, Charles] 1715?–1762.

*The antient and present state of the county of Down. Containing a chorographical description, with the natural and civil history of the same.*
Dublin: printed by A. Reilly for Edward Exshaw, 1744.
Subject: Chorography, medicine, natural history.
G/4228/.04

Chapters 5–9: harbors, rivers, waters, including, in chap. 9, medicinal waters. Chap. 10: medicinal value of goat's whey.

## Smith, John, fl. 1673–1680.

*Horological disquisitions concerning the nature of time, and the reasons why all days from noon to noon are not alike twenty four hours long ... with tables of equations, and newer and better rules than any yet extant, how thereby precisely to adjust royal pendulums.... A work very necessay [sic] for all that would understand the true way of rightly managing clocks and watches.... To which is added the best rules for the ordering and use both of the quick-silver and spirit weather glasses: and Mr. S. Watson's rules for adjusting a clock by the fixed stars.*
London: printed for Richard Cumberland, 1694.
STC II S4106
Subject: Horology, timekeeping.
Case/R/25/.825/no./1

BOUND WITH Smith, John, *Of the unequality*, q.v.

## Smith, John.

*Of the unequality of natural time, with its reason and cavses, together with a table of the true aequation of natvral dayes. Drawn up chiefly for the use of the gentry, in order to their more true adjusting, and right managing of pendulum clocks, and watches.*
London: printed for Joseph Watts, 1686.
STC II S4107
Subject: Horology, timekeeping.
Case/R/25/.825/no./2

## Smith, John, governor of Virginia, 1580–1631.

*New Englands trials. Declaring the successe of 26. ships employed thither within these sixe yeares: with the benefit of that countrey by sea and land: and how to build threescore sayle of good ships, to make a little navie royall.*
London: printed by William Iones, 1620.
STC 22792
Subject: Shipbuilding.
Ayer/150.5/N4/S6/1620

Urges colonization of and fishing etc. in New England.

## Smith, Thomas, 1638–1710.

*Vitae quorundam eruditissimorum et illustrium virorum.*
London: David Mortier, 1707.
Subject: Astronomy, geometry.
E/445/.818

Lives of 8 prominent men. Includes, e.g., astronomers John Bainbridge and John Grave; geometer Henry Briggs; John Dee (with the hour and minute of his birth), with a bibliography appended.

## Snellius, Willebrordus, 1581–1626.

*Willebrordi Snelii à Royen. R. F. Tiphys batavvs, sive histiodromice, de navium cursibus, et re navali.*
Leyden: Elzeviriana, 1624.
Subject: Navigation.
Case/L/995/.828

## Sole, Francesco dal, b. ca. 1490.

*Instrvtioni et regvle di Francesco dal Sole, Francese, ... sopra il fondamento delle alme scientie d'abbacco, arithmetica, geometria, cosmografia, & mathematica.*
Ferrara: Francesco di Rossi da Valenza, 1564.
Subject: Arithmetic, geometry.
Case/L/10/.828

Contains a section, "regula del Zodiaco" (p.65).

## Solinus, Caius Julius, early 3d cent. A.D.

*Delle cose maravigliose del mondo.*
Venice: Gabriel Giolito de' Ferrari, 1559.
Subject: Descriptive geography, natural history.
Case/Y/672/.S664

Trans. Giovan Vincenzo Belprato.

## Solinus, Caius Julius.

*Ioannis Camertus minoritani, artivm et sacrae theologiae doctoris, [polyhistor* in Greek] *ennarrationes. Additus eiusdem Camertis index, tum literarum ordine, tum rerum notabilium copia percommodus studiosis.*
Vienna: printed by Ioannes Singrenius for Lucas Alantse, 1520.
Church 45^ Harrisse 108
Subject: Descriptive geography, natural history.
Ayer/*7/S7/1520

MS note from bookseller gives bibliographical reference as Harrisse 181 ff. BOUND WITH Mela, Pomponius, *De orbis situ*, 1540, q.v.

## Solinus, Caius Julius.

*The excellent and pleasant worke of Iulius Solinus polyhistor contayning the noble actions of humaine creatures, the secretes and prouidence of nature, the description of countries, the manners of the people: with many meruailous things and strange antiquities, seruing for the benefit and recreation of all sorts of persons.*
London: printed by I. Charlewoode for Thomas Hackett, 1587.
STC 22896
Subject: Descriptive geography, natural history.
Case/Y/672/.S66

Trans. from the Latin by Arthur Golding.

## Solinus, Caius Julius.

*De memorabilibvs mvndi.*
Venice: Theodorus de Ragazonibus, 1491.
H-C *14880^ BMC (XV)V:478^ Pr. 5266^ Still. S555^ Goff S-620
Subject: Descriptive geography, natural history.
Inc./5266

MS marginalia passim.

## Solinus, Caius Julius.

*Solinvs de memorabilibvs mvndi.*
Venice [Guilelmus Animamia, Tridinensis], 1493.

H-C *14881^ BMC (XV)V:412^ Pr. 5116^ Still. S556^ Goff S-621
Subject: Descriptive geography, natural history.
Inc./5116

## Solinus, Caius Julius.

*Solinus de memorabilibus mundi diligenter annotatus & indicio alphabetico prenotatus.*
Paris: Jehan Petit [1503?]
Subject: Descriptive geography, natural history.
Greenlee/5100/S68/1503

BOUND WITH Pauli Orosii, *Historiographi clarissimi opus prestantissimum.* Paris: Jehan Petit, 1506. Discusses plague of locusts (pp. 9, 17, 20, 23, 28, 37, 41–43 etc.)

## Solinus, C. Julius.

*De mirabilibvs mondi.*
Brescia: Jacobus Britannicus, 1498.
H-C *14884^ BMC (XV)VII:982–83^ Pr. 7008; Still. S558^ Goff S-623
Subject: Descriptive geography, natural history.
Inc./f7008

## Solinus, Caius Julius.

*Solinus de mirabilibus mundi.*
Brescia: Jacobus Britannicus, 1498.
H-C *14883^ Pr. 7007^ Still. S559^ Goff S-624
Subject: Descriptive geography, natural history.
Inc./f7007

MS marginalia. A reprint [see Goff].

## Solinus, Caius Julius.

*C. Ivlii Solini polyhistor rervm toto orbe memorabilivm thesaurus locupletissimus. Hvic ob argvmenti similitvdinem Pomponii Melae de sitv orbis libros tres, fide diligentiaqve summa denuò iam recognitos, adiunximus.*
Basel: Michael Isingrinivs, 1543.
Harrisse Add. no. 143
Subject: Descriptive geography, natural history.
Ayer/7/S7/1543

## Solis, Giulio Cesare, 16th cent.

*Discorso di cosmografia doue si hà piena notitia di tutte le prouincie, città, castella, popoli, monti, mari, fiumi, & laghi di tutto il mondo.*
Padua: Lorenzo Pasquati, 1602.
Subject: Cosmography, descriptive geography.
G/117/.81

In dialogue form.

## Sorbiere, Samuel, 1615–1670.

*A voyage to England.*
London, 1709.
Subject: Mathematics, medicine, sciences in England.
G/45005/.83

Includes descriptions of English science (pp. 27–39), practice of medicine in England, with description of the quack, Borri (pp. 78–85), mathematics, Hobbes (pp. 39–41), and Royal Society (pp. 47–50).

## Sorel, Charles, 1602–1674.

*La maison des ievx academiques, contenant vn recveil general de tous les ieux diuertissans pour se réjoüir & passer le temps agreablement.*
Paris: Estienne Loyson, 1668.
Subject: Dreams, games and puzzles, physiognomy.
Case/V/16/.478

In 2 pts. Pt. 2: *Les recreations galantes* (1671), which includes "L'explication des songes et un traité de la phisionomie."

## Sorel, Charles.

*Les secrets astrologiqves des figvres ov des anneavx grauez souz certain signe du ciel pour accomplir diuers effects merueilleux. Et de l'unguent des armes, sympathique ou constellé, dont l'on pretend guerir les playes sans les toucher. Ouurages tirez de la derniere partie de la science humaine de M. Ch. Sorel.... a quoy l'on a adjousté des obseruations contre le liure des cvriositez invoyes de M. I. Gaffarel.*
Paris: Anthoine de Sommaville, 1640.
Subject: Astrology, sympathetic medicine, weapon salve.
Case/B/85/.832

"De l'vngvent des armes" (p. 337).

## South, Robert. 1634–1716.

*Musica incantans; or, the power of musick. A poem. Written originally in Latin by Dr. South. Translated: With a preface concerning the natural effects of musick upon the mind.*
London: printed for William Turner, 1700.
STC II S4737
Subject: Music, psychology.
Case/fV/23/.827

## South, Robert.

*Musica incantans, sive poema experimens musicae vires, juvenem in insaniam adigentis et musici inde periculum.*
London: H. Hills [1709]
Subject: Music, psychology.
V/23/.83

## Speed, John, 1552–1629.

*A prospect of the most famous parts of the world. viz. Asia, Africa, Evrope, America. With the kingdomes therein contained.... Together vvith all the provinces, counties, and shires contained in that large theater of Great Brittaines Empire.*
London: printed by John Legatt for William Humble, 1646.
STC II S4882A
Subject: Geography.
Case/+G/117/.82

## Speed, John.

*The theatre of the empire of Great-Britain, presenting an exact geography of the kingdom of England, Scotland, Ireland, and*

*the isles adjoyning; . . . together with a prospect of the most famous parts of the world, viz. Asia, Africa, Europe, America.*
London: printed for Thomas Basset and Richard Chiswel, 1676.
STC II S4886
Subject: Geography.
Ayer/135/S7/1676

5 pts. in 1 vol.

## Spencer, John, fl. 1630–1693.

*A discourse concerning prodigies: wherein the vanity of presages by them is reprehended, and their true and proper ends asserted and vindicated. To which is added a short treatise concerning vulgar prophecies.*
London: printed by J. Field for Will. Graves, 1665.
STC II S4948
Subject: Apparitions and prodigies, prognostication, superstition.
Case/B/863/.83

BOUND WITH *A discourse concerning vulgar prophesies. Wherein the vanity of receiving as the certain indications of any future event is discovered; and some characters of distinction between true and pretending prophets are laid down.* London: printed by J. Field for Timothy Garthwait, 1665.

## Spina, Bartolommeo, 1494(ca.)–1546.

*Qvaestio de strigibvs, vna cvm tractatv De praeeminentia sacrae theologiae, & quadruplici apologia de lamiis contra Ponzinibium.*
Rome: Populi Romani, 1576.
Subject: Witchcraft.
Case/0BF/1565/.S6/1576

## [Spinoza, Benedictus de] 1632–1677.

*Benedictus de Spinoza Opera posthuma.*
[Amsterdam: J. Rieuwertsz] 1677.
Subject: Chemistry, ethics dealt with in a geometrical fashion.
Case/B/246/.817

Contents: "Ethica. More geometrico demonstrata." "Epistola VI continens annotationes in librum nobilissimi viri Roberti Boyle, de nitro, fluiditate, & firmitate. Viro nobilissimo, ac doctissimo Henr. Oldenburgio responsio ad praecedentem."

## Spiriti, Lorenzo, b. ca. 1436.

*T boec van der auōturen ende van tijt cortīge van versinnen.*
Antwerp: Jan van Ghele, 1608.
Subject: Astrology.
Wing/ZP/646/.G345

BOUND WITH *Thuys d fortuņe. Met het huys der doot. En de tien outheden des mentchen.* Rotterdam: widow & children of Jan van Gehle, 1610.

## Spiriti, Lorenzo.

*Le passetemps de la fortune des dez. Ingenieusement compilé par maistre Laurens l'Esprit, pour responses de vingt questions par plusieurs coustumierement faites, & desirees sçauoir.*
Lyon: heirs of François Didier, 1582.
Subject: Astrology, dice, chance, prognostication.
Wing/ZP/539/.D556

## Sprat, Thomas, bp. of Rochester, 1635–1713.

*The history of the Royal-Society of London, for the improving of natural knowledge.*
London: printed by T. R. for J. Martyn, 1667.
STC II S5032
Subject: Academies and learned societies, experimental science.
Case/A/9/.776

MS notes. Selected experiments and observations: weather, dissecting. P. 369: "Experiments not dangerous to the Church of England."

## Sprat, Thomas, Bishop of Rochester.

*The history of the Royal-Society of London.* [Another edition]
London: printed for Rob. Scot, Ri. Chiswell, Tho. Chapman, and Geo. Sawbridge, 1702.
Subject: Academies and learned societies, experimental science.
A/9/.7761

2d ed., corrected.

## Sprat, Thomas, bp. of Rochester.

*The history of the Royal Society of London.* [Another edition]
London: Printed for Samuel Chapman, 1722.
Subject: Academies and learned societies, experimental science.
Case/4A/967

3d ed., corrected.

## Sprat, Thomas, bp. of Rochester.

*The plague of Athens, which happened in the second year of the Pelopennesian War. First described in Greek by Thvcydides; then in Latin by Lvcretivs. Since attempted in English by the right reverend . . . Thomas lord bishop of Rochester.*
London: printed by M. F. for Charles Brome, 1688.
STC II S5044
Subject: Pest tracts, plague.
Case/Y/642/.E/30692

BOUND WITH Walker, 1695, and [Wollaston] 1691 [n.a.]. Prose trans. of Thucydides, bk. 2, by Hobbes, followed by verse (trans. of Lucretius?).

## Sprenger, Jacobus, and Institorus, Henricus.

*Malleus maleficarum.*
[Speier: Peter Drach, 1487]
HC* 9238^ Pr. 526^ BMC (XV)I:26^ Goff I-163
Subject: Demonology, witchcraft.
Inc./f526

According to Goff, an entry in the account book of Peter Drach indicates that this vol. was in print prior to 15 April 1487.

### Sprenger, Jacob, and Institoris, Henry.

*Mallevs maleficarvm, maleficas, & earum haeresim, vt phramea potentissima conterens.*
[Cologne, Ioannes Gymnicvs (colophon: "Anno XX") 1520]
Subject: Demonology, witchcraft.
Case/B/88/.438

MS marginalia. MS note on title page: "1er editio."

### Squire, Jane, 1671?–1743.

*A proposal for discovering our longitude.*
London: printed for the author, 1742.
Subject: Astronomy, longitude, mathematics.
Ayer/8.9/N2/S774/1742

Title also in French. Followed by "Copies of letters written in consequence of the preceding proposal"; "The explanation of a proposal to determine our longitude; or, the longitude discover'd, by Jane Squire, 1731," in English and French.

### Staurophorus, Rhodophilus (pseud.).

*Raptus philosophicus, das ist, philosophische offenbarungen, gantz simpel und einfältig gestellet, und an die hoch-löbliche und berühmte Fraternitet R. C. unterthänig geschrieben: durch Rhodophilum Staurophorum, ejusdem sapeintissimae atque divinitus excitatae fraternitatis, SS. ordinis R. C. indignum clientem; amore tamen penitus languentem, adeoque desiderio summè flagrantem.*
[N.p.] 1619.
Subject: Rosicrucians.
K/975/.839

### Stevin, Simon 1548–1620.

*Materiae Politicae. Burgerlicke stoffen.*
Leyden: A. Rosenboom, [1650?]
Subject: Architecture, city planning, military art.
Case/J/O/.8413

Contains a chapter on the usefulness mathematics in a citizen's public and private life, in bookkeeping, town planning, mechanical devices (see Hans Baron, *Newberry Library Bulletin* 4, no. 2 [May 1955]: 54). Includes Hendrik Stevin's "Lochening van een ewich roerstel gesecht perpetuum mobile" [at beginning of vol.], is not in the issue of 1649. BOUND WITH *Verrechting van domeine,* Leyden: Livius, 1650.

### [Stobniczy, Jan ze] [Ioannes de Stobnicza] d. 1530.

*Introductio in Ptolomei Cosmographiam cum longitudinibus & latitudinibus regionum & ciuitatum celebriorum.*
[Cracow: Hieronymus Vietor, 1519.]
Harrisse 95
Subject: Cosmography.
Ayer/*6/P9/S8/1519

2d ed. With facsimiles of 2 woodcut maps. Among tracts are "Epitoma Europae Eneae Siluij"; "Siriae compendiosa descriptio ex Isidoro"; "Terrae sanctis & urbis Hierusalem"; "Africae breuis descriptio ex Paulo Orosio."

### Stoeffler, Johann, 1452–1531.

*Calendarivm Romanvm magnum, . . . D. Ioanne Stoeffler iustingensi mathematico authore. . . . A strigeros quisquis punctium disquirere tractus optat, & obliqui sidera zodiaci, siue etiam erronum varios expendere cursus, constet vt articulis noxque diesque suis, hanc adeat chartam mox, egregiamque mathesin lecturus doctis omnia scripta modis. Haec etiam aegroto dispensat phármaca membro, haec etiam venis tempora tuta parat, deniquae venturos lunae solisquae labores foelix auspicijs vnica charta docet, hanc igitur toto chartam, doctissime, coelo felicem, lector consule, disce, cole.*
Oppenheim: Jacob Köbel, 1518.
Subject: Astrology, astronomy.
Wing/ZP/547/.K808

MS marginalia. A group of astrological and astronomical tracts, including "Schemata eclypsivm lvminarivm," and "Tabvla cycli solaris." With (unassembled) volvelles at back of vol, including "Instrvmentvm Horarvm."

### Stoeffler, Johann.

*Cosmographicae aliqvot descriptiones Ioannis Stöfleri Iustingensis mathematici insignis. De sphaera cosmographica, hoc est de globi terrestris, artificiosa structura. De dvplici terrae proiectione in planum, hoc est, qua ratione commodius chartae cosmographicae, quas mappas mundi uocant, designari queant. Omnia recens data per Io. Drandrum medicum & mathematicum.*
Marburg: Eucharius Ceruicornus, 1537.
Harrisse, p. 353.
Subject: Cosmography, mapmaking.
Ayer/*7/S8/1537

### Stow, John, 1525?–1605.

The abridgement of the English chronicle.
London: printed for the Company of Stationers, 1611.
STC 23331
Subject: Natural history, technology.
Case/F/45/.8496

MS notes recto and verso of last leaf of epistle dedicatory and verso of dedication (to Craven, lord mayor of London). P. 178: "In 1459 the science of printing was found in Germany at Magunce. William Caxton of London brought it into England about the yeare 1471"; p. 288: tempests at Chelmsford; p. 286: earthquake.

### Stow, John.

*A svrvay of London. Conteyning the originall, antiquity, increase, moderne estate, and description of that city, written in the yeare 1598. by Iohn Stow. . . . Since by the same author increased, with diuers rare notes of antiquity.*
London: Iohn Windet, 1603.
STC 23343
Subject: Topography.
Case/G/455/.8507

"Hospitals in this cittie, and suburbes thereof, that haue beene of old time, and now presently are" (p. 497).

## Stow, John.

*The svrvay of London.* [Another edition]
London: printed by George Purslowe, 1618.
STC 23344
Subject: Topography.
Case/G/455/.8508

P. 122: medical and mathematical lectures.

## Stow, John.

*The survey of London. Afterwards inlarged by the care and diligence of A. M. in the yeere 1618. And now completely finished by the study and labour of A. M. H. D. and others, this present yeere 1633.*
London: printed by Elizabeth Pvrslovv, 1633.
STC 23345
Subject: Topography.
Case/G/455/.8509

"Of the ancient and present brookes" (chap. 3); "Of the ancient and famous River of Thames" (chap. 4).

## Stow, John.

*A survey of the cities of London and Westminster. Written at first in the year MDXCVIII by John Stow. . . . Now lastly corrected, improved, and very much enlarged. . . . by John Strype.*
London: printed for A. Churchill [etc.], 1720.
Subject: Topography.
fG/455/.85

In 2 vols., 6 bks. Pp. 28 ff.: tides, floods, fish in the Thames, pollution of the river; pp. 78–79: building of the London wall; chap. 22: of schools, etc., with descriptions of chirurgery lecture, Hood's mathematical and military lecture; chap. 23: College of Physicians.

## Strabo, 63 B.C.–?24 A.D.

*Geographia.*
[Venice] Wendelin of Speier, 1472.
H-C *15087^ BMC (XV)V:161^ Pr. 4042^ Still. S705^ Goff S-794
Subject: Geography.
Inc./f4042

On front flyleaf: "Strabonis rerum geographicarum libri XVII ea interpretatione Guarini Veronensis et G. Typhernati . . . sans chiffre signature in relieures. Commençant par l'epitre dédicatoire au Pape Paul." Ed. Joannes Andreae, bp. of Aleria. MS marginalia passim.

## Strabo.

*Geographia.*
[Treviso] Joannes (Rubeus) Vercellensis, 1480.
H-C *15089^ BMC (XV)VI:896^ Pr. 6493^ Still. S706^ Goff S-796
Subject: Geography.
Inc./f6493

## Strabo.

*Geographia.*
[Venice] Joannes Rubeus, 1494.
H-C *15090^ BMC (XV)V:418^ Pr. 5135^ Still. S707^ Goff S-797
Subject: Geography.
Inc./f5135

Half title: "Strabo de sitv orbis." Ed. Antonius Mancinellus.

## Strabo.

*Strabonis de sitv orbis libri XVII. Grecè & Latinè simul iam, in eorum qui pariter & geographiae et utriusque lingue studiosi sunt, gratiam editi: olim quidem, ut putatur, à Gvarino Veronensi, & Gregorio Trifernate in Latinum conuersi.*
Basel: Henricvs Petri, 1549.
Subject: Geography.
Case/6A/54

MS notes on front flyleaf (subject matter of each book) said to be in Pietro Angelio's hand. In double columns, Greek and Latin. MS marginalia passim.

## Strabo.

*La prima [–seconda] parte della geografia di Strabone, di Greco tradotta in volgare Italiano da M. Alfonso Bvonaccivoli.*
Venice: Francesco Senese, 1562–65.
Subject: Geography.
Ayer/*6/S8/1562

In 2 vols.

## Strabo.

*Strabonis nobilissimi et doctissimi philosophi ac geographi rervm geographicarvm commentarij libris XVII contenti, Latini facti.*
Basel: Henricpetrina [1571].
Subject: Geography.
Ayer/*6/P9/S7/1571

Maps designed by Sebastian Münster (d. 1552) for, and originally appeared in, his eds. of Ptolemy dated 1540, 1542, 1545, and 1552, qq.v. Used here in a different work by the same printer.

## [Straet, Jan van der], 1536?–1605.

*Nova reperta.*
[Antwerp, 1600?]
Subject: Inventions, technology.
Case/Wing/fZ/412/.85

Designed by van der Straet; engraved by Philipp Galle. Plate 1 depicts Columbus landing on American shores; plate 4 is the "printing house plate"; other subjects treated are astronomy, clock making, medicine, distillation, harness making, etc. On title page: "Aloysio Alamannio Flor[enti]no. I. Strad. invent. D. D."

## Strangehopes, Samuel.

*A book of knowledge. In three parts. The first, containing a brief introduction to astrology, showing the nature, quality, and effects of the twelve signs, and seven planets, their dominion over bodies, with the fortunes of those calculated, who are born*

*under them; also a delightful wheel of fortune. The second, a treatise of physick, the anatomy of man's body, the diseases incident to the body of man: rules & receipts for the curing of them; also rules for sweating, bathing, conserving, and preserving, and the way to make cordial-waters. Also the principal rules of arithmetick, very plain and easie. The third, the countrey-man's guide to good husbandry: Rules for setting and planting of orchards, gardens, and woods; also rare receipts for curing diseases in horses, sheep, cows, and oxen; also an almanack for ever, and other variety of inventions very profitable and advantagious.*
London: printed for Tho. Passinger, 1679.
STC II S5926A
Subject: Almanacs, astrology, household encyclopedias, popular medicine.
Case/B/8635/.84

## Stubbe, Henry [or Stubbs], 1632–1676.

*The Indian nectar, or a discourse concerning chocolata: wherein the nature of the cacao-nut, and the other ingredients of that composition, is examined, and stated according to the judgment and experience of the Indians, and Spanish writers, who lived in the Indies, and others; with sundry additional observations made in England: The ways of compounding and preparing chocolata are enquired into; its effects, as to its alimental and venereal quality, as well as medicinal (especially in hypochondriacal melancholy) are fully related. Together with a spagyrical analysis of the cacao-nut, performed by that excellent chymist, Monsieur le Febure.*
London: printed by J. C. for Andrew Crook, 1662.
STC II S6049
Subject: Chocolate, diet, spagyric medicine.
Ayer/792.5/C5/S8/1662

## Stubbs, Henry.

*Legends no histories; or A specimen of some animadversions upon the history of the Royal Society. Wherein, besides the several errors against common literature, sundry mistakes about the making of salt-petre and gun-powder are detected, and rectified; whereunto are added two discourses, one of Pietro Sardi, an another of Nicolas Tartaglia relating to that subject. Translated out of the Italian by Henry Stubbe, physician at Warwick.*
London, 1670.
STC II S6053
Subject: Chemistry, pyrotechnics.
Case/U/4966/.852

## [Stubbs, Henry]

*Lex talionis; sive vindiciae pharmacoporum: or A short reply to Dr. Merrett's book; and others written against apothecaries: wherein may be discovered the fravds and abvses committed by doctors professing and practicing pharmacy.*
London: printed and are to be sold by Moses Pitt, 1670.
STC II S6055
Subject: Materia medica, medicine.
Case/I/045/.88/no./3

Binders title: *Tracts on physick, 1665–70.*

## Sturm, Johann Christoph, 1635–1703.

*Cometa nuperus an, et qvae, mala terris aut illaturus ipsemet influxu physico, aut aliunde justo dei judicio inferenda portendere saltem & praesignificare, credendus sit? . . . respondendo publicè tuebitur Wolffg. Lvdovicvs Andreae.*
Altdorf: Literis Schönnerstaedtianis, 1681.
Subject: Influence of comets.
Case/B/863/.84

## Sturm, Leonhard Christoph, 1669–1719.

*Architectura civili-militaris. Oder: wollständige anweisung staat-thore brucken, zeug-häuser, lasenmatten, und andere souterrains der wälle, casernen, baraquen, corps de gardes, und proviant-hauser behörig anzugeben.*
Augsburg: printed by Peter Detleffsen for Jeremiah Wolff, 1719.
Subject: Civil and military architecture, bridge building, engineering.
W/2/.853

## Sturm, Leonhard Christoph.

*Freundlicher wett-streit der Französischen Holländischen und Teutschen kriegs-bau-kunst worinnen die befestigungs-manier des hrn. von Vauban an neu-Breisach . . . und zweyerley vorstellungen der von L. C. Sturm publicirten. und nach des weit-berühmten hrn. George Kimplers maximen eingerichteten manier.*
Augsburg: printed by Peter Detleffsen for Jeremiah Wolff, 1718.
Subject: Arithmetic, surveying.
fU/2647/.853

## Sturm, Leonhard Christoph.

*Mathematischer beweiss von dem heil Abendmahl das I. Die worte der enfessung nie recht aus dem Griechischen übersetzet worden. II. An der art wie es von den Lutheranern gehalten wird.*
Frankfurt & Leipzig, 1714.
Subject: Mathematics, witchcraft.
3A/180

BOUND WITH several tracts by Jacob Brill (n.a.), Fabricius, Johann Albert, q.v., and *Promotoris edlen ritters von orthoptera theosophische bedancken von der nacht der finsterniss oder von der gewalt des teufels in der lufft. Aus dem göttlich-und natürlich-magischen central-licht den kindern der weissheit vorgestellet.* 1709.

## Sturm, Leonhard Christoph.

*Le veritable Vauban se montrant au lieu du faux vauban, qui a couru jusqu' ici par le monde, & enseignant par le moien d'une arithmetique & d'une geometrie courte, & aisée, non seulement les regles pour tracer proprement cette maniere célebre de fortifier, mais aussi ses maximes fondamentales et plusieurs autres régles utiles qu'on ya ajoûtées.*
The Hague: Nicolas Wilt, 1708.
Subject: Arithmetic, geometry, military science.
U/26/.8546

1: "Abregé d'arithmetique qu'un ecolier diligent peut aprendre en deux mois" (1st 45 pp.); 2: "De la pratique geometrique."

## Sturm, Leonhard Christoph.

*Vollständige anweisung wasser-kunst wasserleitungen brunnen und listernen wohl anzugeben worinnen Nic. Boldmans text nach lib IV cap X angeführet und durch anmerckungen erkläret. Hernach durch mehreren zufass vermehret. Auch zugleich was goldmaren von mühlen.*
Augsburg: printed by Peter Detleffsen for Jeremiah Wolff, 1720.
Subject: Engineering, water power.
Case/Wing/fZ/30535/.848

## Sturm, Leonhard Christoph.

*Der wahre Vauban, oder der von den Teutschen und Holländern verbesserte Französische ingenieur, vorinnen I. die arithmetic, II. die geometrie, III. die off- und deffensiv-kriegs-bau-kunst, nach den grundsäzen des berühmten herrn von Vauban.*
Nuremberg: Peter Conr. Monath, 1737.
Subject: Arithmetic, geometry, military science.
U/26/.85465

## Sturmy, Samuel, 1633–1669.

*The mariners magazine, stor'd with these mathematical arts: The rudiments of navigation and geometry. The making and use of mathematical instruments. The doctrine of triangles, plain and spherical. The art of navigation, by plain-chart, Mercator's-chart, and the arch of a great circle. The art of surveying, gauging, and measuring. Gunnery and artificial fire-works. The rudiments of astronomy. The art of dialling. Also with tables of logarithms, and tables of the suns declination; of the latitude and longitude, right ascension, and declination of the most notable fixed stars; of the latitude and longitude of places; of meridional parts.*
London: printed by Anne Godbid for William Fisher, 1679.
STC II S6097
Subject: Astronomy, mathematics, navigation, surveying.
Ayer/*f8.9/N2/S936/1679

2d ed. "revised and carefully corrected by John Colson."

## Suarez, de Figueroa, Cristobal, fl. 1613.

*Plaza vniversal de todas ciencias, y artes, parte tradvzida de Toscana, y parte compuesta por el doctor Christoual Suarez de Figueroa.*
Perpignan: Luys Roure, 1630.
Subject: Anatomy, astronomy, botany.
sc 116

Leaf 90 (Disocvrso [*sic*] 22): "De los simpicistas y herbolarios"; leaf 144 (Discvrso 36): "De los anatomistas"; leaf 191v (Discvrso 39): "De los astronomos y astrologos."

## Summenhart, Konrad, d. 1501?

*Conradi Summenhart commentaria in summa physice Alberti Magni.*
[Hagenau: Henricus Gran, 1507.]
Subject: Elements, liquids and solids, metals, natural history.
Case/B/181/.A334

One of the Albertus Magnus spuriae. MS notes on front flyleaf (recto and verso), on title page, and MS marginalia passim (in several hands).

## [Swadlin, Thomas] 1600–1670.

*Divinity no enemy to astrology. Intended to have been delivered in a sermon to the students in that art, but prevented by the sickness of the Author T. S. D.D.*
London: printed by J. G. for Nathaniel Brooke, 1653.
STC II S6215
Subject: Astrology.
Case/C/99/.266

Binders title: *English sermons.* "To the reader. . . . It is astrology I seeke to maintaine, and the question is not whether jugling be lawfull . . . whether witchcraft be lawfull, whether calculation of princes nativities be lawfull . . . but whether astrology is lawfull."

## Swan, John, fl. 1635.

*Specvlvm mundi, or a glass representing the face of the world; shewing both that it did begin, and must also end: . . . whereunto is joyned an hexameron, or a serious discourse of the causes, continuance, and qualities of things in nature; occasioned as a matter pertinent to the work done in the six dayes of the worlds creation.*
Cambridge: printed by the printers to the Universitie of Cambridge, 1635.
STC 23516
Subject: Natural history.
Case/C/257/.85

## Swan, John.

*Specvlum mvndi.* [Another edition]
Cambridge: printed by Roger Daniel, for Iohn Williams, 1642.
STC II S6238
Subject: Natural history.
Case/C/257/.852

2d ed. enlarged. Engraved title page gives 1643 as date of publication and omits Williams's name from imprint.

## Swift, Jonathan, 1667–1745.

*The right of precedence between physicians and civilians enquir'd into.*
Dublin: printed 1720; London: reprinted for J. Roberts.
Subject: Medicine compared to law.
Case/K/0451/.85

"Physick is as old as the occasion of it; as old, indeed, within a few Days as Mankind, which can by no means be said of the other (in comparison) upstart Profession" (p. 25); "But I am weary of proving so plain a Point. To me it is clear beyond contradiction that the Antiquity and Dignity of Physick do give it the precedence of the Civil Law and its Friend [common law]" (p. 29).

## Sydenham, Thomas, 1624–1689.

*The whole works of that excellent practical physician, Dr. Thomas Sydenham: wherein not only the history and cures of acute diseases are treated of, after a new and accurate method; but also the shortest and fastest way of curing most chronical diseases.*

London: printed for R. Ware, R. and B. Wellington, etc., 1740.

Subject: Disease, medicine.

oR/128.7/.S8813/1740

11th ed. Corrected from the original by John Pechey, M.D. Deals with epidemic diseases by year.

## Sydenham, Thomas, 1624–1689.

*Praxis medica. The practice of physick: or, Dr. Sydenham's processus integri . . . containing the names, places, signs, causes, prognosticks, and cures, of all the most usual and popular diseases afflicting the bodies of human kind, according to the most approved modes of practice. Among which you have the pathology, and various methods of curing a clap, or virulent running of the veins, and the French pox, with all their attendant symptoms, beyond whatever was yet publish'd on this subject by any other author ancient or modern, since the disease first appeared in the world, to this day.*

London: printed for J. Knapton, etc., 1716.

Subject: Materia medica, medicine.

oR/128.7/.S8713/1716

In 3 books. 3d ed. enlarged throughout by William Salmon.

## Sykes, Arthur Ashley 1684?–1756.

*A defence of the dissertation on the eclipse mentioned by Phlegon: wherein is further shewn, that that eclipse had no relation to the darkness which happened at our saviour's passion.*

London: James, John, and Paul Knapton, 1733.

Subject: Astronomy, the rational (science) and the miraculous (religion).

C/588/.8581

## Syrianus [Syriani Philoxeni] ca. 380–450.

*Syriani antiqvissimi interpretis in II. XII. et XIII Aristotelis libros metaphysices commentarius, a Hieronymo Bagolino, praestantissimo philosopho, Latinate donatvs.*

[Venice] in Academia Veneta, 1558.

Subject: Metaphysics.

Wing/ZP/535/.A/3636

# T

## Tachard, Gui, 1651–1712.

*Voyage de Siam des pere Jesvites envoyés par le roy, aux Indes & à la Chine. Avec levrs observations astronomiques, & leurs remarques de physique, de géographie, d'hydrographie, & d'histoire.*
Amsterdam: Pierre Mortier, 1687.
Subject: Astrology, astronomy, voyages.
Greenlee/4850/F8/T11/1687

Bk. 5, p. "139" [i.e., 239]: astronomical observations.

## [Tachard, Gui]

*A relation of the voyage to Siam. Performed by six Jesuits, sent by the French king to the Indies and China, in the year 1685. With their astrological observations, and their remarks of natural philosophy, geography, hydrography, and history.*
London: printed by T. B. for J. Robinson and A. Churchil, 1688.
STC II S95A
Subject: Astrology, astronomy, voyages.
Case/Eames/G/688/.859

## Tachard, Gui.

*Second voyage du pere Tachard et des Jesuites envoyez par le roy au royaume de Siam. Contenant diverses remarques d'histoire, de physique, de geographie, & d'astronomie.*
Paris: Daniel Horthemels, 1689.
Subject: Astronomy, natural history, voyages.
Case/G/688/.86

## Tagliacozzi, Gasparo, 1545–1599.

*Gasparis Taliacotii Bononiensis . . . De curtorum chirurgia per insitionem in quibus ea omnia, quae ad huius chirurgiae, narium scilicet, aurium, ac labiorum per insitionem restaurandorum cum theoricen, tum practicen pertinere videbantur, clarissima methodo cumulatissimè declarantur. Additis cutis traducis instrumentorum omnium, atque deligationum iconibus, & tabulis.*
Venice: Gaspar Bindonus junior, 1597.
NLM 4310
Subject: Plastic surgery, skin grafting.
UNCATALOGUED

## Tagliente, Giovanni Antonio, 16th cent.

*Libro de abacho, ilquale isegna fare ogni ragione mercantile: & pertegare le terre con larte della geometria & altre nobilissime ragione straordinarie con la tariffa come respondeno li pesi et monede de molte cittade et paesi con la inclita citta di Vinegia. Ylquale libro se chiama thesoro vniuersale.*
Venice: heirs of Giovanni Padouano, 1557.
Subject: Commercial arithmetic, geometry.
Wing/ZP/535/.P133

## Taillepied, Noel, 1540–1589.

*Traicté de l'apparition des esprits. A sçauoir, des ames separees, fantosmes, prodiges, & autres accidens merueilleux, qui precedent quelquefois la mort des grands personnages, ou signifient changement de la chose publique.*
Paris: Franç. Targa, 1627.
Subject: Prodigies, prognostication, supernatural.
B/893/.861

## Tanner, Robert, fl. 1583–1592.

*A mirror for mathematiques: A golden gem for geometricians: A sure safety for saylers, and an auncient antiquary for astronomers and astrologians. Contayning also an order howe to make an astronomicall instrument, called the astrolab, vvith the vse thereof. Also a playne and most easie instruction for the erection of a figure for the 12 houses of the heauens.*
London: printed by J[ohn] C[harlewood], 1587.
STC 23674
Subject: Astrology, astronomy, instrumentation, mathematics.
Ayer/*8.9/M2/T2/1587

## Tartaglia, Niccolò, 1500–1557.

*Opere del famosissimo Nicolo Tartaglia cioè quesiti, trauagliata inuentione, noua scientia, ragionamenti sopra Archimede. Nelle qvali copiosamente si spiega. . . . E varie maniere arificiose, di cauare ogni gran vasello affondato, & fabricar due vase co i quali si possa descendere nel fondo del mare, & à suo piacere ritornar di sopra.*
Venice: at the sign of the lion, 1606.
Subject: Inventions, mathematics, submarine.
Case/L/0/.862

Included are (1) "Regola generale da suleuare con ragione, e misura no solamente ogni affondata naue: ma vna torre solida di metallo. . . . Insieme con vn artificioso modo di poter andare, & stare per longo tempo sotto acqua, a ricercare le materie affondate & in loco profundo"; and (2) "Ragionamenti di Nicolò Tartaglia sopra la sua travagliata inventione. Nelli quali dechiara volgarmente quel libro di Archimede siracusano intitolato."

## Tartaglia, Niccolò.

*La noua scientia de Nicolo Tartaglia con vna gionta al terzo libro.*
[Venice, 1558.]
Subject: Geometry, physics.
Case/L/35/.862

## Tartaglia, Niccolò.

*Quesiti et inventioni diverse.*
[N.p., n.d.]
Subject: Chemistry, mechanics, physics.
U/O/.862

Title page wanting. First published in Venice, 1546. Libro terzo: sopra del salnitrio et delle varie compositioni. Libro settimo: sopra gli principii delle questioni mechanice di Aristotile (leaf 37).

## Tartaglia, Niccolò.

*Three bookes of colloqvies concerning the arte of shooting in great and small peeces of artillerie, variable ranges, measure, and waight of leaden, yron, and marble stone pellets, mineral saltpeeter, gunpowder of diuers sortes, and the cause why some sortes of gunpowder are corned, and some sortes of gunpowder are not corned: written in Italian by Nicholas Tartaglia. . . . And now translated into English by Cyprian Lvcar. Also the said Cyprian Lvcar hath annexed vnto the same three bookes of colloquies a treatise named Lvcar Appendix collected by him out of diuers authors in diuers languages to shew vnto the reader the properties, office, & dutie of a gunner, and to teach him to make and refine artificial saltpeeter . . . to measure altitudes, longitudes, latitudes, and profundities.*
[London: printed by Thomas Dawson for Iohn Harrison the elder, 1588.]
STC 23689
Subject: Chemistry, pyrotechnics, surveying.
Case/fU/444/.86

## Taust, Johann Gottfried.

*Cometa redivivus, das ist der aus der aschen viel enstetzlicher als zuvor hervor flammende und aufs neue sich unserm geschichte praesentirende unglücks prophete oder der nach gemeiner art genannte comet und schwantz stern welcher seinen curs und lauff geändert und nach dem er unter der sonnen strahlen 3.*
Halle: printed by heirs of Christoph Salfeld for Simon Johann Hübner, 1681.
Subject: Comets.
Case/B/863/.863

## Taust, Johann Gottfried.

*Der von abend gegen morgen lauffende unglücks prophete oder nach gemeiner art benahmte comet oder schwantz-stern, welcher abermahls dieses 1680sten jahres von 6 November an bisz auf den 24 hujus leuchtend sich hat sehen lassen, nebst andern bey dessen leuchtung vorgegangenen himel-wundern.*
Halle: Printed by heirs of Christoff Salfeld for Simon Johan Hübner, 1681.
Subject: Comets.
Case/B/863/.8637

## Taxil, Jean.

*L'astrologie et physiognomie en levr splendevr.*
Tournon: R. Reynaud, 1614.
Subject: Astrology, medicine, physiology.
Case/B/8635/.864

## Telin, Guillaume.

*Bref sommaire des sept vertus, sept ars liberault, sept ars de poesie, sept ars mechaniques, des philosophies, des quinze ars magicques. La louenge de musique. Plusieurs bonnes raisons a confondre les Juifz qui nyent laduenement nostre seigneur Jesuchrist. Les dictz et bonnes sentences des philosophes: auec les noms des premiers inuenteurs de toutes choses admirables & dignes de scauoir faict par Guillaume Telin de la ville de Lusset en Auuergne.*
Paris: printed by Nicolas Cousteau for Galliot du Pre, 1533.
Subject: Encyclopedia of liberal arts.
Case/3A/702

Includes astrology, chiromancy, geometry, medical aphorisms of Galen, Hermes, Hippocrates, natural magic, navigation, among numerous others.

## Temple, Sir William, 1628–1699.

*Works.*
London: printed for J. Round [etc.], 1740.
Subject: Ancients vs. Moderns, disease.
Case/6A/371

In 2 vols. Pt. 1, no. 6: "An essay upon the cure of the gout"; pt. 2: on ancient and modern learning; pt. 3, no. 2: on health and long life; with other essays on gardening and domestication of plants. Each part has separate title page, with slightly different imprint.

## Temple, Sir William.

*Les oeuvres mêlées de Monsievr le chevalier Temple.*
Utrecht: Antoine Schouten, 1694 [1693].
Subject: Ancients vs. Moderns, disease.
Y/12/.T238

In 2 pts. Pt. 1: "Essai du Moxa contre la goute: Of all the ailments to which the dreadful condition of age contributes, especially in this northern land, I must say there is none more common, nor more pernicious than the gout." In pt. 2: "Du scavoir des anciens & des modernes."

## Temple, Sir William.

*Miscellanea.*
London: printed by A. M. and R. R. for Edw. Gellibrand, 1680.
STC II T646
Subject: Disease.
Case/Y/145/.T24

Six essays "by a person of honour." No. 5: upon the excesses of grief; no. 7: "An essay upon the gout by Moxa."

## Temple, Sir William.

*Miscellanea.*
London: printed by J. C. for Edw. Gellibrand, 1681, 1690.
Subject: Ancients vs. Moderns, disease.
STC T647
Case/Y/145/.T241

2 pts. in 1. 2d ed. corrected and augmented. Pt. 2, with "An essay upon ancient and modern learning," printed by J. R. for Ri and Ra Simpson, 1690.

## Temple, Sir William.

*Miscellanea.*
London, [printed & sold by the booksellers of London & Westminster] 1709.

Subject: Ancients vs. Moderns, disease.
Y/145/.T2415

The first part, 5th ed. P.54: "An essay upon the cure of the gout."

### Temple, Sir William.

*Miscellanea.*
London: Printed for Benjamin Tooke, 1701.
Subject: Ancients vs. moderns, health.
Case/Y/145/.T242

3d pt. Published by Jonathan Swift. Has an essay "upon health and long life" and "A defence of the essay upon antient and modern learning."

* * *

*Theatrum sympatheticum, in quo sympathie actiones variae, singulares & admirandae tàm macro- quam microcosmicae exhibentur, & mechanicè, physicè, mathematicè, chimicè & medicè, occasione pulveris sympathetici, ita quidem elucidantur, ut illarum agendi vis & modus, sine equalitatum occultarum, animaeve mundi, aut spiritus astralis magnive magnalis, vel aliorum commentariorum subsidio ad oculum pateat. Opusculum lectu jucundam & utilissimum; Digbaei, Papinii, Helmontii, aliorumque recentiorum scriptorum prolata exhibens & trutinans, atque ipsius pulveris sympathetici germanam & optimam descriptionem simul exponens.*
Nuremberg: heirs of Joh. And. & Wolffg. Jun. Endter, 1660.
Subject: Sympathetic medicine, sympathy and antipathy.
B/529/.231

Also in this vol.: "Nicolaus Papinus Blaesensis, M.D. De pulvere sympathico" (pp. 1[2]53–335); "Erycius Mohyus Eburonis Pvlvis sympatheticvs qvo vulnera sanantur absque medicamenti ad partem affectam applicatione & superstitione. Galenicarum, Aristotelicarumque rationum cribro eventilatus" [pp. 336–(381)].

### Themistus, fl. 384.

*Themistii Evphradae in libros quindecim Aristotelis, commentaria.*
Paris: Simon Colinaeus, 1528.
Subject: Dreams, soul.
Case/6A/41

Title page and an introductory page wanting and supplied in facsimile. MS marginalia passim. Includes "De auscultatione natural"; "De anima"; "De somno et vigilia"; "De diuinatione in somno. Hermolao Barbaro interprete. Ad haec, Alexandri Aphrodisiensis in libros de anima, commentaria."

### Themistus.

[Author's name & title in Greek] *Omnia Themistii opera, hoc est paraphrases, et orationes. Alexandri Aphrodisiensis libri dvo de anima, et de fato vnvs.*
[Venice: heirs of Aldus Manutius & Andreas Asulanus, 1534]
Subject: Dreams, soul.
Wing/fZP/535/.A3623

Except for table of contents (in Greek and Latin) and dedication (in Latin), text is in Greek.

### Theodorus, Jacobus, d. 1590.

*D. Iacobi Theodori Tabernaemontani neu vollkommen kräuterbuch, darinnen vber 3000 kräuter, mit schönen und kunstlichen figuren, auch deren underscheid und würkung samt ihren namen in mancherley sprachen, beschrieben: desgleichen auch, wie die selbige in allerhand kranckheiten, beyde der menschen und des biehs, sollen angewendet und gebraucht werden, angezeigt wirb. Erstlichen durch Casparum Bauhinum.*
Basel: Joh. Ludwig König and Johann Brandmyller, 1687.
Subject: Botany, herbals.
Case/7A/2

In 2 vols. Title page of vol. 2 (which is in 2 parts): *Neu & vollkommen kräuter buch mit schönen künstlichen und leblichen figuren und conterfeyten, allerhand vortrefflichster und fürnemmer, so vol fremder als einheimischer gewachs, kräuter blumen, stauden, hecken, und bäumen, auch köstlicher ausländischer wurzeln, rinden, früchten, &c. . . . wundartzen—apothekern . . . durch langwirige und gewisse erfahrung beschrieben durch Nicolavm Bravn . . .* [and Caspar and Hieronymus Bauhin].

### Theodosius of Tripolis, no later than 1st cent. B.C.?

*Sphaericorvm libri tres, nvnqvam antehac Graece excusi. Idem Latinè redditi per Ioannem Penam.*
Paris: Andreas Wechel, 1558.
Subject: Spherical geometry.
Wing/ZP/539/.W413

Editio princeps of original Greek text followed by Latin translation.

### Theophrastus of Eresos, d. 287 B.C.

*De historia et causis plantarum.*
Venice: Aldus Manutius, the elder, and Andreas Torresanus, de Asula,
1513.
Subject: Botany.
Case/fY/642/.A8479

BOUND WITH Aristotle, *Habentvr hoc volvmine Theodoro Gaza interprete*, and Codrus, qq.v.

* * *

*Thesaurus exorcismorum atque conjurationum terribilium, potentissimorum, efficacissimorum; cum practica probatissima: quibus spiritus maligni, daemones maleficiaque omnia de corporibus humanis obsessis, tanquam flagellis fustibusque fugantur, expelluntur, doctrinis refertissimus atque uberrimus.*
Cologne: Lazarus Zetzner, 1608.
Subject: Exorcism, demonology, witchcraft.
Case/B/88/.87

This vol. contains the follwing 6 tracts: *F. Valerii Polydori Patavini Practica exorcistarum ad doemones & maleficia de Christi fidelibus pellendum* (pt. 1); *Ejusdem, Dispersio daemonum quae est practicae exorcistarum* (pt. 2); *F. Hieronymi*

*Mengi Vitellianensis, Flagellum daemonum; Ejusdem Fustis daemonum; F. Zachariae Vicecomitis, Complementum artis exorcisticae; Petri Antonii Stampae, Fuga Satanae.*

### Thévenot, Jean de, 1633–1667.

*The travels of Monsieur de Thevenot into the Levant.*
London: printed by H. Clark for H. Faithorne [etc.], 1687.
Subject: Discovery, travel.
STC II T887
Case/fG/131/.87

In 3 pts.: Turkey, Persia, East Indies.

### Thevet, André, 1502–1590.

*Cosmographie de Levant, par F. André Theuet d'Angoulesme.*
Lyon: Ian de Tovrnes, and Gvil Gazeau, 1554.
Subject: Cosmography.
Ayer/*109/T4/1557

Includes descriptions of Greece, Egypt, Slavonia, Sicily, Malta, Corsica, Judea. BOUND WITH his *Les singvlaritez de la France antarctiqve*, q.v.

### Thevet, André.

*La cosmographie vniverselle d'André Thevet cosmographe dv roy illvstree de diverses figvres des choses plvs remarquables vevës par l'auteur, & incogneuës de noz anciens & modernes.*
Paris: Pierre L'Huilier, 1575.
Subject: Descriptive geography.
Case/G/117/.88

In 2 vols.Vol. 1: Africa, Asia; vol. 2: Europe, Near East, Yucatan, Peru.

### Thevet, André.

*The new found vvorlde, or antarctike, wherein is contained wonderful and strange things, as well of humaine creatures, as beastes, fishes, foules, and serpents, trees, plants, mines of golde and siluer: garnished with many learned authorities, trauailed and written in the French tong, by that excellent and learned man, master Andrevve Thevet. And novv nevvly translated into Englishe, wherein is reformed the errours of the auncient cosmographers.*
London: Henry Bynneman for Thomas Hacket, 1568.
STC 23950
Subject: Exploration, natural history, travel.
Ayer/*109/T4/1568

Winsor, vol. 4, p. 31. "Of our ariuall to Fraunce antartike, otherwise named America, to the place named Caape Defria." (chap. 24).

### Thevet, André.

*Les singvlaritez de la France antarctiqve, autrement nommée Amerique: & de plusieurs terres & isles decouuertes de nostre temps.*
Paris: heirs of Maurice de la Porte, 1557.
Subject: Discovery, exploration, travel.
Ayer/*109/T4/1557

BOUND WITH his *Cosmographie de Levant*, q.v.

### Thibault, Girard, 17th cent.

*Academie de l'espeé.*
[Leyden: Elzevier] 1628.
Subject: Anatomy, emblems, mathematics of fencing, proportion.
Wing/+ZP/646/.E51874

Bk. 1: "Habitude & cognoissance du corps." At beginning of bk. 2: "emblema tabula" 1 and 11 (proportions & emblems & symbols).

### Thienis, Gaietanus de, Canon of Padua, 1387–1465.

*Expositio super libros de anima Aristotelis.*
Vicenza: Henricus de Sancto Ursio, Zenus, 1486.
C [R] 617 var^ BMC (XV)VII:1046^ Pr. 7168^ Goff G-26
Subject: Soul.
Inc./f7168/Inc./5110.5/Inc./4785.5/Inc./5017.5

Contains his *De sensu agente*. BOUND WITH Burley, Walter;Thomas Aquinas; Gratia Dei de Esculanus; and Joannes de Janduno, qq.v.

### Thomas Aquinas, Saint, 1225?–1274.

*Opuscula.*
Milan: Benignus & Joannes Antonius de Honate, 1488.
HC 1540^ BMC (XV)VI:742^ Pr. 5908^ Pell. 1092^ Still. T 234^ Goff T-259
Subject: Natural science.
Inc./f5908

Includes (opus 14) "Diuinum opus contra Aueroystas dicentes vnum esse intellectum omnium hominum"; (opus 34) "De occultis operationibus nature"; (opus 35) "De motu cordis." MS marginalia.

### Thomas Aquinas, Saint.

*Commentarius in librum Aristotelis de anima.*
Pavia: Martin de Lavalle, 1488.
H 1521(1)^ BMC (XV) VII:1012^ Pell. 1083^ Still. T216^ Goff T-239
Subject: Psychology, soul.
Inc./f7093/.7

MS marginalia on scattered leaves. Bound in vellum music MS.

### Thomas Aquinas, Saint.

*Commentaria diui Thome Aquinatis predicatorum in libri perihermenias.*
Venice: Guilielmus Tridinensis de Monteserato [Guilelmus Anima Mia Tridinensis] 1489.
BMC (XV)V:411^ Goff T-254
Subject: Philosophy, syllogistic reasoning.
Inc./f7168/Inc./5110.5/Inc./4785.5/Inc./5017.5

With his *Commentaria . . . predicatorum in libros posteriorum Aristotelis; Commentaria in libros posteriorum perihermenias Aristotelis;* and *Expositio super libris de interpretatione et posteriorum fallaciae.* BOUND WITH Burley, Walter; Gratia Dei de Esculanus; Joannes de Janduno; and Thienis, Gaetano de, qq.v.

**Thomas Aquinas, Saint.**

*De ente et essentia.*
[Venice] Otinus de Luna, 1496.
H-C 1504^ BMC (XV)V:567–68^ Pr. 5599B^ Still T 264^ Goff T-290
Subject: Form and matter, philosophy.
Inc./f5599B

MS marginalia.

**Thomas Aquinas, Saint.**

*Expositio super libris de interpretatione et posteriorum fallaciae.*
Venice: Joannes de Colonia & Johannes Manthem, 1477.
H* 1497 + HC *1496^ BMC (XV)V:228^ Pr. 4314^ Goff T-252
Subject: Philosophy.
Inc./4314

MS marginalia.

**Thomas Aquinas, Saint.**

*Quaternarius.*
Paris: George Mittelhus, 1493.
H*1395^ Pell. 991^ Goff T-340
Subject: Philosophy, soul.
Inc./8115.5

**Thomas Aquinas, Saint.**

*Summa theologica.*
[Strassburg: Johann Mentelin, not after 1463]
H-C *1454^ BMC (XV)I:51 (var?)^ Pr. 199^ Goff T-208
Subject: Integration of divine and human knowledge, natural science, natural theology.
Inc./+ 199

2d pt.

**Thomas Aquinas, Saint.**

*Summa theologica pars secunda.*
Mainz: Peter Schöffer, 1467.
H* 1459^ BMC (XV)I:24^ Still. T 188^ Pell. 1049^ Pr. 83^ Goff T-209
Subject: Integration of divine and human knowledge, natural science, natural theology.
Inc./+ 83

**Thomas, Corbinianus, 1694–1767.**

*Mercvrii philosophici firmamentvm firmianvm descriptionem et vsum globi artificialis coelestis, ac asterismos ejusdem ad ineuntem [sic] annum 1730.*
Frankfurt & Leipzig, 1730.
Subject: Astronomy.
Wing/ZP/7471/.30

**Thomasius, Christian, 1655–1728.**

*Kurze lehr-gäze von dem laster der zauberey.*
Frankfurt & Leipzig, 1717.
Subject: Witchcraft.
Case/B/88/.872

*Mit dessen eigenen vertheidigung vermehret worbey Johann Kleins, J.U.D. . . . juristiche untersuchung was von der hexen bekäntniss zu halten; dass solche aus schändlichem beyschlaff mit dem teuffel kinder gezeuget.*

**[Thou, Jacques Auguste de] 1553–1617.**

*Hieracosophioy sive de re accipitraria libri III.*
Paris: Mamertus Pattison for Robert Stephan, 1587.
Subject: Falconry, hawking, natural history.
Case/Y/682/.T 39

2d ed. In verse. Has explanatory page (107), which may relate toanimal husbandry, natural history, veterinary studies.

**Thyraeus, Petrus Pierre.**

*Daemoniaci, hoc est: de obsessis a spiritibvs daemoniorvm hominibvs, liber vnvs. In qvo daemonvm obsidentivm conditio: Obsessorum hominum status: rationes & modi, quibus ab obsessis daemones exiguntur; causae item tum difficilis exitus ipsorum, tum signorum quae exituri relinquunt: Loca denique, quò egressi tendunt, & his similia, discutiuntur & explicantur, denuò omnia repurgata & aucta.*
Lyon: Ioannes Pillehotte, 1603.
Subject: Demonology, possession.
Case/B/88/.875

In 4 pts. 1: "Dispvtationis in qva daemonvm obsidentivm conditio, operationes, atque in obsessos potestas"; 2: "Qvi sint obsessi"; 3: "In qva de varia potestate . . . et multiplicibus modis daemones ex humanis corporibus eiiciendi"; 4: "De daemonum ex hominum corporib. egressu, egrediendi difficultate, & crudelitatis, signis quae in egressu, relinquunt."

**Tindal, Matthew, 1657–1733.**

*Christianity as old as the creation: or, the Gospel, a republication of the religion of nature.*
London, 1730.
Subject: Natural theology, reason vs. revelation.
C/525/.877

**Tindal, Matthew.**

*Christianity as old as the creation.* [Another edition]
London, 1732.
Subject: Natural theology, reason vs. revelation.
Case/4A/3146

Chap. 13: "The bulk of mankind, by their reason, must be able to distinguish between religion and superstition" (p. 209). Chap. 14: The law of nature.

**Titelmans, Franciscus, 1502–1537.**

*Philosophiae natvralis compendivm, libri XII. De consideratione rerum naturalium, earumque ad suum creatorem reductione.*
Lyon: printed by Godefridvs and Marcellvs Beringvs bros. for Antonius Vincentivs, 1545.
Subject: Natural philosophy, natural science, metaphysics, physics.
Case/B/42/.876

Includes "De principiis rerum naturalium"; "De motu & accidentibus eius"; "De infinito, loco, uacuo, & tempore"; "De generatione & corruptione rerum naturalium"; "De meteorologicis impressionibus"; "De coelo & mundo."

### Toland, John 1670–1722.

*Adeisidaemon, sive Titus Livius a superstitione vindicatus.*
The Hague: Thomas Johnson, 1709.
Subject: Superstition, witchcraft.
Y/672/.L5387

Together with "Origines Judaicae, sive, Strabonis de Moyse et religione Judaica historia," which has some material on divination by dreams.

### Toland, John.

*Christianity not mysterious: or, a treatise shewing, that there is nothing in the gospel contrary to reason, nor above it: and that no Christian doctrine can be properly call'd a mystery.*
London: printed for Sam Buckley, 1696.
STC II T1763
Subject: Natural theology, reason vs. revelation.
C/525/.88

2d ed. enlarged.

### Toland, John.

*A collection of several pieces of Mr. John Toland.*
London: J. Peele, 1726.
Subject: Astronomy, Hermeticism, occult, supernatural.
Bonaparte/Coll./No/11,494

2 vols. Vol. 1: "Account of the Druids"; "Conjectura verosimilis de prima typographiae inventione" (p. 297); "De genere, loco, & tempore mortis Jordani Bruni Nolani" (p. 304); "An account of Jordano Bruno's book, of the infinite universe and innumerable worlds" (p. 316); vol. 2: "Physic without physicians" (p. 273).

### Toland, John.

*A collection of several pieces of Mr. John Toland.* [Another copy]
Y/145/.T57

### Toland, John.

*Letters to Serena.*
London: printed for Bernard Lintot, 1704.
Subject: Natural philosophy, soul.
B/245/.87

Sec. 5: Motion essential to matter.

### [Toland, John]

*Pantheisticon. Sive formula celebrandae sodalitatis Socraticae, in tres particulas divisa; quae pantheistarum, sive sodalium, continent 1, Mores et axiomata; 2, Numen et philosophiam: 3, Libertatem et non fallentem legem.*
Cosmopoli, [London] 1720.
Subject: Natural philosophy.
B/74/.88

Part 2 in dialogue form.

### Toland, John.

*The theological and philological works of the late Mr. John Toland, being a system of Jewish, Gentile, and Mahometan Christianity.*
London: printed by W. Mears, 1722.
Subject: Astronomy, natural theology.
C/515/.881

Possibly relevant treatises: no. 5: "Hypatia: or, the history of a most beautiful, most vertuous, most learned, and every way accomplish'd lady" [praising her skill and wisdom in astronomy], and "Clidophorus, or, of the exoteric and esoteric philosophy; that is, of the external and internal doctrine of the ancients."

### Tommai, Petrus, Ravennas, b. ca. 1448–1509.

*Foenix domini Petri Rauennatis memoriae magistri.*
Venice: Bernardinus de Choris de Cremona, 1492.
BMC (XV)VII:1511^ R 707^ Pr. 7420^ Still. P487^ Goff P-532
Subject: Mnemonics.
Inc./6579/.7

A pirated edition, attributed (by BMC) to Bazalerius de Bazaleriis, Bologna, 1492. 1st page: "Artificiosa memoria clarissimi iuris vtriusque doctoris & militis domini Petri Rauennati iura canonica ordinarie de sero legentis in celeberrimo gymnasio Patauino in hoc libello continentur."

### Tommai, Petrus, Ravennas.

*Artificiosa memoria clarissimi iuris vtriusque doctoris & militis domini Petri Rauennatis perquam facillime multa memoriter teneri & dici possunt.*
Erffurdt: Wolfgang Schenck, 1500.
H-C *13698^ BMC (XV)II:593^ Pr. 3110^ Still. P488^ Goff P-533
Subject: Mnemonics.
Inc./3110

MS marginalia.

### Topsell, Edward, d. 1638?

*The historie of fovre-footed beastes. Describing the true and liuely figure of euery beast, with a discourse of their seuerall names, conditions, kindes, vertues (both naturall and medicinall) countries of their breed, their loue and hate to mankinde, and the wonderfull worke of God in their creation, preseruation, and destruction. Necessary for all diuines and students, because the story of euery beast is amplified with narrations out of Scriptures, fathers, phylosophers, physitians, and poets: wherein are declared diuers hyerogliphicks, emblems, epigrams, and other good histories, collected out of all the volumes of Conradvs Gesner, and all other writers to this present day.*
London: printed by William Iaggard, 1607.
STC 24123
Subject: Natural history, zoology.
Case/O/500/.88

2 vols. in 1. Vol. 1: "Foure footed beasts"; vol. 2: *The historie of serpents. Or, the second booke of liuing creatures.* London:

printed by William Jaggard, 1608 [STC 24124]. Both parts have index, "Names of all the beastes contayned in this history in diuers languages" (including "Hebrue, Caldey, Arabian, Sarcen, Persian, Greek, Latin, Italian, Spanish, French, German, Illirian"). P. 595: woodcut of rhinoceros by Dürer.

## [Torre, Alfonso de] d. 1460.

*Sommario di tvtte le scienze, dal qvale si possono imparar molte cose appartenenti al uiuere humano, & alla cognition di Dio.*
Venice: Gabriel Giolito de' Ferrari, 1565.
Subject: Encyclopedias, magic, marvels, mathematics, supernatural.
Case/Y/722/.T637

"Della geometria" (p. 48); "Dell' aritmetica, de' svoi inuentori, utilità, & altri secreti" (p. 43); "Tratta delle arte magiche & della douinationi" (p. 139).

## Torreblanca Villalpando, Francisco, d. 1645.

*Epitome delictorum; sive, de magia: in qva aperta vel occvlta invocatio.*
Lyon: Joannis Antonij Huguetan, & partners, 1678.
Subject: Divination, magic, occult.
B/88/.882

Divination (judicial "astronomy," physiognomy, chiromancy, oracles, cabala, necromancy, "aurispicinia," dreams, prodigies, earthquakes, oracles) and operational magic.

## Torrentinus, Hermanus, 1450?–1520?

*Opusculum perutile.*
Deventer: Richardus Pafraet, ca. 1500.
C 5840^ Goff H-74
Subject: Etymology, taxonomy.
Inc./9044.5

Contains "De generibus nominum"; "De heteroclitis"; "De nominum significationibus."

## Tortelli, Giovanni, 1400–1466.

*Ioannis Tortellii Arretini commentariorvm grammaticorvm de orthographia dictionvm e graecis tractarvm.*
Venice: Nicolavs Ienson, 1471.
HC *15564^ BMC (XV)V:170–71^ Pr. 4081^ Still. T 360^ Goff T-395
Subject: Dictionary encyclopedias, etymology.
Inc./+4081

Some MS marginalia. Has tract, "Horlogium."

## Tortelli, Giovanni.

*Commentariorvm grammaticorvm de orthographia dictionvm e graecis tractarvm.*
Vicenza: Stephan Koblinger, 1479.
H-C *15566^ BMC (XV)VII: 1043^ Pr. 7160^ Still. T 32^ Goff T-397
Subject: Dictionary encyclopedias, etymology.
Inc./f7160

MS inscription of ownership on front flyleaf. See s.v., "astronomia, astrologia, horlogium."

## Tortelli, Giovanni.

*Ioannis Tortellii Arretini commentariorvm grammaticorvm de orthographia dictionvm e Graecis.*
Vicenza: Herman Liechtenstein, 1480.
H-C *15567^ BMC (XV) VII:1037–38^ Pr. 7158^ Still. T 363^ Goff T-398
Subject: Dictionary encyclopedias, etymology.
Inc./f7158

Gives definitions and pronunciation. See s.v., "Hippocrates, Horoscopes."

## Tortelli, Giovanni.

*Ioahannis Tortellii Arretini commentariorvm grammaticorvm de orthographia dictionvm e graecis.*
Venice: Herman Liechtenstein, 1484.
H-C *15569^ BMC (XV)V:357^ Pr. 4787^ Still. T 364^ Goff T-399
Subject: Dictionary encyclopedias, etymology.
Inc./f4787

Some MS marginalia.

## Tortelli, Giovanni.

*Ioannis Tortellii Arretini commentariorvm grammaticorvm de orthographia dictionvm e Graecis.*
Venice: Andreas de Paltasichis, 1488.
H-C *15571^ BMC (XV)V:355^ Pr. 4780^ Still. T 365^ Goff T-400
Subject: Dictionary encyclopedias, etymology.
Inc./f4780

## Tortelli, Giovanni.

*Ioannis Tortellii Arretini commentariorvm grammaticorvm de orthographia dictionvm e Graecis.*
Venice: Philippus Pincius, 1493.
H-C* 15572^ H 15577^ BMC (XV)V:494^ Pr.5296^ Still. T 366^ Goff T-401
Subject: Dictionary encyclopedias, etymology.
Inc./f5296

Ed. Pyrrhus Pincius.

## Tournefort, Joseph Pitton de, 1656–1708.

*Josephi Pitton Tournefort Aquisextiensis, doctoris medici Parisiensis, Academiae regiae scientiarum socii & in horto regio botanices professoris, institutiones rei herbariae.*
Paris: Typographia Regia, 1700.
Subject: Botany, herbal, plant classification and description.
N/405/.88

"Editio altera, Gallica longe auctior, quingentis circiter tabulis aeneis adornata." In 3 vols. Vol. 1 text consists of 22 plant classifications, divided into sections and subdivided into genera. BOUND WITH *Corollarium institutionum rei herbariae, in quo plantae 1356, munificentiâ Ludovici Magni in Orientalibus regionibus observatae recensentur, & ad genera sua revocantur.* Plant names in Latin and French. Colophon, vol. 1: "Curante Joanne Anisson, 1703." Vols. 2 and 3 consist of plates only. Colophon, vol. 2: Paris: Anisson, 1700. No colophon in vol. 3.

* * *

*Traitté de la svperstition, composé par Plvtarqve & traduit par Mr. Le Fèvre.*
Savmvr: Iean Iesnier, 1666.
Subject: Superstition.
Y/642/.P689

## [Treachard, John] 1662–1723.

*The natural history of superstition.*
[London] A. Baldwin, 1709.
Subject: Superstition, sympathy and antipathy.
B/57/.886

"There is a certain sympathy and antipathy in nature . . . so agreeable or contrary contexture of different bodies, as by a sort of natural mechanism do incline to or avoid another" (p. 26).

* * *

*The treasvrie of avncient and moderne times. Containing the learned collections, iudicious readings, and memorable obseruations: not onely diuine, morrall and phylosophicall. But also poeticall, martiall, politicall, historicall, astrologicall, &c. Translated out of that worthy Spanish gentleman, Pedro Mexio. And M. Francesco Sansouino, that famous Italian.*
London: printed by W[illiam] Iaggard, 1613.
STC 17936
Subject: Encyclopedias.
Case/fA/15/.88

In 2 vols. Vol. 2: [Greek word] *Containing the following books to the former treasvrie of avncient and moderne times.* Printed by Iaggard, 1619. STC 17936.5. STC attributes this work to Thomas Milles. Vol. 1 includes "Of Amber-Greece, from whence it is brought, &c." (bk. 3, chap. 19); "That brute beastes have instructed many men in sundry medicines for hurts received &c." (bk. 7. chap. 15); "Of many beasts and other creatures, that by instinct of nature, haue foreknowledge of things to come" (chap. 16); vol. 2: "Of the excellencie of the art of physicke" (bk. 8, pp. 755–85).

## Trew, Christoph Jacob, 1695–1769.

*Plantae selectae qvarvm imagines ad exemplaria natvralia Londini in hortis cvuriosorvm nvtrita. . . . occasione havd vvlgari pvblici vsvs ergo collegit et a tabvla prima ad septvagesimam secvndam nominibvs propriis notisqve illvstravit D. Christophorvs Iacobvs Trevv.*
[Nuremberg, 1750–53]
Subject: Botany, herbals.
Case/+W/765/.88

Hand colored plates. Plant names, description, history.

## Trithemius, Johannes, 1462–1516.

*Chronica. Ein an überauss lustig warhaffitg histori von der Franckeu [sic] ankunfft narung auffwachsung das sie nahendt herrn diss gantzen Europe warden durch lxiii künig biss auff Carolum magnum auff den das gantz Römisch imperium.*
[N.p. and n.d.]
Subject: Cosmography, geography.
Case/F/3912/.888

Leaf 46v: "In den namen der länder so gemert is worden nemlich in den buchstaben A.B.C.rc[?] nach ausweisung de folia auch nach den artickeln alpes hochgebirg rc.[?] fol. ccxci. volgt." See leaf 47v for mention of Ptolemy.

## Trithemius, Johannes.

*Polygraphiae libri sex.*
Oppenheim: Ioannis Haselberg, 1518.
Subject: Arcana, cabala, secret writing, word symbolism.
Wing/fZP/547/.H347

## Trithemius, Johannes.

*Polygraphiae libri sex. . . . Accessit clauis polygraphiae liber vnus, eodem authore. . . . Additae svnt etiam aliqvot locorvm explicationes.*
Cologne: Ioannes Birckmann & Werner Richwin, 1564.
Subject: Arcana, cabala, secret writing, word symbolism.
Wing/ZP/547/.B53

## Trithemius, Johannes.

*Polygraphie et vniverselle escriture caballistique. . . . Avec des tables & figures concernants l'effaict & l'intelligence de l'occulte escriture.*
Paris: Iaques Kerver, 1625.
Subject: Arcana, cabala, secret writing, word symbolism.
Case/Z/216/.88

Traduicte par Gabriel de Collagne.

## Trithemius, Johannes.

*De septem secvndeis, id est, intelligentijs siue spiritibus orbes post Deum mouentibus, reconditissimae scientiae & eruditionis libellus.*
Cologne: Ioannes Birckmann, 1567.
Subject: Magic, supernatural.
Case/B/88/.888

Biographies of angels; relation of angels and divinities to history; some astrology. Followed by letters, e.g., "Ioannes Tritemivs . . . Domino Ioanni Vuestenburgh . . . de tribus naturalis magiae principils [*sic*] sine quibus nihil in ipsa ad effectum produci potest" (p. 81).

## Trithemius, Johannes.

*Steganographia: Hoc est: Ars per occvltam scriptvram animi svi volvntatem absentibvs aperiendi certa.*
Darmstadt: Balthasar Aulaeandri for Ioannis Berner, 1621.
Subject: Occult, secret writing.
Case/Z/216/.882

BOUND WITH *Clavis steganographiae* and *Clauis generalis triplex in libros steganographicos*, both with the same imprint. MS notes on back flyleaf.

## Troili, Giulio, 1613–1685.

*Paradossi per pratticare la prospettiva senza saperla, fiori, per facilitare l'intelligenza, frvtti per non operare alla cieca.*

*Cognitioni necessarie à pittori, scultori, architetti, ed à qualunque si diletta di disegno.*
Bologna: Gioseffo Longhi, 1683.
Subject: Geometry, perspective.
Wing/fZP/635/.T845

In 3 parts: definitions and principles; plane surfaces; light and shadow.

* * *

*A true coppie of a prophesie which was found in [the] old ancient house of one Master Truswell, sometime recorder of a towne in Lincoln-shire. Which in all mens judgements was not unwritten these 300 yeares. And supposed to be seene still in a writing of parchment, at Stow. . . . Whereunto is added Mother Shipton's prophesies.*
London: printed for Henry Marsh, 1642.
STC II T2633
Subject: Astrology.
Case/B/867/.886

[Binders title] *Tracts on astrology.* With this are bound Melton, John; Holwell, John; and Ferrier, Oger [Auger] qq.v.

* * *

*A true and exact relation of the severall informations, examinations, and confessions of the late witches, arraigned and executed in the county of Essex. . . . Wherein the severall murthers, and devillish witchcrafts, committed on the bodies of men, women, and children, and divers cattell, are fully discovered.*
London: printed by M. S. for Henry Overton, and Benj. Allen, . . . 1645.
STC II F23
Subject: Witchcraft, confessions, examinations, and informations.
Case/B/8845/.89

Confessions of Elizabeth Clarke and others suspected as witches.

* * *

A *true & faithful relation of what passed for many yeers between Dr. John* Dee and some spirits.

See Casaubon, Meric.

* * *

*A true and impartial relation of the informations against three witches, . . . who were indicted, arraigned, and convicted at the assizes holden for the county of Devon at the castle of Exon, Aug. 14. 1682.*
London: printed by Freeman Collins, 1682.
STC II T2502
Subject: Witchcraft, confessions, examinations, and informations.
Case/B/8845/.891

Trials of Temperance Lloyd, Mary Trembles, Susanna Edwards.

## Tryon, Thomas, 1634–1703.

*A new method of educating children: or, rules and directions for the well ordering and governing them during their younger years. Shewing that they are capable at the age of three years, to be caused to learn languages, and most arts and sciences. Also, what methods is to be used by breeding women, and what diet is most proper for them and their children to prevent wind, vapours, convulsions, &c.*
London: printed for John Salisbury and J. Harris, 1695.
STC II T3190
Subject: Diet for pregnant women, pedagogy, prenatal care and influence.
Case/K/79/.891

## Tryon, Thomas.

*Some memoirs of the life of Mr. Tho. Tryon . . . written by himself.*
London: printed by T. Sowle, 1705.
Subject: Astrology, spagyric medicine.
Case/E/5/.T 782

Pages misnumbered throughout. Keith Thomas, in *Religion and the decline of magic* (London, 1971) refers to p. 24 (possibly p. 22 in this copy): "I cannot therefore hold it unlawful or vain to study astrology." See also sig. C6v about the spagyrical art.

## [Tryon, Thomas]

*A treatise of dreams and visions, wherein the causes natures and vses of nocturnal representations, and the communications both of good and evil angels, as also departed souls, to mankinde, are theosophically vnfolded; that is, according to the word of God, and the harmony of created beings. To which is added, a discourse of the causes, natures, and cure of phrensie, madness or distraction. By Philotheos Physiologus* [pseud.].
[N.p., 1689?]
STC II T3197
Subject: Dreams, humors, psychology.
Case/B/578/.89

Chap. 4: Causes of dreams (the planets, diet, medicine, visions from God). Relation of dreams to the 4 humors and complexions.

## Tryon, Thomas.

*The way to health, long life, and happiness: or, a discourse of temperance, and the particular nature of all things requisite for the life of man; as, all sorts of meats, drinks, air, exercise, &c. With special directions how to use each of them to the best advantage of the body and mind. Shewing from the true ground of nature, whence most diseases proceed, and how to prevent them. To which is added, a treatise of most sorts of English herbs with several other remarkable and most useful observations, very necessary for all families.*
London: printed for H. Newman, 1697.
Subject: Astrological medicine, diet, herbals, humors, quackery.
STC II T3202A
Case/oRA/775/.T78/1697

3d ed. to which is added a discourse of the philosophers stone, or vniversal medicine, discovering the cheats and abuses of those chymical pretenders.

## Tryon, Thomas.

*Wisdom's dictates: or, aphorisms & rules, physical, moral, and divine. For preserving the health of the body, and the peace of the mind, fit to be regarded and practised by all that would enjoy the blessings of the present and future world. To which is added, a bill of fare of seventy five noble dishes of excellent food, far exceeding those made of fish or flesh, which banquet I present to the sons of wisdom, or such as shall decline that depraved custom of eating flesh and blood.*
London: printed for Tho. Salusbury, 1691.
STC II T3205
Subject: Diet, hygiene, medicine, vegetarianism.
Case/B/692/.89

## Tschirnhaus, Ehrenfried-Walther van, 1641–1708.

*Medicina mentis, sive tentamen genuinae logicae, in quâ disseritur de methodo detegendi incognitas veritate.*
Amsterdam: Albertus Magnus & Joannes Rieuwerts, Jr., 1687.
Subject: Mental and physical health and hygiene.
Case/oB/2609/.M4

BOUND WITH *Medicina corporis, seu cogitationes admodum probabiles de conservandâ sanitate.* Amsterdam: Albertus Magnus & Johannes Rieuwertz, Jr., 1686. Each tract is in 3 parts. In "To the reader," preceding *Medicina corporis,* author states that the style of bodily medicine is different from the style of mental medicine.

## Tschudi, Aegidius, 1505–1572.

*De prisca ac vera alpina Rhaetia cvm caetero alpinarum gentium tractu, nobilis ac erudita ex optimis quibusque ac probatissimis autoribus descriptio, autore Aegidio Schudo, . . . cui hac editione accessit regula inuestigationis omnium locorum in tabula Heluetiae contentorum, cum indice,... per Conradum Lycosthenem.*
Basel: [widow of Michael Isingrin (b. 1500), printer] 1560.
Subject: Descriptive and historical geography.
Case/F/38/.892

## Tuccaro, Arcangelo, b. ca. 1535.

*Trois Dialogves de l'exercice de savter, et voltiger en l'air. Par le Sr. Argange Tvccaro, de l'Abruzzo*
Paris: Clavde de Monstr'oeil, 1599.
Subject: Exercise, gymnastics, health.
Case/V/132/.89

In 3 dialogues. 1.: "Traicte des exercices gymnastiques, dont les anciens vsoient auec leur declaration & distinction, & vne dispute du blasme de la loüange du balon de la dance"; 2.: "Plusieurs beaux discours du saut appellé par les anciens cubistique"; 3.: "Au troisieme & dernier est fort amplement discouru des exercices que l'homme peut faire, tantost plus, tantost moins, selon sa nature & complexion, & comme pour se maintenir en santé, il doit vser d'vn exercice, qui est la vraye medicine pour rendre le corps agile, gaillard, vigoureux & sain.

## Turnebe, Adrien 1512–1565.

*Viri clariss. Adriani Tvrnebi regii qvondam Lvtetiae professoris opera.*
Strassburg: Lazarus Zetzner, 1600.
Subject: Disease, natural history.
Case/fX/0163/.894

Bk. 1: "Commentarius"; bk. 2: "Versiones selectorum quorundam librorum Theophrastus: odoribus; lapidibus; igne; ventis. Plutarch: primo frigido; fluuiorum & montium nominibus. Demetrij Pepagomeni de podagra."

## Turner, Daniel, 1667–1741.

*The art of surgery: in which is laid down such a general idea of the same as is founded upon reason confirm'd by practice, and farther illustrated with many singular and rare cases medico-chirurgical.*
London: printed for C. Rivington, 1729.
Subject: Medicine, surgery.
oRD/30/.T88/1729

2 vols. Vol. 1: 3d ed. corrected; vol. 2: 5th ed., printed for C. Rivington, 1736.

## Turner, William, 1653–1701.

*A compleat history of the most remarkable providences,... both of judgment and mercy, which have happened in this present age. Extracted from the best writers, the author's own observations, and the numerous relations sent to him from divers parts of the three kingdoms. To which is added, whatever is curious in the works of nature and art.*
London: printed for John Dunton, 1697.
STC II T3345
Subject: Medicine, natural history.
Case/B/86/.89

3 pts. in 1 vol. Pts. 2 and 3 contain 7 separate tracts with individual title pages. Pt. 1: "History of the divine"; pt. 2: "Wonders of nature" (including sympathy and antipathy); pt. 3, chap. 35: "The excellence of women in the arts. A woman of Kenly in Shropshire, Nurse Corfield, was noted for her skill in chirurgery and physick." Pt. 3: "Curiosities of art, including improvements in astronomy, navigation, mechanicks, and the hydratilick art, or water-works."

## Tyard, Pontus de, 1521–1605.

*Devx discovrs de la natvre dv monde, & de ses parties. A sçavoir, le premier cvrievx, traittant des choses materielles: & le second cvrievx, des intellectuelles.*
Paris: Mamert Patisson, for Robert Estienne, 1578.
Subject: Astronomy, macrocosm-microcosm, natural philosophy.
Case/Y/762/.T955

"Quel autre chemin (ie vous prie) plus droit nous meine à la théologie, que l'astronomie & ses seruantes?" (p. 3).

## Tyard, Pontus de.

*Discovrs dv temps, de l'an, et de ses parties.*
Paris: Mamert Patisson, for Robert Estienne, 1578.
Subject: Astronomy, natural philosophy, time.

Case/Y/762/.T956

"Et si le temps est l'espace du mouuement, comme peut estre le mouuement sans temps, ou le temps sans mouuement?" (p. 2v).

## Tyard, Pontus de.

*Mantice, ou discours de la verité de diuination par astrologie.*
Paris: Galiot du Pré [1573?]
Subject: Astrology.
Case/B/8635/.898

2d ed. enlarged.

## Typotius, Jacobus, 1540–1601.

*Symbola diuina & humana pontificvm imperatorvm regvm. Accessit breuis & facilis isagoge Iac. Typotii. Ex mvsaeo Octavii de Strada civis Romani.*
Prague: Egidius Sadeler, 1601–3.
Subject: Emblems.
Case/fF/0711/.896

3 vols. in 2; vols. 1 and 2 bound together. Vol. 2: *Symbola uarie diuersorum principvm sacrosanc ecclesiae & sacri imperij Romani.* Prague, 1602. Vol. 3: *Symbola varia diversorvm principvm cvm facili isagoge D. Anselmi de Boodt Brvgensis.* Prague, 1603. See also Boodt, Anselm Boece de.

# U–V

*Unmassgebliches bedencken ob die cometen zu künsstige anglucks-hälle als krieg, theuerung, pestilentz, grosser herrn, tod, etc. verkündigen aus beranlassung des jüngsthin neu-erscheinenen cometen auff vielfältiges begehren und anhalten kürtzlich eylfertig, und einfältig entworffen. Quid sibi stella velit nova, non ego judico, at opto: sit crinis finis, meta cometa mali!*
[n.p.] 1681.
Subject: Comets.
Case/B/863/.912

* * *

*Unterricht wegen des cometsterns und dessen lauffs in seinem eigenem circul, nebenst einem kupffer stücke, worinne klärlich vor augen gestellet, dass er seinen schweiff eigentlich nicht umbgewendet sondern dem selben also veränderlich exstrecken müssen.*
Hanover: Georg Friederich Grimmen, 1665.
Subject: Comets.
Case/B/863/.92

### Vadianus, Joachim, 1484–1551.

*Epistola Vadiani, ab eo penè adulescente ad Rudolphum Agricolam iuniorem scripta.*
Basel: Andreas Cratander, 1522.
Subject: Geography.
Ayer/*6/M5/1522

BOUND WITH Mela, Pomponius, q.v.

### Vadianus, Joachim.

[*Epitome topographica*] *Epitome trivm terrae partivm, Asiae, Africae, et Evropae compendiariam locorum descriptionem continens, praecipue autem quorum in actis Lucas, passim Euangelistae & Apostoli meminere. Cvm addito in fronte libri elencho regionum urbium, amnium insularum, quoram nouo testaemnto fit mentio, quo expeditius pius lector quae uelit, inuenire queat.*
Zurich: Christophorvs Frosch[ouer], 1534.
Subject: Descriptive geography, topography.
Ayer/*7/V2/1534a

Some MS marginalia.

### Vadianus, Joachim.

[*Epitome topographica*] [Another copy]
Ayer/*7/V2/1534b

Some MS marginalia passim.

### Vadianus, Joachim.

*Epitome topographica totius orbis, conferens ad ea potissimum loca . . .*
[Antwerp] Ioan. G[rapheus], 1535.
Subject: Geography, topography.
Greenlee/4890/V12/1535

MS marginalia and MS notes in section, "Nvmidia."

### Vairus, Leonardus.

*De fascino libri tres, auctore Leonardo Vairo Beneventano. In quibus omnes fascini species, & causae optima describuntur. . . . Necnon contra praestigias, imposturas, allusionesque daemonorum, cautiones & amuleta praescribuntur: ac denique nugae, quae de ijsdem narrari solent, dilucide confutantur.*
Venice: Aldus, 1589.
Subject: Charms, witchcraft.
Case/B/88/.92

### Vairus, Leonardus.

*Trois livres des charmes, sorcelages, ov enchantemens. Esquels toutes les especes, & causes des charmes sont methodiquement descrites, & doctement expliquees selon l'opinion tant des philosophes que des theologiens; auec des vrais contrepoisons pour rabattre les impostures & illusions des demons: & par mesme moyen les vaines bourdes qu'on met en auant touchant les causes de la puissance des sorceleries y sont clairement refutees.*
Paris: Nicolas Chesneav, 1583.
Subject: Charms, witchcraft.
Case/B/88/.93

Trans. Iulian Bavdon.

### Valentine, Basil.

*The last vvill and testament of Basil Valentine, monke of the order of St. Bennet. Which being alone, he hid under a table of marble, behind the high-altar of the cathedral church, in the imperial city of Erford: leaving it there to be found by him whom God's providence should make worthy of it. To which is added two treatises the first declaring his manual operations. The second shewing things natural and supernatural.*
London: printed by S. G. and B. G. for Edward Brewster, 1671.
STC II B1018.
Subject: Alchemy.
Case/B/8633/.078

Translated by J. W. [John Webster]. 7 pts. in 1 vol.

### Valeriano Bolzani, Giovanni Pierio, 1477–1558.

*Hieroglyphica sive de sacris AEgyptiorvm literis commentarii Ioannis Pierii Valeriani Bolzanii Bellvnensis.*
Basel: [M. Isingrin], 1556.
Subject: Emblems, fauna.
Case/W/1025/.933

Animals and their symbolism, sacred and profane. 7 leaves of Latin index (beginning "Sirius stella") and the 2 first leaves of Greek index in MS.

### Valeriano Bolzani, Giovanni Pierio.

*Ioannis Pierii Valeriani Bellvnensis. Hieroglyphica, seu de sacris AEgyptiorum, aliarumque gentium literis commentarij, libris quinquaginta octo digesti: quibus additi sunt duo hieroglyphicorum libri Caeli Augustini Curionis: Eiusdem Pierii pro sacerdotum barbis declamatio, & poemata varia cum diuersis hieroglyphicis collectaneis, in sex libros ordine alphabetico dispositis, & nunc diligentur expurgatis. Accesserunt in hoc postrema editione, Hori Apollonis hieroglyphicorum libri duo: Hieroglyphicorum emblematumque medicorum.*
Lyon: Paul Frellon, 1626.
Subject: Emblems.
Case/fZ/141/.93

Several tracts bound together. Emblems relating to the zodiac, medicine, the humors, and a list of "variorum phantasmatvm quibus agitantur aegrii melancholici."

### Valeriano Bolzani, Giovanni Pierio.

*Les hieroglyphiques de Ian-Pierre Valerian vvlgairement nomme Pierivs. Avtrement, commentaires des lettres et figvres sacrées des Aegyptiens & autres nations. Oeuvre reducte en cinquante hiuct liures ausquels sont adjoincts deux autres de Coelivs Cvrio, touchant ce qui est signifié par les diuerses effigies, et pourtraicts des dieux et des hommes.*
Lyon: Paul Frellon, 1615.
Subject: Emblems.
Case/W/0120/.93

Arranged by subject (e.g., the eye). Translated by I. de Montlyart.

### Valeriano Bolzani, Giovanni Pierio.

*Ieroglifici, overo commentari delle occulte significationi de gli Egittij, & d'altre nationi. . . . Accresciuti di due libri dal Sig. Celio Augustino Curione.*
Venice: printed by Gio. Antonio, and Giacomo de'Franceschi, 1602.
Subject: Emblems, hieroglyphics.
Case/fZ/141/.929

### Valesius, Hadrianus [Valois, Adrien de, 1607–1692].

*Notitia Galliarum ordine litterarum digesta. In qua situs, gentes, opida, castella, vici, montes, silvae, maria, flumina, fontes, lacus, paludes, insulae maritimae & amnicae, paeninsulae, pagi provinciaequae Galliae illustrantur; locorum antiquitates varium eorum nomina, vetera ac nova, episcopatuum ac monasteriorum origines, aliaque ad historiam Francicam pertinentia notantur; geographi & historici Graeci, Romani ac nostri explicantur, & emendantur.*
Paris: Frederic Leonard, 1675.
Subject: Descriptive geography.
fG/39001/.933

Names of cities and towns, Latin and vernacular, with topographical and historical description.

### Valla, Giorgio, d. 1500.

*Georgii Vallae Placentini, viri clariss. De expetendis, et fvgiendis rebvs opus.*
[Venice: Aldus Romanus for the sons of Ioannes Petri Valla, 1501.]
Subject: Astrology, mathematics, medicine.
Case/+A/251/.93

In 2 vols. Vol. 1: "De arithmetica, de musica, de geometria, de tota astrologia, de physiologia"; vol. 2 heading at top of 1st page (title page wanting): "Georgii Placentini expetendorvm, ac fvgiendorvm primvm qvomodo inventa medicina et in qvot partes sit distribvta" (in 7 pts.).

### Valla, Lorenzo, 1406–1457.

*De libero arbitrio.*
[Strassburg: Georg Husner, ca. 1475.]
Hain 15830*^ Pr. 355^ BMC (XV)I:83^ Goff V-70.
Subject: Natural theology.
Inc./355

Incipit: Lavrencii Vallensis oratoris clarissimi de libero arbitrio et providencia divina tractatvlvs. In dialogue form. Mystery can only be accepted, not rationalized. "We stand by faith, not by the probability of reasons." Penciled note on front pastedown: "In this book Valla defends the thesis that God's presence does not contradict man's free will (against Boethius)."

### Valla, Lorenzo.

*De voluptate ac vero bono libri III.*
Basel: Andreas Cartandrvm [*sic*], 1519.
Subject: Natural theology, philosophy.
Case/B/652/.931

Bk. 1: the natural world; bk. 2: laws (of nature, philosophy); bk. 3: philosophy and religion.

### [Valle, Battista della] d. 1535

*Vallo, libro continente appertenentie ad capitanii, retenere & fortificare una citta con bastioni, con noui artificii de fuoco aggionti, come nella tabola appare, & de diuerse sorte poluere, et de expugnare una citta con ponti, scale, organi, trombe, trenciere, artigliarie, caue, dare auisamenti senza messo alla amico, fare ordinanze, battaglioni, et ponti de disfida con lo pingere, opera molto utile con la experientia de larte militare.*
[Venice: Sessa? 1524]
Subject: Engineering, gunpowder, military science.
Case/F/057/.714

"Far vno arlogio" (leaf 19); modello per cavare aqva (28v).

### Vallemont, Pierre le Lorrain, abbé de, 1649–1721.

*La physique occulte, ov traité de la baguette divinatoire, et de son utilité pour la découverte des sources d'eau, des miniéres, des tresors cachez, des voleurs & des meurtriers fugitifs. Avec des*

*principes qui expliquent les phénoménes les plus obscurs de la nature.*
Paris: Jean Anisson, 1693.
Subject: Divination, dowsing rod, occult, supernatural.
Case/B/8645/.935

## Valsalva, Antonio Maria, 1666–1723.

*De aure humana tractatus, in quo integra ejusdem auris fabrica, multis novis inventis, & iconismis illustrata, describitur; omniumque ejus partium usus indagantur. Qvibvs interposita est musculorum uvulae, atque pharyngis nova descriptio, et delineatio.*
Bologna: printed by Constantini Pisarius for S. Michael, 1704.
Subject: Anatomy of the ear, medicine.
Case/QM/507/.V35/1704

1st ed.

## Valverde, Giovanni [Juan de Valverde].

*Anatomia del corpo humano composta per M. Giouan Valuerde di Hamusco, & da luy con molte figure di rame, et eruditi discorsi in luce mandata.*
Rome: Ant. Salamanca & Antonio la frerj, 1560.
Subject: Anatomy.
UNCATALOGUED

Colophon: "In Vinegia appresso Nicolò Beuilacqua Trentino."

## Vanini, Lucilio [afterward Giulio Cesare], 1585–1619.

*Amphitheatrvm aeternae providentiae divino magicvm. Christiano-physicvm, nec non astrologo-catholicvm. Aduersus veteres philosophos, atheos, epicureos, peripateticos, & stoicos.*
Lyon: widow of Antonius de Harsy, 1615.
Subject: Astrology, fate, providence, will.
Case/C/558/.935

Includes "Autoris sententiam de monstris ex physicis & astronomicis principiis deducit" (sec. 40); "Apud peripateticos quid sit necessitas, fatum, natura, fortuna casus, & fors accurate explicat" (sec. 42); "Respondet argumento secundo probans ex ipso Ptolomaeo astra inclinare tantum, non cogere nostras voluntates. In contrariam tamen subtilissimas rationes adducit quibus ex astrologorum arcanis respondet" (sec. 45).

## Vanini, Lucilio.

*De admirandis naturae reginae deaeque mortalium arcanis.*
Paris: Adrianvs Perier, 1616.
Subject: Natural history.
Case/B/235/.9242

In 4 books, in dialogue form. The heavens, the earth and the waters, generation of animals, and demons and the supernatural.

## Vargas Machuca, Bernardo de, 1557–1622.

*Milicia y descripcion de las Indias, por el Capitan don Bernardo de Vargas Machuca.*
Madrid: Pedro Madrigal, 1595.
Subject: Exploration, geography, New World.
Ayer/*108.5/V2/1599

Compendio de la sphera (leaf 180).

## Vargas y Toledo, Alfonso, abp., d. 1366.

*Quaestiones super libris De anima Aristotelis.*
Florence: Nicolaus Laurentius, Alamanus, 1477.
H-R 877^ BMC(15C)VI:626^ Pr. 6113^ Still. V84^ Goff V-92
Subject: Psychology, soul.
Inc./f6113

MS marginalia.

## Vaughan, Henry, 1622–1695, trans.

*Olor Iscanus. A collection of some select poems, and translations, formerly written by Mr. Henry Vaughan, Silurist.*
London: printed by T. W. for Humphry Moseley, 1651.
STC II V123
Subject: Disease, medicine, psychology.
Case/Y/185/.V456

4 works in 1 vol., including, in addition to the poetry, "Of the diseases of the mind"; "Of the diseases of the mind and the body. A discourse in the Greek by Plutarchus Chaeronensis"; "Of the diseases of the mind, and the body, and which of them is most pernicious"; "The praise and happinesse of the countrie-life; written originally in Spanish by Don Antonio de Guevara." Trans. into Latin by John Reynolds and into English by Vaughan.

## Vaughan, Thomas [Eugenius Philalethes, pseud.], 1622–1666.

*Lumen de lumine, oder ein neues magisches liecht geoffenbahret der welt mit getheilet durch Eugenium Philalethen.*
Hamburg: Gottfried Liebezeit, 1693.
Subject: Alchemy, magic, Rosicrucians.
B/88/.933

"Ein brieff von den brüdern des Rosen-Creutzes. Betreffend den unsichtbahren magischen berg und den darinnen verwahrten schatz." Translated from English by J. R. S. M. C.

## [Vaughan, Thomas]

*Magia Adamica: Or the antiquitie of magic, and the descent thereof from Adam downwards, proved whereunto is added a perfect, and full discoverie of the true coelum terrae, or the magician's heavenly chaos, and first matter of all things. By Eugenius Philalethes* [pseud.]
London: printed by T. W. for H. Blunden, 1650.
STC II V151
Subject: Alchemy, natural magic.
Case/B/88/.935

Bound with The man-mouse taken in a trap. London, 1650 (a diatribe against Henry More).

## Veen, Octavio van, known as Otto Vaenius, 1560?–1629?

*Amoris divini emblemata, stvdio et aere Othonis Vaeni concinnata.*
Antwerp: Plantiniana Balthasar Moreti, 1660.
Subject: Emblems.
Case/W/1025/.944

## Veen, Octavio van.

*Amorvm emblemata, figvris aeneis incisa stvdio Othonis Vaeni.*
Antwerp: printed by Henricus Swingenius, 1608.
Subject: Emblems.
Case/W/1052/.937

In Latin, Italian, and French.

## Veen, Octavio van.

*Emblemata sive symbola a principibus, viris eeclesiasticis[sic]ac militaribus, alijsque vsurpanda. Deuises ou emblems pour princes, gens d'eglise, gens de guerre, & aultres.*
Brussels: Ex officina Huberti Antonii, 1624.
Subject: Emblems.
Case/W/1025/.938

## Veer, Gerrit de, fl. 1600.

*Tre navigationi.*
Venice: Ieronimo Porro, & company, 1599.
Subject: Navigation, voyages.
Ayer/124/D91/V4/1599a

At end of pt. 3 (3d voyage): "I nomi ueramente di quelli che sono ritornati da questa nauigatione sono questi" (list of 12 names follows, including de Veer).

## Veer, Gerrit de.

*Tre navigationi fatte dagli Olandesi, e Zelandesi al settentrione nella Norvegia, Moscovia, e Tartaria . . . Descritte in Latino da Gerardo Di Vera, e nuouamente da Giouan Ginnio Parisio tradotte nella lingua Italiana.*
Venice: Gio. Battista Ciotti, 1599.
Subject: Navigation, voyages.
Ayer/*124/D91/V4/1599b

De Veer was historian of the 3d voyage.

## Veer, Gerrit de.

*Trois navigations admirables faictes par les Hollandois & Zelandois au septentrion lesquelles ont descouuert la mer Vueygats, la nouuelle Zemble & le païs qui est dessous la huictantiesme degré, que l'on estime estre Groenlandie, où iamais personne parauant n'auoit abordé: plusieurs cruels ours & autres monstres marins; auec grands dangers & incroyables difficultez: vtiles & fort necessaires à tous pilotes, nautonniers & autres gens de marine.*
Paris: Gvillavme Chavdiere, 1599.
Subject: Navigation, voyages.
Ayer/*124/D91/V4/1599

## Vegius, Mapheus, d. 1458.

*De educatione librerorum & eorum claris moribus libri VI.*
Milan: Leonard Pachel, 1491.
H-R 15920^ Still. V98^ Goff V-111
Subject: Education of children, pedagogy.
Inc./5991.5

Mothers should nurse their own infants (sig. bii); development of memory (sig. ciii v).

## Vegius, Mapheus.

*Maphei Vegij Laudensis poete et oratoris praeclari dialogus tum festiuus tum elegans, nec non summa ingenii solertia atque industria concinnatus, mores, vitamque hominum certe peruersam complectens, cui nomen Philalethes. Appendices loco annexa est succincta & breuis omnium fere dictionum quae hoc libello continentur, interpretatio, cum praesation veritatis laudem ex uariis scriptoribus complectente per Vdalricum Fabri Thorenburengensem.*
Vienna: Ioannes Singrenius, 1516.
Subject: Philosophy.
Case/Y/682/.V516

Some MS marginalia.

## Vegius, Mapheus, d. 1458.

*Philalethes.*
[Strassburg: Heinrich Knoblochtzer, ca. 1480.]
H-C *15296^ BMC (XV)I:88^ Pr. 368^ Goff V-115
Subject: Philosophy.
Inc./368

## Vegius, Mapheus.

*Philalethes.*
[Basel: Michael Furter, not after 1492.]
H-C *15927^ BMC (XV)III:782^ Pr. 7644^ Still. V103^ Goff V-117
Subject: Philosophy.
Inc./7722/.5

In dialogue form. With table: "Tabula declaratiua quorundam terminorum ac instrumentorum quibus alethia. i. veritas se afflictam et propulsam indicat: vt in precedenti dialogo notatur." Subheads of table include Nautarum instrumenta; rusticorum instrumenta; feminarum arma et ornamenta quedam; artificum instrumenta; instrumenta lalonum: et omnis generis fabrorum; instrumenta tonsorum."

## Venn, Thomas.

*Military and maritime discipline in three books.*
London: printed by E. Tyler and R. Holt for Rob. Pawlet, 1672.
STC II V192
Subject: Gunpowder, metallurgy, pyrotechnics, military science.
Case/fU/O/.9377

Bk. 1: "Military observation or tacticks"; bk. 2: "An exact method of military architecture, the art of fortifying towns . . . rendred into English by John Lacey out of the works of the late learned mathematician Andrew Tacquet"; bk. 3: "The compleat gunner in three parts, shewing the art of founding of great ordnance; making gun-powder; the taking of heights and distance either with or without instruments; with the nature of fire-works. Translated out of Casimir, Diego, Vssano and Hexam, &c. To which is added the doctrine of projects applied to gunnery by Galilaeus and Torricellio. And observations out of Mersennus and other authors."

## Venner, Tobias, 1577–1660.

*Via recta ad vitam longam: or, a plaine philosophicall demonstration of the nature, faculties, and effects of all such things as by way of nourishments make for the preseruation of health, with diuers necessary dietical* [sic] *obseruations; as also of the true vse and effects of sleepe, exercise, excretions and perturbations, with iust applications to euery age, constitution of body, and time of yeere: By To. Venner, doctor of physicke in Bathe.*
London: printed by Felix Kyngston for Richard Moore, 1628.
STC 24645
Subject: Diet, health, hygiene.
Case/Q/.942

* * *

*Vera relatione del terremoto segvito nella Romagna, e Marca il Giouedì Sanctù 14 Aprile del corrente anno 1672.*
In Bracciano [n.p., 1672].
Subject: Earthquakes.
Greenlee/4504/P855

## Vergilius, Polydorus, 1470?–1555?

*Adagiorvm aeqve hvmanorvm vt sacrorvm opvs.*
Basel [M. Ising.] 1550.
Subject: Adages, chance, prodigies, witchcraft.
Case/Y/0567/.942

BOUND WITH his *Dialogorum*, q.v.

## Vergilius, Polydorus.

*I dialoghi di Polidoro Verigilio.*
Venice: Gabriel Giolito di Ferrari, 1550.
Subject: Prodigies.
Case/Y/682/.V5952

Translated by Francesco Baldelli. In 4 parts. Contents include "De' prodigij, libri tre." In dialogue form.

## Vergilius, Polydorus.

*Dialogorum, De iureiurando & periurio, Lib. I. De veritate & mendacio. De prodigij & sortibus lib. III.*
Basel: M. Ising[rinius], 1553.
Subject: Prodigies.
Case/Y/0567/.942

BOUND WITH his *Adagiorum opus*, 1550, q.v..

## Vergilius, Polydorus.

[*De rerum inventoribus*]
*De inuentoribus rerum libri tres. M. Antonii Sabellici De artium inuentoribus ad Baffum carmen elegantissimum. Ex secunda recognitione.*
[Strassburg: Mathis Schurerius Selsetensis, 1512]
Subject: Encyclopedias.
Case/Y/0567/.94

A history of the liberal arts, including discoveries and inventions in agriculture, archaeology, architecture, arithmetic, astrology, divination, engineering, geometry, magic, medicine, metallurgy, necromancy, technology, and others. BOUND WITH his *Prouerbiorum liber*, 1511.

## Vergilius, Polydorus.

[*De rerum inventoribus*]
*Polydori Vergilii Vrbinatis adagiorvm liber. Eiusdem De inuentoribus rerum libri octo, ex accurata autoris castigatione, locupletationéque non uulgari, adeo ut maxima ferè pars primae ante hanc utriusque uoluminis aeditioni accesserit.*
[Basel: Ioan Frobenius, 1524]
Subject: Encyclopedias.
Case/Y/0567/.941

## Vergilius, Polydorus.

[*De rerum inventoribus*]
*Polydori Vergilii Vrbinatis De rervm inuentoribus libri octo, per autorem summa cura recogniti & locupletati. Dices supremam manum impositam.*
Basel: Ioan[nes] Frob[enius], 1525.
Subject: Encyclopedias.
Safe/Case/*fY/682/.V585

Bound for Jean Grolier. Stamped on cover: "Io. Grolierii et amicorvm." MS leaf signed by Grolier mounted inside front cover.

## Vergilius, Polydorus.

[*De rerum inventoribus*]
*Von den Erfyndern der dyngen. Wie und durch wölche alle ding nämlichen alle künsten, handtwercker auch all andere händel, geystliche und weltliche sachen als polliceyen, religiones, orden, ceremonien, vnnd anders &c. betreffende von dem maysten biss auff das mynnste nichts aussgelassen, von anfang der welt her biss auff die vnsere zeit geübt vnn gepraucht, durch Polydorum Vergilium von Vrbin, in acht bücheren aygentich im Latein beschriben.*
Augsburg: Heynrich Steyner, 1537.
Subject: Encyclopedias.
Case/fY/682/.V5924

Trans. Marcus Tatius Alpinus.

## Vergilius, Polydorus.

[*De rervm inuentoribus*]
*De rerum inventoribus libri octo.*
Basel: Isingrinius, 1546
Subject: Encyclopedias.
Case/3A/1286

## Vergilius, Polydorus.

[*De rervm inventoribvs*]
*De rervm inventoribvs libri octo.*
Lyon: Seb. Gryphivs, 1546.

Subject: Encyclopedias.
Case/Y/682/.V586

### Vergilius, Polydorus.

[*De rerum inventoribus*]
*An abridgement of the notable woorke of Polidore Vergile conteignyng the deuisers and firste finders out as well of artes, ministeries, feactes and ciuill ordinaunces, as of rites and ceremonies, commonly vsed in the churche: and the originall beginnyng of the same compendiously gathered by Thomas Langley.*
London: printed by Richard Grafton, 1546.
STC 24656
Subject: Encyclopedias.
Case/Y/682/.V588

### Vergilius, Polydorus.

[*De rerum inventoribus*]
*De rervm inventoribvs libri octo. Denuo recogniti & expurgati.*
Rome: Bartholomaei Grassi, 1585.
Subject: Encyclopedias.
Case/Y/682/.V5861

### Vergilius, Polydorus.

[*De rerum inventoribus*]
*De rervm inventoribvs libri octo. Denuò Romae recogniti & expurgati. Item Sardi Ferrariensis de rerum inuentoribus libri II.*
Lyon: Ant. Gryphivs, 1586.
Subject: Encyclopedias.
Wing/ZP/539/.G/917

### Vergilius, Polydorus.

[*De rerum inventoribus*]
*De rerum inuentoribus libri octo.*
[Geneva] printed by Iacobus Stoer, for Nicolaus Bassej, 1590.
Subject: Encyclopedias.
Case/-Y/682/.V5862

Bindery misprint: call number should be "Case -X" instead of "Case -Y."

### Vergilius, Polydorus.

[*De rerum inventoribus*]
*Di Polidoro Virgilio de Vrbino De gli inventori delle cose, libri otto.*
Florence: Filippo Givnti, 1592.
Subject: Encyclopedias.
Case/Y/682/.V593

Trans. Francesco Baldelli.

### Vergilius, Polydorus.

[*De rerum inventoribus*]
*Polydorvs Vergilius von Vrbin von erfindung vnd erfindern der dinge wie vnd durch welche alle ding, nemblich alle kunst handwercke, allerley händel, geistliche vnd weltliche sachen, als policeyen, religiones, orden, ceremonien, &c. betreffendt von dem grösten biss auff das geringste nichts aussgelassen von anfang der welt der biss auff diese vnsere zeit angefangen geübt vnd gebraucht worden.*
Frankfurt am Main: Wolfgang Richter for Ioannis Theobald Schönwetter, 1603.
Subject: Encyclopedias.
3A/534

Trans. Marcus Tatius.

### Vergilius, Polydorus.

[*De rerum inventoribus*]
*De rervm inventoribvs libri VIII et de prodigiis, libri III.*
Leyden: Franciscus Hegervs, 1644.
Subject: Encyclopedias, prodigies.
Wing/ZP/646/.E524

### Vergilius, Polydorus.

[*De rerum inventoribus*]
*The works of the famous antiquary, Polidore Virgil.*
London: printed for Simon Miller, 1663.
STC II V596
Subject: Encyclopedias.
Case/Y/682/.V591

"Compendiously English'd by John Langley late master of Paul's School, London." Book 9, p. [305], chap. III: "Of alchymy."

### Vergilius, Polydorus.

[*De rerum inventoribus*]
*De inventoribvs rerum libri VIII. Et de prodigiis libri III.*
Amsterdam: Daniel Elzevir, 1671.
Subject: Encyclopedias, prodigies.
Case/Y/682/.V5867

2 pts. in 1 vol.

### Vergilius, Polydorus.

[*De rerum inventoribus*]
*De gli inventori delle cose. Con due tauole, vna de capitoli, e l'altra delle cose più notabili.*
Brescia: Domenico Gromi, 1680.
Subject: Encyclopedias.
Y/682/.V594

Trans. Francesco Baldelli. Book I, chap. 17: "Chi furono i primi, che trouarono l'astrologia, ouero di certe stelle il corso, e la sfera, e la ragione de i venti; e quanti i venti siono, e l'oseruazione delle stelle nel nauicare."

### Vergilius, Polydorus.

[*De rerum inventoribus*]
*Pleasant and compendious history of the first inventers and instituters of the most famous arts, misteries, laws, customs and manners in the whole world. Together, with many other rarities and remarkable things rarely known, and never before made publick. To which are added, several curious inventions, peculierly attributed to England & English-men. The whole work alphabetically digested, and very helpful to the readers of history.*
London: John Harris, 1686.

STC II V598
Subject: Encyclopedias.
Case/Y/682/.V592

### Verheyen, Philippe, 1648–1710.

*Corporis hvmani anatomica in qva omnia tam veterum qvam recentiorvm anatomicorvm inventa methodo nova & intellectu facillima describuntur.*
Leipzig: Thomas Fritsch, 1699.
Subject: Anatomy.
Case/QM/21/.V47/1699

### Verheyen, Philippe.

*Corporis humani anatomiae . . . in quo tam veterum, quàm recentiorum anatomicorum inventa.*
Brussels: t'Serstevens bros., 1726.
Subject: Anatomy.
oQM/21/.V47/1726

In 2 books.

### Verien, Nicolas, fl. 1685–1724.

*Livre curieux et utile pour les sçavans, et artistes composé de trois alphabets de chiffres simples, doubles & triples, fleuronnez et au premier trait. Accompagné d'un tres grand nombre de devises, emblêmes, médailles et autres figures hieroglyfiques. Ensemble de plusieurs supports et cimiers pour les ornemens des armes. Avec une table tres ample par le moyen de laquelle on trouvera facilement tous les noms imaginables. Le tout inventé, dessiné et gravé par Nicolas Verien maistre graveur.*
Paris [1685]
Subject: Emblems.
sc 217

### [Verien, Nicolas]

*Recueil d'emblêmes, devises, medailles, et figures hieroglyphiques au nombre de plus de douze cent, avec leurs explications.*
Paris: Jean Jombert, 1696.
Subject: Hieroglyphics, symbolism.
Case/W/1025/.9459

"Par le sieur Verrien, Maître graveur." First 62 plates have explanations of symbols or Latin aphorisms and French translations.

### Verini, Giovanni Battista.

[Esempi di calligrafia]
*Alla illvstra: et eccell. s. marchesa Del Guasto. Giouambattista Verini Fiorentino suo dedicatissimo saruitore che insegna abbacho, & de ogni sorte littre scriuere . . . i Milano.*
Brescia: Ludouico Britannico for Giouambattista Verini, 1538.
Subject: Calligraphy, ciphers.
Wing/ZW/535/.V58

MS note laid in at back of vol.: "Extremely rare and apparently undescribed by bibliographers. It is a different work from the Liber elementorum published at Florence about 1526. Probably it was composed by the author at Milano while teaching there."

### Verini, Giovanni Battista.

*Elementorum litterarum Ioannis Baptiste de Verinis Florentini nouiter impressus.*
[Florence, 1527?]
Subject: Applied geometry (geometrical analysis of lettering).
Wing/ZW/14/.V58

Running head: "Luminario." In 4 books. "Nel secondo si dimostra la moderna cauare col sesto per geometricha ragione & con grandissima breuita. Nel terzo si dimostra la litera anticha per geometricha ragione.

### Verini, Giovanni Battista.

*Luminario. Libro 7. Opera di Giovambattista Verini Fiorentino che insegna abbacho.*
[Milan: P(ietro) Paulo Verini, 1536]
Subject: Education, lettering.
Wing/ZW/14/.V6

Mutilated fragment.

### Verini, Giovanni Battista.

*El trivmopho di ricette & secreti bellissimi composto per Giouambattista Verini Fiorentini, a comune beneficio di ciascheduno [s]pirito gentile.*
[Milan: Vincentio da Medda for Giouambattista Verini, 1535.]
Subject: Chemistry, iatrochemistry.
Wing/ZP/535/.M463

Contains recipes, formulas, directions for making inks, and colors; treating minor ailments ("a chi non posse orinara") whitening teeth, etc. BOUND WITH Antonio, Milanese, q.v.

### Verini, Giovanni Battista.

*La vtilissima opera da imparare a scrivere di varie sorti lettere di Gioumbattista Verini Fiorentino che insegna al Rialto abbaco & scriuere.*
[N.p., 153–?].
Subject: Ciphers, lettering.
Wing/ZW/14/.V585

With a cipher of "Giouambattista."

### Versor, Johannes, d. ca. 1485.

*Expositio super summulis Petri Hispani.*
Naples [Henricus Alding], 1477.
H-C-R 16032^ BMC (XV)VI:867^ Pr. 67016^ Still. V220^ Goff V-245
Subject: Logic, philosophy.
Inc./6716

MS notes. 7 tracts, e.g., Periarmenius (Aristotle); Predicabilium (Porphyrius); Predicamentorum (Aristotle); and others.

### Versor, Johannes.

*Expositio versoris prestantissimi doctoris parisiensis super summulis magistri Petri Hyspani.*

[Toulouse: Heinrich Mayer, ca. 1494?]
C-R 6182^ Still. V221^ Goff V-246
Subject: Logic, philosophy.
Inc./f8722/.5

Rubricated; MS notes. Edited by Petrus de Sancto Johannes.

## Versor, Johannes.

*Quaestiones super libros Aristotelis.*
[Cologne: Heinrich Quentel] 1489.
H-C *16047^ Voullième 1228, 1231, 1235^ Pr. 1295^ Still. V228^ Goff V-253
Subject: Natural science.
Inc./f1295

Half title: "Questiones magistri Johannis Versoris super libros de celo & mundo cum textu Arestotelis." Tracts include, among others, "De generatione & corruptione, de metheororum, de somno et vigilia," and "De divinationibus."

## Vesalius, Andreas, 1514–1564.

*De humani corporis fabrica libri septem.*
Basel [Ioannes Oporinus, 1543.]
Subject: Anatomy, dissection, medicine.
Osler 567.
Case/6A/156

## Vesalius, Andreas.

*De hvmani corporis fabrica libri septem.*
Venice: Franciscus Franciscius Senensis & Ioanes Criegher Germanus, 1568.
Subject: Anatomy, dissection, medicine.
NLM 1568
UNCATALOGUED

## Vesling, Johann, 1598–1649.

*The anatomy of the body of man; wherein is exactly described every part thereof, in the same manner as it is commonly shewed in publick anatomies. And for the further help of young physitians and chyrurgions, there is added very many copper cuts, far larger than is printed in any book written in the English tongue. Also explanations of every particular expressed in the copper plates. Published in Latin, by Joh. Veslingus, reader of the public anatomy in the most famous university of Padua.*
London: printed for George Sawbridge, 1677.
STC II V287
Subject: Anatomy.
Case/folio/oQM/21/.V4913/1677

Englished by Nich. Culpeper gent. student in physick and astrology.

## Vespucci, Amerigo, 1451–1512.

[*Mundus novus*]
*Mundus novus.*
Augsburg: Johannes Otmar, 1504.
Church 20
Subject: Discovery, New World, voyages.
Ayer/*112/V5/1504a

## Vespucci, Amerigo.

[*Mundus novus.*]
*Mundus nouus.*
[Venice, 1504]
Church 19^ Harrisse 30
Subject: Discovery, New World, voyages.
Ayer/*112/V5/1504b

## Vespucci, Amerigo.

[*Mundus novus.*]
*Mundus nouus. Albericus Vespvtivs Lavrentio Petri de Medicis salvtem plvrimam dicit.*
[Paris, 1504]
Church 17^ Harrisse 23
Subject: Discovery, New World, voyages.
Ayer/*112/V5/1504c

"Ex italica in Latinam linguam focundus interpres hanc epistolam vertit. vt Latini omnes intelligant que multa miranda indies reperiantur. . . . Quando a tanto tempore quo mundus cepit ignota sit vastitas terre, et que contineantur in eo."

## Vespucci, Amerigo.

[*Mundus novus.*]
*Mente retrouati au + poesi no et mondo nouo da Alberico Vesputio Florentino intitulato.*
[Vicenza: Henrico Vicentino, 1507.]
Subject: Discovery, New World, voyages.
Ayer/*11/F8/1507

## Vespucci, Amerigo.

[*Mundus novus.*]
*Diss büchlin saget wie di zwey durch lüchtigsten herren her Fernandus. K. zü Castilien un herr Emanuel K. zu Portugal haben das weyte mör ersüchet vnnd funden vil insulen vnnd ein nüwe welt von wilden nackenden zeüten vormas vnbekant.*
Strassburg: Johannes Grüniger, 1509.
Subject: Discovery, New World, voyages.
Ayer/*112/V5/1509

Missing leaves (following sig. c iii) have been replaced with facsimiles.

## Vespucci, Amerigo.

[*Mundus novus.*]
*Paesi nouamente ritrouati poer la nauigatione di Spagna in Calicut. Et da Albertutio Vesputio Fiorentino intitulato mondo nouo. Nouamente impresso.*
Venice: Zorzo de Rusconi Millanese, 1521.
Subject: Discovery, New World, voyages.
Ayer/*111/F8/1521

Chap. 29: "Varij animali presertim elefanti & Ziraffe."

## Vespucci, Amerigo.

[*Mundus novus.*]

*Le noueau monde et nauigacions faictes par Emeric de Vespuce florentin des pays et isles nouuellement trouuez auparauant a nous incongeuz. Tant en l'Ethiope que Arabie Calichut & aultres plusieurs regions estranges. Translate de italien en langue francoyse par Mathurin.*
Paris: Galliot du Pre, 1516.
Subject: Discovery, New World, voyages.
Ayer/*111/F8/1516

Chap. 42, fol. 39: "des elephans sauuaiges."

### Vespucci, Amerigo.

*Be* [i.e., *De*] *ora antarctica per regem portugallie pridem inuenta.*
Strassburg: Mathias Hupfuff, 1505.
Subject: Navigation by stars.
Ayer/*112/V5/1505

### Vespucci, Amerigo.

*Vita e lettere di Amerigo Vespvcci raccolte e illvstrate dall' abate Angelo Maria Bandini.*
Florence, 1745.
Subject: Exploration, voyages.
E/5/.V636

### Vespucci, Amerigo.

*Vita e lettere di Amerigo Vespvcci.* [Another copy]
Ayer/112/V5/B22/1745

### Vetancourt, Augustin de, 1620–1700.

*Teatro Mexicano. Descripcion breve de los svccessos exemplares, historicos, politicos, militares, y religiosos del nuevo mundo occidental de las Indias.*
México: printed for Doña Maria de Benavides, widow of Iuan de Ribera, 1698 [1697]
Subject: Anthropology, natural history, New World.
Ayer/*655/.52/V5/1698

Includes "svccessos naturales": "De la naturaleza, temple, sitio, nombre, longitud, fertilidad, y otras grandezas de el nuevo mundo. De la fertilidad, y riqueza en commun de este nuevo mundo. "Successos politicos": "De los que habitaron la tierra de la Nueva España antes de diluvio, de origen de sus naciones despues, y de sus primeros pobladores."

### Vicecomes, Hieronymus [Visconti, Girolamo] fl. 1490–1512.

*Lamiarum siue striarum opusculum.*
Milan: Leonardus Pachel, 1490.
C 3210=C 6200^ BMC (XV)VI:788^ Pr. 5986^ Still. V 244^ Goff V-272
Subject: Witchcraft.
Inc./5986

Edited by Aluisius de la Cruce.

### Vieri, Francesco de', fl. 1547–1590.

*Discorso dell' eccellentiss. filosofo M. Francesco De' Vieri cognominato il secondo Verino. Intorno a' dimonii, volgarmente chiamati spiriti.*
Florence: Bartolomeo Sermartelli, 1576.
Subject: Spirits, supernatural.
Case/B/88/.938

Pt. 1: Nella qvale si ragiona, qual fosse il parere d'Aristotile, sopra tal soggetto; pt. 2: Nella qvale si dimostra qual fosse l'opinione di Platone intorno à gli spiriti; pt. 3: View of the Catholic church on the subject. 1st. ed., with imprimatur leaf of Florentine Inquisition.

### Vieri, Francesco de.

*Trattato di M. Francesco de' Vieri, cognominato il Verino secondo nel qvale si contengono i tre primi libri delle metheore. Nvovamente ristampati, & da lui ricorretti con l'aggiunta del quarto libro.*
Florence: Giorgio Marescotti, 1582.
Subject: Alchemy, humors.
Case/M/3/.944

"Alcune verità & regole ò conclusioni de corpi similari come humidi, & come secchi" (p. 392); "alcune verità ò regole, ò conclusioni de corpi similari, come caldi & come freddi" (p.400). Discusses generation and corruption, digestion, maturation, coagulation.

### Vieussens, Raymond, 1641–1715.

*Oeuvres Françoises de M. Vieussens.*
Toulouse: Jean Guillemette, 1715.
Subject: Anatomy, medicine.
UNCATALOGUED

Two treatises, each with its own title page, both illustrated. 1: *Traité nouveau de la structure et des causes du mouvement naturel du coeur* [1st ed., 1715]; 2: *Traité nouveau de la structure de l'oreille divisé en deux parties* [1714], internal and external ear.

### Vigenere, Blaise de, 1523–1596.

*Traicté des chiffres ov secretes manieres d'escrire.*
Paris: Abel L'Angelier, 1586.
Subject: Ciphers.
Case/Z/216/.943

Lacking index, errata, and privileges.

### Vignola, Giacomo Barozzio, called, 1507–1573.

*Le dve regole della prospettiva pratica di M. Iacomo Barozzi da Vignola. Con i commentarii del reuerendo padre maestro Egnatio Danti dell' ordine de predicatori mattematico dello studio di Bologna.*
Bologna: Gioseffo Longhi, 1682.
Subject: Perspective.
fW/5922/.944

### Villa, Manuel Angelo.

*Lista noticiosa instrvmentos, e artefactos phisicos, e mathematicos, que se fabricaõ, e se vendem nesta cidade de Lisboa em casa de Manoel Angelo Villa.*
Lisbon: Antonio Isidoro da Fonseca, 1745.

Subject: Instruments.
Greenlee/4505/P855

### [Villano, Giovanni fl. ca. 1390, supposed author]

[*Chroniche de la inclyta cita de Napole emendatissime*] [Title page lacking, text begins] "Tavola de li capitoli de le chroniche de Napolii & de li capitoli deli bagni de Puzolo & Ischia."
Naples: M. Evangelista di Presenzani de Pauia, 1526.
Subject: Medicinal baths.
Bonaparte/Coll./No./4740

Attributed to Bartolommeo Caracciolo called Carafa (fl. ca. 1445) cf. Graesse. First tract is followed by "Sequita tractato utilissimo de li bagni" (baths, their medicinal uses and locations). Cap. 20, car. viii: "Come se uno cauallo sub certa constellatione che sanaua la infirmita delli caualli."

### [Villars, Nicholas Pierre Henri de Montfaucon, abbé de], 1635–1673.

*The Covnt of Gabalis or, conferences about secret sciences.*
London: printed for Robert Harford, 1680.
STC II V386A
Subject: Demonology, Rosicrucians.
Case/B/88/.94

Rendered out of French into English with an advice to the reader by A. L. A.M. Translated by Archibald Levell.

### [Villars, Nicholas Pierre Henri de Montfaucon, abbé de]

*The Count of Gabalis; or, the extravagant mysteries of the cabalists exposed in five pleasant discourses on the secret sciences.*
London: printed for B. M., 1680.
STC II V386
Subject: Demonology, Rosicrucians, witchcraft.
Case/Y/1565/.V688

Done into English by P. A., gent. [Philip Ayers] with short animadversions. STC: "Printed for B. M. printer to the Cabalistical Society of the Sages, at the sign of the Rosy-crusian."

### Vincent de Beauvais, d. 1264.

*Bibliotheca mvndi. Vincenti Bvrgvndi, ... Specvlvm qvadrvplex, natvrale, doctrinale, morale, historiale.*
Douai: Balthazar Beller, 1624.
Subject: Encyclopedias, natural history.
Case/A/251/.944

In 4 vols. Vol. 1: creation of the world; vol. 2: mechanical arts, e.g., architecture, navigation, alchemy, medicine, natural philosophy; vol. 3: psychology, with table of the passions; vol. 4: history.

### Vincent de Beauvais.

*Speculum naturale.*
[Strassburg: Printer of the Legenda Aurea, 1483?]
C 6257^BMC (XV)II:10, III:860^Pr. 2056^Still. V 625^Goff V-293
Subject: Encyclopedias, natural history.
Inc./417.5

In 2 vols.

### Violante, Philipp de.

*De variolis et morbillis tractatvs physico-mechanicvs.*
Dresden: Georg Conrad Walther, 1750.
Subject: Medicine.
oRB/115/.V65

With half-title only; one of a group of medical tracts. Binders title: *Varia medica.* BOUND WITH Werlhof, q.v., and 3 other medical tracts (published after 1750).

### Virling, Georg. Samuel.

*Der wackere stab des herren oder frisch-grünende und zur straffe bereitete ruthe über die in sünden und sicherheit schlaffende welt-menschen: vorgestellet durch den am 18 (28) Decembr. dieses zu end laufenden 1680 jahres mit viel grösserem als erstmals; daher desto mehr erschrecklicherm schwanz oder schweiff hervorstrahlenden cometen.*
Erffurt: Benjamin Hempel, 1681.
Subject: Comets.
Case/B/863/.93

### Visscher, Roemer, 1547–1620.

*Sinnepoppen.*
Amsterdam: Willem Iansz [n.d.]
Subject: Emblems.
Case/-W/1025/.9475

Emblems are representations of contemporary mechanical devices. Engraved by C. J. Visscher. Privilege dated 1614, 's Gravenhage.

### Vitruvius Pollio, Marcus, 1st cent. B.C.

*I dieci libri dell'architettvra di M. Vitrvvio.*
Venice: printed by Francesco de' Franceschi Senese, & Giouanni Chrieger, Alemano associates, 1567.
Subject: Architecture.
Case/5A/202

### Vitruvius Pollio, Marcus.

*Di Lucio Vitruuio Pollione de architectura libri dece traducti de latino in vulgare affigurati: commentati: & con mirando ordine insigniti: per il quale facilmente potrai trouare la multitudine de li abstrusi & reconditi vocabuli a li soi loci & in epsa tabula utilitate de ciascuno studiose & beniuolo di epsa opera.*
Como: Gotardus de Ponte, 1521.
Subject: Architecture, astronomy, engineering, surveying.
Wing/+ZP/535/.P77

Edited by Agostino Gallo. In 10 books, including (bk. 9) "De la norma emendata inuentione di Pythagora de la deformatione del trigono hortogono"; "De le gnomonice ratione da li radii del sole per lumbra trouate: & del mundo & de li planeti"; "Del curso del sole per li duodeci signi." Chapters on catapults, aqueducts. Text surrounded by commentary. Typed note laid in: "This is the handsomest edition of the great handbook of classical architecture which transmitted rules of classical architecture and engineering to the Middle Ages and the Renaissance and which formed the basis for the work of Bramante, Michaelangelo and Palladio, among others." Illustrations are attributed in part to Leonardo da Vinci.

## Vives, Juan Luis, 1492–1540.

*De concordia & discordia in humano genere.*
[Antwerp] Michael Hillenius [1529].
Subject: Philosophy, psychology.
Case/B/686/.942

In 4 books.

## Vives, Juan Luis.

*De conscribendis epistolis Joannis Lvdovici Viuis libellus uerè aureus. Des. Erasmus Roterodami compendium postremo iam ab eodem recognitum. Conradi Celtis methodus. Christopherii Hegendorphini methodus.*
Basel [B. Lasius & T. Platterus], 1536.
Subject: Astrology, astronomy, calendar.
Case/Y/682/.V826

BOUND WITH Joachim Camerarius, Theodor Gaza, qq.v., and other tracts on various aspects of humanistic philosophy, not all within scope.

## Vives, Juan Luis.

*De disciplinis libri XX, in tres tomos distincti, quorum ordinem uersa pagella indicabit.*
Cologne: Ioannes Gymnicus, 1532.
Subject: Educational reform, quadrivium.
Case/I/40195/.9372

Pt. 1: "De cavsis corruptarum artium"; bk. 5: "De philosophia natvrae, medicina, & artibus mathematicis corruptis."

## Vives, Juan Luis.

*De disciplinis libri XX.* [Another edition]
Cologne: Ioannes Gymnicus, 1536.
Case/I/40195/.9374

## Vives, Juan Luis.

*De disciplinis libri XII. 7 de corruptis artibus, quinque de tradendis disciplinis.*
Leyden: Joan. Maire, 1636.
Subject: Educational reform, quadrivium.
Case/I/40195/.937

## Vives, Juan Luis.

*L'institvtion de la femme chrestienne, tant en son enfance, que mariage & viduité. Auec l'ofice du Mary. Le tout composé en Latin, par Loys Viues. Et nouuellement traduict en langue Francoise, par Pierre de Changy esquyer. . . . Le tout reueu & corrigé.*
Paris: Charles L'Angelier, 1555.
Subject: Health, hygiene, psychology, women.
Case/3A/2815

## Volpi, Giovanni Antonio, 1686–1766.

*Academicorum et scepticorum philosophiae rationem non esse in physica omnino repudiandum, oratio Jo. Antonii Vulpii.*
Padua: Josephus Cominus, 1732.
Subject: Natural philosophy.
B/018/.948

## Voltaire, François Marie Arouet de, 1694–1778.

*Elémens de la philosophie de Neuton, mis à la portée de tout le monde.*
Amsterdam: Etienne Ledet & associates, 1738.
Subject: Astronomy, gravity, motion, popular science.
L/35/.945

## Voltaire, François Marie Arouet de.

*The elements of Sir Isaac Newton's philosophy.*
London: printed for Stephen Austen, 1738.
Subject: Astronomy, gravity, motion, popular science.
L/35/.946

Revised, corrected, and trans. John Hanna, M.A.

## Vossius, Gerardus Joannes, 1577–1649.

*De qvatvor artibvs popvlaribvs, grammatisce, gymnastice, musicae, & graphice, liber.*
Amsterdam: Ioannis Blaeu, 1650.
Subject: Mathematics.
L/10/.948

BOUND WITH *De philologia* and *De vniversae mathesios natvra & constitvtione liber; cui subjungitur chronologia mathematicorvm.*

## Vossius, Gerardus.

*De studiorum ratione opuscula.*
Utrecht: Theodor ab Ackersdyck, & Gisberti à Zyll, 1651.
Subject: Medicine, natural science.
Case/I/409/.94

Includes a tract by Johan van Heurne (1543–1601), *De studio medicinae bene instituendo. Dissertatio* (on the teaching of medicine), pp. 485–513.

## Vossius, Isaac, 1618–1689.

*Isaaci Vossii dissertatio de vera aetate mundi, quâ ostenditur natale mundi tempus annis minimum 1440 vulgarem aeram anticipare.*
The Hague: Adrian Vlacq, 1659.
Subject: Universal chronolgy.
Case/A/91/.948

BOUND WITH *Catalogvs librorum in diversis linguis orientalibus—partim manvscriptorvm, partim typis editorvm, bibliothecae celeberrimi . . . Thomae Erpenii* (Arabic, Persian, Turkish, Hebrew, Syriac); Gerardus Vossius, *De philosophorum sectis;* and other tracts.

* * *

*Voyages and discoveries in South America. The first up the river of the Amazons to Quito in Peru, and back again to Brazil. . . . By Christopher d'Acvgna. The second up the river of Plata, and thence by land to the mines of Potosi. By Mos. Acarete. The third from Cayenne into Guiana, in search of the Lake Parima, reputed the richest place in the world. By M. Grillet and François Bechamel.*
London: Printed for Samuel Buckley, 1698.
STC II V746
Subject: Discovery, voyages.
Case/G/98/.94

"The beauty of this country, and the abundance of medicinal simples, plants and trees it yields" (chap. 30, p. 73).

## Vulson de la Colombière, Marc, d. 1658.

*Les nouveaux oracles divertissans, où les curieux trouveront la réponce agréable des démandes les plus divertissantes pour se réjoüir dans les compagnies. Augmentées de plusieurs nouvelles questions, avec un traité de la phisionomie, recüeilli des plus graves auteurs de ce siècle. Ensemble l'explication des songes & vision [sic] nocturnes, traduit par le sieur W. de la Colombiere & mis nouvellement dans un meilleur ordre.*
Paris: Gabriel Quinet, 1696.
Subject: Dreams, physiognomy, prognostication.
Case/oBF/1852/.V85/1696

Cf. Dorbon Ainé, *Bibliotheca esoterica* no. 5221

## Wafer, Lionel, 1660?–?1705.

*A new voyage and description of the isthmus of America, giving an account of the author's abode there, the form & make of the country, the coasts, hills, rivers, &c. Woods, soil weather, &c. fishe, &c. The Indian inhabitants, their features, complexion &c. their manners, customs, employments, marriages, feasts, hunting, computation, language, &c.*
London: printed for James Knapton, 1699.
STC II W193
Subject: Anthropology, natural history, voyages.
Ayer/1269/P12/W2/1699

## Wagstaffe, John, 1633–1677.

*The question of witchcraft debated. Or a discourse against their opinion that affirm witches, considered and enlarged.*
London: printed for Edw. Millington, 1671.
STC II W199
Subject: Witchcraft.
Case/B/88/.95

2d ed. 2 MS leaves tipped in, containing a biographical sketch of Wagstaffe. Meric Casaubon is said to have been his principal adversary.

## [Walkington, Thomas], d. 1621.

*The optick glasse of hvmors; or, the touchstone of a golden temperature, or the philosophers stone to make a golden temper. Wherein the foure complections sanguine, cholericke, phligmaticke, melancholicke are succinctly painted forth and their externall intimates laid open to the purblind eye of ignorance itselfe, by which euery one may iudge, of what complection he is, and answerably learne what is most sutable to his nature.*
London: printed for I. D., 1639.
STC 24969.
Subject: Humors, psychology, sympathy and antipathy.
Case/B/597/.94

"That the soule sympathizeth with the body, and followeth her crasis and temper" (chap. 2); "Whether the internall faculty may be known by the externall physiognomy and visage" (chap. 3); "Of the diversity of wits, according to the divers temperature of the body" (chap. 7); "the close to the whole work, in verse."

## Walkington, Thomas.

*The optick glasse of hvmors.* [Another edition]
London: printed for G. Dawson, 1664.
STC II W459
Case/B/597/.95

## Wanley, Nathaniel, 1634–1680.

*The wonders of the little world: or, a general history of man. In six books. Wherein by many thousands of examples is shewed what man hath been from the first ages of the world to these times. In respect of his body, senses, passions, affections; his virtues and perfections, his vices and defects, his quality, vocation and profession; and many other particulars not reducible to any of the former heads. Collected from the writings of the most approved historians, philosophers, physicians, philologists and others.*
London: printed for T. Basset, ... R. Chiswel, ... J. Wright, ... and T. Sawbridge, 1678.
STC II W709
Subject: Anthropology, prodigies.
Case/A/15/.954

## Ward, Samuel, d. 1643.

*The wonders of the load-stone or, the load-stone newly reduc't into a divine and morall vse.*
London: printed by E. P. for Peter Cole, 1640.
STC 25030
Subject: Lodestone, magnetism.
Case/3A/2268

Imperfect copy, with photocopies of missing portions inserted. These include preliminary matter, pp. 125–26, pp. 267 to end. Contains a description of the lodestone and the lodestone as Christian metaphor. Trans. by H. Grimeston.

## Ward, Seth, bp. of Salisbury, 1617–1689.

*Vindiciae academiarum containing, some briefe animadversions vpon Mr. Websters book, stiled, the examination of academies. Together with an appendix concerning what M. Hobbs, and M. Dell have published on this argument.*
Oxford: printed by L. Lichfield for T. Robinson, 1654.
STC II W832
Subject: Educational reform.
Case/I/701/.952

A rebuttal to John Webster, q.v.

## Waterhouse, Edward, 1619–1670.

*A declaration of the state of the colony and affaires in Virginia.*
London: G. Eld for Robert Mylbourne, 1622.
STC 25104
Subject: Descriptive geography, exploration, navigation, Northwest Passage.
Ayer/*150.5/V7/W3/1622

With "A treatise annexed, written by that learned mathematician Mr. Henry Briggs, of the northwest passage to the south sea through the continent of Virginia... And a note of the charges of necessary prouisions fit for euery man that intends to goe to Virginia.

## Webster, John, 1610–1682.

*Academiarum examen, or the examination of academies. Wherein is discussed and examined the matter, method and customes of academick and scholastick learning, and the insufficiency thereof discovered and laid open; As also some expedients proposed for the reforming of schools, and the perfecting and promoting of all kind of science.*
London: printed for Giles Calvert, 1654.
STC II W1209
Subject: Educational reform.
Case/I/701/.96

"This school philosophy is altogether void of true and infallible demonstration, observation, and experiment, the only certain means, and instruments to discover, and anatomize natures occult and central operations" (p. 68).

## Webster, John.

*The displaying of supposed witchcraft. Wherein is affirmed that there are many sorts of deceivers and impostors, and divers persons under a passive delusion of melancholy and fancy. But that there is a corporeal league made betwixt the devil and the witch, or that he sucks on the witches body, has carnal copulation, or that witches are turned into cats, dogs, raise tempests, or the like, is utterly denied and disproved. Wherein is also handled, the existence of angels and spirits, the truth of apparitions, the nature of astral and sydereal spirits, the force of charms and philters; with other abstruse matters.*
London: printed by J. M., 1677.
STC II W1230.
Subject: Rational approach to occult, supernatural, witchcraft.
Case/fB/88/.963

## Wecker, Johann Jacob, 1528–1586.

*De secretis libri XVII. Ex variis authoribus collecti, methodiceque digesti & aucti.*
Basel: Conrad Waldkirch for Episcopius, 1598.
Subject: Astrology, magic, physics, quadrivium.
Case/A/15/.964

## Weigel, Christoph, 1654–1725.

Collection of scattered leaves from German writing books.
Nuremberg: Christoph Weigel [1709?]
Subject: Ciphers.
Wing/F/ZW/7471/.091

"In viler arten curiose inventionen." Year 1709 appears on 2 leaves.

## Weigel, Erhardt, 1625–1699.

*Himmels-zeiger der bedeutung aller dinge dieser velt insonderheit derer sterne. Sampt dessen fort-setzung, nechts einem muster wörnach ein gottseelig nativitet zu stellen. auff veranlassung des ungemeinen cometen im 1680 und 1681 sten jahre.*
Jena: Johann Bielcken, 1681.
Astronomy, astrology, comets.
Case/B/863/.935

## Weigel, Valentin, 1533–1588.

*Theologia VVeigelii. Das ist: Oeffentliche glaubens bekändtnüss des weyland ehrwürdigen durch die dritte mentalische oder intellectuallische pfingst-schule erleuchteten mannes.*
Frankfurt: Samuel Müller, 1699.
Subject: Natural theology.
C/58/.412

"Dass diese zahl des thiers 666. alle ding beschliesse und begreiffe ja aller ding wesen und leben sey" (chap. 18).

## [Weigel, Valentin].

*Y dYas mystica ad monadis simplicitatem. Ein nutzbares zwei faches tractätlein, so einem einfeltigen Christlichen hertzen den weg weiset zur ewigen seeligkeit. Darinnen erinnert wird I. Des menschen composition auss dreyen unterschiedlichen wesendlichen theilen. II. Der hochwichtige unterscheid der beyden vornehmsten spiecierum fidei gratiae & naturae, das ist, des irsdischen natürlichen vnd himlischen Christlichen seligmáchenden glaubens auss betrachtung des spruchs Pauli I Cor. 2. v. ult: Der natürliche mensch (das mittere theil des menschen die seele) verstehet (für such vnd alleine) nichts vom geiste gottes. Zu bewhärung der ewigen warheit, erbamung vnd vermehrung des wahren Christen thumbs, vnd erweckung rechter gottseliger gedancken, auss wolmeinendem Christlichen hertzen, é collegio spiritus sancti der gemeinen im reich Christi herfür gegeben per Christianum Theophilum* [pseud.] *Christianopoli.*
[Erffurdt] 1620.
Subject: Rosicrucians
K/975/.962

## [Welling, Georg von] 1652–1727.

*Tractatus mago-cabbalistico-chymicus, et theosophicus, von des saltzes uhrsprung und erzeugung, natur und nutzen wobey zugleich die erzeugung derer metallen, mineralien und anderer salien, aus dem grunde der natur bewiesen wird. . . . Alles, nach einem systemate magico universi, nebst andern in kupffer gestochenen problematibus.*
Salzburg, 1729.
Subject: Alchemy, cabala, magic.
L/0114/.959

Letterpress title page is preceded by 2 MS leaves.

## Werckmeister, Andreas, 1645–1716.

*Musicalische temperatur, oder deutlicher und warer mathematischer unterricht vie man durch anweisung des monochordi ein klavier sonderlich die orgel-wercke, positive, regale, spinetten, und dergleichen wol temperirt stimmen könne, damit nach heutiger manier alle modificti in einer angenehm- und erträglichen harmoniem mögen genommen werden.*
Frankfurt & Leipzig: Theodorus Philippus Calvisius, 1691.
Subject: Mathematics and music.
Case/4A/1029

## Werlhof, Paul Gottlieb, 1699–1767.

*Observationes de febribvs praecipve intermittentibvs et ex earvm genere continvis deqve earvm pericvlis ac reversionibvs praenoscendis et praecavendis per medelam tempestivam efficacem adaeqvatam candide et perspicve propositam ad viros clarissimos et experientissimos avctores commercii literarii Norimbergensis qvi problema proposverant de febribvs intermittentibvs soporosis et apoplecticis scribebat Pavl. Gottlieb Werlhof.*
Hanover: for Foerster heirs, 1745.
Subject: Medicine, disease, fevers.
oRB/115/.V65

2d ed. BOUND WITH Violante, q.v., and 3 other medical tracts (published after 1750). Binders title: *Varia medica.*

## [Weston, Sir Richard] 1591–1652.

*A discourse of husbandrie used in Brabant and Flanders: Shewing the wonderful improvement of land there; and serving as a pattern for our practice in this common-wealth.*
London: printed by William Du-gard, 1652.
STC II W1483
Subject: Agriculture.
Case/R/51465/.968

2d ed., corrected and enlarged.

## Weston, Thomas, fl. 1723.

*Veteris arithmetica elementa, sive de symbolicis practicis partibus arithmeticae, ab antiquis Hebraeis, Graecis et Romanis usurpatae, (omnibus quibus potui vestigiis indagatae) tractatus: in usum studiosae juventutis in academia Grenovici a Tho. Weston.*
[London? 1725]
Subject: Arithmetic.
Wing/+ZW/745/.W522

BOUND WITH *A copy-book written for the use of the young-gentlemen at the academy in Greenwich by Thomas Weston* and *A drawing-book* [in 2 parts: human figure and architecture, landscape]. Leaf giving arabic numerals and roman numeral counterparts is signed "Thomas Weston scripsit 1725."

## Wharton, Sir George, 1617–1681.

*An astrologicall jvdgement vpon his Majesties present march: begun from Oxford, May 7, 1645.*
Oxford: H. Hall, 1645.
STC II W1542
Subject: Astrology.
Case/B/8635/.972

With a postscript by William Lilly, who takes exception to Wharton's prophecies.

## Whiston, William, 1667–1752.

*Astronomical lectures read in the publick schools at Cambridge; by William Whiston, M.A. Mr. Lucas's professor of mathematicks in that university. Whereunto is added a collection of astronomical tables; being those of Mr. Flamsteed, corrected; Dr. Halley; Monsieur Cassini; and Mr. Street.*
London: printed for R. Senex and W. Taylor, 1715.
Subject: Astronomy.
L/9/.97

Back flyleaf has some MS notation.

## Whiston, William.

*A new theory of the earth, from its original, to the consummation of all things. Wherein the creation of the world in six days, the universal deluge, and the general conflagration, as laid down in holy scriptures, are shown to be perfectly agreeable to reason and philosophy.*
Cambridge: printed for Benj. Tooke, bookseller . . . in . . . London, 1708.
Subject: Bible and science, natural theology, reason and religion.
C/257/.972

Bk. I, Lemmata. 1: "All bodies will persever for ever in that state, whether of rest or motion, in which they once are, if no other force or impediment act upon them, or suffer by them"; 2: "All motion is of it self rectilinear"; bk. II, Hypotheses. 1: "The ancient chaos, the origin of our earth, was the atmosphere of a comet"; 3: "Tho' the annual motion of the earth commenc'd at the beginning of the Mosaick creation; yet its diurnal rotation did not till after the fall of man"; bk. III, Phaenomena.

## Whiston, William.

*The testimony of Phlegon vindicated: or, an account of the great darkness and earthquake at our Saviour's passion, described by Phlegon*
London: sold by Fletcher Gyles and by J. Roberts, 1732.
Subject: Earthquakes, reconciling of history, science, and Biblical accounts.
pC/588/.973

## [White, John], 17th cent.

*A rich cabinet with variety of inventions: unlock'd and open'd, for the recreation of ingenious spirits, at their vacant hours. Being receits and conceits of severall natures, and fit for those who are lovers of natural and artificial conclusions. As also variety of the recreative fire-works both for land, air, and water. And fire-works of service, for sea and shore. Whereunto is added divers experiments in drawing, painting, arithmetic, geometry, astronomy, and other parts of the mathematicks.*
London: printed for William Whitwood, 1668.
STC II W1791
Subject: Encyclopedia of inventions, miscellany, and materia medica.
Case/R/2/.972

4th ed. with many additions. Receit no. 52: "How to make a candle diall whereby you may know the hours of the night." Verse prologue (or epigraph), "The author to his book. As in a glass herein you may behold/ A goodly cascate set with peares and gold;/ Not petty gugau's to adorn the brest,/ The neck, the arm, but jewels of the best,/ And choicest learning, such herein you'l find/ Will please your fancy & content your mind;/ Some for delight and recreation/ And some for serious contemplation;/ Some in arithmetick, that lofty art,/ Some likewise in geometry are taught,/ Some in astronomy that art most high,/ Others teach how to decorate the skie,/ With splendent stars, silver & golden showres,/ Which are th' effects of philosophick powers;/ Mind well therefore what's in this book, and let/ It hence be called the Artists Cabinet."

## White, Thomas, 1593–1676.

*Peripateticall institutions. In the way of that eminent and excellent philosopher Sir Kenelm Digby. The theoreticall part. Also a theologicall appendix of the beginning of the world.*
London: printed by R. D., 1656.
STC II W1839
Subject: Natural science, physics.
Case/B/245/.97

In 5 books. Subjects include nature of bodies, metaphysical essences of bodies, time and local motion, planets, light, plants, dreams.

## Whithorne, Peter, fl. 1550–1563.

*Certaine vvayes for the ordering of soldiours in battleray, . . . : And more ouer howe to make salt peter, gunpowder, and diuers sortes of fireworkes or wilde fyre, with other thinges appertayning to the warres.*
London: printed by VV. VVilliamson for Iohn VVight, 1573.
Subject: Chemistry.
Case/U/2/.537

BOUND WITH Machiavelli, *The arte of warre*, trans. Peter VVithorne, 1573. Fol. 23v: Nature of saltpeter, etc.

## Whitney, Geoffrey, 1548?–1601?

*A choice of emblemes, and other devises, for the most part gathered out of sundrie writers, Englished and moralized. And divers newly devised, by Geffrey Whitney.*
Leyden: Francis Raphelengius for Christopher Plantyn, 1586.
Subject: Emblems.
Case/*W/1025/.968

## Wier, Johann, 1515–1588.

*Cinq livres de l'imposture et tromperie des diables: des enchantements & sorcelleries.*
Paris: Iaqves du Puys, 1569.
Subject: Witchcraft.
Case/B/88/.9724

Translated by Jacques Grévin.

## Wier, Johann.

*Histoires, dispvtes et discovrs des illvsions et impostures des diables, des magiciens infames, sorcieres & empoisonneurs: Des ensorcelez & demoniaques, & de la guerison d'iceux: Item de la punition que meritent les magiciens, les empoisonnerus, & les sorcieres. Le tout comprins en six livres (augmentez de motié en ceste derniere edition) Par Iean Wier, medecin du duc de Cleues. Devx dialogves de Thomas Erastvs, touchant le pouuoir des sorcieres: & de la punition qu'elles meritent.*
[Paris?]: for Iaqves Chovet, 1579.
Subject: Witchcraft.
Case/B/88/.9735

Binders title: *Histoires disputes et discours des diables.*

## Wier, Johann.

[*De lamiis*]
*De lamiis liber: item De commentitiis ieivniis.*
Basel: Oporiniana, 1577.
Subject: Witchcraft.
Case/B/88/.971

## Wier, Johann.

*De lamiis liber.* [Another edition]
Basel: Oporiniana, 1582.
Case/B/88/.9722

BOUND WITH his *De praestigiis daemonum*, q.v.

## Wier, Johann.

[*De lamiis*]
*De lamiis, das ist von teuffelsgespenst zauberern vnd gifftbereytern kurtzer doch gründtlicher bericht was für unterscheidt vnter den hexen vnd unholden vnd den gifftbereytern im strassen zuhalten darmit beydes die richter im urtheil fällen vnd verdammen nicht zu viel thun ihr gewissen beschweren und das unschuldiges blut zuvergiessen verhütet werde. Sampt einem angehängten kleinen tractätlein von dem falchen vnd erdichten fasten alles mit vielen nutzlichen vnd glaubwirdigen historien aussgeführet.*
Frankfurt: Nicolaus Basseus, 1586.
Subject: Witchcraft.
Case/fB/88/.974

Trans. Peter Rebenstock von Giessen.

## Wier, Johann.

*De praestigiis daemonvm et incantationibus ac ueneficijs, libri V.*
Basel: Joannes Oporinus, 1566.
Subject: Witchcraft.
Case/B/88/.972

3d ed.

## Wier, Johann.

*De praestigiis daemonum, & incantationibus ac ueneficiis libri sex, postrema editione sexta aucti & recogniti. Accessit liber apologeticvs, et psevdomonarchia daemonvm.*
Basel: Oporiniana, 1583.
Subject: Witchcraft.

BOUND WITH his *De lamiis*, q.v.

## Wigan Eleazar, fl. 1695.

*Practical arithmetick: An introduction to ye whole art wherein the most necessary rules are fairly describ'd in the usuall hands adorn'd with great variety of flourishes perform'd by command of hand design'd to be interleav'd for more speedy fitting of youth for merchandize or trade. By Eleazar Wigan writing master on Great Tower Hill London.*

[London: sold by the author, 1695]
Subject: Commercial and tradesmen's arithmetic.
Wing/fZW/645/.W632

Date, 1695, on both frontispiece portrait and list of roman and arabic numerals.

## Wilkins, John, 1614–1672.

*The discovery of a world in the moone. Or, a discovrse tending to prove, that 'tis probable there may be another habitable world in that planet.*
London: printed by E.G. for Michael Sparks and Edward Forrest, 1638.
STC 25640
Subject: Astronomy.
Case/L/9116/.974

## [Wilkins, John]

[*The discovery of a world in the moon*]
*The discovery of a new world: or, a discourse tending to prove that 'tis probable there may be another habitable world in the moone. With a discourse concerning the possibility of a passage thither.*
London: printed by Iohn Norton for Iohn Maynard, 1640.
STC 25641
Subject: Astronomy
Case/L/976/.97

2 parts in 1 vol. 3d impression, corrected and enlarged.

## Wilkins, John.

[*The discovery of a world in the moon*]
*A discovery of a new world, or, a discourse, tending to prove, that 'tis probable there may be another habitable world in the moon.*
London: printed by T. M. & J. A. for John Gillibrand, 1684.
STC II W2186
Subject: Astronomy.
Case/L/9116/.976

In 2 parts. 4th ed. corrected and amended. Proposition 2: "That a plurality of worlds does not contradict any principle of reason or faith"; proposition 5: "That the moon hath not any light of her own"; proposition 9: "That there are high mountains, deep vallies, and spacious plains in the body of the moon"; proposition 11: "That as their world is our moon, so our world is their moon"; proposition 14: "That 'tis possible for some of or [*sic*] posterity to find out a conveyance to this other world, and if there be inhabitants there, to have commerce with them."

## Wilkins, John.

*An essay towards a real character and a philosophical language.*
London: printed for Sa. Gellibrand, and for John Martyn, 1668.
STC II W2196
Subject: Philology.
Bonaparte/Coll./No./11,544

With index, "An alphbetical dictionary, wherein all English words according to their various significations, are either referred to in their places in the philosophical tables, or explained by such words as are in the tables."

## Wilkins, John.

*An essay.* [Another copy]
Case/fX/186/.974

Penciled in following index title: "By William Lloyd, bp. of Worcester, 1627–1717."

## Wilkins, John.

*Mercury: or the secret and swift messenger. Shewing, how a man may with privacy and speed communicate his thoughts to a friend at any distance.*
London: printed by Rich. Baldwin, 1694.
STC II W2203
Subject: Ciphers, codes, secret writing, signals.
Case/Z/216/.97

Communication by sight (smoke signals), sound (musical notes), conveyancy by bodies (carrier pigeons, shooting messages in cannisters), emblems (hieroglyphs, invented characters), invisible ink. Verso of frontispiece lists "mathematical works of the Right Reverend Father in God John Wilkins, late bishop of Chester. In 3 treatises, viz. A discovery of a new world; Mathematical magic; Mercury."

## Willis, John (stenographer), d. 1628?

*Mnemonica; or, the art of memory, drained out of the pure fountains of art & nature. Digested into three books. Also, a physical treatise of cherishing natural memory; diligently collected out of divers mens writings.*
London: printed by Leonard Sowersby, 1661.
STC II W2812
Subject: Mnemonics.
Case/C/7445/.833

## Willis, Thomas, 1621–1675.

*Opera omnia, nitidius quàm unquam hactenus edita, plurimum emendata, indicibus rerum copiosissimis, ac distinctione characterum exornata.*
Amsterdam: Henricus Wetstenius, 1682.
Subject: Anatomy, medicine, practice.
Case/R/128.7/.W55/1682

## Willis, Thomas.

*De anima brutorum quae hominis vitalis ac sensitiva est, exercitationes duae. Prior physiologica ejusdem naturam, partes, potentias & affectiones tradit. Altera pathologica morbos qui ipsam, & sedem ejus primariam, nempe cerebrum & nervosum genus afficiunt, explicat, eorumque therapeias instituit.*
Oxford: Sheldonian theatre for Ric. Davis, 1672.
Subject: Diseases of the nervous system, physiological psychology.
Case/QP/354/.W6/1672

## Willughby, Francis, 1635–1672.

*The ornithology of Francis Willughby. In three books. Wherein all the birds hitherto known . . . are accurately described. To which are added three considerable discourses by John Ray.*

London: printed by A. C. for John Martyn, 1678.
STC II W2880
Subject: Ornithology.
Case/P/4/.975

Willughby's 3 divisions: birds, landfowl, waterfowl. Ray's discourses: art of fowling, ordering of singing birds, falconry.

### Wilson, Henry, b. 1678.

*A compleat universal history of the several dominions throughout the known world. Containing an account of their situation, extent, boundaries, climate, air, soil, provinces, cities, towns, curiosities, trade, riches, navigation, ports, seas, rivers, mountains, vegetables, animals, minerals, laws.... Also of the persons, complexions, habits, diet, customs, religion, languages, learning, arts, and sciences; of the inhabitants. Including all that is remarkable in their geography, natural, ecclesiastical, and civil history, and policy.*
London: printed for J. and J. Hazard, etc., 1738.
Subject: Geography, navigation, voyages.
Ayer/folio/oD/18/.W5

"Of the Peruvian animals, vegetables, and minerals" (chap. 15).

### Wilson, Henry [attrib. author]

*Atlas maritimus & commercialis; or, a general view of the world so far as relates to trade and navigation: describing all the coasts, ports, harbours, and noted rivers, according to the latest discoveries and most exact observations.... To which are added sailing directions for all the known coasts and islands on the globe; with a sett of sea-charts, some laid down after Mercator, but the greater part according to a new globular projection, adapted for measuring distances (as near as possible) by scale and compass, ... the use of the projection justified by Dr. Halley. To which are subjoined two large hemispheres on the plane of the equinoctial: containing all the stars in the Britannic catalogues: of great use to sailors for finding the latitude in the night.*
London: printed for James and John Knapton, William and John Innys, etc., 1728.
Subject: Hydrography, navigation.
Ayer/135/A83/1728

### Winchilsea, Henage Finch, 3d earl of, d. 1689.

*A true and exact relation of the late prodigious earthquake & eruption of Mt. AEtna, ... as it came in a letter written to his majesty from Naples.... Together with a more particular narrative of the same, as it is collected out of severall relations from Catania.*
[London]: printed by T. Newcomb, 1669.
STC II W2967
Subject: Earthquakes.
Case/M/525/.976

### Wing, Vincent 1619–1668.

*Harmonicon coeleste: or, the coelestiall harmony of the visible world: conteining an absolute & entire piece of astronomie. Wherein is succinctly handled the trigonometricall part, generally propounded, and particularly applyed in all questions tending to the diurnall motion.... Grounded upon the most rationall hypothesis yet constituted, and compared with the best observations that are extant, especially those of Tycho Brahe, and other more modern observators... and commended as usefull to all scholers, astronomers, astrologers, divines, physitians, historiographers, polititians, and poets, &c.*
London: Printed by Robert Leybourn, for the Company of Stationers, 1651.
STC II W2993
Subject: Astronomy.
Case/Wing/fL/9/.977

### [Winslow, Jacques Bénigne] 1669–1760.

*Dissertation sur l'incertitude des signes de la mort, et l'abus des enterremens & embaumemens précipités.*
Paris: Morel the younger, Prault, father & son, Simon, son, 1742.
Subject: Signs of death, medicine.
Q/.977

Traduite et commentée par Jacques Jean Bruhier. Stories of remarkable returns of the dead to life.

### Wither, George, 1588–1667.

*Britain's remembrancer containing a narration of the plagve lately past; a declaration of the mischief present; and a prediction of ivdgments to come; (if repentance prevent not.)*
[London?] Imprinted for Great Britaine and are to be sold by Iohn Grismond, 1628.
STC 25899
Subject: Disease.
Case/Y/185/.W768

For description of the plague, see canto 2, p. 45.

### Wither, George.

*Britain's remembrancer.* [Another copy]
Subject: Disease.
Case/3A/869

### Wither, George.

*A collection of emblemes, ancient and moderne; quickened vvith metricall illvstrations, both morall and divine: And disposed into lotteries, that instruction and good counsell, may bee furthered by an honest and pleasant recreation.*
London: printed by A. M. for Henry Taunton, 1635.
STC 25900d
Subject: Emblems.
Case/*W/1025/.98

With volvelles at back of vol. "A direction, shevving hovv they who are so disposed, shall find out their chance, in the lotteries aforegoing."

### Wolf, Johann Christian, 1683–1739.

*Bibliotheca Hebraea, sive notitia tvm avctorvm Hebr. Cvjvscvnqve aetatis, tvm scriptorvm; qvae vel Hebraice primvm exarata vel ab aliis conversa svnt, ad nostram aetatem,*

*dedvcta. Accedit in calce Jacobi Gaffarelli index codicum cabbalistic. mss. quibus Jo. Picus Mirandulus comes, usus est.*
Hamburg & Leipzig: Christian Liebezeit, 1715–1733.
Subject: Astrology, bibliography, cabala.
Z/961/.98

4 vols. Vol. 1 (1715); vol. 2 (Theodor Christoph. Felginer, 1721); vols. 3 and 4 (widow Felginer, 1727 and 1733, respectively). See Archer Taylor, *Book catalogues, their varieties and uses.* The 24 pp. at end of vol. I lists the astrological and cabalistic materials in Giovanni Pico della Mirandola's library.

## Woodward, John, 1665–1728.

*An essay toward a natural history of the earth: and terrestrial bodies, especially minerals: as also of the sea, rivers, and springs. With an account of the universal deluge and of the effects that it had upon the earth.*
London: printed for Ric. Wilkin, 1695.
STC IIW3510
Subject: Earth sciences.
Case/C/257/.983

## Woodward, John.

*The natural history of the earth, illustrated, inlarged, and defended. To which are added, physical proofs of the existence of God.*
London: printed by Tho. Edlen, 1726.
Subject: Earth sciences.
M/0/.984

In 2 parts, both trans. Benj. Holloway.

## [Woodward, John]

*A catalogue of the library, antiquities, &c. of the late learned Dr. Woodward . . . which will be sold by auction on Mon. 11 Nov. 1728 . . . by Mr. Christopher Bateman . . . and Mr. John Cooper.*
[n.p., n.d.]
Subject: Archaeology, bibliography, earth sciences.
Z/491/.W857

Binders title: *Woodward library catalogue.* BOUND WITH several additional tracts, most by Woodward, including "Remarks upon the antient and present state of London, occasion'd by some Roman urns, coins, and other antiquities lately discover'd" (London: printed for A. Bettesworth, etc., 1723).

## Worm, Ole, 1558–1654.

*Antiqvitates Danicae, literatura runica. Lexicon runicum. Monumenta runica.*
Copenhagen: [various publishers] 1643–1651.
Subject: Archaeology, antiquities, calendar.
Bonaparte/Coll./No./8,554

5 tracts, including "Danicorum monumentorum"; "Fasti Danici universam tempora computandi rationem antiqvitatus Dania et vicinis regionibus observatam libris tribus exhibentes."

## Worrall, John, d. 1771.

*Bibliotheca topographica anglicana: or, a new and compleat catalogue of all the books extant relating to the antiquity, description, and natural history of England, the counties thereof, &c., to the present year, 1736, alphabetically digested in an easy method.*
London: printed for J. Worrall, 1736.
Subject: Topographical dictionary.
Case/ḡ/45/.986

Sale catalogue, with items listed by format and alphabeticaly by author (e.g., folio, A; quarto, A; octavo, A). Includes Dee's conference with spirits (1659); Gesner, four-footed beasts (1607); Thevet, Antarcticke (1568); Platt's Jewel house of art and nature (1653), this last priced at 3s. 6d.

## Wotton, Edward, 1492–1555.

*Edoardi VVottoni Oxoniensis de differentiis animalivm libri decem. Cum amplissimis indicibus . . . deinde omnium animalium nomenclaturae, itémque singulae eorum partes recensentur, tam Graecè, quàm Latinè.*
Paris: Michael Vascosanvs, 1552.
Subject: Natural history, zoology.
Case/6A/131

## Wotton, William, 1666–1726.

*A defense of the reflections upon ancient and modern learning, in answer to the objections of Sir W. Temple, and others. With observations upon the* Tale of a Tub.
London: printed for Tim. Goodwin, 1705.
Subject: Ancients vs. Moderns.
Case/A/91/.987

## Wotton, William.

*Reflections upon ancient and modern learning.*
London: printed by J. Leake for P. Buck, 1694.
STC II W3658
Subject: Ancients vs. Moderns.
Case/A/91/.984

1st ed. Contains a passage on M. Servetus's lesser circulation of the blood (p. 211).

## Wotton, William.

*Reflections upon ancient and modern learning.*
London: printed by J. Leake for Peter Buck, 1697.
STC II W3659
Subject: Ancients vs. Moderns.
Case/A/91/.985

2d ed. with large additions. Comparison of ancients and moderns with respect to natural philosophy, medicine, alchemy, anatomy, geometry and arithmetic, chemistry, circulation of the blood, astronomy, and biology, among others.

## Wotton, William.

*Reflections upon ancient and modern learning.* [Another copy]
UNCATALOGUED

## Wright, Thomas, D. D.

*The passions of the minde in generall.*
London: printed by Valentine Simmes for Walter Burre, 1604.
STC 26040
Subject: Passions of the mind.
Case/B/529/.978

Corrected, enlarged, and with sundry new discourses augmented. Discusses "the essence of passions"; "foure effects of inordinate passions"; "how passions may be discouered." BOUND WITH *A succinct philosophicall declaration of the nature of the clymactericall yeeres, occasioned by the death of Queene Elizabeth.* London: printed for Thomas Thorpe, 1604 (subject: numerology, medicine).

## Wright, Thomas, D.D.

*The passions of the mind in generall.*
London: printed by Miles Flesher, 1630.
STC 26403
Subject: Passions of the mind.
Case/B/529/.98

In 6 bks. "Corrected, enlarged, and with sundry new discourses augmented." MS notes on front and back flyleaves; MS marginalia pp. 89–108.

## Würtz, Felix, ca. 1517 or 1518.

*Felix Würtzen weiland des berühmten wundartztes zu Basel wundartzney, darinnen allerhand schädliche missbrauche welche bissnero von vnerfahrenen vngeschickten wund-aertzten in gemeinem schwang gegangen seynd aussführlich angedeutet vnd vmb vieler erheblichen ursachen willen abgeschafft werden. Jetzund alles mit grossen fleiss auss des authoris hand-geschriebenen bücheren von neuem übersehen mit vielen schäden-curen wie auch einem schönen vnd sehr nutzlichen hebammen- und kinder-büchlein vermehret durch Rudolff Würtzen, wund-artzt zu Strassburg. Sampt angehencktem anatomischen abriss herrn D. Henrici Schaevii.*
Basel: Emanuel König and son, 1675.
Subject: Medicine, treatment of wounds.
Case/3A/1014

## Wytfliet, Corneille, b. ca. 1550.

*Descriptionis Ptolemaicae avgmentvm, sive Occidentis notitia breui commentario illustrata studio et opera Cornely Wytfliet Louaniensis.*
Louvain: Iohannes Bogardus, 1597.
Phillips 1140^ John Carter Brown vol. 1, no. 516–17
Subject: Geography, New World.
Ayer/*6/P9/W9/1597

1st impression.

## Wytfliet, Corneille.

*Descriptiones Ptolemaicae avgmentvm, siue occidentis notitia. Breui commentario illustrata studio et opera Cornely Wytflit Louaniensis.*
Louvain: Iohannes Bogardus, 1597.
Subject: Geography, New World.
Ayer/*6/P9/W9/1597a

Appears to be a different printing from the 1st impression listed above: different preliminaries, different running heads, lacking list of maps, different spelling of Wytfliet's name. A slip inserted in the vol.: "First distinctively American atlas."

## Wytfliet, Corneille.

*Descriptionis Ptolemaicae avgmentvm.* [Another edition]
Louvain: Gerardus Rivius, 1598,
Ayer/*6/P9/W9/1598

Note on bookplate: "Second edition in Latin of Wytfliet's supplement to Ptolemy's Geography."

## Wytfliet, Corneille.

*Descriptionis Ptolemaicae avgmentvm, siue occidentis notitia breui commentario illustrata, et hac secunda editione magna sui parte aucta Cornelio Wytfliet Louaniensi auctore.*
Douai: Francisco Fabri, 1603.
Subject: Geography, New World.
Ayer/*6/P9/W9/1603

Note on bookplate: "Although called the second edition, . . . this appears really to be the third edition of Wytfliet's supplement to Ptolemy's Geography."

## Wytfliet, Corneille.

*Descriptionis Ptolemaicae avgmentvm. Sive occidentis notitia: Breui commentario illustrata studio & opera Cornely Wytfliet, Louaniensis. De nouveau reveu & corrigé.*
Arnheim: Iean Ieansz, 1615.
Subject: Geography, New World.
Ayer/*6/P9/W9/1615

## Wytfliet, Corneille.

*Histoire vniverselle des Indes, orientales et occidentales. Divisée en devx livres le premier par Cornille Wytfliet: le second par Ant. M. & avtres historiens.*
Douai: François Fabri, 1605.
Phillips 1143^ John Carter Brown, vol. 1
Subject: Discovery and exploration, East and West Indies.
Ayer/*6/P9/W9/1605

"Livre premier . . . avqvel est representé tant par discovrs qve cartes l'entiere et parfaicte description des Indes Occidentales." "Livre second . . . representant l'entiere histoire dv decovvrement des Indes Orientales, et levr descriptions." Note on bookplate states that this is a translation of Wytfliet's supplement to Ptolemy with the addition of a 2d part relating to the East Indies by Giovanni Antonio Magini of Padua and others.

## Wytfliet, Corneille.

*Histoire vinverselle des Indes occidentales, diuisée en deux liures, faicte en Latin par Monsieur Wytfliet: nouuellement traduicte. Où il est traicté de leur descouuerte, description, & conqueste faicte tant par les Castillans que Portugais, ensemble de leur moeurs, religion, gouuernemens, & loix.*
Douai: François Fabri, 1607.

Subject: Discovery and exploration, East and West Indies.
Ayer/*6/P9/W9/1607

Much enlarged from 1605 ed. with a new treatise, "La suite de l'histoire des Indes orientales."

## Wytfliet, Corneille.

*Histoire vniverselle des Indes occidentales et orientales, et de la conversion des Indiens. Diuisee en trois parties, par Cornille VVytfliet, & Anthoine Magin, & autres historiens.*
Douai: François Fabri, 1611.
Subject: Discovery and exploration, East and West Indies.
Ayer/*6/P9/W9/1611

2d part is in 2 books (with individual title pages): "Contenant la descouuerte, nauuigation, situation & conqueste, faicte tant par les Portugais que par les Castillans. Ensemble leur moeurs & religion." A 3d part (with separate title page): "De la conversion des Indiens."

# Y–Z

## [Yves, de Paris], Francisci Allaei [pseud.], d. 1678.

*Astrologiae nova methodvs.*
[N.p.] 1658.
Subject: Astrology.
Case/fB/8635/.998

1st section: Astrologiae nova methodvs; 2d section [with separate title page]: Fatvm Vniversi; observatvm a Francisco Allaeio. Arabis Christiano. With volvelles.

## Zabarella, Giacomo, 1533–1589.

*De rebvs naturalibvs libri XXX. Quibus quaestiones, quae ab Aristotelis interpretibus hodie tractari solent, accuratè discutiuntur.*
Cologne: Lazarus Zetzner, 1597.
Subject: Natural history, natural science.
Case/L/0/.992

3d ed. Bk. 1.: "De naturalia scientiae contstitutione"; bk. 2.: "De prima rerum materia"; bk 4.: "Libri de inuentione aeterni motoris."

## Zabarella, Giacomo.

*De rebvs naturalibvs libri XXX.*
Treviso: Robertus Meietti, 1604.
Subject: Natural history, natural science.
Case/fL/0/.993

10th ed. With sections on sky, elements, parts of soul, motion; "Libri de mistione," (p. 231), containing "Auicennae opinio, & argumenta; Auerrois opinio, & adversus Auicennam disputatio; Scoti opinio, & argumentum."

## Zacutus, Abraham Lusitanus, 1576–1642.

*Opervm* [tomus primvs] [tomvs secvndvs] *in qvo de medicorvm principvm historia, libri sex: vbi medicinales historiae, de morbis internis, quae passim principes medicos occurrunt, concinno ordine disponuntur, paraphrasi, & commentariis illustrantur: necnon quaestionibus, dubiis, & obseruationibus exquisitissimis exornantur.*
Lyon: Ioannes-Antonius Hvgvetan, & Gvillielmvs Barbier, 1667.
Subject: Anatomy, history of medicine.
UNCATALOGUED

In 2 vols.: 1: anatomy, disease; 2: history of the practice of medicine; cure of disease.

## Zarlino, Gioseffo 1517-1590.

*Dimonstrationi harmoniche del R. M. Gioseffo Zarlino da Chioggia. . . . Nelle quali realmente si trattano le cose della musica: & si risoluono molti dubij d'importanza.*
Venice: Francesco de i Franceschi Senese, 1571.
Subject: Geometry, harmonics, mathematical theory, numerology.
Case/fV/5/.998

Numerological theory of music.

## Zeisold, Johann, 1599–1667.

[Greek words] *Brevis consideratio qvaestionis: an in generatione hominis anima prolis emanet ab anima parentis? Sub praesidio Johannis Zeisoldi . . . ad disputandum propositae à Nicolao Klopfleisch.*
Jena: Jacob Bauhöffer, 1662.
Subject: Genesis of the rational soul.
Case/3A/985

A group of tracts, binders title: *Disputationes ludicrae.* penciled note on front flyleaf: "Eight German dissertations on absurd and curious themes." Other relevant dissertations: "Brevis considerationis . . . pars posterior . . . praeside Johanne Zeisoldo." Jena: Bauhöffer, 1642; "Controversiae illustres philosophicae, ex Arriaga & Mendoza." Tubingen: Gregorius Kerner, 1664.

## Zerbi, Gabriele, d. 1505.

*Gabrielis Zerbi Veronensis in quaestionibus metaphysicis quas edidit prologus.*
[Bologna: Johannes of Nördlingen and Henricus of Harlem, 1482.]
H *16285^ BMC (XV)VI:820^ Pr. 6556A^ Still. Z27^ Goff Z-27
Subject: Metaphysics.
Inc./f6556A

First 5 bks. only, (234 lines of a total of 512).

## Ziegler, Jacob, 1480–1549.

*Terrae sanctae, qvam Palaestinam nominant, Syriae, Arabiae, Aegypti & Schondiae doctissima descriptio, unà cum singulis tabulis earundum regionum topographicis, authore Iacobo Zieglero Landauo Bauaro. Terrae sanctae altera descriptio ivxta ordinem alphabeti, quae ad scripturam proxime directa est, utilissima etiam plebeio lectori, authore Wuolffgango Vueissenburgio.*
Strassburg: Vuendelinus Rihelius, 1536.
Harrisse B.A.V. 217; John Carter Brown I:1:121
Subject: Navigation, topography, travel.
Ayer/*7/Z6/1536

## Zinkgref, Julius Wilhelm, 1591–1635.

*Emblematvm Ethico-Politicorvm centvria.*
[Frankfurt?]: sold by Iohann Theodor de Bry, 1619.
Subject: Emblems.
Case/W/1025/.994

## Zinkgref, Julius Wilhelm.

*Emblematum ethico-politicorum centuria.*
Heidelberg: printed by Adriani Wyngaerden for Clement Ammonivs, 1666.
Subject: Emblems.
Case/*W/1025/.995

"Editio ultima, auctior et emendatior annexo indice."

## Zwinger, Theodor, 1533–1588.

*Theatrvm vitae hvmanae, omnium ferè eorum, quae in hominem cadere possunt, bonorum atque malorum exempla historica, ethicae philosophiae praeceptis accomodata, & in XIX . . . à Conrado Lycosthene Rubeaquense.*
Basel: Ioan. Oporinvm, Ambrosivm et Avrelivm Frobenius brothers, 1565.
Subject: Encyclopedias, natural science.
Case/fA/251/.996

In 1 vol. Bk. 2: "De mechanicis artibvs"; bk. 18: "De bonis atqve malis corporis."

## Zwinger, Theodor.

*Theatrvm vitae hvmanae à Theodoro Zvingero Basiliense post primam Conr. Lycosthenis Rubeacensis manum plus myriade exemplorvm auctum, methodicè digestum, accuratè recognitum.*
Basel: Frobeniana, 1571.
Subject: Encyclopedias, natural science.
Case/fA/251/.998

4 vols. Includes "De philosophia naturali, & medicina" (vol.I, bk. 3, chap. 2); "Mathematicis disciplinis" (with geometry, cosmography, astronomy, astrology, arithmetic) (chap. 3).